MEANS
FACILITIES
MAINTENANCE
STANDARDS

ROGER W. LISKA, PE, AIC

Means
Facilities Maintenance Standards

Roger W. Liska, PE, AIC

RSMeans

Copyright 1988
Construction Publishers & Consultants
63 Smiths Lane
Kingston, MA 02364-0800
(781) 422-5000

20 19 18 17 16 15 14 13 12

Library of Congress Catalog Number 88-139375

ISBN 0-87629-096-9

This book is dedicated to Judy; her patience, understanding, and encouragement have brought a vision to reality.

TABLE OF
CONTENTS

FOREWORD

The escalating costs of new construction and equipment point to the increasing necessity of assuring that existing buildings, systems, and equipment are maintained on a regular basis and repairs are made to last over a long term. An effective and efficient maintenance program is the only answer to lengthening the life of existing capital investments.

There is hardly a structure in existence today in which there are not defects that could have been prevented by better design and more careful construction. These defects are not usually serious, but they exist and can shorten the life of the structure. Defects are seldom detected before they become so serious that repairs must be made on a rush basis, and are, therefore, more costly. The result, at best, is that management planning and maintenance budgets are upset, and, at worst, that the facility is put out of service while repairs are being carried out.

The purpose of this book is to provide the reader with the capabilities to supervise and administer the functions of a preventive maintenance program for basic building components and materials. The book is oriented toward the inspection of buildings in order to locate defects and potential material failure problems. Recognition of building defects and their effects on the building and its occupants and contents are stressed. Special attention is given to the causes and correction or repair of common defects.

The book is divided into five parts. The first part examines the methods of building inspection, location of deterioration, evaluation of the need for repairs in the light of parameters such as cost, and the behavior of basic materials, providing a better understanding of the causes of deterioration.

The second part of the book covers the causes of deterioration and methods of repair and housekeeping of reinforced concrete, structural steel, and timber building components, which are the major materials used in building structures. The third part focuses on the causes of deterioration, repair, and maintenance of all building materials found both on the exterior and interior of the building, including roofs, floors, walls, ceilings, windows, doors, and hardware. In addition, moisture control and ground maintenance are discussed in this part.

The fourth section examines methods of inspection and maintenance of the basic mechanical and electrical building systems, including heating and air conditioning systems, fresh and waste water plumbing systems and facilities, and common electrical systems.

The last part of the book covers the subjects of planning, estimating, scheduling, and controlling maintenance operations. This section includes how to establish maintenance policy and its implementation and evaluation. Also included is an examination of in-house maintenance programs versus contractual maintenance programs.

Here are just a few of the many items contained in this book that, when utilized, have a potential of saving thousands of dollars in maintenance and repair costs for one's company:

1. Makes accurate judgments to determine structural failures or weaknesses; pinpoints causes of the deterioration based on knowledge of material components and thus saves costs during the investigation stage of repair.
2. Provides a step-by-step procedure of determining the cause of material failures relative to the type of material, the exposure conditions, and the location of the material in the building.
3. Presents step-by-step methods of evaluating whether or not maintenance and/or repairs are needed and thus eliminates much unneeded and unplanned maintenance.
4. Sets forth proven methods of increasing the life of building materials.
5. Helps utilize a maintenance budget more efficiently by aiding in the selection of the correct maintenance and/or repair procedure for the specific type of deterioration and thus eliminates guesswork or trial by error procedures.
6. Details methods of building and material inspection; with inspection checklists included.
7. Serves as the basis for a training program for new employees and as a review for existing ones.
8. Helps new companies establish maintenance programs; along with serving as a tool to up-date existing programs.
9. Presents step-by-step methods of performing housekeeping and repair procedures, along with the required tools and equipment.
10. Aids in eliminating many of the built-in maintenance items when planning a new building, or altering an existing one.
11. Provides the reader with recommended intervals between specific maintenance operations.
12. Provides suggested man-hours and crew sizes to do specific activities, along with methods of developing one's own man-hour allocation records.
13. Illustrates the economic parameters which must be considered when deciding whether to perform a repair procedure now or at some time in the future.
14. Includes many charts and tables for the selection of finish building materials, such as floor and wall coverings, for specific exposure and environmental conditions.
15. Presents specific procedures for determining whether or not to abandon, repair, or replace deteriorated materials.
16. Outlines methods of locating possible 'hidden' deterioration, along with presenting the causes.
17. Presents flow charts to be used in evaluating the need for repair, along with determining the cost of the repair.
18. Lists causes of concrete, steel, and timber deterioration, along with specific methods of diagnosing the causes in order to get at the problem faster.
19. Shows how the repair of one item may cause the failure or deterioration of an adjacent item; and how to safeguard against this problem.
20. Presents methods of determining whether or not material deterioration is potentially hazardous to the extent that it can cause significant structural or mechanical damage, and thus necessitate the partial or complete shut-down of operations.
21. Spells out major repair procedures for structural damage in a checklist manner, along with required tools, equipment, and labor.
22. Outlines the types of repair procedures that can be done by in-house personnel and those which require outside specialty contractors.

23. Presents means of alleviating moisture problems and a discussion of the different types of moisture that can occur within a building.
24. Tables, charts, checklists, and other similar tools are provided to cut down the amount of time in planning for, and performing, maintenance and repair operations.
25. Pinpoints guidelines which must be considered when deciding whether to utilize an in-house maintenance staff or contract the service to an outside company.
26. Presents job descriptions, organization charts, staff requirements, and other managerial tools used at all levels of an organization in setting up a maintenance group.
27. Simplifies methods of estimating the required labor, equipment, and tools, as well as the plant facilities needed to perform specific maintenance and repair operations, along with useful charts and formats for records.
28. Sets forth the "How-To's" of planning and scheduling maintenance and repair operations that minimize interruption of the on-going operations in the building.
29. Presents tested control systems, with formats, for minimizing maintenance costs by following up, on a regular basis, the cost of maintenance and revising the organization to incorporate the potential cost savings.
30. Shows how to utilize managerial tools such as work-study analysis to get the most out of the maintenance dollar.
31. Highlights proven methods of hiring, training, and evaluating in-house and contractual maintenance personnel.
32. Includes various personnel, accounting, evaluation, and other types of management forms to save time in developing or revising forms for recordkeeping and control in one's company.
33. Presents a step by step procedure to be utilized in the inspection of mechanical and electrical systems in the building, including checklists.
34. Suggests routine maintenance items that need to be done to the building support systems (mechanical and electrical) and the frequency at which these items should be performed.
35. Includes a comprehensive list of sources to eliminate the "search and seek" time needed for specialized equipment and materials. Many of these sources provide cost-free information and consulting services.
36. Features fingertip data to aid in selecting the correct cleaning and preserving agents (such as wax and sealers) for specific exposure conditions and material types in a building.
37. Enumerates cost-saving methods of evaluating the quality of cleaning and repair procedures, along with expected life of specific repairs.
38. Includes instant reference tables which outline what substances are harmful to specific types of materials.
39. Reviews key construction and maintenance terms.

The material contained in this book is directed toward both those people with a limited knowledge of building design and maintenance and those who have experience in this field. This book presents the subject matter in a new light. Furthermore, this book will also aid the design-build group in any company by showing them how to design out potential maintenance problems with little additional costs.

The type of building maintenance used to date has always been a very localized matter. It is up to the reader to use the information in this book as an aid in making maintenance a more controllable, and thus less costly, budget item.

Individual maintenance problems, unfortunately, cannot be specifically categorized relative to cause, effect, prevention, and repairs. Each problem presents its own set of unique circumstances. However, through a program with a planned sequence of steps and incorporating a process of elimination, the specific maintenance problem can be solved and remedied quicker and for less cost. The material presented in this book will be a definite asset in attaining this goal.

TABLE OF FIGURES

BEHAVIOR AND EVALUATION OF MATERIAL DETERIORATION

The answers to maintenance problems are not always clear and are never easy. In order to understand the need for maintenance, one must understand the basic properties of materials and how they behave individually and together with other components of the building. Chapter 1 describes the major properties of materials and how they behave individually and together, which can cause specific maintenance problems, especially mutual deterioration.

Once the behavior and properties of material are understood, the next step is to evaluate the extent of any material deterioration and whether or not to abandon, replace, or repair the particular material in question. Chapter 2 examines the following:

- Methods of locating deterioration
- Inspection checklist formats
- Methods of determining causes of deterioration
- Evaluating effects of deterioration
- Cost considerations in abandoning, replacing and/or repairing the deterioration now or in the future.

CHAPTER 1

BASIC BEHAVIOR OF A BUILDING AND ITS COMPONENTS

Today, everyone is looking for some way to improve maintenance. Plant managers are trying to justify expenditures, while management is attempting to trim costs. Everyone involved in the maintenance program wants to be sure that each dollar expended is justified. Those involved in maintenance are forever searching for a program, a concept, or an approach that will satisfy the common objective of improving maintenance. In order to determine the most effective maintenance program, the facilities manager should first understand the basic behavior of a building and its components. This process begins with an analysis of the maintenance function.

The Why's of Maintenance

A starting point is needed to develop a plan of action for any organization that will fit its objectives for improving maintenance. This common beginning point, toward improvement of maintenance or the establishment of a new maintenance program, starts with the understanding of *the why's of maintenance*. A maintenance program is necessary for numerous reasons, some of which are listed below.

- Increase the life of a building and its support systems.
- Insure the safety of the building's occupants and capital equipment.
- Insure that the building's occupants are exposed to sanitary conditions.
- Make the building acceptable for sociological and psychological reasons.
- Insure that work flow of the building's occupants and equipment is not impeded, thus insuring the highest rate of return for the productive activity being carried on in the building.

These reasons and others should be kept in mind when establishing the objectives for a maintenance program.

Repairs are an important part of a maintenance program. However, distinction must be made between maintenance and repairs, since the objectives of each are different.

Maintenance is the day-to-day, or periodically scheduled work required to preserve or restore a facility to a condition in which it may be effectively utilized for its designated purpose. This includes preventive maintenance which is less costly than repairing or replacing damaged facilities.

Repair is the restoration of a material or facility so that it may be utilized effectively for its designated purpose. This may be accomplished by overhaul, reprocessing, or replacement of constituent parts or materials that have deteriorated by action of the elements or by usage, and which have not been corrected through maintenance.

3

Types of Maintenance

There are three general types of maintenance: maintenance-free, built-in maintenance, and built-on maintenance. The basic differences between these three types and instances in which they are used are described in the following section.

Maintenance-Free

The first type of maintenance is known as *maintenance-free*. Ideally, if given enough time and money, this could be attained. Realistically, the unlimited resources necessary to develop maintenance free materials, equipment, and components are not normally available. This type of maintenance, however, should always be an ultimate goal in the minds of those developing new materials and systems for industry. If we lose sight of this goal, newly developed products will mean a higher expenditure than necessary in any one company's budget.

Built-in Maintenance

The second type of maintenance is *built-in maintenance*. There are certain items in any building that must be included in order for the building to meet the functional needs of the owner, and these items must be maintained. For example, when designing a manufacturing facility, the owner specifies certain types of machinery needed to manufacture the items for which he has set up the plant. These items must be maintained on a regular basis to insure that they function properly in order to guarantee the realization of the objective for which the facility was built—its profit-making potential. During the planning and design stage, it is important that an item-by-item value engineering analysis takes place, at which time each item, whether a door, window, piece of machinery, and/or area should be evaluated as to its need within the building and the function it will play. By doing this, the owner may be able to eliminate certain items and thus reduce his proposed maintenance budget.

Built-on Maintenance

The last type of maintenance is *built-on maintenance* which is performed on materials which have been included in the building, but are not necessary for its function. As an example, consider the need for rugs on the floor of a warehouse area. A value engineering analysis would certainly uncover the necessity to eliminate the carpet. This type of maintenance is one which could be definitely eliminated through careful planning and thus reduce the scope and cost of the building's maintenance program.

Stages of Maintenance

Maintenance, whether built-in or built-on, is a necessary aspect in the day-to-day life of a building. There are essentially three stages of maintenance: the planning and design stage, the construction stage, and the maintenance stage. The following sections describe these stages with suggested methods of dealing with each one.

Planning and Design Stage

The first stage occurs during the planning and designing of the facility. This is known as the *design stage*. During this stage, building owners and managers have the opportunity to save thousands of dollars by planning and designing the building to be maintenance-efficient. It is ironic that the same problems consistently occur in similar structures under similar conditions of exposure. This reoccurrence appears to be the result of the use of unsuitable details and/or practices in the design phase of the building. It indicates that architects (or other design professionals) are not aware that such details and practices have proven unsatisfactory relative to the maintenance of the building.

When Does Maintenance Begin?: Maintenance considerations for any facility should begin the day the new building is planned. The design of any facility should be based on the identified function and be as maintenance-free as possible. Considerations relative to the scope of maintenance should always be present in the design process. It is of utmost importance that the personnel who will be actively involved in the planning and managing and/or operating of the maintenance program are involved during the design and planning stage. These people have experienced the shortcomings of similar buildings and have greater on-the-job insight as to the problems that might be encountered and possibly alleviated or built out of the new similar facility. Their recommendations should be incorporated into the final specifications and drawings, if they are economically and aesthetically feasible.

Built-out Maintenance: The term *built-out maintenance* refers to the ease of maintaining the building once it is constructed and occupied. It is of utmost importance that the owner of the building devote time, manpower, and budget during the planning stage of the building to ease its maintenance once built. If this is not done, it is safe to assume that the maintenance budget will be more costly, especially in today's world of ever-increasing wage rates, fringe benefits, and the difficulty of finding and training personnel for maintenance work. The ease of maintaining any material refers to the condition of an item or a surface that permits its repair, adjustment, or cleaning, with reasonable effort and cost. Effort and cost to maintain a particular item should not require abnormal or unusual working skills or exceedingly expensive equipment that is seldom used and requires specialized maintenance itself.

The amount of money wastefully expended each year in the maintenance of facilities due to improper design and planning is phenomenal. Lifetime maintenance costs will probably equal the original cost of construction. Millions of dollars can be saved by properly planning and designing the building for ease of maintenance. This, in itself, is a reason why building managers and maintenance personnel should be consulted during the early stages of the building design. Only they can bring with them the experience required to design or build out maintenance problems for the future.

The opportunity to build out unnecessary maintenance items comes but once, and once the opportunity is lost, it is gone forever. A new building can become old very rapidly, earlier than planned, without a well-planned maintenance program.

Material Selection: During the design stage, it is important that the architect, owner, and maintenance personnel select the proper construction materials to withstand environmental exposure and functional use within the structure. Attention to the details of design and their importance with respect to the overall facility must always be of concern during the design stage.

The designer must remember that what is obvious to him may not be understood by the contractor and/or building owner. Basic economy of design takes place in the selection of the type of structure, the materials, and the building support systems. Normally, the details have little effect on the actual cost, but they are, by far, the primary source of difficulty not only in the construction process but also after the facility has been built and is being occupied. Details not designed and/or constructed with maintenance in mind can only lead to future problems.

Construction Stage

The meticulous process and additional expenditures of designing a maintenance-efficient building, are of no value if the materials and equipment are not assembled and installed properly. For this reason, the second stage of maintenance is the *construction stage* must be performed with the highest quality of workmanship to create a facility requiring a minimum level of maintenance. The contractor selected for the construction of a facility should be well-experienced in constructing the type of facility being planned. This is the responsibility of the owner and the architect. Once the contractor has been selected, a clear and concise set of drawings and specifications is prepared by the architect and followed by the contractor. This effective planning is realized in the finished facility.

Frequent Inspections: Once construction begins, every pipe line, valve, floor tile, and electrical outlet should continually be observed as construction progresses so as to insure proper location and the highest quality of material. This, in turn, can help to minimize the amount of maintenance required once the building is completed and ready for occupancy. The owner should realize that he has made a large investment which he needs to protect through the best quality control mid inspection process should be implemented to monitor construction progress. Although this may add time and cost to construction of the new facility, the owner should realize that he must protect his investment by utilizing the best quality control and inspection process possible.

Maintenance Personnel Involvement: It is interesting to note that it has been found that the best maintenance programs result when maintenance personnel are allowed to observe the actual on-site construction and are involved in checking the locations of many of the "hidden" maintenance items, such as underground piping.

Changes: During the construction stage, if a particular maintenance problem is anticipated in the future due to factors such as poor location of equipment, the owner or architect should immediately consider revising its location to make the equipment more accessible for maintenance purposes. This can be done through the utilization of *change orders* to the contract documents. Change orders often result in a cost increase in the original price of construction, but may, in the end, result in potential savings over the lifetime of the building. The reader will find, throughout the balance of this book, many checklists which can be utilized during the design, and especially the construction, stages that will insure a more maintenance efficient building.

Maintenance Efficient Buildings: The design, construction, and maintenance of any facility are, for the most part, performed by separate departments and frequently by separate firms. Moreover, the maintenance problems often do not appear for several years after the construction of the building. By this time the persons responsible for the design and construction are likely to be engaged in other duties, deceased, or otherwise detached from the need to be concerned about the building. As a result, liaison between the architect, the constructor, and the maintenance personnel is poor. The architects and constructors, however well-intentioned, have little opportunity to learn from the poor performance of their work. In the same light, there is a natural reluctance on the part of those who have performed the deficient work to admit that they have made a mistake. This is one reason among many why it is often so difficult, and thus such a challenge, to construct a maintenance-efficient building. However, if those involved are aware of these potential obstacles, they can meet this challenge by using available funds, manpower, and resources to design and build out possible future maintenance problems. Use of the "Reminder List for the Planning and Construction Stage" in Figure 1.1 can help build out potential maintenance problems.

Reminder for Planning and Construction Stage

Exterior Grounds
____ Be sure backfill is sloping away from the building.
____ Be sure the soil is compacted according to specifications.
____ Protect trees and shrubbery during construction.
____ Consider soil poisoning where insect infestation can become a problem.
____ Specify soil suitable for lawn growth.
____ Use ground cover where possible.
____ Do not plant grass on steep slopes.
____ A built-in sprinkler system will reduce maintenance costs.
____ Avoid small grassy spots.
____ Provide adequate walkways to avoid lawn damage.
____ Use shrubbery, trees, and grasses which are common to the geographical area.
____ Consult with a landscape professional.
____ Provide an unplanted space around trees to avoid mowing damage.
____ Do not plant trees and shrubs near curbs.

Exterior Materials
____ Protect existing walkways or paving from damage.
____ Protect all surfaces which are susceptible to damage during exterior cleaning.
____ Install concrete or asphalt paving over a well-compacted base.
____ Apply protective sealers to flooring and pavement.
____ Asphalt should not be used on grades exceeding 10%.
____ Crown all walkways and roadways for drainage.
____ Consider the exposure condition relative to the type of material being considered for placement in a specific area.
____ Use maintainable surfaces such as polished stone, stainless steel, and glass for exterior surfaces.
____ Construct exterior walls of brick, cast stone, natural stone, and other permanent materials (including certain woods).
____ Avoid using porous stone in climates which have sub-freezing temperatures.
____ Do not use precast concrete, plastic, or metal panels in horizontal plane.
____ Install rustproof fasteners on exterior.
____ Do not use ferrous materials for flashing.
____ Provide adequate flashing where columns and other items penetrate the roof.
____ Items penetrating the roof should not be too close together.
____ Avoid flat slope tops of walls and parapets.
____ Concrete block should be stored under cover on the site as should other materials.
____ Provide adequate flashing and sealing for sky lights.
____ Use safety glass on all ground floor large glass areas.

Exterior Systems
____ Protect catch basins and storm drains from accumulation of construction sediment.
____ Roof drains must be at low points.
____ Provide storm drainage in loading and parking areas.
____ Install lighting in parking areas.
____ Install an adequate number and size of roof drains and downspouts.
____ Protect downspouts where they may become damaged.
____ Provide interior downspouts in cold weather climates.
____ Provide for roof-mounted mechanical equipment.

Figure 1.1

Exterior—General

_____ Remove soil from exterior pavements to prevent its being tracked into the building.

_____ Consider the size of grounds-care equipment in the design of sidewalks, entry ways, etc.

_____ Install metal corner guards to protect the exterior corners of all walls.

_____ Use a gravel or paved splash area to prevent mid splatter on outside walls.

_____ Install ramps, where needed.

_____ Provide exterior egress to storage closets.

_____ Avoid as many changes of surface indentations as possible.

_____ Install roof walkways for maintenance and inspection purposes.

_____ Do not use parapets and other decorative roof feathers, if at all possible.

_____ Provide ramps for wheel chair victims.

_____ Alter the design of window ledges that are near floor level so that they cannot be sat or stood on.

_____ Consider specifying windows with built in guides for window washers in high places.

Interior Materials

_____ Use building paper to protect finish surfaces.

_____ Use a vacuum attachment device for the installation of terrazzo and concrete flooring when grinding.

_____ Remove any foreign materials from walls prior to applying finish coverings.

_____ Replace marked or soiled tile.

_____ Be sure all finish building materials such as tile floors are given their initial preservative treatment as per manufacturer's recommendations.

_____ Consider the exposure condition relative to the type of material being considered for placement in a specific area.

_____ Provide expansion space between concrete slabs and masonry walls.

_____ Carpet should be of a dense pile, multi-colored light tweed and have a durable backing.

_____ Specify finish flooring, wall and ceiling materials which will endure the conditions to which they are exposed (see Chapter 9).

_____ Install concrete for industrial and other uses where appearance is not a prime factor and specify it to be sealed.

_____ Do not use resilient tile, slate, or marble for stair treads and landings.

_____ Stairwell walls should be durable and not made of easily marked materials.

_____ Handrails should be simply designed, continuous, and well attached.

_____ Burlap, natural grass paper, and felt wall coverings are not recommended.

_____ Flat paints mark easily and are difficult to wash.

_____ Use plastic-finished wood panels.

_____ Minimize wall marks by using both materials that are difficult to mark and provide good illumination.

_____ Install lift-out acoustical tile.

_____ Specify plastic-faced mineral acoustic tile.

_____ Install adequate number and size of expansion joints.

Interior Systems

_____ Temporary heating should not produce soot.

_____ Disposable air conditioning filters should be used during construction.

Figure 1.1 (continued)

_____ Be sure that the floor drains are at the lowest point of the floor.
_____ Test and balance all ventilation and hydraulic systems.
_____ Install trenches and/or drains in floors which are exposed to water.
_____ Be sure elevator is of the correct size and capacity for facility needs.
_____ Install an electrical outlet in every elevator.
_____ Install a wall-mounted cigarette urn/waste-receptacle unit in each elevator.
_____ A large freight elevator should be provided, preferably one that goes to the roof and can be entered from the exterior of the building.
_____ The elevator shaft should contain proper lighting for maintenance.
_____ Lights should not be located over stairs.
_____ Provide an adequate number of wall-mounted cigarette urn/waste containers.
_____ Install an electric receptacle on each stair landing.
_____ Provide glazed tile wall material for areas such as those meant for food processing, health care, and high moisture utilization areas, such as rest rooms.
_____ Specify easily cleaned surfaces, such as plastic laminates for elevator walls and doors.
_____ Fire extinguishers should be placed in recessed enclosures.
_____ Surfaces adjacent to diffusors should be smooth tile or metal.
_____ Minimize installation of equipment in overhead areas.
_____ Be sure that all piping, wiring, conduit, equipment, etc., are readily accessible.

Interior—General
_____ Provide for thorough cleaning before final occupancy.
_____ Do not paint windows and doors closed.
_____ Plan for the immediate proper disposal of all waste materials.
_____ Do not use open-slot type expansion joints.
_____ Provide a hanging strip, if many wall hangings are to be used.
_____ Main corridors should be at least eight feet wide.
_____ Install stainless steel corner guards or rounded corners for walls and columns which are exposed to high traffic.
_____ Minimize the use of windows and consider fixed, rather than operating ones.
_____ Specify windows that can be washed from the inside in high places.
_____ Attempt to eliminate venetian blinds.
_____ Install heavy-duty, standard size hardware.
_____ Avoid the use of floor-mounted door stops.
_____ Use plastic laminates for push plates and kick plates.
_____ Specify the type, pattern, material, and operation of doors and windows which will most efficiently meet the needs for the location in which they will be installed and result in a minimum amount of maintenance.
_____ Do not use full-length glass doors.
_____ Do not use louvered doors for decorative purposes.
_____ Specify hard swinging doors in industrial warehouses.

Figure 1.1 (continued)

Inventory for Pre-Maintenance: During the construction stage, a written inventory should be devised to establish the following information:

- Elements in the building (wood flooring, chalk boards, etc.).
- Types of materials of which elements are composed.
- Manufacturers of all building elements, including systems, machinery, and structure.
- Manufacturers' and/or constructors' recommendations relative to maintenance and repair of their products.
- Date material and/or system is installed.

The source for the information for the first three items are the contract documents including as-built drawings and specifications. The job construction schedule can provide some information when the various material and systems were installed.

This inventory should become the foundation of the maintenance and repair program for the facility. It would serve as documentation for any future repair and troubleshooting. Finally, all guarantees and warranties should be part of the documentation in a complete inventory system. The reader should note that there are many commercially available computer software programs which can be used to inventory all the building components.

Maintenance Stage

The third stage of maintenance is the maintenance of the building itself, once it has been constructed and occupied. As stated earlier in this chapter, maintenance is a very complex subject. In order to grasp the concept and bring it into a more controllable focus, cost-wise, we must ask the question, "What are we basically maintaining?" No matter what type of building or facility being referred to, we are speaking of maintaining materials. These materials take the form of surfaces (such as tiled floors), items that are auxiliary to the building (such as sidewalks and landscaping), the structural elements supporting the building (such as beams and columns), systems that provide air conditioning and heating for the building, and other equipment which allows the building to be functional. No matter what the material, maintenance is called for to prevent premature material failure. Some types of localized failures, such as peeling of paint or staining of carpets, will not lead to disaster, such as the death of an occupant. However, other failures, such as structural cracks in a concrete beam, or interior rotting of structural timbers due to insect infestation, could result in structural collapse and, thus, deaths. Therefore, a list of priorities should be established based on the maintenance of specific materials within the building.

There are five basic parameters of any material which one should consider when a) selecting and implementing a maintenance procedure, b) selecting a material for a specific use and c) determining the cause of the deterioration of the material. These five parameters, which are discussed in the subsequent sections, are listed below:

- Forces acting on the material.
- Chemical and physical properties of the material.
- Location of the material.
- Support of the material.
- Coexistence with other materials.

Forces A force is any effect which tends to change the state of motion or the shape of any material. Examples are listed below:

- Dead weight of a material i.e., gravity.
- Moving objects e.g., impact.
- Push or pull produced by expansion or contraction due to temperature change or swelling, or shrinkage due to moisture change.
- Internal distortions caused by yielding of supports, such as foundation settlement.

Forces are either *static* or *dynamic*. Static forces are those forces that are always present, such as the dead weight of the element itself. Dynamic forces are moving forces, such as a truck hitting a tree or an ocean wave hitting a pier.

Forces are induced by *external loads*. The two basic types of loads are *dead loads* and *live loads*. A dead load is one that is permanent, such as the weight of an air conditioning unit on a roof member. Live loads are occasionally applied loads, such as people walking on a floor.

Every material must resist both dead and live loads. An example of this principle is an office building, where on any one floor there exists a live load comprised of the people and furnishings occupying the floor space and the dead load comprised of the weight of the finish floor material. Refer to Figure 1.2 for a summary of the various types of forces, their origin, and how they are applied.

The magnitude of the minimum loads allowed can be found in one of the four regional building codes used in the United States, specifications and/or manufacturers' material codes. It is the structural engineer's responsibility to use the minimum specified loads according to local codes and owner's needs. It is the maintenance department's responsibility to insure that the structural members are functioning only for what they were designed to do and nothing additional without further analysis.

Load Types

The first step in determining whether or not a structural failure is impending is to be able to recognize the types of loads that are acting on the building element. Loads are applied to a structure in two ways. The first type is referred to as a *distributed load*. This type of load is either *uniformly* or *non-uniformly* applied. Most floor loads are examples of uniformly distributed loads. This means that it is assumed (for design purposes) that there exists the same magnitude of load everywhere on the floor. This type of distributed load, sometimes called an *area load*, is given in units of force or weight per unit area (such as pounds per square foot). Uniform loads on structural elements, which support floor areas, are expressed in units of so many pounds per linear foot of length.

Summary of Forces

1. **Gravity**
   Weight of materials and permanently attached items—dead loads. Weight of building occupants, moveable contents, and snow—live load applied in downward direction.
2. **Wind**
   Wind velocity. Applied horizontal from any direction.
3. **Earthquake**
   Produced by movement of internal earth mass producing shock waves, resulting in shaking of the ground. Random vertical and horizontal application.
4. **Blast**
   Due to explosions. Horizontally applied.
5. **Hydraulic and Pneumatic Pressures**
   Liquids and gas forms applied horizontally and/or vertically.
6. **Thermal Expansion and Contraction**
   Due to temperature variations in materials. Internal forces applied in vertical and horizontal directions.

Figure 1.2

Non-uniform distributed loads exist when analyzing such structures as liquid storage tanks or below-ground foundations such as retaining walls. The units of non-uniform loads are essentially the same as for uniform loads. The only difference is the method of design.

The second way a load is applied is as a *concentrated* or *point load*. An example of this is a piece of machinery sitting on a floor member. The units of a concentrated load are measured in weight units such as pounds or kips (one kip equals 1,000 pounds). See Figure 1.3 for an example of load types.

Stress and Strain

A structure, and thus the materials making up the structure react to the applied load by resisting the effect of the external force. In resisting external force action, internal forces called *stresses* are set up within the material as they resist the attempt of the element to change shape. A stress in a body is an internal resistance to an external force and may be measured in pounds per square inch (see Figure 1.4).

Internal stresses are always accompanied by deformation of the material, known as *strain*. Strain is defined as the total change in length divided by the total length of the member. The units assigned to strain are inches per inch.

When subjected to a load, a structure or structural element twists, curves, stretches, shortens, and/or sags, thus assuming some new configuration as the strains accumulate in an overall shape change of the structure. The visualization of what is actually occurring is important. Some types of deformation can be observed visually, while others can be observed only by using scientific measuring devices depending on such parameters as material size and shape, length between supports, and others.

The type of deformation can be a clue as to the types of stresses that are being produced to cause possible material deterioration. It is normal for stresses and deformations to occur up to a certain limit, depending on the type of material being utilized. As with human beings, only so much stress can be endured before a breakdown occurs. When an overstressed

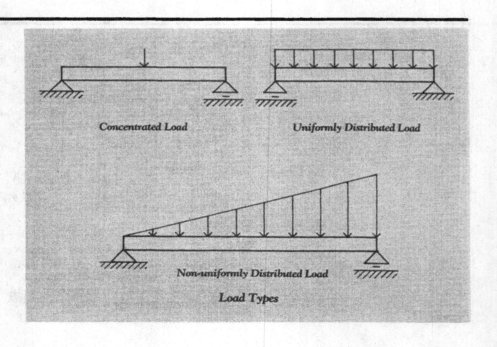

Concentrated Load   Uniformly Distributed Load

Non-uniformly Distributed Load

Load Types

Figure 1.3

condition is induced in any material, it reacts by breaking down physically, thus impairing its ability to carry the induced load. Localized deterioration of the material is evidence of the breakdown process taking place.

Tension and Compression
Two basic types of forces, and thus induced stresses, cause material deterioration. These are *tension* and *compression*. Tensile stresses are produced by pulling-type forces, while compressive stresses are produced by pushing- or crushing-type forces (see Figure 1.5).

When compression and tension are occurring simultaneously on a structural element, the result is known as *bending* (see Figure 1.6). Other types of internal stresses that the structural engineer is concerned about are *shear*, a type of *slicing action*, and *torsion* or *twisting* (see Figure 1.7).

Types of Stresses
It is important for maintenance and inspection personnel to be able to relate the type of deformation and its location on the structural element to the basic type of stress causing it in order to evaluate whether or not the resulting deterioration is evidence of a potential structural failure. Following is a summary of the basic types of stresses and the resulting types of deformation along with possible resulting deterioration clues that appear when the member is in an overstressed condition. This is followed by Figure 1.8 which can be used by the maintenance department in their evaluation of the seriousness of deterioration relative to possible structural failure. The outline below lists the characteristics of basic types of stresses.

- **Tension-pulling force:** Produces tearing at holes and notches; induces straightening of bent elements.
  1. Steel: Produces elongation; Produces tears in structure.
  2. Concrete: Causes cracking.
  3. Wood: Causes elongation, then splitting.
- **Compression—pushing, crushing:** Produces crushing of stocky elements, buckling of slender elements; can be transferred by simple contact bearing with no connection as such.
  1. Steel: Shortening—stocky members (relatively short and large in cross-section); buckling—slender members (relatively long and small in cross-section).
  2. Concrete: Crushing, spalling—stocky members; Buckling, crushing, and spalling—slender members.

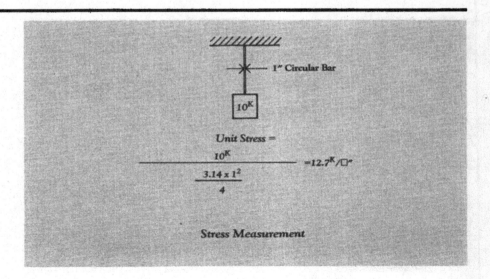

Figure 1.4

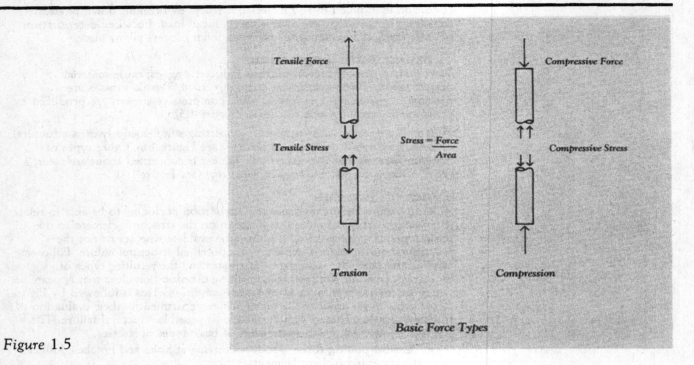

Figure 1.5

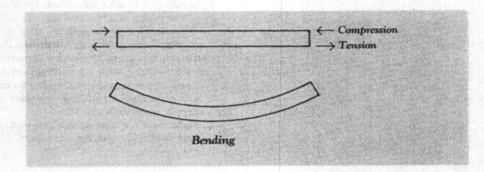

Figure 1.6

- **Shear—slicing action:**
  1. Steel: Tearing separation; localized buckling at point of load.
  2. Concrete: Diagonal cracking beginning at point of maximum shear.
  3. Wood: Tearing with cutting separation.
- **Bending:** Causes curving, tension, compression, and shear forces in a section. Bending stress causing possible failure will be maximum at the outermost fibers of a member, such as the flanges of an open shell member. Buckling is a form of bending.

Detecting Structure Deterioration

Some of the general points that must be considered in determining the cause of deterioration and, thus, the need for maintenance of any material. These general points are outlined below.

1. If a potential failure is being caused by overstress, one must examine the location at which the tension and/or compressive stress have become maximum to determine if this location corresponds to the point of deterioration.
2. Maximum stresses will occur at the outermost fibers or faces of any material first and, therefore, any deformities or deterioration due to overstress will appear on the surface of the material first.
3. A structural designer should be consulted for any specific required material and design analysis.
4. All materials and structures have some inherent strength reserves. These are:
   a. Many of the design assumptions inherent in the mathematical procedure of design are often conservative, resulting in a margin of safety.
   b. Deterioration of a member may not be located at points of maximum stress in the member.
   c. If original design is based on elastic analysis, there is an inherent factor of safety.
   d. Full design live loads seldom are fully realized in practice.
   e. When selecting specific size members to satisfy design requirements, larger than required members an often used.

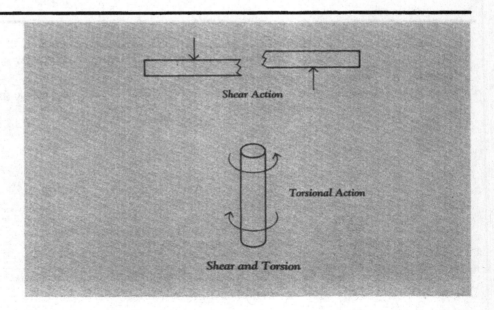

Figure 1.7

f. Strength of material may increase with age.
g. Actual structural action may differ from that assumed in the design stage.
h. Often times a "construction or temporary load" is calculated into the design and no longer exists once the construction is completed.
i. Conventional design procedures do not consider all actions which take place under actual service loads.
j. Some structural elements are designed based on stiffness rather than strength, resulting in a reserve of strength.

Chemical and Physical Properties of Materials

The chemical and physical properties of a material are important. For instance, the material must be of a certain size in order for it to resist the induced loadings and fit into the standard scheme or pattern of the system of which it is a part. It would be ideal to have materials which would require no maintenance. However, this is impossible, and the sooner one realizes the specific type of maintenance needed for the specific material and follows up with the required program, the less costly it will be to keep up the material. Not only must a material stand up to the load and also the exposure conditions, but it must also convey a certain feeling and give a certain appearance, depending upon its placement within the building. Therefore, such characteristics as color, texture, form, and appearance are also of importance and must be considered in the establishment of a maintenance program. See Figure 1.9 for a complete listing of general properties of materials. Specific entries under each one of these classifications that are important or refer to a particular type of material, such as floor tile, will be covered in the specific chapter of this book.

Summary Material vs. Failure Mechanisms					
Material	Tension	Compression	Shear	Torsion	Bending
Wood	Splitting	Crushing — short, stocky members Buckling — long, slender members	Cutting Action Some splintering with plane of wood	Cracking and splintering	A combination of tension and compression
Steel	Necking down	Same as wood	Local distortion at concentrated loads and supports	Twisting and local buckling	Compression and tension
Reinforced Concrete	Cracking (usually without spalling)	Crushing and spalling	Diagonal cracking from support or concentrated loads	Cracking and local crushing	Dependent on design. Need to have some cracking and no spalling.

Figure 1.8

A working knowledge of both the general and structural properties of materials is essential to anyone involved with their maintenance. A brief summary of the advantages and disadvantages of the primary materials used in construction today is presented in Figure 1.10.

Material Location

The third parameter, location of material, actually can be incorporated in the other parameters listed earlier in this chapter. However, the location of the material, with respect to the role that it plays within the building, is important enough to warrant its own place in the list. Locations refer to the exposure conditions of the material, specifically such things as impact, chemical attack, freeze-thaw action, operation misuse, and other such parameters.

Material Support

It may be found that it is not the material that is the cause of its own deterioration, but what supports it. For this reason, the method of material support is very important. The connection of one material to another (the support means) is always a source of potential maintenance problems because it is usually inaccessible once the building is completed. An example of this would be the mastic that holds the tile to the concrete floor. The support of one material by another must be considered and should be a priority item on any inspection list to be used within a maintenance department.

Basic Classifications of Materials

Structural Properties
1. Strength varies for different types of forces, in different directions, and at different ages or temperatures.
2. Strain Resistance is the degree of rigidity, elasticity and ductility.
3. Hardness is the resistance to surface indentation and scratching.
4. Fatigue Resistance is the time loss of strength, shape change with time, and progressive fracture.
5. Uniformity of Physical Structure relates to such things as grain and knots in wood, cracks in concrete, and shear planes in stone.

General Properties
1. Form is being natural or man-made.
2. Weight.
3. Fire Resistance relates to combustibility, conductivity, performance at high temperatures, and melting point.
4. Coefficient of Thermal Expansion is utilized in evaluating movements of structure due to temperature changes.
5. Durability is the resistance to weathering, rot, insects, and wear.
6. Workability in producing shaping, assembling, and altering.
7. Appearance refers to whether natural or finished.
8. Availability.
9. Cost.

Figure 1.9

A Review of Major Building Materials — Advantages and Disadvantages

Material	Advantages	Disadvantages
Aluminum	Lightweight High resistance to corrosion Low electrical resistance Good conductor	Softness Limited strength for structural uses Low stiffness High rate of thermal expansion Low rate of fire resistance Relatively high cost
Concrete	High compressive resistance Durable Resistance to moisture, rot, insects, fire and wear Low bulk cost Water tight (depending on water-cement ratio)	Workability Lack of tensile strength Lack of resistance to many types of chemical exposure, such as salt Hard to remove stains
Copper	High resistance to corrosion Good electrical conductivity Workable Forms its own surface protection Thermal contraction and expansion not high	Costly Strength varies with treatment and mechanical working
Epoxy	(can have varied properties depending on composition) Liquid use very applicable Controllable Excellent strength properties Small creep during curing Hard Tough Resistant to abrasion Resistant to corrosion, salts, acids, petroleum products, solvents and other chemicals Adhesion to surfaces good	Color Form
Glass	Considerably strong Large sizes expensive Non-corrosive	Brittle, subject to shattering under shock Transmits energy at a rapid rate
Lead	Good resistance to corrosion	Heavy High coefficient of thermal expansion Difficult to hold in place
Masonry	Available in small units Appearance (available in many textures, sizes and colors) Good insulator	Stains hard to remove Porous (absorption rate high) Shrinkage of mortar Thermal and expansion cracking Others — see concrete
Paper	Readily available Low in cost Used in conjunction with other materials	Susceptible to water damage Susceptible to rot Relatively weak Highly combustible
Plastics — general	Applicable to many uses Relatively low in cost Workable into many shapes High strength Lightweight Non-corrosive Others — See specific plastic entry	Lack of resistance to fire Low stiffness High rate of thermal expansion Low thermal conductivity Some cases of chemical or physical instability with time Non-salvageable Others — see specific plastic entry
Plastics — specific Acrylics	Transparent Hard Weather resistant Shatter resistant	Easily scratched

Figure 1.10

A Review of Major Building Materials — Advantages and Disadvantages (continued)

Material	Advantages	Disadvantages
Polyethylene	Flexible Tough Translucent Low cost	Easily scratched
Polystyrene	Hard Clear Water and chemical resistance Low cost	Brittle
Vinyls	Tough Wear resistant Stain resistant	
Polyamides (nylon)	Hard Tough Wear resistant	Costly
Alkyds	Water Resistant Tough Good adhesive properties	
Melamines	Hard Durable Abrasive resistant Chemical and heat resistant	
Polyesters	Weather and chemical resistant Stiff Hard	
Plywood	Stronger than standard lumber Durable High resistance to impact and load Not as affected by moisture changes as standard wood Workability See Wood for others	Dimensional stability not as good as that of standard structural lumber Poor quality glues are possible Thermal expansion and contraction more of a problem
Steel	Strong Most resistant to aging Most reliable in quality, non-combustible, non-rotting, dimensionally stable with time and moisture change Resistant to staining	Costly Resultant loss of strength when exposed to intense heat, rapid heat gain and loss Corrosive when exposed to moisture and air or other corrosive conditions
Tin	Extremely workable Very resistant to corrosion	Heavy Loss of strength when exposed to heat
Wood	Readily available Relatively low in cost Simple to work with Available in many shapes and forms Good insulating properties	Paintability changes with moisture content Combustible Susceptible to rot and insect infestation Soft and easily damaged (porous) Dimensional changes due to changes in temperature and moisture Strength changes with changing moisture content
Zinc	Fairly high resistance to corrosion Forms its own protective surface Workable	Brittle

Figure 1.10 (continued)

Coexistence of Materials

Not only do materials have to support each other, but they also must coexist in the same environment. This can cause complications. An example of this is what happens when steel sheet metal screws are used on aluminum gutters. As anyone in maintenance well knows, corrosive action is set up due to the chemical action of water and air as a catalyst between the chemicals within the steel and aluminum, causing rusting and rapid deterioration. Steps must be taken during the design and construction stage to identify potential problems in the connection or installation of incompatible materials and make appropriate substitutions.

Another example of materials coexisting can be found in the act of breathing. Just as human beings do, materials also need to breathe (expand and contract). If the act of breathing is restrained due to a change in temperature, internal stresses are set up causing possible material deterioration. These internal stresses, in turn, create a need for maintenance. One must consider the fact that different materials breathe at different rates. Allowance must be provided in the building for this to happen in the form of expansion and contraction joints. The reader is referred to any engineering handbook for a list of Coefficients of Thermal Expansion of different materials. These coefficients are a measure of the amount of linear changes the materials will undergo with increases and decreases in temperature. The reader should note that each material has its own unique coefficient, which means that each material will breathe at a different rate. Dissimilar materials should not be fastened together so as to inhibit their differential movements. Provision must be made for expansion and contraction of all materials, from the basic vinyl floor tile to the complex concrete structure.

Each of the five parameters discussed above will be explained in more depth in the following chapters which examine basic materials as they are found in any building.

CHAPTER 2

EVALUATION OF MATERIAL DETERIORATION

Once the behavior and properties of material are understood, the facilities manager must learn to detect and control these properties. This can be successfully accomplished by implementing a formal inspection program. Allocating time and money to inspection is important for the following reasons:

- to locate any new deterioration that has taken place since the last inspection.
- to review the existing conditions of the materials in the building, including equipment.
- to follow up on repairs that have been accomplished since the last inspection.
- to evaluate present housekeeping procedures.

The inspection process should be structured as a set of formal guidelines or procedures. These general guidelines should establish the following criteria:

- How often to make an inspection, which is dependent upon the available manpower and items being inspected (see specific chapter for this information).
- Where to begin the inspection process.
- The order of inspection (establish a set of inspection priorities and follow the list).
- How to inspect any one specific component or material.
- Types of things to look for relative to degree of deterioration (refer to checklist in respective chapters).
- Determining probable causes of deterioration.
- How to record, on standard format sheets, what is observed.
- Types of follow-up action recommended.
- Other general information relative to the inspection such as:
  —Time of day.
  —Name of inspector.
  —Hour of inspection.
  —Weather conditions, if outside.
  —Location of inspection, if not a general overall building inspection.

Figure 2.1 is a suggested inspection format. The most effective inspection format, or checklist, however, is one which is developed by the company for its own specific use. Figure 2.2 shows recommended inspection frequencies and Figure 2.3 shows time estimates to make inspection.

Means Forms
BUILDING INSPECTION
CHECKLIST (Interior and Exterior)

BUILDING	SHEET OF
DATE	TIME OF DAY
WEATHER	INSPECTOR

I. Exterior Maintenance	Comments
A. Roof	
___ Cracks	
___ Blisters (vapor)	
___ Wrinkles	
___ Ponding	
___ Gravel Coming Off	
___ Wet Insulation	
___ Dry Felts	
___ Foreign Objects on Roof	
___ Other (Comment)	
B. Flashings and Counterflashings	
___ Composition Flashings	
___ Cracks	
___ Open Joints or End Laps	
___ Pulled Away From Wall	
___ Other (Comment)	
C. Metal Counterflashings	
___ Pulled Out of Reglets	
___ Loose Wedges	
___ Deteriorated Caulking	
___ Rusted Galvanized Iron	
___ General Deterioration	
___ Other (Comment)	
D. Expansion Joints	
___ Cracks	
___ Mislocated or Loose Blocking	
___ Deteriorated Elastic Membrane	
___ Other (Comment)	
E. Gravel Stops and Gutters	
___ Peeling Paint	
___ Broken Solder Joints	
___ Separation From Roof Piles	
___ Other (Comment)	
F. Skylights	
___ Secured to Supports	
___ Cracks	
___ Other (Comment)	

Page 1 of 4

Figure 2.1

22

⚓ Means Forms

Items	Comments
G. Plumbing Vents	
Secured to Supports	
Cracks	
Other (Comment)	
H. Antennas, Signs and Other Similar Items	
Rust and Peeling Paint	
Loose Guys	
Roof Seal	
Other (Comment)	
I. Parapets	
Open Mortar Joints	
Open Joints in Copings	
Other (Comment)	
J. Chimneys	
Deteriorated Brick, Joints	
Flue Liner	
Flashing	
Other (Comment)	
K. Walls	
Cracks	
Paint Condition	
Evidence of Moisture	
Efflorescence	
Areaways	
Caulking	
Other (Comment)	
L. Doors	
Closers	
Exit Devices	
Locks and Misc. Hardware	
Deteriorated Surface	
Other (Comment)	
II. Interior Maintenance	
A. Structural System	
Rot, Splitting, Rust	
Termites	
Deflection	
Cracks	
Other (Comment)	

Page 2 of 4

Figure 2.1 (continued)

23

Items	Comments
B. Stairs and Handrails	
Loose Handrails	
Worn Treads	
Paint or Varnish	
Other (Comment)	
C. Interior Walls	
Cracks, Bulges, Etc.	
Paint Conditions	
Dirt	
Other (Comment)	
D. Ceilings	
Cracks, Bulges, Etc.	
Paint Condition	
Dirt	
Other (Comment)	
E. Floors	
Wood	
Deteriorated Areas	
Finish Condition	
Other (Comment)	
Resilient Floor Covering	
Damaged Areas	
Finish Condition	
Other (Comment)	
Carpet	
Stains	
Dirt	
Other (Comment)	
F. Trim	
Stains	
Dirt	
Paint Condition	
Other (Comment)	

Page 3 of 4

Figure 2.1 (continued)

24

⚓ Means Forms

Items	Comments
G. Doors	
Closers	
Exit Devices	
Hinges	
Locks and Latch sets	
Keying	
Misc. Door Hardware	
Surface Condition	
Other (Comment)	
H. Windows	
Putty and Caulking	
Broken and Cracked Glass	
Sash Operation	
Hardware	
Decay and Termites	
Sills and Stools, Paint, Etc.	
Other (Comment)	
I. Shades and Blinds	
Hardware	
Blades (Ditty or Paint Cond.)	
Fabric	
Rollers	
Other (Comment)	

III. Items Not Specifically Covered (List Item and Comment)

Figure 2.1 (continued)

Recommended Inspection Frequencies	
Item	**Frequency**
Large Electrical Power Plants	6
Disconnecting Switches	5
Electrical Grounds & Grounding Systems	5
Electrical Instruments	5
Electrical Potheads	4
Electrical Relays	4
Lightning Arresters	5
Power Transformers, Deenergized	6
Power Transformers, Energized	3
Safety Fencing	4
Steel Poles and Structures	6
Vaults and Manholes (Electrical)	5
Wood Poles and Accessories	6
Cathodic Protection System	5
Electrical Motors and Generators	
Running Inspection	4
Shutdown Inspection	5
Electrical Systems (Buildings)	5
Lighting (Buildings)	5
Pier Circuits and Receptacles	5
Switchgear (Buildings)	5
Distribution Transformers, Deenergized	6
Distribution Transformers, Energized	5
Aerial Telephone Cable	
From Ground	
From Pole	6
Buried or Underground Telephone Cable	6
Open Wire Telephone Lines	6
Telephone Substations	6
Kitchen Equipment	3
Ventilating & Exhaust Air Systems	
Roof Ventilators — Motor Operated	4
Roof Ventilators — Natural Draft or Wind Operated	6
Ventilating Systems — Galleys and Bakeries	4
Ventilating Systems — Other	4
Exhaust Air Systems — Shower and Toilets	5
Air Velocity (Motor-operated Fans)	3
Cranes and Hoists	4
Heating Equipment (Buildings)	5
Plumbing (Buildings)	3
Water Heaters	
Domestic type & integral storage tank units (20 to 100 gal)	4
Other	
Fuel Facilities (Distribution)	4
Aeration Equipment (Water)	4
Aeration System (Sewage)	4
Chemical Feed Equipment for Water Supply	3
Chlorinators and Hypochlorinators	3
Fresh Water Supply & Distribution Systems	
Backflow prevention devices	3
All Other	4

Figure 2.2

Recommended Inspection Frequencies (continued)

Item	Frequency
Gas Distribution Systems	4
Pumps (Sump and Bilge)	4
Sewage Screening, Grinding, & Grit Removal Equipment	4
Septic Tank Systems	4
Sewage Collection & Disposal Systems	4
Special inspection during or after prolonged rains	
or severe storms	
Sewage Pumps	4
Sludge Pumps (Reciprocating)	4
Steam Traps	4
Turbine Surface Condensers	4
Turbines (Large)	5
Turbines (Small)	4
Air-Conditioning Equipment	4
Chimneys and Stacks	6
Buildings (Except Roofs and Trusses)	6
Furniture	6
Roofs	5
Trusses	6
Swimming Pools	6
Antenna-Supporting Towers and Masts	6
Fuel Facilities Receiving and Issues)	6
Camels and Separators	5
Piers, Wharves, and Bulkheads	6
Seawalls, Groins, and Breakwaters	6
Fuel Facilities (Storage)	5
Bridges and Trestles	6
Fences and Walls	
Perimeter Barriers	3
Other	6
Fresh Water Storage	6
Grounds	6
Incinerators	5
Pavements	
Airfield Pavements	4
Roads, Walks, Parking Areas	
Concrete	5
Flexible	5
Other	5
Railroad Trackage	
Heavy and Fast Moving Traffic	3
Other	6
Refuse and Garbage Disposal (Sanitary Fill)	5
Retaining Walls	6
Storm Drainage Systems	6
Tunnels and Underground Structures	6

Key to Frequency:	1. Daily	4. Quarterly
	2. Weekly	5. Semi-annually
	3. Monthly	6. Annually

Figure 2.2 (continued)

Inspection Time Estimate			
		Estimated Inspection Time Per Unit	
Item	**Unit**	**Hours**	**Minutes**
Air-Conditioning Equipment			
Under 3 tons	Each		30
3-5 tons	Each	1	
15 tons	Each	2	
800 ton	50 ton	2	
Aircraft Operating Areas			
Drainage	1,000 Lin Ft		15
Expansion Joints	1,000 Lin Ft		8
Field Lighting	1,000 Lin Ft		8
Pavements	1,000 Sq Yds		5
Traffic Markings	1,000 Lin Ft		8
Boilers			
Under 30 HP	Each	2	
40–50 HP	Each	8	
175 HP	Each	14	
480–52 HP	Each	18	
Breakwaters	1,000 Lin Ft		8
Bridges	100 Lin Ft	2	
Buildings			
Roofs	10,000 Sq Ft.	1	
Hangar and Storage Space			
Electrical	10,000 Sq Ft floor area		30
Mechanical	10,000 Sq Ft floor area	1	
Structural	10,000 Sq Ft floor area	1	30
Office Space			
Electrical	10,000 Sq Ft floor area		40
Mechanical	10,000 Sq Ft floor area		20
Structural	10,000 Sq Ft floor area		30
Shop Space			
Electrical	10,000 Sq Ft floor area	1	
Mechanical	10,000 Sq Ft floor area	2	
Structural	10,000 Sq Ft floor area	2	15
Cathodic Protection System	Each	8	
Causeways	1,000 Lin Ft		8
Chimneys and Stacks	Each	4	
Communication Systems			
Fire Alarm			
Overhead	1,000 Lin Ft		30
Underground	1,000 Lin Ft	1	
PA and Intercom	Station		15
Telephone and Telegraph			
Overhead	1,000 Lin Ft		30
Underground	1,000 Lin Ft	1	
Docks	Each	40	
Dredging	1,000 Sq Yds surface area	6	40
Electric Power Generation Equipment			
Auxiliaries	All	8	
Generator	Each	8	
Switchboard	Each	8	
Electric Power & Light Distribution			
Distribution Lines			
Overhead	1,000 Lin Ft		30
Underground	1,000 Lin Ft	1	

Figure 2.3

Inspection Time Estimate (continued)

Item	Unit	Estimated Inspection Time Per Hours	Estimated Inspection Time Unit Minutes
Overhead	1,000 Lin Ft	1	
Underground	1,000 Lin Ft	1	20
Transformers			
Pole	Each		15
Substation	Each		30
Elevators, Platform			
Lines, Dumbwaiters	Each	2	40
Fences	1,000 Lin Ft		8
Fresh Water System			
Piping	1,000 Lin Ft		8
Pumps	Each		15
Tanks	10,000 Gal		24
Fuel Storage & Distribution			
Piping	1,000 Lin Ft		12
Pumps	Each		15
Tanks	1,000 Bbl	1	45
Gas Distribution System			
Gages	Each		15
Meters	Each		15
Piping	1,000 Lin Ft		8
Regulators	Each		10
Valves	Each		6
Grounds	Acre		24
Hot Water Heaters			
30 Gallon	Each		15
300 Gallon	Each	1	
Piers	100 Lin Ft berthing space	1	
Pneumatic Power Distribution	1,000 Lin Ft		12
Refrigeration Equipment			
Under 3 tons	Each		30
3–5 tons	Each	1	
15 tons	Each	2	
800 ton	50 Ton	2	
Salt Water System			
Piping	1,000 Lin Ft		8
Pumps	Each		15
Tanks	10,000 Gal		24
Seawalls	1,000 Lin Ft		8
Sewerage Pumping &			
Treatment Plants	100,000 Gal per day		48
Sewage Pumps	Each		15
Sewage & Storm Drainage			
System	1,000 Lin Ft	1	20
Steam Distribution System			
Above Ground	1,000 Lin Ft	1	
Tunnels	1,000 Lin Ft	2	
Streets, Roads, &			
Parking Areas	1,000 Sq Yds		5
Towers	100 Ft Vertical Height	2	30
Trackage	1,000 Lin Ft (2 rail)		12
Trestle	100 Lin Ft	2	
Walls	1,000 Lin Ft		8
Wharves	100 Lin Ft berthing space	1	

Figure 2.3 (continued)

Locating Deterioration

As discussed in Chapter 1, included in one of the three stages of maintenance was *prevention*. If potential maintenance problems were not prevented, and are occurring to such an extent as to cause deterioration, the maintenance organization, in cooperation with plant management, must make a decision whether to abandon, replace, or repair the deteriorated building elements.

What motivates the need to make such a decision? The answer to this question is the existence of material deterioration. But before such a decision can be made, the deterioration must be observed or located. Some deterioration will show itself during the inspection process, while other deterioration will remain hidden to the inspector, and must be found by careful examination.

Time, manpower, and other resources must be delegated for the purpose of searching out and examining all possible areas of deterioration in a building. It is often the so-called "hidden deterioration" that creates future problems in terms of material and/or component failure. An example of this is connections made between roof beams and columns. These are most often inaccessible to the naked eye from the ground. And the only way the connection can be effectively inspected is by ascending to the elevation of the connection and examining it.

The process of locating deterioration is often a very difficult activity to carry out effectively, in terms of time, money, and manpower. To locate deterioration, one must know the following:

- Where to look.
- How to look.
- What to look for.

To become efficient in these areas, one needs to know how the building was constructed, the property of the materials in the building, and the location of all mechanical, electrical, plumbing, and other equipment. This comes from a combination of technical training and on-the-job experience. Thus, the exact process of location depends on:

- One's basic experience.
- Material or materials being inspected.
- Types of construction.
- Types of equipment and structural elements.
- Knowledge of types of deterioration.
- Knowledge of causes of deterioration.

Determining Causes of Deterioration

In order to arrive at an economically feasible decision, one must first determine the cause of the deterioration. This is often a difficult thing to accomplish. In fact, many times the exact cause may never be found and the best anyone can do is to eliminate the problem down to two or more causes and repair for all of them. The actual method used in the determination process can either be:

- Direct method.
- Process of elimination.

Direct Method

The first method is the one in which the cause is known directly from the type and extent of deterioration. This method requires little, if no, guesswork as to what the cause is. An example of this would be a localized failure of a steel building column caused by impact from a piece of equipment.

Process of Elimination

The second method is, as stated, a process of elimination. Much time, money, and manpower can be expended in locating the cause by this method. However, in many cases, it is the only way.

The charts and tables presented in the following chapter can be used as tools in the process of elimination. Their utilization will aid one to arrive more quickly at the cause or causes of deterioration. Some tips that the reader may find helpful in locating the cause or causes are listed below:

1. Inspect the structure.
2. Study it carefully through observation.
3. Compare it with other constructions in the area or elsewhere.
4. Analyze what there is in this case that is abnormal.
5. Speak with all persons working near or on the structure.
6. If necessary, bring in consultants who specialize in these problems.

When inspecting a building for causes of deterioration, one should use their senses to detect signs of things that do not belong where they are. Develop a sense of smell, a sense of feel, and a sense of sound when carrying out the above steps.

Decisions to Repair Deterioration

If the deterioration appears to have affected the material, component, or member to such an extent that it creates an unsafe condition, a professional engineer should be called in to evaluate the strength of the existing member. His recommendations will usually fall into one of the following categories:

- Continual usage—no need for structural repair.
- Restricted usage—no need for structural repair.
- Usage prohibited—until repair has been made.
- Abandon—the member or structure, repair impossible.

The reader should note that the final decision of whether or not to repair is to be made by the management of the company with input from a structural engineer. Whether the strength of a member is adequate or inadequate is not the only criterion which should be considered when deciding on the need for a repair. Other criteria, such as appearance and economical feasibility, should be considered.

If the anticipated decision is to repair the item, and the item has adequate strength, one of the following conclusions may be reached:

- If the present deterioration has produced an objectionable appearance, repair it.
- If appearance is not objectionable, determine if the deterioration is *dormant* or *active*:
  —If dormant, leave it alone.
  —If active, study the future effects of the repair. Will it cause the member or adjacent members to lose their strength?
  —If so, repair now, taking these items into consideration. If not, consider the possibility of repair now or in the near future.

On the other hand, if repair is called for and the material has, or shortly will have, inadequate strength, one of the following conclusions must be reached:

- Repair or rebuild immediately.
- Abandon all or part of the system.
- Consider changing the use of the area or item.

Repairing Deterioration

If repair is needed, a procedure must be selected and implemented. Some of the things which must be considered when selecting a repair procedure are:

- Cost—initial cost, future maintenance, and investment value of deferred costs
- Time of repair
- Extent of repair
- Future effect of the repair
- Appearance
- Operations in vicinity of repair
- Possible strength changes of existing members due to repair.

Keep in mind that repairing for the incorrect cause will be an incorrect repair procedure and can possibly cause additional problems. Relative to the economics affecting the decision of making a repair, the reader is referred to references on engineering economics.

Deferring a Repair

Often, it is possible to defer a repair if the strength, now and in the near future, is adequate. If funds, manpower and/or time are not immediately available, making a repair at a later date may be acceptable. If this is the case, the organization should set up an effective inspection program to check on the deterioration in question on a regular, frequent basis. Keep in mind that deferred action often results in a larger, more costly repair procedure later. Many times one is apt to want to take so-called immediate quick steps to slow down the deterioration, such as by the application of a preservative treatment on rotting wood. This type of action will only add to the ultimate cost of the repair.

Replacing vs. Repair

The question of whether to repair or replace is often a complicated one to answer. A thorough economic cost study is the only effective way of attaining an answer. However, as a general rule of thumb, if the repair costs 50 percent or less of the replacement cost, repair; and replace if the percentage is greater. This rule of thumb is generally effective if the facility can be put out of service during replacement construction. In selecting a repair procedure, the goal is to select the least expensive method to do the job effectively. For any cause or causes, there may be a number of acceptable repair procedures. A very methodical process must be followed in choosing a method. The first step is to list the facts above the deterioration and its causes. Do not include assumptions or value judgments, only facts. Involve all persons who will be affected by the repair (or who have been affected by the deterioration).

Second, based solely on the facts, list all the repair procedures possible that will alleviate the deterioration. From the list of acceptable procedures, select one and implement it. Some repair procedures are easy to implement, as when a worker is given the tools and equipment and is asked to make the repair. Such a procedure would be for a small isolated repair when there is no threat of structural collapse, as in the repair of peeling paint.

However, when the repair job becomes of such magnitude as to require close and careful supervision, design calculations, and incorporating various crafts and equipment, a complete set of specifications and drawings should be prepared prior to making the repair. The drawings and specifications help to insure that the repair will be done both effectively and efficiently, utilizing workers and materials of high quality.

It is useful to know the expected lifetime of the component under consideration when deciding whether to replace or repair it. Refer to Figure 2.4 for average useful life of many building components.

Once the repair has been made, it is of utmost importance that a routine follow-up inspection be made to determine if the selected procedure was, in fact, effective or ineffective. If ineffective, another one of the possible repair procedures selected in the earlier stage should be tried. A repair procedure can become a trial-and-error process, especially if there are many causes of the deterioration.

Once an effective repair has been made, frequent follow-up inspections of the repair areas should be done. The reader is referred to Figure 2.5, which can be utilized in the decision-making process of whether to make a repair or not.

Suggested Average Useful Life of Building Components

Item	Years	Item	Years
I. Major Construction		b. Fire Pumps	20
A. Reinforced Concrete Frame		a. Hose Housings	
1. Masonry Exterior		1) Wood	15
a. Heavy	45	2) Steel	20
b. Light & Medium	40	3) Masonry	30
B. Steel Frame		5. Sump Pumps	
1. Masonry Exterior		a. Small	10
a. Heavy	45	b. Large	15
b. Medium	35	6. Water Heaters — gas & electric	10
c. Light	30	7. Water Wells	25
2. Metal Exterior		D. Service Systems	
a. Heavy	30	1. Elevators (all types)	20
b. Medium	25	2. Fire Alarm	20
c. Light	20	3. Intercom	15
C. Wood Frame		4. Telephone	15
1. Masonry Exterior		III. Miscellaneous Items	
a. Heavy	35	A. Bulkheads	
b. Medium	25	1. Concrete	30
2. Metal Exterior		2. Steel	25
a. Heavy	30	3. Timber	20
b. Medium	25	B. Chimneys	
c. Light	20	1. Brick or concrete	35
3. Wood Exterior		2. Steel-lined	25
a. Heavy	25	3. Steel-unlined	20
b. Light & Medium	20	C. Culverts	
II. Electrical and Mechanical Equipment		1. Concrete	35
A. Electrical Systems		2. Galv. Steel	20
1. Lighting Systems		D. Curbing	
a. Conduit & Wire	20	1. Concrete	25
b. Fixtures	15	E. Fencing	
c. Flood Lighting	15	1. Brick or stone	30
2. Power Feed Wiring		2. Chain Link	20
a. Bus Duct	25	3. Concrete	30
b. Capacitor	20	4. Wire	10
c. Power Feed Wiring Mains	25	5. Wood	10
d. Switchboards	20	F. Flag Poles	25
e. Switch Units	20	G. Incinerators	
3. Transformers		1. Commercial type, steel fire brick lined	20
a. Wet Type	20	2. Concrete block or brick	20
b. Dry Type	15	3. Steel	15
B. HVAC Systems		H. Paving and Walks	
1. Air Conditioning Systems		1. Asphalt on gravel or stone	15
a. Central, including ducts & piping	15	2. Brick	20
b. Window Type	10	3. Concrete	20
c. Cooling Towers	15	4. Gravel, stone, cinders	10
2. Heating Systems		5. Parking area guard rails	10
a. Furnaces & Boilers	20	I. Platforms	
b. Radiators, Convectors, Piping	25	1. Reinforced concrete	35
c. Unit Heaters, gas & steam piping	20	2. Wood frame on concrete piers	20
d. Unit Heaters — Electrical	15	3. Wood frame on wood posts	15
3. Ventilating Systems including fans		J. Railroad sidings	25
& exhausters	15	K. Reservoirs, concrete	35
C. Plumbing Systems		L. Retaining Walls	
1. Drinking Water System	15	1. Brick	30
2. Fixtures	20	2. Concrete	40
3. Piping		3. Steel	25
a. Cast Iron Waste	35	4. Stone	40
b. Concrete	30	5. Wood	15
c. Copper	30	M. Sheds	
d. Plastic	20	1. Brick, tile or concrete block with wood frame	25
e. Steel	20	2. Brick, tile or concrete block with steel frame	35
f. Vitrified Tile	30	3. Metal clad, steel frame	27
4. Sprinkler Systems		4. Metal clad, wood frame	20
a. Wet & Dry Systems	30	5. Wood siding and frame	20

Figure 2.4

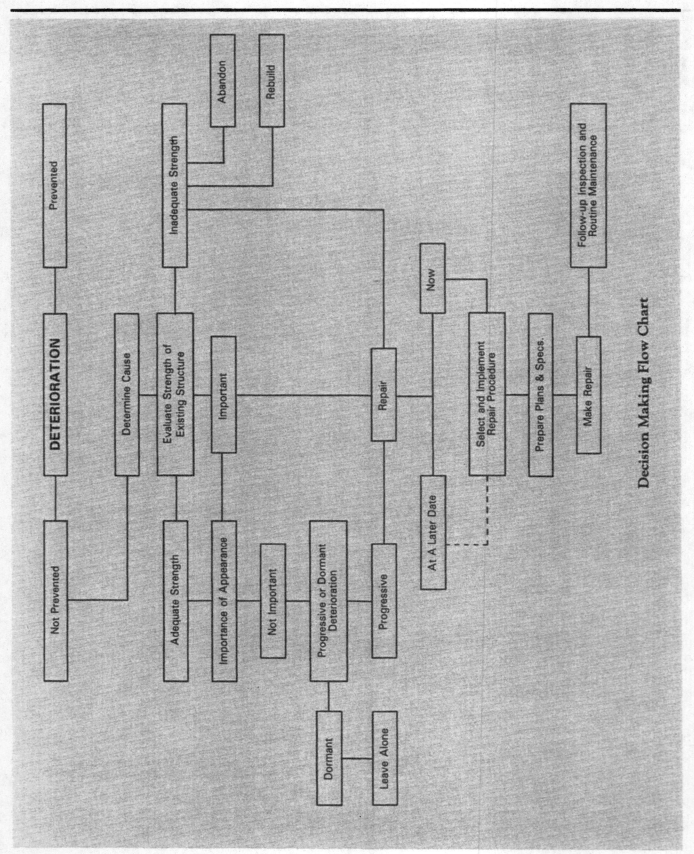

Decision Making Flow Chart

Figure 2.5

PART II
MAINTENANCE AND REPAIR OF MAJOR STRUCTURAL MATERIALS

The maintenance and repair of reinforced concrete are the subject of Chapter 3. Because reinforced concrete is a non-homogeneous material, that is, it is comprised of more than one type of material, it is susceptible to more types of deterioration than homogeneous materials such as steel and wood. In Chapter 3, reasons for concrete failures, as well as how to recognize and eliminate problems, and step-by-step methods of maintaining and repairing reinforced concrete, such as floors, walls, curbs, and ceilings are covered in detail.

Structural steel is one of the most versatile of structural materials. However, it, too, presents problems of deterioration which call for maintenance and repair. Chapter 4 examines problems and methods of maintaining and repairing structural steel.

Wood (or timber) has been used for hundreds of years in the construction of homes, light and heavy commercial buildings, and many other types of facilities. Since wood is not man-made, as structural steel and concrete are, it has its own peculiar types of maintenance problems. These problems, along with methods of prevention and repair, and causes of deterioration, are covered in Chapter 5.

CHAPTER 3

REINFORCED CONCRETE

Because it is comprised of more than one type of material, reinforced concrete is a non-homogenous material. Unlike steel and wood, it is susceptible to more types of deterioration.

Prior to repairing any deterioration, the causal factor(s) must be located and eliminated. If this problem cannot be eliminated, repair must be made so that further deterioration is not allowed to continue. For example, if the cause of concrete deterioration is building movement, install expansion-contraction joints. This chapter contains descriptions of concrete deterioration, repair, and maintenance methods.

Inspection of Concrete

The most important tool in determining types of deterioration is visual inspection. Establish formal procedures and utilize formal formats to record observed condition of the concrete. The American Concrete Institute (ACI) has developed a "Survey of Existing Concrete," which is an effective tool in the inspection process (Figure 3.1).

The inspector must be familiar with the terminology and visual evidence of the various types of deterioration. Refer to the bibliography at the end of this book for a list of ACI resources that contain information on reinforced concrete. Also refer to Figure 3.2 for a summary of types, causes, and diagnoses of deterioration. Use this table during and after the inspection process.

Survey of Existing Concrete

1. Description of structure or pavement
   Name, location, type, and size
   Owner, project engineer, contractor, when built

   Design
       Architect and/or engineer
       Intended use and history of use
       Special features

   Photographs
       General view
       Detailed close-ups of condition of area

   Sketch map—orientation showing sunny and shady walls, and well and poorly drained regions

2. Present condition of structure
   Overall alignment of structure
       Settlement

   Deflection
       Expansion
       Contraction

   Portions showing distress (beams, columns, pavement, walls, etc. subjected to strains and pressures)

   Surface condition of concrete
       General (good, satisfactory, poor, etc.)
       Cracks
           Location and frequency
           Type and size
           Leaching, stalactites
       Scaling
           Area, depth
           Type
       Spalls and pop-outs
           Number, size and depth
           Type
       Extent of corrosion or chemical attack
       Stains
       Exposed steel
           Previous patching or other repair
       Interior condition of concrete
           Strength of cores
           Density of cores
           Moisture content (degree of saturation)
           Evidence of alkali-aggregate or other reaction
           Bond to aggregate, reinforcing steel, joints
           Pulse velocity
           Volume change
           Air content and distribution

3. Nature of loading and detrimental elements
       Exposure
           Environment—arid, sub-tropical, marine, freshwater, industrial, etc.
           Freezing and thawing
           Wetting and drying
           Drying under dry atmosphere
           Chemical attack—sulfates, acids
           Abrasion, erosion, cavitation
           Electric currents

Figure 3.1

Drainage
    Flashing
    Weep holes
    Contour

Loading
    Dead
    Live
    Impact
    Vibration
    Traffic index
    Other

Soils (foundation conditions)
    Stability
    Expansive soil
    Settlement
    Restraint

4. Original condition of structure
    Condition of formed and finished surfaces
    Smoothness
    Air pockets
    Sand streaks
    Honeycomb
    Soft area

    Early structural defects
    Cracking
        Plastic shrinkage
        Settlement
        Cooling
        Curling
        Structural settlement

5. Materials of construction
    Hydraulic cement
        Type and source
        Chemical analysis (obtain certified test data if available)
        Physical properties

    Coarse
        Type, source and mineral composition (representative
          sample available)
        Quality characteristics
            Percentage of deleterious material
            Percentage of potentially reactive materials
            Coatings, texture, and particle shape
            Gradation, soundness, hardness
            Other properties

    Fine aggregate
        Type, source, and mineral composition (representative
          sample available)
        Quality characteristics
            Percentage of deleterious material
            Percentage of potentially reactive materials
            Coatings, texture and particle shape
            Gradation, soundness and hardness
            Other properties

    Mixing water
        Source and quality

Figure 3.1 (continued)

Air-entraining agents
    Type and source
    Composition
    Amount
    Manner of introduction
Admixtures
    Mineral admixture
        Type and source
        Physical properties
        Chemical properties
    Chemical admixture
        Type and source
        Physical properties
        Chemical properties
        Composition
        Amount
Concrete
    Mixture proportions
        Cement content
        Proportions of each size aggregate
        Water-cement ratio
        Water content
        Chemical admixture
        Mineral admixture
        Air-entraining agent

    Properties of fresh concrete
        Slump
        Percent air
        Workability
        Unit weights
        Temperature

    Type
        Cast-in-place
        Precast
        Prestressed

    Reinforcement
        Yield strength
        Thickness of cover
        Presence of stirrups
        Use of welding

6. Construction Practices
Storage and processing of materials
Aggregates
    Grading
    Washing
    Storage
        Stockpiling
        Bins

Cement and admixtures
    Storage
    Handling

Reinforcing steel and inserts
    Storage
    Placement

Forming
    Type
    Bracing
    Coating
    Insulation

Figure 3.1 (continued)

Concreting operation
    Batching plant
        Type—automatic, manual, etc.
        Condition of equipment
        Batching sequence
    Mixing
        Type—central mix, truck mix, job mix, shrink mix, etc.
        Condition of equipment
        Mixing time
    Method of transporting—trucks, buckets, chutes, pumps, etc.
        Equipment—buckets, elephant trucks, vibrators, etc.
        Weather conditions—time of year, rain, snow, dry wind,
            temperature humidity, etc.
        Site conditions—cut, fill, presence of water, etc.
        Construction joints
    Finishing
        Type—slabs, floors, pavements, appurtenances
        Method—hand or machine
        Equipment—screeds, floats, trowels, straight-edge, belt, etc.
        Additives, hardeners, water, dust coat, coloring, etc.
        Curing procedures
            Method—water, covering, curing compounds
            Duration
            Efficiency
        Form removal (time of removal)

7. Initial physical properties of hardened concrete
    Strength—compressive, flexural, elastic modulus
    Density
    Percentage and distribution of air
    Volume change potential
        Shrinkage or contraction
        Expansion or swelling
        Creep
    Thermal properties

8. Additional items pertaining to pavements
    Structural section (sketch and thickness of pavement
        layers—base, subbase, etc.)
    Joints
        Type, spacing, design
        Condition
        Filling material
        Faulting—(measured in mm.)
    Cracks
        Type (longitudinal, transverse, corner), size (measured in mm.),
          frequency
        Patching
        Riding quality
        Condition of shoulders and ditches

Reprinted with permission of the American Concrete Institute, Committee 201.

Figure 3.1 (continued)

Common Causes of Concrete Deterioration

Damage	Cause	Diagnosis
Alkali-Aggregate Expansion	Chemical reaction between aggregate and cement paste	Slight cracking to complete breakup
Cavitation	Rapid movement of water or other liquids across the surface	Spalling around projections. Honey-combing. Popping and cracking noises when water moves over the surface
Cracks (Active and Dormant) Before hardening of the concrete:	Construction movement, settlement, shrinkage around reinforcement. Setting shrinkage due to inadequate finishing and curing	Random, isolated or patterned cracks
After hardening of the concrete:	Chemical reactions, such as corrosion	Patterned cracking, protruding aggregate, popouts. Chemical analysis indicates deterioration of cement paste
	Physical, such as drying shrinkage.	Surface cracking, patterned
	Thermal changes (subjected to temperature extremes, such as from freezing and thawing cycles).	Extreme change in measured temperatures between inner and outer surfaces. Shallow cracking.
	Stress concentration	Localized cracking
	Structural Design	Cracks, usually isolated
	Accidents from overload, vibration, fatigue and earthquake	Cracks can be isolated or patterned depending on crack-producing agent
Corrosion of Reinforcing Steel	Insufficient cover of steel. Quality of concrete. Over-use of calcium chloride as admixture.	Cracks will occur at the level of the reinforcement and parallel to it. Rusting or discoloration will be evident.
Crazing	Surface shrinkage more rapid than interior of concrete mass. Too high a slump. Too rich a mix. Poor timing on finishing. Too rapid absorption of moisture	Shallow cracks forming a hexagonal pattern
Dusting	Too wet of a concrete mixture. Premature or excessive working of the surface. Organic materials in the aggregate. Inadequate curing.	Appearance of a powdery substance on the surface of the concrete
Efflorescence	Water migrating from the interior mass of the concrete to the surface and depositing salts	Appearance of crystaline salts on the concrete surface
Fire	Fire	Charred and spalled surfaces
Form Scabbing	Form oil improperly applied	Difficult to remove forms. Uneven, spalled areas
Holes (Small and Large)	Chemical reaction. Inadequate construction and design	Popouts, holes, random pattern or isolated in extreme
Honeycombing	Placing concrete aggregate with insufficient type of mortar. Improper placing techniques, such as inadequate vibration	Surface defects-voids. Coarse aggregate broken away from the surface
Popouts	Breaking away of a particle near the surface	Depressions left by material popping out
	Excessive amount of moisture or temperature changes in the region	Presence of disintegrated material near the popout

Figure 3.2

Common Causes of Concrete Deterioration (continued)		
Damage	Cause	Diagnosis
Sand Streaking	Concrete mixed with a high water content or a deficiency of finer sand sizes are placed in a formwork that is not water-tight	Vertical streaks of sand which appear on the surface, most noticeable when forms are immediately stripped
Scaling	Severe free-thaw conditions. Improper use of deicing salts. Repeated wetting and drying of concrete. Improper finishing. Chemical attack of concrete. Heat blast	Flaking or peeling away of thin layers of concrete
Spalling	Corrosion of reinforcement. Mechanical damage. Incorrect form removal. Shock-waves	Fragments of concrete that have been broken from the surface. Corrosion of reinforcement
Stain and Uneven Color	Chemical action of foreign materials on the surface. Mixing of different types of cement with each other. Reaction of materials comprising the concrete mixture	Discoloration or lacking uniformity in appearance

Figure 3.2 (continued)

Prevention of Deterioration

Probably 50 percent or more of the existing concrete deterioration that exists today in buildings could have been prevented by effective design and construction techniques. The existing specifications as developed by manufacturers' associations are adequate. However, if not followed to the letter, even by using effective construction techniques, poor concrete, which is susceptible to deterioration, will result.

There are many ways in which concrete deterioration can be prevented. Some of the major ways are listed below. The reader should note that this list can be used as a checklist in the review of existing specifications on concrete construction.

- Properly compact subgrades.
- Provide adequate drainage of subgrade.
- Eliminate traffic on the compacted subgrade.
- Delay the final finishing as long as the concrete surface remains workable.
- Maintain controlled rate of concrete placement as recommended by the ACI.
- Establish and maintain effective inspection and quality control procedures on the job site.
- Begin curing operations immediately after finishing.
- Utilize adequate vibrating methods, as recommended by the ACI.
- Protect curing concrete from external loads due to impact and ground vibrations.
- Use stiff concrete mixes.
- Use recommended quantities, types of construction, and expansion joints.
- Provide required reinforcing steel.
- Place all reinforcing steel in strict accordance with contract drawings.
- Insure that all materials in the concrete mixture are at correct temperatures for specific environmental conditions according to ACI recommendations.
- Utilize the highest quality of materials as per ACI recommendations.
- Do not use curing compounds on surfaces that are to receive protective treatments.
- Do not use form oils or waxes on forms to be placed against surfaces to be coated.
- Use minimum cement and water contents in the design mix (do not permit over-wetting for workability).

Routine Maintenance

The extent and frequency of routine (housekeeping) type maintenance of concrete will depend upon the exposure, location, desired appearance, extent of any existing deterioration, and available maintenance funds. Columns, walls, and other vertical surfaces that are exposed to the environment are given a minimum of routine maintenance. The reason for regular housekeeping would be to retain a specified or desired appearance. If appearance is not a problem, the surface could be sealed or left as is and inspected at regular intervals for deterioration.

If, however, vertical surfaces need to be cleaned for the sake of appearance, utilization of mechanical means, such as air hoses, or light sand blasting may be used. Chemical etching is also a possibility and should be performed under safe conditions. Whatever means are used in the cleaning process, it should be emphasized that the existing concrete surface not be removed, because doing so would decrease the amount of cover on the concrete reinforcement which, in the long run, would induce further and more dangerous deterioration.

Maintenance of concrete floors includes regular dusting and, possibly, damp mopping, as discussed in the chapter on floors. At regular intervals, a cleaning solution can be applied to the floor, scrubbed in, picked up, and rinsed. Most floors that receive any amount of traffic should be sealed. The sealant should not be placed on the floor until about six months after the concrete has been cured. The clear sealer is more desirable than painting or using colored sealer. If a maintenance department is considering painting the floor, it should be noted that the painting process must be repeated periodically.

Coatings

The durability of concrete can be improved by preventive maintenance in the form of surface treatments. Care must be taken not to use a surface sealer which will seal moisture into the concrete by preventing evaporation of water penetrating from unsealed surfaces. This is especially so for floors, which receive the most abuse in any building. Coatings can also be placed to add characteristics to the concrete that do not otherwise exist, such as the ability to resist attack from potentially disintegrating types of elements. The magnitude of protection needed in any given exposure will depend upon the following:

1. Temperature.
2. Concentration of potentially disintegrating chemical agent.
3. Nature of the potentially disintegrating chemical agent.
4. Whether the agent is flowing or stationary.
5. Whether the exposure is continuous or intermittent.

Coating Types

The coatings can be secured to the concrete surface by *bonding*, *percolation* into the pores, or *chemical reaction*. It should be noted that coatings give either permanent or temporary protection. In the latter case, the coating surface has to be retreated. The ACI has developed a list of substances harmful to concrete (see Figure 3.6 at end of chapter) along with recommended coatings and/or surface treatments to protect against damage from the respective substance. Prior to treating an entire floor with any one of the recommended treatments, it should be tried on a small patch of the surface and tested for its effectiveness.

Coating Application

The most important aspect of applying coatings to concrete surfaces is proper preparation. The following general guidelines should be adhered to in preparation:

- Surfaces should be well-cured and free from moisture.
- Eliminate objectionable voids in the surfaces. Fill any voids with a neat grout.
- Remove all grease, oil, waxes, efflorescence, laitance, dust, and loose particles.

The method used is dependent upon the surface conditions and the type of coating to be used. Acid etching or sand blasting are the methods most utilized today; however, manufacturers' recommendations should be referred to in order to insure proper application.

Hardening Agents

Effective coatings are available for most degrees of required protection. It should be remembered that no one material will serve best for all conditions. Coatings are usually categorized in broad areas. The first is a hardening agent. This includes such substances as sodium silicate, sodium sulphate, and fluorosilicate hardeners. Sodium silicate will not react chemically with the concrete, but leaves a sheen-like deposit on top of the surface. This deposit eventually wears off and creates dusting conditions. Sodium sulphate and silicate types provide better hardening protection; however, dusting will occur when traffic wears through the thin surface.

Seals and Finishes

The other type of general coating classification is referred to as seals and finishes. Concrete floor seals and finishes will help eliminate dust by filling the pores of the concrete surface. Soilage does not penetrate these coatings. By applying additional coats, protection can be obtained from possible excessive wear and thus prolong the life of the floor. Sealed floors will also be easier and less costly to maintain. Conditions to be considered when selecting a protective sealant and/or finish are as follows:

- The standard of maintenance required.
- The amount of traffic on the floor.
- The type of soilage that will occur.

- Existing floor conditions.
- Application requirements.
- The degree of maintenance required after applying the finish or sealer.

Available seals include chlorinate rubber, oleoresin, one-component epoxy esters, oil-modified urethanes, and acrylic resins. Chlorinate rubber and oleoresins offer light wear, darken over time, and collect dirt.

One-component epoxy esters and oil-modified urethanes wear well, yellow with age and collect dirt, to a moderate degree. Acrylic resin seals provide good wear, light color, and only moderate dirt collection.

Floor finishes offer the best protection for any surface. The finish locks in the concrete and forms a second surface on top of an existing floor. Floor finishes, however, must be occasionally reapplied. Different types of floor finishes available are one-and two-component epoxies and oil-free urethanes. Single-component epoxies have fairly good wear, but tend to yellow with age. They also are moderate dirt collectors. The two-component type provides good wear and color retention. The disadvantages are brittleness and a tendency to scratch. Both types of oil-free urethanes give excellent wear, flexibility, and scratch resistance. For mildly eroded concrete, a resinous resurfacer will do an effective job. Severe erosion will require the use of toppings. The major types available are two-component type (epoxy resin and curing agent), three-component type, and urethane-based.

Two-component are tough, wear-resistant, and flexible. However, good wear characteristics can be adversely affected by moisture during application. Three-component toppings are tough and wear-resistant, but they can chip from the edges, if brittle. They also have good impact resistance. Urethane-based toppings are normally more flexible than epoxy types. They are, however, affected by moisture, during application.

Other types of toppings available are latex modified concrete, which gives good adhesion and is somewhat flexible, and asphalt toppings, which are soft and pliable and, therefore, not as durable as other toppings. The reader should consult with manufacturers for information on surface preparation, coating thickness, and application techniques.

Stain Removal

It is important that stains be removed, not only for the sake of appearance, but also to prevent future material surface damage. The first step in removal of stains, whether caused by construction activities or through exposure of the concrete during service, is to determine the source of the stain and then select the proper method for removal.

Common mechanical methods for removing some stains are sand blasting, grinding, steam cleaning, brushing, and light blowtorch application. Steel brushes, when used by themselves, wear at times in a manner that leaves iron deposits which will eventually rust and may later stain the concrete.

Chemical cleaning is more involved and requires application of specific chemicals. The action takes place either by dissolving the staining substance, which can then be blotted from the concrete surface, or by bleaching the discoloring agent chemically into a product having a color that blends with the concrete. Many chemicals can be applied to concrete without appreciable damage to the surface, but strong acids or chemicals having a highly active reaction should be avoided; even weak acids may etch the surface, if left for any length of time. It is advisable to saturate the surface with water before application of an acid so that the acid will not be absorbed too deeply into the concrete. The acid used should be completely flushed from the surface with water.

Some stains can be removed by more than one method. No attempt should be made until one is sure that the method or solvent selected will do the job. Experimentation with different bleaches or solvents is helpful. Some experimentation should be done in an inconspicuous small area of the concrete surface. With careful experimentation, the most effective method and materials can be selected. The reader is referred to Figure 3.3 for a summary of the major types of stains and recommended methods of their removal.

With either mechanical or chemical methods, care should be taken to protect surrounding areas and materials other than concrete, such as glass and wood, from the effects of any treatment.

Many of the methods described for stain removal make reference to a *poultice*. A poultice is a paste containing a solvent or reagent and a powdery inert absorbent material. The inert material may be diatomaceous earth, calcium chloride lime or talc. The selection of the solvent depends on the stain to be removed. The paste is usually spread over the stain with a trowel to a thickness of about one-half inch.

Concrete Repair

As discussed in Chapter 2, prior to making a repair, determination of the cause and the type of deterioration is mandatory. Tables presented earlier in this chapter can be utilized in this process.

Once the cause or causes have been determined, the next step is to evaluate the extent of damage and whether or not an effective repair can be made to the remaining sound structural concrete.

If an inadequate amount of existing sound concrete is evident, consideration should be given to the possible complete replacement or abandonment of the element or structure. However, if repair can be effectively made, the reader is referred to Figure 3.4 for the different types of repair procedures that can be used for the various modes of deterioration that were presented in Figure 3.2. Many of these procedures are briefly discussed in the balance of this chapter. The reader should note that many of these procedures require consultants who specialize in these areas. Maintenance personnel should not attempt these procedures without some prior experience.

Pre-repair Considerations

When selecting an applicable repair procedure, the resulting appearance must be considered. The following precautions should be taken when appearance is an important aspect for patches:

- Use concrete having qualities similar to those of the surrounding concrete.
- Finish the patch in the same manner as the surrounding concrete.
- Try a sample patch before patching the actual structure.
- Most times it is best not to patch.

For extensive repair areas, the following precautions should be considered:

- Large areas may be broken up into small units, each unit at a different level. This gives an ideal appearance.
- All repair work must be uniform in both materials and workmanship.
- Use construction joints in approved areas only.
- Consider the use of a brick or stone facing after the repair is made.

Admixtures

The second factor that must be considered during the pre-repair stage is that many repairs incorporate the use of a concrete admixture. A concrete admixture is a chemical which adds some desired effect to the concrete. The reader should refer to the summary of the different types of admixtures commonly used in repairs listed below:

- **Air-entrainment Agents:**
  Provide for a more workable material.
  Definitely used when repair will be exposed to freeze-thaw cycles.
- **Retarder and Densifying Agents:**
  Retarders will retard the set and are suggested in warm weather to reduce cracking due to setting up too fast.
  Increase the workability of the mix.
  Allow for delayed finishing, resulting in a less permeable concrete.
- **Acceleration:**
  Useful in winter (cold weather concreting).
  Useful when working to seal against the flow of water.
  Should be used to a minimum since the admixture tends to increase shrinkage.

Recommended Stain Removal Procedures for Concrete

Stain Type	Recommended Methods for Removal
Air, Smoke and Wood Tar	Apply a trichlorethylene poultice. Scrape off when dry and repeat the application, if needed. Scrub with clear water and let dry. As an alternate method, use bleach.
Aluminum	Wet area well and scrub with a solution of 10% hydrochloric acid (use weaker solutions on colored concrete). Rinse thoroughly with clear water and allow to dry.
Bitumen	
1. Molten Bitumen	Cool with ice until it is brittle and chip off with a chisel. Scrub the surface with an abrasive powder to remove residue and rinse with clear water.
2. Emulsified	Scrub the stained area with scouring powder and water. Do not use solvents.
3. Cutback	Reduce the intensity of stain by application of a poultice impregnated with toluene or benzine. Scrub surface with scouring powder and water.
Blood	Wet the stain with clear water and cover it with a thin layer of sodium peroxide powder. Sprinkle it with water and allow to stand for five minutes. Wash with clear water and scrub vigorously. Brush a 5% solution of muriatic acid on the surface to neutralize any alkaline traces. Rinse with clear water.
Caulking Compound	Scrape off from the surface and apply a poultice impregnated with denatured alcohol. Leave until dry. Allow caulking to become brittle and brush off with a stiff brush. Wash the surface with hot water. Rinse, let dry.
Chewing Gum	See Caulking Compound.
Coffee	Use bleach as described for Smoke stains
Copper and Bronze	Dry mix one part ammonium chloride with four parts fine-powdered inert materials. Add household ammonia to make a smooth paste; apply over the stain and let dry. Scrub with clear water and allow to dry.
Curing Compound	This type of stain will eventually be removed with time. Old stains can be removed by mechanical abrasion methods, such as light grinding or sand blasting.
Epoxy	Burn off small areas with a blow torch and use light sand blasting for large areas.
Grease Stains	Scrape off excess grease from the surface and scrub with scouring powder, soap trisodium phosphate or detergent.
Ink Stains	Ink may vary in chemical composition and therefore, a trial-error solution is needed.
Iodine	Apply an alcohol impregnated poultice. After the paste has been allowed to dry, scrape it off and wash the surface with clear water.
Iron Rust Stains	Mop the surface with a solution containing one pound of oxalic acid powder per gallon of water. After three hours, rinse with clear water. Apply a second application for bad spots. Use mechanical means, such as sand blasting, if the above method doesn't work.
Linseed Oil	Soak up with an absorbant material. Do not use a wiping action as it will rub it into the surface. Cover the spot with a dry, powdered absorbant inert material and let stand. Apply a poultice impregnated with refined naphtha to remove any deep stain.
Lubricating Oil	Soak up. Scrub with a strong soap for use on concrete. Make a poultice with a solution of 5% caustic soda. Let dry for twenty to twenty-four hours, remove and wash the surface with clean water.
Mildew	Use a mixture of one ounce of commercial laundry detergent, three ounces of trisodium phosphate, one quart of commercial laundry bleach, and three quarts of water. Apply to the area with a soft brush. Rinse with clear water.
Moss	Utilize a commercially available compound.

Figure 3.3

Recommended Stain Removal Procedures for Concrete (continued)

Stain Type	Recommended Methods for Removal
Paint Stains	
1. Wet Paint	Soak up the wet paint. Do not use a wiping action. Scrub the surface with scouring powder and water.
2. Dried Paint	Scrape off as much as possible of the dried paint. Apply a poultice impregnated with a commercial paint remover. Allow to stand for twenty to thirty minutes. Scrub the stain and wash off.
Perspiration Stain	Use bleach as for smoke stain removal.
Petroleum Oil	See Lubricating Oil.
Soybean Oil	See Linseed Oil.
Tobacco Stain	Apply a poultice of hot water and grit scrubbing powder. Mix to a mortar consistency. Apply to the surface in a thin layer. When dry, scrape off and scrub with clear water.
Tung Oil	See Linseed Oil.
Urine Stain	Use bleach as for fire stain removal.
Wood Stains	Scrub the surface with one part glycerol diluted to four parts water.

Figure 3.3 (continued)

Figure 3.4

Concrete Repair Procedures

Repair Procedure	Alkali Aggregate Expansion	Cavitation	Cracks – Active	Cracks – Dormant	Crazing	Dusting	Efflorescence	Fire	Form Scabbing	Holes – Small	Holes – Large	Honeycombing	Permeability	Popouts	Sand-streaking	Scaling	Spalling	Stains and Uneven Color
Total Replacement	NA	NA	NA	NA	NA	X	X	X	X	NA	X	X	X	X	X	X	X	NA
Sand Blasting	NA	NA	NA	X	X	NA	NA	X	NA	NA	NA	NA	NA	NA	NA	X	X	X
Sack Run	NA	NA	NA	NA	X	NA	NA	X	NA	NA	NA	NA	NA	NA	NA	NA	NA	X
Stressing	NA	NA	X	X	NA	NA	NA	X	NA	NA	NA	NA	NA	NA	NA	NA	NA	NA
Stitching	NA	NA	X	NA	NA	NA	NA	X	NA	NA	NA	NA	NA	NA	NA	NA	NA	NA
Prepacked Concrete	NA	X	NA	NA	NA	NA	NA	NA	X	X	X	X	X	NA	X	NA	X	NA
Pneumatically Applied Mortar	NA	X	NA	X	NA	NA	NA	X	NA	X	NA	X	NA	NA	X	X	X	X
Overlays	NA	NA	X	X	X	X	NA	X	X	X	NA	NA	NA	X	NA	X	X	X
Mortar Replacement	NA	NA	NA	NA	NA	NA	NA	X	X	NA	X	NA	NA	NA	NA	NA	NA	NA
Grouting	NA	NA	NA	X	NA	NA	NA	NA	NA	NA	NA	NA	NA	NA	NA	NA	NA	NA
Jacketing	X	X	X	X	NA	X	NA	X	NA	NA	NA	X	NA	NA	X	NA	X	NA
Grinding	NA	NA	NA	X	X	X	NA	X	NA	NA	NA	NA	NA	NA	NA	X	X	X
Epoxy Bonding	NA	NA	X	X	NA	NA	NA	NA	NA	NA	NA	NA	NA	NA	NA	NA	NA	NA
Dry Pack	NA	NA	NA	X	X	NA	X	X	X	X	NA	X	NA	NA	NA	NA	NA	NA
Autogenous Healing	NA	NA	NA	NA	NA	NA	NA	NA	NA	NA	NA	X	NA	NA	NA	NA	NA	NA
Concrete Replacement	X	NA	NA	X	NA	NA	NA	X	X	NA	X	NA	NA	X	NA	X	X	NA
Blanketing	NA	NA	NA	X	X	X	NA	X	X	NA	NA	NA	X	NA	X	X	X	X
Caulking	NA	NA	X	X	NA	NA	NA	X	NA	NA	NA	NA	NA	NA	NA	NA	NA	NA
Acid Etching	NA	NA	NA	X	NA	X	X	X	NA	X	NA	NA	NA	NA	NA	NA	NA	X
Coatings	X	X	NA	X	X	X	NA	X	X	X	X	NA	X	X	NA	X	X	X

X = Applicable Procedure   NA = Not Applicable

- **Expanding Mortars:**
  Useful when attempting to attain a tight fit, with no shrinkage cracks.
- **Waterproofing Admixtures:**
  Provide for a denser concrete.
  Not a replacement for good, dense concrete.

Preparation

All damaged, loosened or unbonded portions of existing concrete must first be removed by chipping hammers or other approved equipment. The surfaces may then be prepared by wet sand blasting, water blasting with approved water blasting equipment, bush hammering or any other approved method and then cleaned and allowed to dry thoroughly. During the process, care should be taken to prevent undercutting aggregate in the existing concrete. Replacement of the deteriorated concrete should be delayed several days until a re-examination of excavated surfaces confirms the soundness of the remaining concrete. It is far better to remove too much concrete than too little because affected concrete generally continues to disintegrate and, while the work is being done, it costs more to excavate to ample depths. Surface cleaning should be done by air water jets. Surface drying must be complete and may be accomplished by air jet. Compressed air used in cleaning and drying must be free from oil and all other contaminating materials.

After surfaces have been prepared and properly cleaned, they need to be kept in a clean, dry condition until the placing of concrete has been completed (except for dry pack repairs). Dry pack repairs require the application of a mortar bond coat prior to placement of repair material. There are other special preparation requirements relative to special repair procedures that will be discussed in the balance of this chapter. When a portion of the reinforcing bars are exposed during the preparation process, all the concrete encasing the ban (at least one inch all around) should be removed.

Cracks in Concrete

The causes of cracking are poor structural design, movement due to temperature differentiation, overload, shrinkage during the curing stage, etc. The following are factors that can be utilized in finding out the cause of a crack:

- Location of the crack.
- Pattern of the crack.
- Depth and width of the crack.
- Presence of foreign materials.
- Difference in elevation between two contiguous cracked concrete masses.

Another factor to be determined is whether the crack is active or dormant. A dormant crack can be repaired permanently after the full extent of cracking has taken place. An active crack is due to continuing movement of the structural element for some reason, such as foundation movement. Methods of determining whether the crack is active or dormant are shown in Figure 3.5 and are as follows:

1. To determine if the crack is progressing along and parallel to its length, mark the end of the crack. After a few days, return to see if the crack has progressed beyond the mark.
2. Place a piece of tape, that has been notched parallel to the crack, across the crack. If the tape tears or compresses (wrinkles), the crack is active. If there is no apparent change in elongation of the tape, after some time, the crack is dormant.
3. Apply gauge points, such as pins, on either side of the crack. At regular intervals of time, measure the distance between the gauge points, using vernier calipers, to determine if the crack is moving.

Prior to repairing a crack, it must be determined if the crack is of a type that only requires sealing (coating) against intrusion of foreign material and if restoration of structural integrity is required. If the crack is an active crack, a stress analysis must be performed to determine if strengthening requirements are needed. If the crack is dormant, a factor that must be considered is the occurrence of water in the cracked area and whether or not it must be controlled.

Methods of crack repair include acid etching, autogeneous healing, caulking, protective coatings, or concrete replacement. These methods and others are described in the following sections.

Acid Etching
Use a commercially available diluted (to about 10 to 30 percent) acid. Apply by vigorously scrubbing with a stiff wire or fiber bristle brush or broom. Follow this by completely and thoroughly scrubbing with clear water to remove all traces of acids and reactive products.

Autogenous Healing
This is a self-healing type process in that the cement has the ability to re-establish continuity over a crack by itself. This method is effective for concrete that is in a submerged environment or is exposed to a continual wash of water.

Caulking
This method utilizes a high-grade plastic caulk and is applied with a gun or in strips, according to manufacturers' recommendations. This method should be utilized only for dormant or nearly dormant type deterioration that is not exposed to abrasive type loading.

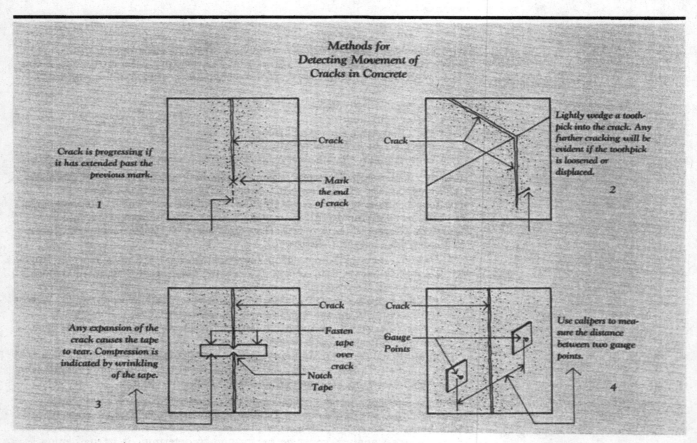

Figure 3.5

Coatings

See previous section for a discussion on coatings.

Concrete Replacement

This method consists of replacing the defective concrete with ready-mixed or job-mixed concrete of quality equal to that of the existing concrete. This method is used when the cavity or large crack goes through the entire member or the concrete is defective beyond the reinforcing steel. This method is also used when the depth of the area exceeds six inches and the repair will be appreciably continuous. If this procedure is to be used for the repair of holes, the top edge of the hole at the face should be cut to a horizontal line. If the shape of the defect makes it advisable, the top of the cut should be stepped down and continued on a horizontal line. The top of the hole should be cut on a 1:3 upward slope from the back toward the face from which the concrete will be placed. When the hole extends through the concrete section, spalling and feather edges may be avoided by having chippers work from both faces. All interior corners should be rounded to a minimum radius of one inch.

Use of Forms

The construction and setting of forms are important steps in the procedure for satisfactory concrete replacement where the concrete must be placed from the side of the structure. To obtain a tight and acceptable repair, the following requirements must be observed:

1. Front forms for wall repairs more than 18 inches high are to be constructed in horizontal sections so the concrete can be conveniently placed in lifts of not more than 12 inches deep. The back form may be built in one piece. Sections to be set as concreting progresses are to be fitted before placement is started.
2. To exert pressure on the largest area of form sheathing, tie bolts should pass through wooden blocks fitted snugly between the walers and the sheathing.
3. For irregularly shaped holes, chimneys may be required at more than one level. When beam connections are required, a chimney may be necessary on both sides of the wall or beam. For such construction, the chimney should extend the full width of the hole.
4. Forms should be substantially constructed so that pressure may be applied to the chimney cap at the proper time.
5. Forms must be mortar tight at all joints between adjacent sections and between the forms and concrete and at tie-bolt holes to prevent the loss of mortar when pressure is applied during the final stages of placement.
6. Structural concrete placements are to be made with an oversanded mix containing about a 3/4 inch maximum size aggregate; a maximum water-cement ratio of .47 by weight; 6 percent air, by volume of concrete; and a mixture slump of 4 inches. This special mix should be placed several inches deep on the joint at the bottom of the placement. A mortar layer is not to be used on the construction joints.

Concrete for repairs should have the same water-cement ratio as used for similar structures, but should not exceed .47 by weight. The concrete should contain 3 percent to 5 percent entrained air. Where surface color is important, the cement should be carefully selected, or blended with white cement, to obtain the desired result. To minimize shrinkage, the concrete should be as cool as practicable when placed, preferably at 70 degrees or lower.

Best repairs are obtained when the lowest practicable slump is used. This is about three inches for the first lift in an ordinarily large form. Subsequent lifts can be dryer, and the top two inches of concrete in the hole and that in the chimney should be placed at about zero dump. It is usually best to mix enough at the start for the entire hole; thus, the concrete will be up to 1-1/2 hours old when the successive lifts are placed. Such pre-mixed concrete, provided it can be vibrated satisfactorily, will have less settlement, less shrinkage, and greater strength than freshly mixed concrete. The quality of a repair depends not only on the use of low dump concrete, but also on the thoroughness of the vibration during and after depositing the concrete. There is little danger of overvibration. Immediately after the hole has been completely filled, pressure should be applied to the fill and the form vibrated. The operation should be repeated at 30-minute intervals until the concrete hardens and no longer responds to vibrations.

When placing concrete in lifts, placement should not be continuous; a minimum period of 30 minutes should elapse between lifts. When chimneys are required at more than one level, the lower chimney should be filled and allowed to remain for 30 minutes between lifts. When chimneys are required on both faces of a wall or beam, concrete should be placed in one chimney only until it flows to the other.

Forms for concrete replacement repairs usually may be removed the day after casting unless form removal would damage the fresh concrete, in which event stripping should be postponed for another two days. The projections left by the chimney normally should be removed the second day. If the trimming is done earlier, the concrete tends to break back into the repair. These projections should always be removed by working up from the bottom, because working from the top tends to break concrete out of the repair. The rough area resulting from trimming should be filled and finished to produce a surface comparable to that of the surrounding areas. Plastering of these surfaces should never be permitted.

Concrete Replacement Without Forms

Concrete replacement techniques are used without forms only for slabs on grade or holes. The same type of procedures, materials, and workmanship as described above can be incorporated. If holes are less than six inches square, they should be filled using the dry-pack procedure as described in another section, or the epoxy, bonding-epoxy mortar method. If they are larger, low slump concrete should be used. Slump of the concrete should not exceed two inches for slabs that are horizontal or nearly horizontal and three inches for all other concrete. The net water-cement ratio should not exceed .47 by weight. A water-reducing, set-controlling admixture may be used. Set-retarding agents should be used only when the interval between mixing and placing is quite long. If practicable, the replacement concrete should be preshrunk by letting it stand as long as possible before it is tamped into the hole.

Dry Packing

This process involves the placement of a stiff mortar by tamping it into place. Dry pack should be used for filling holes having a depth equal to or greater than the least surface dimension of the repair area (minimum depth = one inch). Dry pack should not be used for relatively shallow depressions where lateral restraint cannot be obtained, for filling around reinforcement, or for filling holes that extend completely through a concrete section.

For the dry-pack method of repair, holes should be sharp and square at the surface edges, but corners within the holes should be rounded, especially when watertightness is a requisite. The interior surfaces of holes should be roughened to develop an effective bond.

Care must be taken when using this technique to insure that no free water is on the surface when the previously mixed mortar bond coat is applied. To apply the bond coat, the surface should be thoroughly brushed with a stiff mortar or grout. After this, the dry pack material should be immediately packed into place before the bonding grout has dried. The mix for the bonding grout is a 1:1 cement and fine sand mixed with water to a fluid paste consistency. Under no circumstances should the bonding coat be so wet or applied so heavily that the dry pack material is more than slightly rubbery. Dry pack is usually a mix (by dry volume or by weight) having a 1:2-1/2 cement to sand ratio. A quantity of water should be used to produce a mortar that will adhere together while being molded into a ball with the hands and will not exude water, but will leave the hands damp.

The proper amount of water will produce a mix that will be rubbery when solidly packed. Less water will not yield a sound, solid pack; additional water will result in excessive shrinkage and loose repair.

Dry pack material should be placed and packed in layers having a compacted thickness of about 3/8 inch. Thicker layers will not be well-compacted at the bottom. The surface of each layer should be scratched to facilitate bonding of the next layer. One layer may be placed immediately over another unless an appreciable rubbery quality develops. If this occurs, work on the repair should be delayed 30 to 40 minutes. Under no circumstances should alternate layers of wet and dry materials be used.

Each layer should be solidly compacted over the entire surface by a hardwood plank and a hammer. Much of this tamping should be directed at a slight angle and toward the sides of the hole to insure maximum compaction in these areas. The hole should not be overfilled. Finishing may be completed at once by laying the flat side of the hardwood piece against the fill and striking it with several good blows. Steel finishing tools should not be used and water must not be used to facilitate fishing.

Grinding

This method can be used to remove deep-seated stains and irregularities in the surface and as a preparation procedure to receive a repair. Consideration should be given to etching and hammering before incorporating a grinding process. When grinding is used, it should be after the surface is hardened sufficiently to prevent the dislodgement of aggregate and should be continued until the course aggregate is exposed. The machines used should be approved types with stones that cut freely and rapidly. The surface should be kept wet during the grinding process and the cuttings removed with a squeegee and then the surface flushed with water. After the surface is ground, air holes, pits and other blemishes are filled with a thin grout composed of one part cement and one part #80 grained sand and grit. After the fillings have hardened for seven days, the surface should receive a final grinding to remove the film and give the finish a polished appearance.

Jacketing

This process involves applying and fastening a material over the concrete. Covering materials used today are metal, rubber, plastic and concrete. In water flow environments, the jackets should be streamlined to prevent scouring. There are various brand-named cementatious materials which can be used to coat concrete that has been stained or has minor surface spalling. These materials are applied by floating or spraying the material onto the concrete base according to manufacturers recommendations. Jacketing is used principally for the repair of deteriorated columns, piers and piling and when all or a portion of the section to be repaired is under water. This method can be used for protecting members from further deterioration, as well as a means of strengthening them. The following items should be considered when developing the specifications for the repair procedure:

1. Provide a form for the jacket. The form should be provided with spaces to assure clearance between it and the existing concrete surface. Types of forms are:
   A. **Wood:** used for marine environments or elsewhere where it is desired to protect the concrete from chemical action. The lumber should be given a preservative treatment and should be tongue and groove. Corrosion-resistant metal bands are used to tie the formwork together. They must be spaced close enough to prevent bulging of the formwork. A seal must be provided at the bottom of the form.
   B. **Wrought Iron:** This is an expensive type of permanent form, but can be used when it is justified from a cost standpoint.
   C. **Precast Concrete:** Used where timber forms constitute a fire hazard or the lumber is susceptible to attack by borers, termites, or other insect action.

D. **Gauge Metal:** Used as a temporary form where:
  —Repair will be in the dry and corrosive attack is mild.
  —The forms are to be stripped.
  —The life of the repair is limited.
  —Timber forms are not recommended.
2. Concrete surfaces are prepared as discussed previously in this chapter.
3. Any reinforcement required is usually in the form of wire-welded mesh.
4. Placement, vibration, and curing of the material should be done in accordance with the discussion contained in the chapter on steel repair or as specified in the job specifications.

Mortar Replacement

Portland cement mortar may be used for repairing defects on surfaces not prominently exposed where the defects are too wide for dry pack filling, too shallow for concrete filling, or where they are not deeper than the far side of the reinforcement that is nearest the surface. Repairs may be made either by use of shotcrete or by hand methods. If a mortar gun is used for the replacement mortar method, comparatively shallow holes should be flared outward at about a 1:1 slope so that rebound will fall free. Corners within the holes should be rounded. Shallow imperfections in new concrete may be repaired by mortar replacement if the work is done promptly after removal of the forms and while the concrete is still green. In the repair of old concrete, the importance of removing all traces of disintegrated material cannot be overemphasized. All areas to be repaired should be chipped to a depth not less than an inch. Wherever hand-placed mortar replacement is used, the edges of chipped areas should be squared with the surface, leaving no feather edges.

Best results with replacement mortar are obtained when the material is pneumatically applied, using a small gun. Equipment commonly used for shotcreting is too large to be satisfactory for the ordinarily small-sized repair of new concrete. Neat work is difficult in small areas and cleanup costs are high because cleanup is seldom done properly. However, small-sized equipment has been used satisfactorily for small-scale repair work. After the areas to be repaired have been cleaned, roughened, and surface-dried, the mortar should be applied immediately. No initial application of cement, cement grout, or wet mortar should be made.

If the repairs are more than one inch deep, the mortar should be applied in layers not more than 3/4 of an inch thick to avoid sagging and loss of bond. After completion of each layer, there should be a lapse of 30 minutes before the next layer is placed. Scratching or otherwise preparing the surface of a layer prior to addition of the next layer is unnecessary, but the mortar must not be allowed to dry.

In completing the repair, the hole should be filled slightly more than level full. After the material has partially hardened, but can still be trimmed off with the edge of a steel trowel, excess material should be shaved off, working from the center toward the edges. Extreme care must be used to avoid impairment of bond. Neither the trowel nor water should be used in finishing. A satisfactory finish may be obtained by rubbing the surface lightly with a soft rag.

For minor restorations, satisfactory mortar replacement may be performed by hand. This should be followed by a weatherproof coating.

Slab Jacking

This procedure is used to raise a slab that has settled. Holes are drilled through the slab based on a predetermined pattern established by consultants experienced in this type of work. Each hole is then fitted with a pressure fitting. This is followed by pumping a concrete slurry under the slabs to raise it to the desired elevation. For more details on this technique, the reader should consult with a professional, experienced in this type of work. Do not use this method for a severely cracked slab or a very heavy slab.

Pneumatically Applied Mortar (Gunite)

This method is used for restoration of concrete surfaces where the deterioration is shallow. It is also used for restoring surface spall due to corrosion of reinforcement. Some of the possible problems one should be aware of when working with this method are as follows:

- The material is porous.
- It shrinks more than conventional concrete.
- Coating may inadvertently include pockets of rebound or may overlay hollow spots, leaving unfilled depressions.
- This method may be expensive.

There are two processes of producing shotcrete. In the dry mix process, the dry materials are thoroughly mixed with enough moisture to prevent dusting. This dry mixture is forced through the delivery hose by compressed air. Water is added at the nozzle. In the wet process, all materials and water are mixed to produce mortar or concrete. The product is then forced through the delivery hose to the nozzle where air is injected to increase the velocity.

The optimum mix for shotcrete contains a little less water than will cause sloughing and just enough cement for the desired water/cement ratio. For large jobs, the optimum mix for sand shotcrete (as discharged at the nozzle) is one part cement to 4.5 parts sand and coarse aggregate by weight; this gives proportions in place of $1 : 3.2$ to $1 : 3.8$. The water cement ratio of fresh shotcrete in places should be about .57 for sloping and .54 for overhanging surfaces. These are approximately the maximum ratios that can be used without causing sloughing. To obtain optimum mix proportions, it is suggested that a trial batch be run and applied to sample surfaces with follow-up quality control procedures implemented.

For proper application, the nozzle should be held normally at about three feet from the surface to be coated. The most favorable velocity for material leaving the nozzle depends upon the size of the nozzle. With the dry process, it is essential that water pressure is greater than the air pressure to insure complete wetting of the materials at the nozzle and to give the nozzle operator a quicker, more positive control.

When coatings one inch thick or more are to be applied to vertical or overhanging surfaces, shotcrete without coarse aggregate should be applied in several layers to prevent sloughing. For level or slightly sloped surfaces, the thickness of a single layer may vary up to a maximum of 3-1/2 inches. When more than one layer is applied, a delay of 30 minutes to one hour between applications is usually sufficient to prevent sloughing. For shotcrete containing coarse aggregate and accelerator, no delay is necessary, since initial set takes place almost immediately and it gains strength rapidly. Layers should be applied before the previously placed shotcrete has set completely; otherwise, a glaze coating will form on the surface of the previous layer. There is no apparent difference between finished placements started at the top and those placed from the bottom upward. It is essential that the surface to be coated be free of rebound. Personnel placing shotcrete should be experienced in this method.

Prepacked or Preplaced Aggregate Concrete

This technique involves the filling of the area to be repaired with gap-graded aggregate. This is followed by filling the voids with water and then pumping in a sand-cement grout to replace the water-filled voids. This method is primarily used for:

- Refacing structures.
- Jacketing.
- Filling of cavities in and under structures.

The advantage is the ease with which preplaced aggregate concrete can be placed in certain locations where placement of conventional concrete would be extremely difficult. Preplaced aggregate concrete is especially adaptable to underwater construction, to concrete masonry repairs and, in general, to certain types of new structures. Since preplaced aggregate concrete is adaptable to special types of construction, it is essential that the work be undertaken by well-qualified personnel, experienced in this method of concrete repair.

Preplaced aggregate should be washed and screened to remove dirt immediately before placing in the form. The maximum size of aggregate should be an inch and a half. Grout for preplaced aggregate concrete may consist of sand of specified gradation mixed with water and cement at high speeds to a creamy consistency. A water-reducing, set-controlling agent is added to inhibit early stiffening of the grout; also, it enhances the fluidity and holds the solid constituents in suspension. Consistency of grout for preplaced aggregate concrete should be uniform from batch to batch and should be such that it may be readily pumped, under reasonably low pressure, into the voids of the preplaced mass of aggregate.

Forms for preplaced aggregate concrete may be of wood, steel, or other materials suitable for conventional concrete. The form workmanship should be of better quality than is normally suitable for conventional concrete.

The grout pipe system is used to deliver and inject grout into the preplaced aggregate, to provide means for determining grout elevations within the aggregate mass, and to provide vents in enclosed forms for water and air escape. Proper design and arrangement of the pipe system are essential for successful placement. The simplest and most reliable system consists of a single pipeline connected to insert pipes positioned during the placing of the aggregate. The pumping systems should have a bypass for returning grout to an agitating tank.

The length of delivery line should be kept to a practicable minimum. Pipe sizes should be such that during operation, under normal conditions, the grout velocity ranges between two to four feet per second, or at a pumping rate of about one cubic foot of grout per minute through a one-inch diameter pipe. Higher velocities require excessive pumping pressure. The recommended velocity range is for delivery pipes up to 300 feet long. From 300 to a maximum of 1,000 feet, the diameter must be increased about one pipe size to avoid excessive pumping pressure.

Grout insert pipes are normally 3/4 to one inch in diameter and may be placed vertically or at various angles so as to inject grout at the proper point. The pipe should be in sections of five and one-half feet long for easy withdrawal. For depths below 15 feet, they should be flush-coupled. For shallower depths, standard pipe couplings may be used.

Connections between the grout delivery line and insert pipe should be quick-opening fittings. Quick-disconnect pneumatic fittings are not suitable because of the reduction in cross-section of the flow area. Valves should be a quick-opening, plug type which can be readily cleaned.

Spacing of insert pipes depends on aggregate gradation, void ratio, depth and area of work, and location of embedded items. Spacing of insert pipes may range from four to twelve feet; five or six feet spacing is commonly used. For the purpose of insert pipe layout, it is assumed that the grout surface will be on about a 1:4 slope in a dry placement, and 1:6 underwater; however, actual grout surfaces may be considerably flatter. It is helpful to color code or number and record locations of each insert pipe so there is no question where the outlet pipe ends.

The pump should be of a positive displacement type, such as a piston pump or a progressive cavity type. Although a well-proportioned grout mix will retain solids in suspension within a piping system, pumps shut down for prolonged periods will permit the sand to settle within the pump and lines. As pumps normally require a period of maintenance on each shift, one or more standby pumps should be provided or quick changeover to maintain continuous operations. The pump should have a pressure gauge on the outlet line to indicate any incipient line blockage.

Sounding wells, slotted pipes, are used to indicate the level of the grout. The ratio of sounding wells to insert pipe ranges from 1:4 to 1:10. These sounding wells, through which a sounding line (equipped with a one-inch, diameter float, weighted so that it will sink in water and float on grout) may be lowered into the mass. There should be no burrs or obstructions inside the sounding well on which the float will catch.

Fundamentally, grout injections should start at the lowest point within the form and be continued until the placement is completed. Usually a sufficient quantity of grout is pumped through the insert pipe to raise the grout level from six to twelve inches. The insert pipe outlet is set initially six inches from the bottom and progressively raised as the grouting proceeds. The lower end should remain embedded twelve inches below the grout surface. The grout surface should be kept relatively level, although often a gentle slope is maintained. Care should be exercised not to permit grout to cascade on a steep slope through the aggregate, causing separation of sand from grout. Adequate venting should be provided to insure complete filling and prevent entrapment of air or water in enclosed spaces.

Internal vibration cannot be employed with this method of placing, but external vibration of the forms can be, and is, beneficial in improving surface appearance. If it is not done, a spotty appearance will develop where coarse aggregate particles have been in contact with the form.

Stitching

This procedure is incorporated when major cracks must be repaired and structural integrity re-established. Holes are drilled into the concrete on either side of the crack, on a predetermined design, founded on a structural analysis. Shaped metal binders, called "stitching dogs," are installed across the cracks. The dogs must be of correct size and orientation as determined by the stress analysis. They should be placed close together at the ends of the crack. If water is a potential problem, seal the crack first before placing the dogs. If the crack is active, first stabilize the crack with the dogs, then seal. Dogs are installed into the holes with a non-shrink or expanding grout. The holes should be completely filled. Also drill out the ends to relieve stress concentration. It must be noted that if the repair method is incorrectly done, the concrete will become cracked in another direction. Encasement of the dogs with an overlay is beneficial, and a must when water is present.

Stressing

This method is used when structural cracks have occurred and the entire area must be strengthened and cracks closed. The process utilizes stressing cables or rods to apply compression force, which closes the cracks and at the same time increases the structural capacity of the member. As for stitching, this procedure should be performed only by personnel experienced in this subject and only after a stress analysis has been performed.

Thin Resurfacing (Bonded or Unbonded)

This procedure sometimes is known as an overlay. A bonded type of resurfacing is used if the reason for doing this repair is that wear, such as from abrasion, has taken place. Do not use bonded resurfacing for active deterioration. In this case, unbonded resurfacing should be incorporated. This method of repair is not a means of structural stabilization, but is a means for prolonging the life of the underlying concrete section. This method is also used to give areas which have not been repaired an appearance similar to those which have been repaired. This technique is dependent on the type of deterioration. Techniques for specific deterioration types are detailed below:

- **Active Deterioration:** The overlay must be flexible (bendable) and extensible (stretchable). If the cracks are subjected to longitudinal movement parallel to their axis, do not use an overlay; use blanketing. Use a two-or three-ply built up surface of roofing felt layed in a top coat of tar and cover the entire area with gravel.
- **Dormant Deterioration:** Use any type of overlaying material, as long as it can stand up to traffic and can be well bonded to the surface.

Proper preparation of the surface of the concrete is of essential importance. Any loose or poorly bonded concrete should be removed. The use of a scarifying machine is a satisfactory method for preparing large, flat areas. Small areas of loosened material can be removed by jackhammering, chipping tools, or rough grinding. After the area has been scarified, the surface should be dry swept or vacuumed to clear it of debris and dust. It should then be washed with water, thoroughly brushed, and finally inspected for loose areas and off drippings that may have been mined. In areas where the old floor surface is sound and where scarifying is not necessary, the surface can be chemically etched. Factors that must be kept in mind in repair of concrete floors and pavements are as follows:

- Changing the level or grade of the floor surface for reasons of:
  —Drainage.
  —Curbing.
- Traffic Patterns—Will they be affected?
- Joints
  —Maintain expansion joints.
  —Provide control jointing to control joint locations.
  —Seal joints in old surfaces.

When selecting a method, keep in mind that a good bonding of the resurfacing to the existing surface is hard to obtain, and this often is an expensive and problematic operation. Therefore, the unbonded course method is usually selected except in cases where the increased thickness creates further problems, or the method becomes more expensive than the bonded course method.

The procedure for constructing an unbonded course is described below.

1. Remove the damaged areas of the existing slab.
2. Compact or stabilize the subgrade.
3. Use a dry concrete mix, having the same strength as the existing concrete.
4. Use conventional methods of placement.

For constructing a bonded course:

1. Remove damaged areas of the existing slab.
2. Compact or stabilize the subgrade.
3. Roughen the existing surface with the use of an acid etch, sand blast, or scarification.
4. Clean all dirt, debris, oil patches, or other contaminants with a solvent.
5. Thoroughly wet the surface and let dry.
6. A bond coat of grout should be scrubbed into the surface with a stiff bristle broom as soon as the surface is dried. Leave this finish rough.
7. Apply the resurfacing course as soon as the bond coat dries to thumbnail hardness. Use conventional procedures to apply the finish coat.

Relative to expansion joints in bonded courses, the following general outline should be followed:

1. Provide an expansion joint in the new slab over each active expansion joint in the old one.
2. Provide control or contraction joints in the new slab over all cracks in the underlying slab which show movement.
3. For exterior exposures, provide additional joints in the new slab.
4. Place floor slab concrete in accordance with the job specifications, along with providing control joints.
5. Expansion joints in the resurfacing course should be filled and sealed.
6. End dams between pours should go down flush with the old surfaces.
7. Increase the thickness of the edges of the resurfacing course along joints or dowel the joints.
8. Provide an intercourse waterproofing between surfaces, if required.
9. Remember, defects occur over defects.

Asphalt Concrete Overlays

The advantages of asphalt concrete overlays over concrete are lower cost, easier application, and availability of materials and equipment. The disadvantage is that this type of overlay cannot be used when pavement must be raised. It also has poor wear resistance and is non-resistant to fuel and other chemical type spillages.

1. All joint filler and sealer must be raked out of existing joints and cracks larger than one-half inch.
2. Clean the pavement of all excess filler and remove the asphalt patches.
3. Eliminate all high spots, bumps, raised joints or other projections.
4. Remove all disintegrated concrete.
5. Sweep the existing surface clean. Remove all foreign matter that may hinder bond.
6. Apply a light top coat of asphalt emulsion to all surfaces to be bonded.
7. Apply an approved asphaltic concrete leveling course, followed with both a bonder and a surface course. For localized patching, use a surface course only. Provide wire-welded fabric reinforcement if cracking of the old slab is extensive. Epoxies can also be used for overlay materials. The advantages of using epoxy are listed below:
   A. Excellent bonding characteristics.
   B. Provides skid resistance.
   C. Can be used as a water proofing agent (except at joints).
   D. Protects concrete surfaces from chemical disintegration.

Application of Epoxy Overlays

1. Clean concrete surfaces thoroughly.
2. Insure that the surface is rough.
3. Flush the surface with water and allow to dry.
4. Apply a binder (be sure to take precautions not to allow the binder to get into drains, expansion joints, etc). Spray the binder evenly and do not overspray.
5. Spread the aggregate immediately after applying the binder. Use a tough abrasion-resistant aggregate for large areas and pure silica sand for patching.
6. Complete the epoxy surface application according to the manufacturer's directions.

Repair of Water Seepage

When water seeps through the cracks in the floor, it is advisable to place a water proofing membrane between the old and new concrete. This water proofing membrane consists of two layers of roofing felt with hot bituminous material mopped between the layers (as for foundation wall repair). The membrane should cover the entire floor and should extend a distance of about twelve inches up the sides of the wall. The new slab to be placed on top of the membrane should be at least 2-1/2 inches thick. Where there will be a pressure from ground water, it may be necessary to increase the thickness and add reinforcement to the new slab. In extremely bad conditions, the possibility of installing a well point system should be investigated.

Sack Rubbing

This method is used to remove stains or to fill small holes. Spray the concrete with water. Follow this by rubbing mortar over the surface and into the voids with a rubber float or a piece of burlap. When preparing the mortar, be sure to add the correct amount of cement to match the existing concrete color.

Blanketing

This is used for large cracks and can be used for sealing active as well as dormant cracks. This is actually a form of coating or overlay, but used for isolated cracks. If the cracks are active, an extensible material must be used. For dormant cracks, a brittle material such as mortar may be used. There are various methods, depending on the blanketing material.

Using an Elastic Sealant:

1. If the sealant is to be placed below the surface, a chase (groove) needs to be cut into the surface. The width should be at least four to six times the anticipated amount of movement in the joint. The minimum depth is about one inch.
2. Cut the chase square. The bottom should be chipped as smooth, level and as clean as possible. The sides should be rough.
3. Remove all loose and disintegrated concrete adjacent to the crack. If this results in too large a chase, rebuild to a desired dimension with dry pack or replacement concrete.
4. The chase should be dried out. Use a blowtorch, if needed. If there is an inflow of water, a temporary mortar plug made with a flash-set material can be used to stop the flow. The plug is only a temporary measure. If the flow of water is large, install weep pipes to drain the water away.
5. Place the bond breaker at the bottom of the chase so as not to hinder the functional properties of the sealant.
6. Select a sealant that will be capable of providing the required sealing properties (such as tensile value, bonding value, deformation qualities, etc.). Other factors to consider are fatigue, age vs. functionality and abrasion resistance.
7. Place the sealant according to manufacturer's directions.

Using a Mastic-Filled Joint:
This method is similar to an elastic sealant, except that the bond breaker is omitted and the sealant is bonded to the bottom as well as to the sides of the chase. The sealant is of a mastic material.

Using a Mortar-Plugged Joint:

1. Prepare the chase. The depth is the sum of the thicknesses of the mortar plug and sealant. The usual depth is 1 1/2 to 3 inches (depending on the materials used). The width of the chase is about 2 to 3 inches.
2. The edges of the chase should be undercut, as close as possible. This makes for a tighter plug. There is an exception to this, i.e., when the repair is to be covered with an overlay.
3. All surfaces of the chase should be cleaned of deposits and be dry. Remove all loose, deteriorated concrete. Seal the joint, if needed for water problems.
4. Place the plug by first applying the mastic sealant; then a bonding agent followed by the mortar compound.
5. If the water pressure is too great, the plug can be anchored to the bond coat with header bars.

This type of repair is not suitable for active cracks.

Crimped Water Bar-Water Stops:

1. The water stop can be placed directly on the surface and anchored into a chase and anchor holes (no traffic areas) or be placed into a large recess and covered with mastic and mortar cover (for traffic areas). If a mortar cover is used, the depth, width, installation and anchoring of the mortar cover is similar to item 2 above.
2. An epoxy bonding agent is used to bond the wings of the stop into the concrete.
3. Be sure to solder or vulcanize all splices to make the stops continuous (for reasons of possible leaking).

This method is used where there is only simple expansion or contraction and in the presence of small amounts of water pressure.

Thermo Setting Plastic (Epoxy)

Epoxy should be used to bond new concrete or mortar to old concrete whenever the depth of repair is between 1-1/2 to 6 inches. Epoxy-bonded epoxy mortar should be used when the depth of repair is less than 1-1/2 inches to feather edges. Epoxies are useful in special applications such as bonding steel anchor ban in old concrete. The surface of existing concrete to which concrete and epoxy mortar are to be epoxy bonded must be prepared and maintained in a clean condition, as discussed above, except when wet or dry sandblasting is used. In this case, cleaning is by water or air jet. Concrete to be repaired with epoxy materials should be heated when necessary so that the surface temperature will not drop below 65 degrees Fahrenheit during the first four hours after placement of an epoxy bond coat. This may require several hours of preheating with radiant heaters. Concrete temperatures during preheating should never exceed 200 degrees Fahrenheit and the final surface temperature at the time of placing the epoxy material should never be greater than 105 degrees Fahrenheit.

Many epoxy formulations are now available, but some are unsuitable for all applications. When repairing concrete, the epoxy is generally mixed with sand to make an epoxy mortar. The sand used in epoxy mortar must be clean, dry, well-graded, composed of sound particles, and maintained in a dry area not less than 70 degrees Fahrenheit for 24 hours immediately prior to the time of use. Filler materials other than sand, such as portland cement, can be used. However, for general applications, a natural sand is recommended.

Preparation of Epoxy Bonding Agent

Epoxy resin bonding agent is a two-component material which requires a combination of components and mixing prior to use. Once mixed, the material has a limited pot life and must be used immediately. The bonding agent should be prepared by adding a hardener component to the resin component in the proportions recommended by the manufacturer, followed by thorough mixing. Since the working life of the mixture depends on the temperature (longer at low temperatures, much shorter at higher temperatures), the quantity to be mixed at one time should be applied and topped within approximately 30 minutes. The addition of thinners or dilutants to the resin mixture is not permitted.

Preparation of Epoxy Mortar

The epoxy mortar is composed of sand and an epoxy bonding agent suitably blended to provide a stiff workable mix. The epoxy component should be mixed thoroughly prior to addition of sand. Mixed proportions should be established, batched, and recorded by weight, although the dry sand and mixed epoxy may be batched by volume. Suitable measuring containers that have been calibrated by weight can be used. It should be equivalent to a ratio of approximately 4 to 4-1/2 parts to one part epoxy by volume. If equivalent volume proportions are being used, care must be taken to prevent confusing them with weight proportions. Epoxy mortar should be thoroughly mixed with a slow-speed mechanical device. The mortar should be mixed in small-sized batches so that it can be made and placed within a 30-minute period.

Application of an Epoxy Bonding Agent

Immediately after the epoxy resin is mixed, it must be applied to the prepared, dry, existing concrete at a coverage of not more than 80 square feet per gallon, depending upon the surface conditions. The area of coverage per gallon of agent depends on the roughness of the surface to be covered and may be considerably less than the maximum specified above. The epoxy bonding agent may be applied by any convenient, safe method such as a squeegee, brushes, or rollers which will yield an effective coverage. Spraying of the material is permitted if an effective and efficient airless spray is used when the concrete surface is to receive the agent. If spraying is to be used, a small sample should be made to demonstrate its effectiveness.

During application of the epoxy bonding agent, care must be exercised to confine the material to the area being bonded and to avoid contamination of adjacent surfaces. However, the epoxy bond coat should extend slightly beyond the edges of the repair area.

Applied epoxy bonding agent film must be in a fluid condition when the concrete or epoxy mortar is placed. Unless epoxy mortar is placed on steep sloping or vertical surfaces, the agent may be allowed to stiffen to a very tacky condition. Special care must be taken to thoroughly compact the epoxy mortar against the tacky bond coat. If the applied film cures beyond a fluid condition, a second application of bonding agent must be applied while the first coat is still tacky.

Application of Epoxy Bonded Concrete

Use of epoxy bonded concrete in repairs requiring forming (such as on steeply sloped or vertical surfaces) can be permitted only when sufficient time has been allowed to place concrete against the bonding agent while it is still smooth. Immediately after application of the epoxy-resin bonding agent, while the epoxy is still fluid, unformed epoxy bonded concrete should be spread evenly to a level slightly above grade and compacted thoroughly by vibrating or tamping. After being compacted and vibrated, the concrete should be given a wood float or steel finish, as required. Water, cement, or a mixture of dry cement and sand should not be sprinkled on the surface. Troweling, if required, should be performed at the proper time and with heavy pressure to produce a smooth, dense finish, free of defects and blemishes. As the concrete continues to harden, the surface should be retroweled. The final troweling should be performed after the surface has hardened so that no cement paste will adhere to the edge of the trowel. Excessive troweling should not be permitted.

Application of Epoxy Bonded Epoxy Mortar

Surfaces of existing concrete to which epoxy mortar is to be bonded should be prepared as discussed earlier. Epoxy-resin bonding agent should then be applied as outlined above. The agent should be applied to the areas immediately before placing an epoxy mortar. Special care must be taken to prevent the bond coat from being spread over concrete surfaces not properly cleaned or prepared.

The prepared epoxy mortar should be tamped, flattened, and smoothed into place in all areas while the epoxy bonding agent is still in a fluid condition, except that on steep slopes the bond coat can be brought to a tacky condition. The mortar should be worked to grade and given a steel trowel finish. Special care should be taken at the edges of the area being prepared, particularly where there are thin feather edges, to assure complete filling and leveling and to prevent the mortar from being spread over surfaces not having the epoxy bond coat application. Steel troweling should best suit prevailing conditions: in general, it should be performed by applying slow, even strokes. Trowels may be heated to facilitate the finishing, but the use of thinner, dilutants, water, or other lubricants on placing or finishing tools is not recommended. After leveling the epoxy mortar to the finished grade, where precision surfaces are required, the mortar should be covered with plywood panels smoothly lined with polyethylene sheathing and weighted with sandbags otherwise braced by suitable means until the possibility of slumping has passed. When polyethylene sheathing is used, no attempt should be made to remove it from the epoxy mortar repair before final hardening. All areas of repair requiring feather edging should be finished with epoxy mortar.

Surfaces of all epoxy-bonded mortar repairs should be finished to the elevation of surfaces adjoining the repair areas. The final finished surfaces should have the same smoothness and texture of surfaces adjoining repair areas.

Application of Epoxy by Pressure Injection

An effective method for repairing structural members such as walls, piers, floors, ceilings, and pipes is the epoxy pressure injection system. If the cracks to be injected are clean and dry, and the epoxy is properly mixed, placed, and cured, the repaired member will be restored to its original structural integrity. Although damp or wet cracks can be repaired using this method, development of sufficient bond between epoxy and concrete cannot be assured. Small repair jobs can use any system that will successfully deposit the epoxy into the required zones. Epoxy injection jobs generally require a single stage injection technique in which the two epoxy components are pumped independently of one another from the reservoir to the mixing nozzle. At the mixing nozzle, located adjacent to the crack being repaired, the two epoxy components are brought together for mixing and injecting. The epoxy used in the injection technique must have a slow, low initial viscosity and a closely controlled set time. Cracks as small as 0.002 inches in width have been successfully repaired with injected epoxy resin. If this method is to be considered, consult with experienced personnel in this area.

Curing the Repair

Because of the relatively small volume of most repairs, and the tendency of old concrete to absorb moisture from new material, water curing is a highly desirable procedure, at least during the first 24 hours. When forms are used for repair, they can be removed and then reset to hold a few layers of wet burlap in contact with new concrete. The best method of water curing is to do a slow soaking from a hose. If curing compound is used, the best combination is an initial water curing period of seven days followed (while the surface is still damp) by a coat of compound. It is always essential that repairs receive some water curing and be thoroughly damp before the curing compound is applied.

Curing compound will be used whenever there is any possibility that freezing temperatures will prevail during the curing period. Sheet polyethylene must be an airtight, non-staining, waterproof covering that will effectively prevent loss of moisture by evaporation. Edges of polyethylene should be lapped and sealed. The waterproof covering should be left in place for at least two weeks. If a waterproof covering is used and the concrete is subjected to any usage during the curing period that might rupture or otherwise damage the covering, the covering must be protected by a suitable layer of clean wet sand or other cushioning material that will not stain concrete. After curing, the covering (except if curing compound is used) and all foreign materials should be removed as directed.

Relative to curing epoxy bonded concrete, when it has hardened sufficiently to prevent damage, the surface should be moistened by spraying lightly with water and then covering with sheet polyethylene or by coating with an approved curing compound.

Epoxy mortar repair should be cured immediately after completion at not less than 60 degrees F. until the mortar is hard. Post curing should then be initiated at elevated temperatures by heating in depth the epoxy mortar and the concrete beneath the repair. Post curing should continue for a minimum of four hours at a surface temperature of not less than 90 degrees F. nor more than 110 degrees F. The heat should be supplied by use of portable propane fire heaters, batteries of infrared lamp heaters, or other approved sources positioned to attain the required surface temperatures. In no case should epoxy-bonded epoxy mortar be subjected to moisture until the specified post curing has been completed.

Figure 3.6 is a list of substances which are harmful to concrete, as well as recommended coatings and/or surface treatments to protect the concrete against damage from the respective substance.

Effect of Chemicals on Concrete (see end of table for special notations)

Material	Effect	Material	Effect
*Acetic acid, all concentrations	Disintegrates slowly	Ashes	Harmful if wet, when sulfides and sulfates leach out (see sodium sulfate)
Acetone	Liquid loss by penetration. May contain acetic acid as impurity (which see)	Ashes, hot	Cause thermal expansion
Acid waters (pH of 6.5 or less) (a)	Disintegrates slowly. In porous or cracked concrete, attacks steel	Automobile and diesel exhaust gases (n)	May disintegrate moist concrete by action of carbonic, nitric, or sulfurous acid
*Alcohol	See ethyl alcohol, methyl alcohol	*Baking soda	See sodium bicarbonate
Alizarin	Not harmful	Barium hydroxide	Not harmful
*Almond oil	Disintegrates slowly	Bark	See tanning bark
*Alum	See potassium aluminum sulfate	*Beef fat	Solid fat disintegrates slowly, melted fat more rapidly
Aluminum chloride	Disintegrates rapidly. In porous or cracked concrete, attacks steel	*Beer	May contain, as fermentation products, acetic, carbonic, lactic, or tannic acids (which see)
*Aluminum sulfate	Disintegrates. In porous or cracked concrete, attacks steel	Benzol (benzene)	Liquid loss by penetration
*Ammonia, liquid	Harmful only if it contains harmful ammonium salts (see below)	Bleaching solution	See specific chemical, such as hypochlorous acid, sodium hypochlorite, sulfurous acid, etc.
Ammonia vapors	May disintegrate moist concrete slowly or attack steel in porous or cracked moist concrete	*Borax	Not harmful
Ammonium bisulfate	Disintegrates. In porous or cracked concrete, attacks steel	*Boric acid	Negligible effect
Ammonium carbonate	Not harmful	*Brine	See sodium chloride or other salt
*Ammonium chloride	Disintegrates slowly. In porous or cracked concrete, attacks steel	Bromine	Gaseous bromine disintegrates. Liquid bromine disintegrates if it contains hydrobromic acid and moisture
Ammonium cyanide	Disintegrates slowly	*Buttermilk	Disintegrates slowly
Ammonium fluoride	Disintegrates slowly	Butyl stearate	Disintegrates slowly
Ammonium hydroxide	Not harmful	Calcium bisulfite	Disintegrates rapidly
Ammonium nitrate	Disintegrates. In porous or cracked concrete, attacks steel	*Calcium chloride	In porous or cracked concrete, attacks steel. (b) Steel corrosion may cause concrete to spall
Ammonium oxalate	Not harmful	*Calcium hydroxide	Not harmful
*Ammonium sulfate	Disintegrates. In porous or cracked concrete, attacks steel	Calcium nitrate	Not harmful
Ammonium sulfide	Disintegrates	*Calcium sulfate	Disintegrates concrete of inadequate sulfate resistance
Ammonium sulfite	Disintegrates	Carbazole	Not harmful
Ammonium superphosphate	Disintegrates. In porous or cracked concrete, attacks steel	Carbolic acid	See phenol
Ammonium thiosulfate	Disintegrates	*Carbon dioxide	Gas may cause permanent shrinkage (see also carbonic acid)
Animal wastes	See slaughter house wastes	*Carbon disulfide	May disintegrate slowly
Anthracene	Not harmful	*Carbon tetrachloride	Liquid loss by penetration of concrete
Arsenious acid	Not harmful	*Carbonic acid	Disintegrates slowly (c)

Figure 3.6

Material	Effect	Material	Effect
Castor oil	Disintegrates, especially in presence of air	*Cottonseed oil	Disintegrates, especially in presence of air
Chile saltpeter	See sodium nitrate	Creosote	Phenol present disintegrates slowly
China wood oil	Liquid disintegrates slowly.	Cresol	Phenol present disintegrates slowly
Chlorine gas	Slowly disintegrates moist concrete	Cumol	Liquid loss by penetration
Chrome plating solutions (o)	Disintegrates slowly	Deicing salts	Scaling of non-air-entrained or insufficiently aged concrete (b)
Chromic acid, all concentrations	Attacks steel in porous or cracked concrete	Diesel gases	See automobile and diesel exhaust gases
Chrysen	Not harmful	Dinitrophenol	Disintegrates slowly
*Cider	Disintegrates slowly (see acetic acid)	Distiller's slop	Lactic acid causes slow disintegration
Cinders	Harmful if wet, when sulfides and sulfates leach out (see, for example, sodium sulfate)	Epsom salt	See magnesium sulfate
		*Ethyl alcohol	Liquid loss by penetration
Cinders, hot	Cause thermal expansion	*Ethyl ether	Liquid loss by penetration
Coal	Sulfides leaching from damp coal may oxidize to sulfurous or sulfuric acid, or ferrous sulfate (which see)	*Ethylene glycol	Disintegrates slowly (d)
		Feces	See manure
Coal tar oils	See anthracene, benzol, carbazole, chrysen, creosote, cresol, cumol, paraffin, phenanthrene, phenol, toluol, xylol	*Fermenting fruits, grains, vegetables, or extracts	Industrial fermentation processes produce lactic acid. (e) Disintegrates slowly (see lactic acid)
		Ferric chloride	Disintegrates slowly
Cobalt sulfate	Disintegrates concrete of inadequate sulfate resistance	Ferric nitrate	Not harmful
*Cocoa bean oil	Disintegrates, especially in presence of air	Ferric sulfate	Disintegrates concrete of inadequate quality
*Cocoa butter	Disintegrates, especially in presence of air	Ferric sulfide	Harmful if it contains ferric sulfate (which see)
Coconut oil	Disintegrates, especially in presence of air	Ferrous chloride	Disintegrates slowly
*Cod liver oil	Disintegrates slowly	Ferrous sulfate	Disintegrates concrete of inadequate sulfate resistance
Coke	Sulfides leaching from damp coke may oxidize to sulfurous or sulfuric acid (which see)	Fertilizer	See ammonium sulfate, ammonium superphosphate, manure, potassium, nitrate, sodium nitrate
Copper chloride	Disintegrates slowly	Fish liquor	Disintegrates (f)
Copper plating solutions (p)	Not harmful	*Fish oil	Disintegrates slowly
		Flue gases	Hot gases (400–1100 F) cause thermal stresses. Cooled, condensed sulfurous, hydrochloric acids disintegrate slowly
Copper sulfate	Disintegrates concrete of inadequate sulfate resistance		
Copper sulfide	Harmful if it contains copper sulfate (which see)	Foot oil	Disintegrates slowly
*Corn syrup	Disintegrates slowly	*Formaldehyde, 37 percent	Formic acid, formed in solution, disintegrates slowly
Corrosive sublimate	See mercuric chloride	Formalin	See formaldehyde

Figure 3.6 (continued)

Material	Effect	Material	Effect
*Formic acid, 10 percent	Disintegrates slowly	Lignite oils	If fatty oils are present, disintegrates slowly
*Formic acid, 30 percent	Disintegrates slowly	*Linseed oils	Liquid disintegrates slowly. Dried or drying films are harmless
*Formic acid, 90 percent	Disintegrates slowly	Locomotive gases (r)	May disintegrate moist concrete by action of carbonic, nitric or sulfurous acids (see also automobile and diesel exhaust gases)
*Fruit juices	Hydrofluoric, other acids, and sugar cause disintegration (see also fermenting fruits, grains, vegetables, extracts)	Lubricating oil	Fatty oils, if present, disintegrate slowly
Gas water (g)	Ammonium salts seldom present in sufficient quantity to disintegrate	Lye	See sodium hydroxide
Gasoline	Liquid loss by penetration	Machine oil	Fatty oils, if present, disintegrate slowly
*Glucose	Disintegrates slowly	*Magnesium chloride	Disintegrates slowly. In porous or cracked concrete, attacks steel
*Glycerine	Disintegrates slowly	Magnesium nitrate	Disintegrates slowly
*Grain	See fermenting fruits, grains, vegetables, extracts	*Magnesium sulfate	Disintegrates concrete of inadequate sulfate resistance
*Honey	Not harmful	Manganese sulfate	Disintegrates concrete of inadequate sulfate resistance
Horse fat	Solid fat disintegrates slowly, melted fat more rapidly	Manure	Disintegrates slowly
Humic acid	Disintegrates slowly	*Margarine	Solid margarine disintegrates slowly, melted margarine more rapidly
*Hydrochloric acid, all concentrations	Disintegrates rapidly, including steel	Mash, fermenting	Acetic and lactic acids, and sugar disintegrate slowly
Hydrofluoric acid, all concentrations	Disintegrates rapidly, including steel	Mercuric chloride	Disintegrates slowly
Hydrogen sulfide	Not harmful dry. In moist, oxidizing environments converts to sulfurous acid and disintegrates slowly	Mercurous chloride	Disintegrates slowly
Hypochlorous acid, 10 percent	Disintegrates slowly	Methyl alcohol	Liquid loss by penetration
Iodine	Disintegrates slowly	Methyl ethyl ketone	Liquid loss by penetration
Kerosene	Liquid loss by penetration of concrete	Methyl isobutyl ketone	Liquid loss by penetration
*Lactic acid, 5–25 percent	Disintegrates slowly	*Milk	Not harmful. However, see sour milk
*Lamb fat	Solid fat disintegrates slowly, melted fat more rapidly	Mine water, waste	Sulfides, sulfates, or acids present disintegrate concrete and attack steel in porous or cracked concrete
*Lard and lard oil	Lard disintegrates slowly, lard oil more rapidly	*Mineral oil	Fatty oils, if present, disintegrate slowly
Lead nitrate	Disintegrates slowly	Mineral spirits	Liquid loss by penetration
Lead refining solutions (q)	Disintegrates slowly	*Molasses	At temperatures ≥ 120 F, disintegrates slowly
Leuna saltpeter	See ammonium nitrate and ammonium sulfate	Muriatic acid	See hydrochloric acid
		*Mustard oil	Disintegrates, especially in presence of air
		Nickel plating solutions (v)	Nickel ammonium sulfate disintegrates slowly

Figure 3.6 (continued)

Material	Effect	Material	Effect
Nickel sulfate	Disintegrates concrete of inadequate sulfate resistance	Potassium hydroxide, 25 percent or over	Disintegrates concrete
Niter	See potassium nitrate	*Potassium nitrate	Disintegrates slowly
Nitric acid, all concentrations	Disintegrates rapidly	Potassium permanganate	Harmless unless potassium sulfate present (which see)
*Oleic acid, 100 percent	Not harmful	Potassium persulfate	Disintegrates concrete of inadequate sulfate resistance
Oleum	See sulfuric acid, 110 percent	Potassium sulfate	Disintegrates concrete of inadequate sulfate resistance
*Olive oil	Disintegrates slowly	Potassium sulfide	Harmless unless potassium sulfate present (which see)
Ores	Sulfides leaching from damp ores may oxidize to sulfuric acid or ferrous sulfate (which see)	Pyrites	See ferric sulfide, copper sulfide
Oxalic acid	Not harmful. Protects tanks against acetic acid, carbon dioxide, salt water. Poisonous. Do not use with food or drinking water	*Rapeseed oil	Disintegrates, especially in presence of air
		Rock salt	See sodium chloride
Paraffin	Shallow penetration not harmful, but should not be used on highly porous surfaces like concrete masonry (u)	Rosin	Not harmful
		Rosin oil	Not harmful
*Peanut oil	Disintegrates slowly	Sal ammoniac	See ammonium chloride
Perchloric acid, 10 percent	Disintegrates	Sal soda	See sodium carbonate
Perchloroethylene	Liquid loss by penetration	Salt for deicing roads	See text. Also calcium chloride, magnesium chloride, sodium chloride
Petroleum oils	Liquid loss by penetration. Fatty oils, if present, disintegrate slowly	Saltpeter	See potassium nitrate
Phenanthrene	Liquid loss by penetration	*Sauerkraut	Flavor impaired by concrete. Lactic acid may disintegrate slowly
Phenol, 5–25 percent	Disintegrates slowly	Sea water	Disintegrates concrete of inadequate sulfate resistance. Attacks steel in porous or cracked concrete
*Phosphoric acid, 10–85 percent	Disintegrates slowly		
*Pickling brine	Attacks steel in porous or cracked concrete	Sewage	Usually not harmful (see hydrogen sulfide)
Pitch	Not harmful	Silage	Acetic, butyric, lactic acids (and sometimes fermenting agents of hydrochloric or sulfuric acids) disintegrate slowly
*Poppy seed oil	Disintegrates slowly		
*Potassium aluminum sulfate	Disintegrates concrete of inadequate sulfate resistance	Slaughter house wastes (w)	Organic acids disintegrate
*Potassium carbonate	Harmless unless potassium sulfate present (which see)	Sludge	See sewage, hydrogen sulfide
*Potassium chloride	Magnesium chloride, if present, attacks steel in porous or cracked concrete	Soda water	See carbonic acid
		*Sodium bicarbonate	Not harmful
Potassium cyanide	Disintegrates slowly	Sodium bisulfate	Disintegrates
Potassium dichromate	Disintegrates	Sodium bisulfite	Disintegrates
Potassium hydroxide, 15 percent	Not harmful (h)	Sodium bromide	Disintegrates slowly

Figure 3.6 (continued)

Material	Effect	Material	Effect
Sodium carbonate	Not harmful, except to calcium aluminate cement	Sulfurous acid	Disintegrates rapidly
*Sodium chloride	Magnesium chloride, if present, attacks steel in porous or cracked concrete. (b) Steel corrosion may cause concrete to spall	Tallow and tallow oil	Disintegrates slowly
		Tannic acid	Disintegrates slowly
		Tanning bark	May disintegrate slowly if damp (see tanning liquor)
Sodium cyanide	Disintegrates slowly	Tanning liquor	Disintegrates, if acid
Sodium dichromate	Dilute solutions disintegrate slowly	*Tartaric acid solution	Not harmful
*Sodium hydroxide, 1–10 percent	Not harmful (h)	Tobacco	Organic acids, if present, disintegrate slowly
*Sodium hydroxide, 20 percent or over	Disintegrates concrete	Toluol (toluene)	Liquid loss by penetration
Sodium hypochlorite	Disintegrates slowly	*Trichloroethylene	Liquid loss by penetration
*Sodium nitrate	Disintegrates slowly	*Trisodium phosphate	Not harmful
Sodium nitrite	Disintegrates slowly	Tung oil	Liquid disintegrates slowly. Dried or drying films are harmless
Sodium phosphate (monobasic)	Disintegrates slowly	Turpentine	Mild attack. Liquid loss by penetration
Sodium sulfate	Disintegrates concrete of inadequate sulfate resistance	*Urea	Not harmful
Sodium sulfide	Disintegrates slowly	Urine	Attacks steel in porous or cracked concrete
*Sodium sulfite	Sodium sulfate, if present, disintegrates concrete of inadequate sulfate resistance	Vegetables	See fermenting fruits, grains, vegetables, extracts
Sodium thiosulfate	Slowly disintegrates concrete of inadequate sulfate resistance	Vinegar	Disintegrates slowly (see acetic acid)
		Walnut oil	Disintegrates slowly
*Sour milk	Lactic acid disintegrates slowly	*Whey	Disintegrates slowly (see lactic acid)
*Soybean oil	Liquid disintegrates slowly. Dried or drying films harmless	*Wine	Not harmful. Necessary to prevent flavor contamination
Strontium chloride	Not harmful	Wood pulp	Not harmful
*Sugar	Disintegrates slowly	Xylol (xylene)	Liquid loss by penetration
Sulfite liquor	Disintegrates	*Zinc chloride	Disintegrates slowly
Sulfite solution	See calcium bisulfite	Zinc nitrate	Not harmful
*Sulfur dioxide	With moisture forms sulfurous acid (which see)	Zinc refining solutions (x)	Hydrochloric or sulfuric acids, if present, disintegrate concrete
*Sulfuric acid, 10–80 percent	Disintegrates rapidly	Zinc slag	Zinc sulfate (which see) sometimes formed by oxidation
*Sulfuric acid, 80 percent oleum	Disintegrates	Zinc sulfate	Disintegrates slowly

Figure 3.6 (continued)

Sometimes used in food processing or as food or beverage ingredient. Ask for advisory opinion of Food and Drug Administration regarding coatings for use with food ingredients.

a   Waters of pH higher than 6.5 may be aggressive if they also contain bicarbonates. (Natural waters are usually of pH higher than 7.0 and seldom lower than 6.0, though pH values as low as 0.4 have been reported. For pH values below 3, protect as for dilute acid.)

b   Frequently used as a deicer for concrete pavements. If the concrete contains too little entrained air or has not been aged more than one month, repeated application may cause surface scaling. For protection under these conditions, see "deicing salts."

c   Carbon dioxide dissolves in natural waters to form carbonic acid solutions. When it dissolves to extent of 0.9 to 3 parts per million it is destructive to concrete.

d   Frequently used as deicer for airplanes. Heavy spillage on runway pavements containing too little entrained air may cause surface scaling.

e   In addition to the intentional fermentation of many raw materials, much unwanted fermentation occurs in the spoiling of foods and food wastes, also producing lactic acid.

f   Contains carbonic acid, fish oils, hydrogen sulfide, methyl amine, brine, other potentially reactive materials.

g   Water used for cleaning coal gas.

h   However, in those limited areas of the United States where concrete is made with reactive aggregates, disruptive expansion may be produced.

n   Composed mostly of nitrogen, oxygen, carbon dioxide, carbon monoxide, and water vapor. Also contains unburned hydrocarbons, partially burned hydrocarbons, oxides of nitrogen, and oxides of sulfur. Nitrogen dioxide and oxygen in sunlight may produce ozone, which reacts with some of the organics to produce formaldehyde, peracylnitrates, and other products.

o   These either contain chromium trioxide and a small amount of sulfate, or ammonium chromic sulfate (nearly saturated) and sodium sulfate.

p   Many types of solutions are used, including
    (a) Sulfate—Contain copper sulfate and sulfuric acid.
    (b) Cyanide—Contain copper and sodium cyanides and sodium carbonate.
    (c) Rochelle—Contain these cyanides, sodium carbonate, and potassium sodium tartrate.
    (d) Others such as fluoborate, pyrophosphate, amine, or potassium cyanide.

q   Contains lead fluosilicates and fluosilicic acid.

r   Reference here is to combustion of coal, which produces carbon dioxide, water vapor, nitrogen, hydrogen, carbon monoxide, carbohydrates, ammonia, nitric acid, sulfur dioxide, hydrogen sulfide, soot, and ashes.

u   Porous concrete which has absorbed considerable molten paraffin and then been immersed in water after the paraffin has solidified has been known to disintegrate from sorptive forces.

v   Contains nickelous chloride, nickelous sulfate, boric acid, and ammonium ion.

w   May contain various mixtures of blood, fats and oils, bile and other digestive juices, partially digested vegetable matter, urine, and manure, with varying amounts of water.

x   Usually contains zinc sulfate in sulfuric acid. Sulfuric acid concentration may be low (about 6 percent in "low current density" process) or higher (about 22–28 percent in "high current density" process).

Reprinted with permission of the American Concrete Institute, Committee 201.

Figure 3.6 (continued)

CHAPTER 4

STRUCTURAL STEEL

Steel is one of the most versatile structural materials. Some of the many characteristics of steel are listed below:

- Strong
- Stable chemical properties
- Permanent (if maintained)
- Ductile
- Adaptable to prefabrication
- Easily erected
- Weldable
- Reusable

Despite its resilience, however, steel must be continually inspected and repaired in order to preserve its structural integrity. Methods for inspection, repair, and maintenance of steel are described in this chapter.

Need for Inspection

Structural steel has its own peculiar needs in terms of maintenance and repair. An effective plant maintenance program should include a formal inspection on a regular basis, as discussed in Chapter 2. Refer to the Inspection Checklist for Structural Steel (Figure 4.1) to simplify the inspection process.

As with any structural-type members, not only should the members be inspected, but also their connections to other members. When performing the inspection, notes should be made as to the present condition of the structural steel, the observable surface deteriorations (such as defects and rusting), and the dates the particular element was repaired, as noted in the inspection checklist. Refer to Chapter 6 for a complete discussion of corrosion.

Suggested Inspection Frequency

Movable structures or structural components subjected to moving loads should be inspected every three months to one year. Stationary structures or structures not subjected to moving loads should be inspected every one to five years.

Preventive Measures

The strength of structural steel is controllable when accurate requirements are given. Steel failures are seldom caused by any fault of the material, but mainly by the design and construction of a component or facility. Steel that is inaccessible for periodic inspection and maintenance, such as items buried in masonry construction, must be permanently protected, especially against moisture from outside of the environment or from within the surrounding material. Keep in mind that hidden deterioration that is inaccessible, if

73

allowed to continue, eventually causes deterioration of adjacent materials. Therefore, if specific parts are going to become inaccessible in the future, the most economical decision is to encase them.

As discussed in Chapter 1, it is of utmost importance that time and resources be devoted to planning a structural steel building in order to build out those items which cause deterioration. The reader is referred to Figure 4.2 for a listing of the more common types of deterioration, their causes and methods of prevention.

Relative to other means of preventing steel failure, see Figure 4.3 which lists methods that may be utilized in the design stage. By no means is this list complete. It is up to the reader, based on his experience, to expand this list to encompass particular or specific problems in his area.

Inspection Checklist for Structural Steel

1. **Structural Members:**
   A. *Types*:
   - Roof Trusses
   - Open Web Joists
   - Platform Supports
   - Equipment Supports
   - Stairs
   - Bearing and Bass Plates
   - Siding Girts
   - Roof Purlins
   - Bridging
   - Steel Decking
   - Bracing—Vertical and Horizontal
   - Knee Braces
   - Columns
   - Crane Rails
   - Beams and Girders
   - Plate and Shells
   - Stacks
   - Structural members that are part of equipment, such as cranes and conveyers.

   B. *Check for*:
   - Cracks
   - Localized Distortion
   - Paint or other protective coating failures
   - Unusual wear, such as impact failure
   - Misalignment
   - Plumbness
   - Corrosion
   - Stress marks (show up in coat of paint)
   - General Cleanliness

2. **Connections:**
   A. *Types*:
   - Bolts
   - Welds
   - Anchor Bolts
   - Rivets
   - Splices—Beams and Columns
   - Connecting Plates and Angles

   B. *Check for*:
   - Corrosion-General
   - Loosening of bolts and/or rivets
   - Cracks in weld
   - Local distortion
   - Corrosion-Specific (rust ring worn through paint by connection head)
   - Unusual wear, such as destruction of bolt threads
   - Protection coating failure
   - General cleanliness

Figure 4.1

Where a structural member has been badly damaged such as by corrosion and the cross-sectional area has been reduced by more than 30 percent, the member should be repaired or replaced immediately.

Painting Structural Steel

Painting can be employed as a protective coating and/or part of a repair process for most types of structural steel, except the following:

- Areas involving problems of accessibility.
- Thin sections.
- Severe exposure conditions.

The important aspects of painting structural steel begin with the correct preparation of the surface. The surface should be cleaned and free of loose materials, such as grease and rust. The next step is to select the correct type of paint. Be sure the paint and primer are compatible with each other and with the type of steel surface. Once the materials are selected, the primer should be placed on the surface as soon as possible after the cleaning or surface preparation process so that the surface does not become re-contaminated. The paint should be applied with the correct equipment, according to the manufacturer's recommendations.

Galvanized steel is steel coated with zinc and then treated with chemicals to prevent white rust. A combination of the zinc metal and chemical treatment often creates problems of adhesion of applied coatings after exposure. If the incorrect paint system is used, extreme flaking and peeling may take place after a year or so of exposure, especially when wide temperature changes take place. Allow galvanized steel to weather, if at all possible, and use appropriate primers prior to application of paint.

A Summary of the Types, Causes and Prevention of Deterioration of Structural Steel

Type of Deterioration	Cause	Prevention
Abrasion	Members subjected to contact with moving parts.	Replacement of member
	Members subjected to wave action.	Sacrificial Metal
	Members immersed in a moving liquid.	Armoring
Corrosion	Resulting from a chemical or electro-chemical reacton which converts the metal into an oxide, carbonate and sulfides.	Replacement of member
		Coatings, such as painting
		Encasement
		Replacement with alloys
		Sacrificial Metal
		Cathodic Protection
		Armoring
Fatigue	Repetitive, cyclic loading occurring at stresses at or below allowable design values.	Change loading characteristics.
		Correct design for this type of possible failure.
Lamellar Tearing	Incorrect welding process.	Correct and careful welding techniques.
Loosening of Connections	Impact and fatigue loading	Encase connection in concrete.
		Reinforce by welding.

Figure 4.2

Once the required number of coats of paint have been applied, the painting process does not end. The paint should be inspected, not only after the job is completed, but also after the application of each coat, including the primer. During the inspection process, the inspector should check for pinholes, holidays, blisters, and coat thickness.

Check film thickness of existing paint periodically. Repaint when it decreases to four mils. Watch for signs of local rusting or corrosion. Spot paint as soon as possible, before general surface preparation and painting are required. A complete discussion of painting preparation, materials, and methods are found in Chapter 7.

Other Protective Coatings

Further discussion of methods of prevention should consider other coatings besides paint. These coating would include bituminous paints, galvanizing, and other coatings, such as those manufactured from plastics. (Bituminous paints and their application are covered in Chapter 7.) Where longer life protection is required, zinc coating should be considered. Readers should note that zinc coating is not a permanent protection and in some climates the galvanized coating should be covered with a paint application. Relative to the maintenance and repair function, it is important that the zinc coating on the steel not be broken.

Corrosion-Resistant Alloys

Another method of preventing corrosion and the effects of it is to specify corrosion-resistant alloys in the steel. Corrosion of a steel structure may be decreased by the use of corrosion-resistant alloys in place of ordinary carbon steel. The type of alloy used will depend on the atmospheric exposures. One example is the use of a low copper bearing steel to resist corrosion.

Except under extremely severe atmospheric conditions that are corrosive to the metals, it is recommended that corrosion-resistant alloys be used for only critical members, rather than in a general application for the entire structure. Some uses would be:

- Moving parts and assemblies, such as bearing plates, rocker or roller bearings.
- Small member assemblies, such as for gratings, which are expensive to maintain.

Prevention of Steel Deterioration

1. Make parts of structure accessible to maintenance or encase inaccessible parts.
2. Minimize exposure of steel parts.
3. Avoid shapes or details which will catch dirt.
4. Eliminate pockets, low spots and crevices.
5. Protect column bases with concrete encasement.
6. Slots and holes in horizontal surface must be plugged or kept open.
7. Avoid narrow crevices.
8. Draw tight all adjacent metal surfaces.
9. Provide full-penetration weld.
10. Pipe or tubular columns should be concrete filled or sealed air tight.
11. Isolate dissimilar metals.
12. Provide ventilation in spaces around steel members.

Figure 4.3

- For structural elements which would be inaccessible in the final building.
- Items in the building that are susceptible to severe exposure conditions due to wetting or water exposure on a frequent basis.

For buried and subaqueous elements, the use of copper-bearing or high-strength, low-alloy steels has little advantage over the use of ordinary carbon steel. Alloys high in silicone, chromium, or nickel do have better corrosion-resistance properties for such applications and may be used when economically justified. (The reader is referred to Chapter 7 for an in-depth discussion on corrosion.)

Encasement

Another method of prevention is encasing the member in concrete or other similar materials to make it inaccessible to environmental exposure. (Repair procedures are discussed later in this chapter.)

Sacrificial Metal

The term sacrificial metal denotes an over-design of the structural element with respect to its thickness requirements. The purpose of this is to provide an extra amount of material, for expected corrosion loss, in excess of the structural requirements for the section. Initially, a structural design is made of the member and the section selected. The next step would be to anticipate, using the best experience and records possible, the rate of expected corrosion or deterioration. Once this information is obtained, the next step would be to establish an estimated projected lifetime of the structure. From this information, an amount of total expected corrosion over the lifetime of the structure can be calculated. This thickness can be added to the required design thickness of the member resulting in the total thickness required. The extra metal required is known as the sacrificial metal. It is important that the reader realize that any loss of metal will happen on all exposed faces of the structural element. Also, the rate of loss on each side or each face may differ.

Armoring

Armoring is just what it says, a means of protecting the structure against damage, due especially to abrasion and impact, by covering or encasing it with another material. Methods of armoring would be concrete encasement, wrought iron plates, timber facings, the provision of sacrificial metal, of hard metals, and the use of abrasion-resistant paint coatings, such as vinyls, neoprene, and baked phenolic finishes. The most effective of these would be, of course, concrete encasement. The use of wrought iron protection plates also provides an excellent armoring device, but is an expensive procedure. The other means listed above are suitable only for light to moderate exposures and must be frequently repaired and maintained. When armoring is to be provided, streamline the encasement. Streamlining increases the life not only of the member being armored, but also the armoring itself.

The most important aspect of an effective maintenance program for steel is to *keep the material clean*. The reader must keep in mind that dirt holds moisture and moisture, in combination with oxygen, will cause corrosion. Also, dirt often hides defects that are occurring or have occurred in a structural member, which can become a very dangerous situation. Therefore, regular cleaning and inspection are probably most important in the day-to-day maintenance of steel members. The cleaning process may take many forms, from basic dusting of the steel, that is in a relatively clean environment, to rubbing down with a detergent or cleaning with air under pressure or controlled sandblasting in an extremely corrosive environment.

Needs and Types of Repair

Before a correct repair procedure can be selected, one must be able to identify the type of deterioration occurring and alleviate the cause. See Figure 4.4 for symptoms of deterioration of steel.

Many repair procedures that are called for in a building can often be done by maintenance personnel. However, when the integrity of the structure or a structural element is in question, it is of utmost importance that the limitations of the maintenance personnel be realized. In very large companies a house staff may be available to do a structural analysis. However, if such a staff does not exist within the company, outside consultants should be called in. This is especially important since incorrect repair procedures of structural items, such as structural steel, may result in structural collapse or an increased rate of present deterioration. The following procedures are presented only for use as a general checklist or reminder list as to the steps that are followed in the entire repair activity. In the same light, specific labor, material, and equipment requirements are not given as for other repair activities in this text. It would be unrealistic to think that any company would have the budget to retain on its staff personnel who are experienced in the various types of repairs required for the many materials in any one facility. It is the maintenance staff's responsibility to insure that a competent consultant be hired, that drawings and specifications be prepared, that a thorough inspection and quality control program be initiated for the specific repair, and that a follow-up inspection be carried out to keep a watchful eye on the repaired item.

It is suggested relative to any new or trial-by-error type of repair procedure that a prototype or model trial be performed on a sample piece of the deteriorated metal prior to initiating a complete and comprehensive repair on the entire assembly. The three major repair procedures are:

- Replacement.
- Plating.
- Encasement.

Plating and encasement are described in subsequent sections.

Where corrosion or other types of deterioration are severe throughout the structural element, the most economical decision would be to replace it. This decision would probably also be made if appearance is a consideration and if there is no room to add plates or strengthening members.

Figure 4.4

Symptoms of Types of Deterioration of Structural Steel	
Type of Deterioration	**Symptoms**
Abrasion	Worn, smooth appearance, general depression of the abraded area.
Corrosion	Pitted, oxidized surface showing loose flakes, reddish-brown rust colored appearance.
Fatigue	Small fractures oriented perpendicular to the line of stress.
Impact	Local distortion of the member in the form of a sharp crimp. Will occur in a tension member of flange.
Lamellar Tearing	Minute, often times unseen cracking in the weldment. May need microscopic instruments to observe.
Loose Connections	Vibrations and improper tightness.

Plating

In most cases, the repair of steel elements involves some degree of strengthening. Strengthening can be accomplished through plating. Plating is used when abrasion, corrosion, or other deterioration is localized on the steel members. The purpose of plating is to compensate for the loss of the section which has been deteriorated away. Basically, all that is done is to splice in the new metal across the areas of deterioration. Plating can also be used for repairing members which have cracked, buckled, or suffered local crushing due to impact. This technique is not applicable where the appearance of plates would be objectionable. In such a case, replacement is a better solution.

The material used for plating must be of the same grade of material as the original structural elements. It is important that the original building specifications be followed when installing the plating material, allowing for close tolerances. The following general procedure can be used in the repair:

1. Make a visual inspection and take measurements as to the size and thickness of plates required.
2. Check the actual measurements with a structural analysis.
3. Prepare a brief specification and complete drawing, including methods of connections and any other instructions that must be given to field personnel in order for them to perform the repair procedure as efficiently and effectively as possible to insure a long lasting repair.
4. The contact surfaces between the new and existing members should be thoroughly cleaned, utilizing the same methods for preparing steel surfaces for painting, as presented in Chapter 6. It is important that the contact surface between the new and existing materials does not induce the entrapment of rust or dirt; therefore, any contact surfaces should be primed and painted with special care before plating, using additional coats of paint, if needed.
5. Following the drawing and specifications, install the plates.
6. After the plates have been installed, the entire assembly should be primed and painted.
7. If the new assembly is in a very corrosive atmosphere, it would be advantageous to encase the entire assembly. Plating is not a permanent repair. Unless protection against further corrosion is provided, this method is nothing more than a form of providing sacrificial metal.
8. Frequent periodic follow-up inspections should be done to insure that the repair is sound and performing correctly.

Encasement

The third procedure for repair is referred to as encasement. This method is considered an indirect repair procedure for steel in that it is performed for the purpose of preventing any new deterioration or further deterioration on a component which has been repaired by plating, or replaced. This is nothing more than covering the existing steel with a covering or encasing it in concrete, bituminous base material, or other similar substances. Concrete encasement is used when its appearance is not objectionable. This is a very economical and satisfactory method of permanently protecting structural steel. In the protection of steel, concrete has many uses, some of which are listed below:

- Protection of waterfront structures in tide zones and below water line.
- Protection of buried pipe structures.
- Lining pipes.
- Filling of pipes and tubular columns, sometimes known as lally columns.
- Protection of structural elements which will be inaccessible in the final building.
- Protection of structural members subjected to highly corrosive atmospheres, such as those existing in chemical plants.
- Protection from fire damage.
- Protection from abrasive type loads.

Preparation for Encasement

1. The steel should be cleaned of all loose, rust, scale or dirt.
2. Wrap the steel member in material, such as wire mesh or welded-wire fabric.
3. For extreme exposure conditions, paint or coat the first six inches of embedment of the member. In other types of environments, the steel does not need to be painted.
4. Select a concrete that is of a dense, sound, rich mixture. Use at least 4,000 pounds per square inch compressive strength concrete.
5. If forms are to be used to hold the concrete in place during setting, prepare the formwork. Formwork may be of construction materials such as lumber or of such prefabricated items as steel piping. The formwork should be inspected for possible leaks, inadequate bracing, and general overall integrity prior to filling with encasing material.

Filling of Forms

There are various methods available to inject the encasing material into the forms. These are:

- Pumping.
- Pre-pack.
- Tremie.
- Dry pack or dry place.

Each of these will be considered in turn with an explanation and procedural steps for each.

Pumping

Pumping grout can be very economical except for filling large volumes and where it is necessary to insure a minimum of shrinkage in the drying process of the cement.

1. Select a grout which is comprised of two to three parts sand to one part cement by volume. Use the richer grout (2:1) for thin sections and the leaner grout (3:1) for heavier sections.
2. Using approved equipment, begin pumping the grout. The grout does not need to be pumped from the bottom, but it is important that, if pumped from the top, the end of the nozzle be continuously inserted into the top surface of the rising grout. It is important that the grout be pumped in a uniform, smooth, and continuous operation.
3. If the lateral travel of the grout is no more than four feet, and/or the pipe opening is three inches or larger, one grout pipe need only be used. However, if the lateral travel is greater, or if the pipe opening is smaller, more than one pipe will be needed.
4. Fill the forms to overflowing. Allow the grout to settle for about 15 minutes and refill the forms to overflowing.
5. It is important that the forms be continuously vibrated when filling, but do not over-vibrate the forms. (Refer to the American Concrete Institute's Specifications on amounts and intensities of vibrations required.)

Prepacked Concrete

The prepacked method of placing concrete is to be used for underwater installations and where accessibility is difficult.

1. Fill the forms with coarse aggregate conforming to ASTM Specifications. The aggregate must be clean.
2. The forms must be closed and watertight and be vented only at the top. Otherwise, the grout will escape and not completely fill all the coarse aggregate voids.
3. Using a sand-cement grout mixture, fill the forms by pumping the grout into the coarse aggregate.
4. Begin pumping the grout at the lowest point and proceed upward. The pipes should be placed at a maximum spacing of five feet, center to center.
5. Place the grout in a smooth continuous operation.
6. Maintain the grouting pressure until the mortar has set up.

Tremie

Use a tremie grout material for small voids. If larger voids are to be filled, use concrete. The same preparation for grout as for pumping procedures is used for tremie. The only major difference is that the grout pipe must be placed in the fluid during the entire operation. The supply hopper must be kept filled so that the seal between the fluid and the placed material is not broken. This is often a difficult task to achieve.

Placing in the Dry

In this method a cofferdam or other similar structure, needed to keep the water away from the steel, must be designed and constructed. This, in itself, usually makes the repair procedure of encasement a very costly item. However, in many cases, this is the only economical way of encasing steel. This type of work is very specialized and should not be attempted by untrained maintenance personnel. A consultant should be brought in to design and perform the repair procedure.

Handpacked Concrete

Handpacked concrete is usually used below water level. This procedure involves taking concrete with a strength of about 4,000 psi and adding an admixture to the grout which gives it quick-setting properties. The grout is applied pneumatically or by hand. This procedure is especially economical, since no forms are needed.

Finishing

Once the grout or concrete has been placed, the last step is to finish the top of the repair with a cover of grout. The grout should be sloped away from the structural steel member so that the accumulation of water is not a problem.

Pneumatically Applied Mortar (Shotcrete, Gunite)

This procedure is another method of applying a protective casing on the structural steel. It is used for thin coatings (three to four inches thick). This method is also used for encasing relatively large areas where its appearance is not objectionable. Some of the disadvantages of this method are:

- Results in a porous finish.
- High rate of shrinkage takes place causing possible cracking.
- Highly skilled workmen are required.

The procedural steps of this method are as follows:

1. Prepare a mixture of cement and water. A uniformly graded clean sand containing about 4 percent moisture should be used. The mix should be a 3:1 to 3-1/2:1 mix. Use as much water as possible.
2. An air entraining agent is suggested for increased workability. However, keep in mind that too much air will reduce the structural capacity of the encasement.
3. Using qualified personnel, apply the mortar.
4. Adequate curing is needed. Water-spray when the surface appears dry or, better yet, apply two coats of sealing compound. The minimum curing period should be seven days.
5. If shrinkage cracks occur, they should be immediately sealed with coatings of linseed oil, silicone, or similar materials.

The reader should note that encasement of steel with concrete will not protect the steel from corrosion due to electro-chemical action. Where there is the possibility of a discharge of stray currents into the ground, install a grounding system to the steel element (cathodic protection).

Besides being used as a method for preventing deterioration due to corrosion and abrasion, encasement can also be used to reinforce a deteriorated section, provided there is a clearance in the structure for the enlargement of the section, the existing member has adequate strength to carry the added weight, and the appearance of the final encased member is not objectionable.

Reinforced Bituminous Coatings

Reinforced bituminous coatings (wrappings) are used for buried members in highly corrosive soils. They are also used to protect pipes and rods in fittings that anchor retaining structures. The general procedure to be used is as follows:

1. Prepare the surface.
2. Immediately after preparing the surface, coat with a hot-coal-tar primer or enamel.
3. Cover the surface with one or more spiral wound wrappings of felt or other suitable fibrous material. The fiber must be saturated with a waterproofing bitumen mixture.
4. If backfilling is required, protect the coating.
5. Do not allow the protected member to come into contact with clay or silty soils. These types of soils will tend to stick to the coating and cause its eventual deterioration.
6. Do not allow rocks, boards, or any protruding items to cut into the wrapping, which may result in further deterioration. Use a sandy backfill around the encasement.

Other casings or coverings that may be considered and have been used successfully for specific situations are as follows:

- Packing of urethane foam covered with polysulfide liquid.
- A sheathing of 18-gauge nickel-copper alloys.
- A sprayed-on vinyl plastic coating.

Connections

If the metal adjacent to the fasteners has deteriorated, repairs can be accomplished by plating or replacement. If the fasteners have deteriorated, or are extremely corroded, they should be removed, the resulting holes reamed out, and larger or stronger bolts installed. An alternative to bolting is welding the connection.

Loose connections should be given immediate attention. If the connections are bolts, retorque them, except if they are high-strength type bolts; then replace them. When rivets become loose, replace them with bolts of equal or higher strength. It may be necessary for the existing loose rivets to be drilled out prior to the installation of new fasteners.

Connections of Steel Members to One Another

Whenever two or more structural members are connected, there is the possibility of entrapment of dirt creating a possible corrosion condition. This is especially the case when angles and channels are connected back to back.

If deterioration is uncovered during the inspection process, first determine the extent of the damage. If strengthening is not needed, use sandblasting to clean out the deterioration and seal all around the opening.

A more effective repair procedure would be to fill the space between the members with such material as cement grout. However, this type of process often is difficult due to the small confining space between the members.

If the deterioration requires strengthening, replacement of plating should be done, making sure that all resulting voids have been sealed or filled. This is done so that future problems with dirt collection are alleviated.

Beams and Columns—Bearing Plates

Check for pitted surfaces from corrosive action. If such a condition is found, remove the plate, grind it smooth and replace, making up the loss of thickness by metal shims.

Heat Straightening

A relatively new method of repairing members is known as heat straightening. Heat straightening is used mainly to straighten members that have become distorted due to impact or overload. Generally, the method is accomplished through application of heat (by use of torches) and cold (by use of dry ice) on specific locations of a structural steel member, which induces it to straighten out. This is such a specialized method that maintenance personnel should not attempt it. There are only a few companies in the continental United States that perform this specialized work.

Cracking Due to Brittle Fracture

The topic of cracking due to brittle fracture is quite theoretical and requires the services of a registered professional structural engineer to ascertain the damage. This type of cracking occurs when a high stress or strain develops at a point where the material has lost its ductility. The high stress or strain may be due to one or a combination of reasons. Once the crack or cracks have been observed, it is important that they be arrested. Bolted splices have been used to provide increased capacity and a method of placing material in the crack path in hopes of stopping it. Another procedure is to weld strips of material in the fracture paths. However, either of these procedures should not be attempted until a structural analysis has been made to determine if, in fact, the cracking is due to a brittle failure.

Expansion Joints

A major maintenance problem which appears in almost every facility is ineffective expansion joints. The reasons for this problem are many. First, they are usually not fully designed in detail relative to the future maintenance requirements of the joints. They are often hard to get to and usually constructed of many small details which are often hard to clean and maintain. There are basically three types of expansion joints and these are:
- Slotted hole expansion joint.
- Expansion plate expansion joint.
- Finger type expansion joint.

A major objective of any expansion joint is to provide free and unrestrained movement of the structure. Some points to keep in mind when designing and detailing the building relative to the expansion joints are:
- Do not use slotted hole connections, if at all possible (unless they will be inspected and maintained at frequent intervals).
- Use corrosion-resistant methods as much as possible. However, do not allow dissimilar metals to come into contact in such a manner as to set up an electro-chemical corrosion process.
- Use high-strength steels in bolts and bearing surfaces to keep wear to a minimum.
- The threaded portion of any bolt should be outside the connecting plates of the expansion joints.
- Do not install washers, if at all possible.

Relative to the frequent and periodic maintenance of expansion joints, the following points should be kept in mind:
- Inspection should be frequent.
- Frequent cleaning of the entire assembly, preferably at least twice a year, depending upon the environmental exposure conditions. The cleaning can be accomplished with conventional industrial brushes or brooms. Hard-to-get-at areas can be cleaned with pressurized air.
- Take care of any corrosion at first signs of its appearance.
- Do not alter the expansion joints, if repairs are needed, without first considering the structural effect of the repair.
- Check for leaks at expansion joints and, if present, attempt to repair the joint. If the joint cannot be repaired in such a way as to stop the leak, means should be provided for proper drainage without affecting other parts of the structure. By no means should the water be allowed to splatter inside and onto the building materials below.

In summary, the problem of expansion joints seems to be more serious in masonry structures (see Chapter 6 for more information) and continuous structures. The most effective type of expansion joints for these types of structures is the finger or expansion plate assembly and not the sliding or rolling type.

Lamellar Tearing

This is a relatively new-found problem with welded steel. Lamellar tearing is nothing more than the pulling apart of a hot-rolled steel member in its thinnest dimension. The source of the problem is welding-caused contraction in highly restrained connections. Lamellar tearing is a separation within the steel, caused by through-thickness strains induced by shrinkage of hot weld metal as it cools. It occurs mainly in corner and T-joints. It is more serious with thick steel members and with large cross-section welds.

To prevent such problems, the steel members being welded should be free to move closer together as the weld metal cools and shrinks. The American Institute of Steel Construction suggests that electrodes be used which have the lowest acceptable weld-metal yield strength. This will help reduce the tendency for lamellar tearing. Many times this type of deterioration is hard to detect since minute tears in the metal occur under weldments. Therefore, it is of extreme importance that all large welds be 100 percent X-rayed at the steel fabricator's plant and in the field (if in-place field welding is being done).

In an existing building, if the inspector notices signs of lamellar tearing, he should contact the steel fabricator and obtain permission to review the X-ray film. Also, they should have their X-ray supervisor examine the flaw. A meeting should be held to discuss the necessary corrections required to repair the flaw.

CHAPTER 5 TIMBER

Wood has been used for centuries in the construction of homes, light and heavy commercial buildings, and many other types of facilities. The major reasons why wood is used so extensively are listed below:

- It is readily available.
- It is easily manufactured into useful sizes and shapes.
- Its decorative character.
- Its resiliency.
- Its high strength to weight ratio.
- Its low heat conductivity.
- Its high resistance to electric current.
- Its ease of fabrication and fastening.
- Its durability.
- Its good weathering qualities when treated.

Some types of wood are more durable than others; in fact, some parts of the same species are more durable than others (heart wood is more durable than sap wood). Figure 5.1 lists some of the common imperfections in wood. Inadequately maintained timber structures usually result in extensive deterioration which will require replacement instead of repair. In fact, most of the repair of timber structures today is by replacement.

This chapter contains descriptions of the various types of wood, types of deterioration, and alternatives for repair.

Types of Wood Deterioration

The major causes of wood deterioration are decay-causing fungi, marine borers, insects, and fire damage. Each of these types (and others) of decay are discussed in subsequent sections; methods of prevention are presented for each one. See Figure 5.2 for a summary of the causes and the symptoms of wood deterioration that can be used in the inspection process. Types of problems inherent in wood that require eventual maintenance are excessive deflection, checking, splitting due to shrinkage of the timber, and deterioration of the hardware. An example of a checklist used during an annual building inspection is shown in Figure 5.3.

Common Imperfections in Wood

Imperfection	Description	Effects on Strength	Effect on Grading Structural Lumber
Checks and Splits	Split in the wood.	In lumber subjected to bending, checks and splits reduce the resistance to shear; they do not affect the strength for longitudinal compression.	Checks and splits are restricted in those parts of a bending member where shearing stresses are highest.
Holes	Either a knothole or a hole caused by some other means.	Reduces tensile strength somewhat more than compressive and shear strength and affect stiffness.	The size, number and location of knots is restricted for structural lumber; cluster knots are prohibited.
Knots	Localized imperfections.	Same as for holes.	Same as for holes.
Pitch Pockets	Opening between growth rings containing pitch or bark.	Little or no effect.	Usually disregarded except if a large number occur; shake may be present or bond between annual growth rings may be weakened.
Shakes	A separation of the wood between the annual growth rings.	Same as for checks and splits.	Same as for checks and splits.
Slope of Grain	Areas where the direction of the wood grain is not parallel to the edges of the piece of lumber.	Will twist with changes in moisture content.	Cross-grained pieces are undesirable; reduction of strength due to cross grain in structure is taken as twice the reduction observed in tests of small clear specimens.
Wane	Bark or lack of wood on the edge or corner of the piece of lumber.	Affects nailing and bearing.	Limited in structural lumber requirements for fabrication, bearing, nailing and appearance and not for effect on strength.

Figure 5.1

Summary of Causes of Timber Deterioration with Symptoms

Cause of Deterioration	Symptoms
Carpenter Ants, Beetles and Carpenter Bees	Similar to termites.
Decay (Rot) Due to Fungi	Softening and discoloration of wood, fluffy or cottony appearance, destruction of wood cells, appearance of fruity bodies in the form of mushrooms, incrustations in the advanced stage.
Excessive Splitting and Checking	Excessive relative amount of members at a joint, bowing of compression members (shown by broken paint lines or newly exposed wood), elongated bolt holes.
Fire Damage	Surface cellular damage, charred surfaces, easily probed with a knife.
Hardware	Loose connections, formation of rust on hardware surfaces, discoloration of wood adjacent to hardware.
Loosened Connections	Loose connections, excessive deflection.
Marine Borers	Minute openings in the timber, hollow sound when struck with a hammer, a myriad of surface grooves, narrowing of a section giving it an hourglass appearance.
Termites	Presence of swarm or winged termites. Discovery of one or more shelter tubes, wood that collapses easily when probed with an ice pick or a pocket knife, or tapped with a hammer.

Figure 5.2

Inspection Checklist for Wood (Timber)

Items
- Sills and Plates
- Beams and Girders (Ledgers)
- Rafters
- Joists
- Columns (Posts)
- Bridging
- Headers
- Studs
- Lintels
- Stairs
- Wood Trusses
- Nailers
- Connection Hardware
- Finish Wall and Ceiling Covering
- Finish Floor Covering
- Subfloors
- Fascia and Related Trim
- Finish Trim
- Other Millwork
- Laminated Wood Members

Check for
- Decay
- Marine Borers
- Insect Damage
- Rodent Damage
- Fire Damage
- Splitting
- Checking
- Misalignment
- Soundness of Wood
- Loosened Connections
- Loosened Laminates

Figure 5.3

Decay

Untreated lumber is subject to wood decay. Wood decay is caused by wood-rotting fungi that grow in damp wood. Fungi attack wood members in contact with damp masonry foundations, moist ground or standing-water, and water pipes that accumulate condensation or on which moisture condenses. Poor ventilation around the wood hastens the process of decay. Common signs of wood decay are:

- A damp, musty odor.
- Opening or crumbling of the wood.
- The presence of fine, dusty, reddish-brown powder under the building or structure.
- A hollow sound when timber is tapped.
- Easy penetration of timber by a sharp-pointed tool.

Since decay (fungi) must have moisture, food, air, and favorable temperature conditions to develop and survive, the elimination of one or more of these will prevent this type of deterioration from taking place. The deprivations must be permanent.

- **Moisture:** Moisture content of wood must be maintained at or below 20%.
- **Food:** Use fungi-resistant woods (see Figure 5.4) or poison the wood by treating.
- **Air:** Cut off the air supply by encasement or embedment.
- **Temperature:** Keep temperature over 110 degrees Fahrenheit or below 35 degrees Fahrenheit.

Decay Resistance of Major Woods

Wood types	High Resistant	Moderate Resistant	Low Resistant	Wood types	High Resistant	Moderate Resistant	Low Resistant
Alder			X	Oak, Chestnut	X		
Ashes			X	Oak, Gambel	X		
Aspens			X	Oak, Oregon White	X		
Baldcypress		X		Oak, Post	X		
Basswood			X	Oak, Red and Black Species			X
Beech			X	Oak, Swamp Chestnut		X	
Birches			X	Oak, White	X		
Buckeye			X	Osage Orange	X		
Butternut			X	Pine, Eastern White		X	
Catalpa	X			Pine, Longleaf		X	
Cedars	X			Pine, Slash		X	
Cherry, Black	X			Pines, Most Other Species			X
Chestnut	X			Poplar			X
Cottonwood			X	Redwood	X		
Cypress, Arizona	X			Sassafras	X		
Douglas Fir		X		Spruces			X
Elms			X	Sweet gum			X
Hackberry			X	Sycamore			X
Hemlocks			X	Tamarack		X	
Hickories			X	Walnut, Black	X		
Honeylocust		X		Willows			X
Junipers	X			Yew, Pacific	X		
Larch, Western		X		Yellow-Poplar			X
Locust, Black	X						
Magnolia			X				
Maples			X				
Mesquite	X						
Mulberry, Red	X						
Oak, Bur	X						

Figure 5.4

See Figure 5.5 for some specific suggested methods of preventing decay.

The following steps should be taken to treat for and remove any further decay:

1. Remove fungus-infested lumber. Spray infested areas with a wood preservative.
2. Eliminate the cause(s).
3. Provide ventilation for affected areas.
4. Replace infested lumber with treated lumber.

Marine Borers

Marine borers are found in brackish waters and are especially active when the water temperature is in the 50 degree range. See Figure 5.6 for methods of prevention. The damage by marine borers can be controlled by natural conditions of salinity, temperature, turbidity, current, or pollution. Means of improving the resistance of existing structures to marine borer attack are expensive. The only effective method is armoring. See Figure 5.7 for various methods of armoring.

Methods of Preventing Decay

1. Shelter member out of weather conditions.
2. Provide adequate ventilation.
3. Do not lay members against each other, if at all possible.
4. Do not place members in contact with moist ground or standing water.
5. Encase member in extreme environmental conditions.
6. Treat with a preservative compound.
7. Use woods having natural fungi resistance properties. (See Figure 5.4)

Figure 5.5

Prevention of Damage by Marine Borers

1. Armoring (see Figure 5.7).
2. Encasement with plastic coatings or wrappings.
3. Impregnating with a preservative treatment prior to installation.
4. The use of immune species of wood.
5. Discouraging the development of the borers by using electrolysis, toxic treatments, or explosives.
6. Remove the wood out of the reach of the borers.

Figure 5.6

Untreated timber used in water infested with marine borers may last only twelve months. Therefore, it is mandatory that timber be treated. Experience has shown that treated timber can have a lifetime of 20 to 30 years.

This text will not describe in detail the method of preventing damage by marine borers, but the reader should be aware of the methods available which can be found in some of the bibliography entries.

Insects

Insects that can cause wood damage are listed below:

Insect Type	Type of Damage
Beetles	Burrow through and excavate channels on surface layers.
Carpenter Ants and Bees	Burrow in the wood and make homes, but do not feed on the wood. Rarely a problem.
Termites	Bore and eat into wood or other cellulose material such as fiberboard.

Termites

Termites are the most destructive insects on our list. They are found in most regions of the United States, but are most active along the Pacific and Gulf Coasts and in some of the South Atlantic and Southwest states. There are three types of termites: the *subterranean*, the *dry wood* and the *damp wood* species.

Summary of Types of Armoring

Types of Armoring	Comments
Leave bark on.	Not effective.
Charring and tarring.	Only effective for light borer activity.
Painting with tar, oil paints, lead and copper paints.	Does not give continuous protection for extended periods of time.
Metallic sheathing (zinc, copper, or zinc-copper alloy).	Effective as long as it is free of punctures, gaps, etc., very expensive.
Pipe casing.	Good protection, but expensive. Usually used for protection of piles only.
Cast-in-place concrete.	Used mainly for protection of pilings. Not to be used where there is a possibility of impact type forces. Otherwise, quite effective.
Plastic coating.	Quite effective, but relatively new.

Figure 5.7

Subterranean: Subterranean termites account for 95 percent of all termite damage in the United States. They live in nests and develop their colonies underground, but build tunnels through the earth to get at wood or other cellulose material such as fiberboards, fabric, and paper. Therefore, the evidence that this species is active is in the form of soil mounds or tubes along or inside of foundations or under porches, etc.

Swarming: The other species of termites will enter the home directly at the time of swarming activity. They frequently attack wood which has a high moisture content and can be found in decaying wood. This type sets up colonies and lives in the wood. There is little, if no, evidence of their existence. They can cause destruction from within the timber.

Preventing Termite Damage

Prevention can begin by utilizing good building practices consistent with the geographical region and the potential type of insect infestation possible. See Figure 5.8 for methods of prevention.

Treatment for Termites

Regular inspection should be made to investigate the possibility of insect damage. When damage is uncovered and repair is needed, the first action should be to destroy the insects by fumigating, poisoning, or removing and destroying all the material which has been infested. The deteriorated member can be then replaced or reinforced.

Corrective measures include the following:
- Excavating the adjacent grade to remove the timbers away from the soil.
- Drilling infected areas and injecting poison.
- Raising the foundation wall or piers.
- Insuring adequate ventilation under the crawl spaces.
- Installing termite shields.
- Painting with protective coatings.
- Encasing member so termites cannot get at it.
- Removing the infected wood.

Methods of Preventing Termite Infestation

1. Grade soil so it slopes away from the foundation.
2. Remove tree stumps, roots, and all untreated wood items from the construction area.
3. Avoid embedding non-pressure treated wood in the concrete.
4. Reinforce the concrete.
5. Cover hollow concrete masonry with concrete cap block, termite shield, or pressure treated wood.
6. Chemically treat the soil both inside and outside the foundation walls and prior to installation of ground slabs. Follow manufacturer's recommendations for applying the treatment.
7. Fill joints and cracks with coal tar pitch or acceptable sealer.
8. Provide adequate ventilation in crawl spaces.

Figure 5.8

Fire

Heavy timber construction is more resistant to fire than exposed metal construction. However, extensive fire damage can result.

Building codes have done a lot to prevent damage due to fires. Fire retardant chemicals are available to increase the fire resistance of wood. The proper fire retardant chemical treatment of wood greatly reduces the magnitude of flammable products released. This, in turn, reduces the amount of flame spread over the surface. Treatment also reduces quantity of heat available or released in the volatiles during the early stages of fire. Another result is that the timber becomes self-extinguishing, once the main source of heat or fire is removed. The treatment does not prevent wood from decomposing under fire conditions. The American Wood Preservers Association's *Standards for Fire Retardant Treatment* should be followed very closely in any treatment process. Another beneficial result of this type of treatment is the inhibition of the formation of decay and termite attack. It has recently been found, however, that fire retardant treatments can cause early and rapid deterioration of the connecting metal hardware and a premature failure of the structure.

Shrinkage

Wood will undergo some shrinkage depending on such items as the moisture content of the wood, the species and grade of wood. This action can cause loosening of connections resulting in splits and checks in the wood. The loosened connections can induce increased deflections and distortion of members, resulting in the weakening of the facility.

Checking (Splits)

By definition, a check is a surface opening caused by seasoning. It does not extend through the thickness of the piece of lumber, but follows the grain of the wood. A split (or through-check) is a lengthwise separation of the wood extending from one surface through the piece to the opposite surface or to an adjoining surface. The more serious checking or splitting may result at the ends, then through the body of the piece. End drying (and thus checking) can be minimized by application of paint coatings to the material while unseasoned.

If checking occurs once the structural member is in place, two potential hazards exist. These are: 1) possible structural failure and, 2) a break in the timber surface which can allow insect infestations and possible decay to begin. Checking that occurs at the end of a member can be controlled by sealing the ends with a commonly available sealer and/or by driving a C-iron into the end. The last method can only realistically be done prior to installation of the timber member. Another method which can be used to control checking, which is effective, not only at member ends, but also within the member itself, is using bolts. The reader should keep in mind that lumber codes allow for some checking to occur without affecting structural integrity of the member. It is checking beyond this amount about which one must be concerned.

At the ends of members loaded in compression parallel to the grain, checks and splits may be disregarded, provided there is no evidence of slip from wedging action of connectors and bolts. For members loaded in tension parallel to the grain, splits outside the connector area, that are about parallel to the grain, may also be disregarded. If checks and splits occur in any other locations, immediate attention should be given to the problem by consulting with a professional registered structural engineer.

Loosening of Connections

Excessive deflection along with the normal action of the loads applied to the structure may cause loosening of connections. As for structural steel, all connections must be a high priority item on the inspection checklist. If loose connections are found, they must be tightened immediately by retorquing the hardware. It is suggested that connections are inspected and tightened three to six months after the erection of the structure and again after one year and three to five years thereafter. It is best to tighten the bolts in the humid season.

Hardware Deterioration

The main problem with hardware, when used in conjunction with timber, is corrosion. If corrosion is anticipated, the size of the connecting hardware should be increased (a form of sacrificial metal). Some other suggestions that will aid in preventing failures due to corrosion are as follows:

- Use galvanized steel for connectors.
- Keep exposed portions of connectors to a minimum by doing such things as minimizing the number of washers.
- Limit the size of the bolt holes to 1/16 over the bolt diameter.
- Secure full bearing of the adjacent timbers when bolts are used as connecting members.

Replacing and Reinforcing Timbers

The majority of repairs of timber members usually are accomplished by replacing or reinforcing the timber. The method utilized will depend upon the type of forces the member carries and its function in the building.

- **Compression Members** may be replaced or a new section spliced into the good portion.
- **Tension Members** may be spliced using side plates. It is very difficult to design a splice in a tension member. However, if it is done, it should be checked to insure that it is, in fact, carrying the tensile load. Replacement is a more effective means of repairing a tensile member.
- **Bending Members** may be replaced or spliced. Splicing should be done where the bending movement is zero, which is usually near a support. A structural analysis can be utilized to locate the point of zero movement.
- **Braces** should be replaced rather than spliced.
- **Corrosion:** Hardware which has corroded or disintegrated to the point of being ineffective should be removed and replaced with an oversize or a higher-strength connector.

The reader must keep in mind that, whether replacing or repairing, a structural analysis must be done to determine the extent of structural damage. This should be done only by a trained professional, such as a registered structural engineer. It is not within the scope of this book to present step-by-step replacement procedures. Since there are too many parameters which must be considered prior to making the repair, that can only be ascertained by a structural engineer.

When the damage done by decay is extensive, the deteriorated member should be replaced with timber having the same strength characteristics and size as the existing element. The decision as to whether the deterioration is extensive enough to warrant replacement must be based upon economics.

If the decision is made to repair the element, the following general items should be kept in mind when developing the repair procedure:

- Remove any live loads and dead loads imposed on the structural member.
- Provide any temporary supports needed to insure that repair procedures don't result in a premature failure of the member.

- Using applicable hand and power tools, remove all decayed material.
- Treat the damaged surface with a fungicide coating to help prevent any further deterioration.
- If further decay is probable, utilize one of the methods listed under prevention to permanently protect the repair area, such as encasement.

Glue-Laminated Members

Glue-laminated members do not ordinarily have the same defects as solid wood because laminated members are kiln-dried before gluing and are made up of small and selected pieces. Besides making the same type of inspection as for solid wood members, glue laminated components should be inspected periodically for delamination or separation between pieces of a member. Minute cracks should be examined carefully; cracks larger than 1/16 inch should be examined by an engineer.

Wood Treatments

Preservative treatments protect wood against decay and insect attack. Research has shown that available preservative treatments will extend the life of wood many times. The reader should note that cost of the initial treatment to prevent wood decay is many times lower than the cost of replacing the member in the future. Therefore, any lumber that will be used: 1) on the exterior of the building; 2) in the interior, but in contact or in close proximity of the ground and; 3) in below-grade installations, should be treated with a preservative treatment. This fact should be specified in the construction documents for the building.

When considering the types of treatment processes one must remember the following points:
- Cost.
- Structural integrity.
- Materials that will be in contact with, or in close proximity to, the treated wood.
- Environment.
- Effect on the health of humans and wildlife.
- Appearance.
- Paint adhesion to surface.
- Odor.

The types of treatment processes are:
- Vacuum Pressure Impregnation.
- Surface Application (non-pressure).

The former is more effective and longer lasting and must be done prior to the installation of the structural member. Once the member is in place, the second process is the only one available to preserve the member.

The reader is referred to Figure 5.9 for a summary of the types of preservative treatments available, along with some advantages and disadvantages of each. See Figure 5.10 for types of preservatives to be used for marine borer attack.

Preservatives — Advantages and Disadvantages

Oil-Based Wood Preservatives

Type of Preservative	Advantages	Disadvantages
Anthracene Oils	High toxity to wood-destroying organisms; insoluable water; low volatility; ease of application; permanence.	Dark brown color, cannot be painted; strong, unpleasant odor; easily ignited when first applied.
Coal-tar creosotes	See Anthracene Oils.	See Anthracene Oils.
Copper Naphthenate	High protection against decay fungi and termites; can be painted; not unpleasant odor; less easily ignited than coal-tar creosotes.	Gives wood greenish or dark color and provides less protection against marine borers than creosote.
Creosotes derived from wood, oil and water gas	Same as Anthracene Oils and Coal-tar Creosotes.	About the same as Anthracene Oils and Coal-tar creosotes, but less effective.
Creosote Solutions	See Anthracene Oils and Coal-tar Creosotes.	About the same as Anthracene Oils and Coal-tar creosotes, but less effective.
Crystal-free-coal-tar Creosotes	See Anthracene Oils and Coal-tar Creosotes.	See Anthracene Oils and Coal-tar Creosotes.
Pentachlorophenol	See Copper.	May alter color as dark-colored petroleum oils are used; provide less protection against marine borers.
Water-repellent Preservatives	Retards moisture changes in wood; good protection against decay and insects.	Cannot be used in contact with ground or areas where continual dampness can occur unless preservative is thoroughly applied.

Water-based Wood Preservatives

Type of Preservative	Advantages	Disadvantages
Acid Copper Chromate	Provides protection against decay and insects; can be painted; no objectionable odor; if thoroughly impregnated has some resistance to marine borers.	Wood can be used in contact with ground or water, but generally not recommended for contact with water.
Ammonical Copper Arsenite	Good protection against decay and insects and some protection against marine borers.	Wood can be used in contact with ground or water, but generally not recommended for contact with water.
Chromated Zinc Chloride	Provides protection against decay, insects and fire; can be painted; no objectionable odor.	Wood cannot be used in contact with ground or water.
Chromated Zinc Chloride (FR)	See Chromated Zinc Chloride.	See Chromated Zinc Chloride.
Copperized Chromated Zinc Chloride	See Chromated Zinc Chloride.	See Chromated Zinc Chloride.
Tanalith (Wolman Salts)	Protects against decay and insects; can be painted; no objectionable odor.	Wood cannot be used in contact with ground or water.
Zinc Meta Arsenite	Good protection against decay and insects; can be painted; no objectionable odor.	Wood can be used in contact with ground, but generally not recommended for contact with water.

Figure 5.9

Preservative Treatments Against Marine Borer Attack

Borer Type	Suggested Preservative
Teredo and Pholad (but not with Limnoria Tripuntata)	Creosote and/or Creosote/Coal Tar
Limnoria Tripuntata and Pholad	Water-borne Salts plus Creosote
Limnoria Tripuntata (but not with Pholad)	Water-borne Salt
Pholad (only)	Creosote

Figure 5.10

PART III

MAINTENANCE AND REPAIR OF BUILDING FINISH MATERIALS

Chapter 6 examines the maintenance and repair of most exterior materials. Since these materials co-exist with other types of materials, special problems result. These problems as well as the means for preventing and maintaining them are discussed in Chapter 6. Some of the materials that are covered in this chapter are listed below:

- Masonry units
- Flashings
- Vents
- Wall Materials
- Canopies
- Fences

Chapter 6 also presents methods for painting and for preventing corrosion.

Chapter 7 presents step-by-step procedures for the maintenance and repair of all ceiling and wall finishes, such as tiles, paints, papers, and paneling, as well as charts giving possible causes and periodic maintenance schedules. Tables are also included to aid in selecting new material coverings.

Repair and maintenance of floor coverings is one of the most important aspects of building operations. Floors must be inspected on a continuous basis and minor defects corrected before they become major problems. Chapter 8 discusses problems related to flooring materials, and methods of maintaining and repairing them.

The roofing system is one of the most important exterior components. It provides resistance to environmental elements and forces that are detrimental to the building and its contents. Roof repair and maintenance are examined in Chapter 9.

Windows and doors needs as much, if not more, attention as walls, ceilings, and floors. Causes and effects of problems, procedures for repair, and routine housekeeping needs are covered in Chapter 10.

Because there are so many types, styles, and manufacturers, maintenance and repair of hardware and miscellaneous materials are complex problems. Maintenance and repair of these products are covered in Chapter 11.

Chapter 12 examines the importance of maintaining landscaping with routine procedures, as well as an outline of potential drainage problems and how to alleviate them. The information is presented in order for a maintenance department to adequately maintain and care for all landscape items, along with maintaining an adequately drained site.

CHAPTER 6 EXTERIOR MATERIALS

Exterior wall coverings are made of a wide variety of materials, including wood (shingles, weatherboard siding, plywood); concrete and masonry (brick, concrete, structural tile, stone, stucco); metal (corrugated iron or steel, aluminum, enamel-coated steel, protected metals); mineral products (asphalt and fiberglass shingles, glass block); and other materials.

General Procedures

When exterior walls are repaired, the original specifications for materials should be followed to match adjacent existing materials. Improved construction techniques and materials should be constantly examined to prevent or offset similar failure. Repaired or replaced structural members must be capable of bearing the full loads specified in the design of the building or structure. Structural changes, particularly those affecting bearing walls, should not be attempted without the consultation of a qualified engineer. When weathering, normal wear, or other deterioration necessitate refinishing of an exterior wall, the existing material should be matched if found satisfactory, or a completely new finish applied over the existing surface.

Wood Exteriors

The main causes of wood exterior failures are:
- Moisture
- Inferior workmanship and materials
- Structural failure

Wood exteriors should be regularly inspected for damage from wear, accidents, and the elements. Painting and surface treatments should be inspected quarterly for deterioration; exteriors should be checked for loose, warped, cracked, or broken boards and shingles.

Exterior wall surfaces should be washed at least once a year, using a neutral detergent. This cleaning process can take place when the outside windows are washed. After a number of years of exposure, repaint or restain the surface.

Make a careful check to determine that existing structural, functional, and material conditions warrant repair to the existing wall rather than complete re-siding, insulation, or other overall repair or rehabilitation. Where existing situations are satisfactory, replace damaged material with a similar material. Cut back sufficient areas beyond the damaged part to obtain good joining and sound nailing. Tighten nails in existing material. Be sure that materials receiving the new nailed pieces of sections are sound. Cover replacement wood with treatment and/or paint matching the original design. When as-built plans are available, it is advisable to examine the original construction detail for out-of-vision construction and utilities that may be damaged. Warped, split, or curled shingles should be removed with a ripper and replaced in a similar manner as roofing shingles. Panel siding should be checked for looseness and faulty caulking. It is usually more economical and satisfactory to replace damaged or deteriorated panels than to attempt patching.

When existing siding does not meet the functional requirements of a building or structure, careful consideration should be given to re-siding over the existing material. A qualified architect or engineer should be consulted for planning the job and preparing the plans and specifications.

Metal Exteriors

Materials commonly used for metal exteriors include aluminum, protected metal, galvanized iron and steel, and prefabricated sheet metal. All require similar maintenance measures.

Precision shop fabrication of metal siding and new methods of fastening panels with interlocking joints and clips generally insure against misalignment in erecting metal buildings. Buildings may sag, lean, or suffer surface damage from foundation settlement, heavy wind pressures, corrosion of vehicles, or impact of heavy objects. If foundation settlement causes misalignment of structural members and consequent displacement or bending of siding, the basic problem should be solved as recommended in Chapter 4 before siding repairs are made. Where severe wind or vibration affect the stability of the siding, additional bracing and fasteners should be applied according to the siding manufacturer's recommendations. Constant vibration makes more frequent inspection of sheeting fasteners necessary.

1. Keep all bolts, clips, rivets, nails, and other ties and fasteners tight. Where corrosion or rust has destroyed the effectiveness of a fastener, replace it immediately. Where stresses have damaged the siding connection with the fastener, patch as necessary with matching material and replace the connector so as to ensure a positive connection. Use stainless steel or aluminum nails and neoprene washers for trouble-free service, as they not only waterproof the joint, but prevent corrosion. It is often advantageous from the standpoint of maintenance to use double-headed nails in placing corrugated metal siding.
2. Where sections of metal siding have been bent or cut, remove and straighten the section, making neat patches where necessary. If damage is severe, replace the panel with a matching substitute. To prevent vehicles (i.e., forklifts) from striking and damaging the siding, provide the building with bumper guards.
3. If metal buildings have interior insulation, take care not to damage existing construction and utilities when repairing or replacing metal wall coverings.
4. Keep ventilators in metal buildings clean and clear of obstruction. Keep door-sliding devices and locks in adjustment and tightly fastened.
5. Periodic washing and coating are required at specific frequencies depending on the environment. Wash using a solution that contains a detergent specifically prescribed for the type of metal.

Protected Metal Sidings

Protected metal roofing and siding are designed to give reasonably long service, but damage from abrasion may expose the metal core to corrosion or reduce the thickness of the protective coating. Unless the damaged areas are repaired, touched up, or recoated as necessary, serious damage may result. If structural changes to a building make it necessary to cut existing protected metal roofing or siding, treat the affected areas to prevent corrosion.

Types: Protected steel roofing and siding are of the standard corrupted or deep-corruption types, with steel core protected at the factory with coatings or a combination of coverings and coatings. The sheets are resistant to flame spread. The weather surfacing is weather-resistant, water-repellant, and resistant to fumes, chemicals, and corrosion. Color pigments used in color coats are heat- and light-stable and chalk-resistant. Roofing and siding sheets are furnished complete with all flashings, fastenings, and accessories. Interior and/or exterior surfaces of the sheets are provided with a bituminous (black) or colored synthetic resin factory-applied weather surfacing. Protective coverings and/or coatings are applied by one of the following methods:

- **Type A:** The cleaned steel core is dipped in a bath of molten zinc, fusing the zinc to the steel. While the zinc is in a molten state, a layer of felt is pressed over it, on each side of the sheet, squeezing the felt fibers into the zinc. The exterior and/or interior surface is finished with a heavy bituminous compound or a color coating of synthetic resin.
- **Type B:** The cleaned steel core is given a phosphatizing treatment, coated with a rust-inhibitive primer, and coated with a resinous adhesive. A layer of impregnated felt is rolled onto both sides of the sheet with the application of heat and pressure. The sheet is then given a protective coating of bituminous compound or alkyd resin color compound.
- **Type C:** The cleaned steel core is coated on all surfaces with a rust-inhibitive coating. The sheet is then heated and coated with adhesive. After application of the second coat, a layer of mineral mica is applied to both sides of the sheet by heat and pressure. A synthetic-resin color coating may be applied over the bituminous compound coatings.

Cleaning

Before repairing, touching up, or recoating, clean the affected areas. Remove all loose or torn felt, flaked coating, and rust with a stiff fiber brush. Remove dust, chemical deposits, grease, and dirt. Use synthetic detergent cleaner to remove chemical deposits and grease. Scrape off thick deposits of grease with a trowel or putty knife before using the detergent cleaner. Remove rust with a wire brush. The coatings on protected metal are usually petroleum asphalt compounds. They can be softened or dissolved by organic cleaners and solvents. Do not use naphtha, bensol, sylol, or carbon tetrachloride for cleaning protected metal.

Repair of Protected Metal

For small areas (under 10 square inches) of metal core exposed to insure compatibility, the manufacturer's recoating compound should be used in recoating protected metal.

- **Type A and B Materials:** Coat the exposed metal, including a width of about one inch of the adjoining felt and bituminous covered surfaces, with asphalt primer. Allow the primer to dry for at least 24 hours. Then apply bituminous plastic cement. Build up the plastic cement to form a continuous plane with adjoining surfaces. When the cement has been exposed to the weather for at least 24 hours, apply a coat of bituminous compound. Thin the compound to finishing consistency, with a suitable solvent and apply it at the rate of about one gallon to 125 square feet. Overlap adjoining area about one inch.
- **Type C Material:** Follow the treatment recommended for Type A and B materials except that the final coat should be of asphalt-base emulsion. While the coating is still tacky, cover it completely with 1 60-mesh mica topping.

For large areas (10 square inches or more) of metal core exposed:

- **Type A and B Materials:** Prime the metal surfaces and about one inch of the adjoining area as recommended for small areas. Then apply a brush coat of bituminous compound, using about one gallon for each 75 square feet. While the compound is still wet, apply 48-by 48-mesh unbleached muslin weighing four ounces per linear yard, or a porous, tough mat of fiberglass reinforced with random continuous glass yarns and bonded with a resinous binder compatible with bituminous coatings. Before using muslin, soak it in clean water until all sizing has been removed, and hang it up until it is damp dry. Apply the damp muslin or the fibrous glass at right angles to the corrugations, using material as wide as practicable and overlapping the edges about 3/4 inch. Press the muslin or glass mat firmly and evenly into the bituminous compound. Provide a continuous pleat in the muslin, in the center of and parallel with each low corrugation of the projected metal roofing and siding. At side or end laps, pleat the muslin or the glass at the edges of the overlapping sheets of protected metal and tuck the muslin or glass into the crevice between the sheets. Immediately after the muslin or glass is installed, apply a brush coat of the bituminous compound recommended for the first coat. Cover the muslin or glass completely. Allow this coat to dry for at least 24 hours. Then apply a final coat of the bituminous compound, at the rate of one gallon for each 75 square feet, over the repaired area.

- **Type C Material:** Prime the metal surface, including a width of about one inch of the adjoining bituminous covering, as recommended for smaller areas. Then apply a brush coat of asphalt-base emulsion at the rate of one gallon for each 75 square feet. While the emulsion is still wet, apply muslin or a fibrous glass mat as recommended for Type A and B materials. Immediately after the muslin or fibrous glass is installed, apply a finish coat of the asphalt emulsion, covering the muslin or fibrous glass completely. Allow the emulsion to dry for at least 24 hours. Then apply a final coat of the emulsion at the rate of one gallon for each 75 square feet. While this coat is still tacky, apply an overall coat of 160-mesh mica topping.

Exposed Felt

Apply a brush coat of bituminous compound over the exposed felt. If the felt is dry and bleached, the compound should be applied at the rate of about one gallon to 100 square feet. If the felt is only a few years old and in good condition, apply the bituminous compound at the rate of about one gallon to 150 square feet. Allow the coating to dry for at least 24 hours. Then apply a final coat of bituminous compound at the rate of about one gallon to 75 square feet.

Rusted and Loosened Bolt and Screw Heads

Replace corrosion-weakened bolts and screws with new bolts or screws. If the old holes are too large for the new fasteners, drill new holes in an adjoining solid portion of the sheet. Use stainless steel sheet metal screws for fastening sheets to each other, and use stainless steel self-tapping screws to fasten sheets to structural steel. Clean the rust from old holes, cover them with muslin or fibrous glass patches, and recoat them as recommended for repairing large areas of exposed metal core.

Caulking Side and End Laps

If water is entering the building through the laps, caulking of the laps is usually necessary. Before caulking, however, be certain that all fasteners in the vicinity of the leaks are tight. Remove all dirt and debris from the laps. Using a caulking gun, lay a continuous bead of plastic cement along the edge of the protected metal sheet. Then force the cement into the lap with a putty knife or a small pointing trowel. After the cement has been worked into the lap, smooth off the exposed cement and remove any excess cement.

Painting

Recoat or paint protected metal if the existing coatings are dry and brittle. Recoating or painting will extend the life of the protected metal for several years. Recoating or painting should be done during warm weather. It should not be done while it is raining or when the surfaces are wet. Be sure the surfaces are clean. Type C material cannot be painted or recoated unless the surface mica is removed or has disappeared.

Unpainted Surfaces: If the existing surfaces have not been painted, coat them with a bituminous compound. Before applying the compound to oily and brittle surfaces, apply a coat of asphalt primer. Allow the primer to dry for at least 24 hours. Then mix the bituminous compound to brushing consistency with a suitable solvent. Apply the compound to roofs at the rate of about one gallon to 80 square feet. Use three-knot roof brushes with long handles to apply tar coating to roofs. Push the roof finishes forward and backward when applying the coating and exert considerable pressure on the brushes. Use four-inch paint brushes with coarse, stiff bristles to apply the coating to siding. Do not use roof brushes to apply the coating to siding because of the danger of splashing and wasting the coating material.

Painted Surfaces: If the existing protected metal has been previously painted with aluminum paint that has deteriorated or flaked off, repaint it with aluminum paint. Be sure to remove all loose and blistered paint before repainting. Use aluminum paint mixed in the proportions of two pounds of aluminum paste-pigment to one gallon of mixing varnish for aluminum paint. Apply the paint at the rate of about one gallon to 140 square feet. Best results are obtained by spray applications, rather than brushing. The use of finishes may cause bleeding of the bituminous coating underneath. If local conditions make brushing rather than spraying necessary, use a three- or four-inch soft, long-bristle brush. Lay the paint on the surface, working in one direction only. Work rapidly, do as little brushing as possible, and do not brush the paint into the surface. Be sure that all areas are covered. Aluminum paint has a tendency to 'leaf' immediately after it is applied and touching up an insufficiently painted area usually results in a spotty job.

Masonry and Concrete Exteriors

Concrete and masonry exteriors, such as brick, granite, limestone, and marble, require less frequent maintenance than most outside materials, but certain failures can occur. For example, leakage through concrete walls is caused by cracks in the concrete and, in rare cases, porosity of the concrete. The cracks may be caused by foundation settlement, excessive floor loadings, temperature expansion and contraction in structural members, or poor materials and poor workmanship in the original construction. Exteriors should be inspected quarterly for structural cracks, open mortar joints, settlement, efflorescence, stains, and deterioration of paint or other surface covering.

Common defects in brick walls are open vertical joints, cracking, spalling, and porosity. Efflorescence and leakage of rainwater through the walls usually result from such defects. Inadequate or improperly designed and constructed flashings also cause serious leakage. Open vertical joints result from failure to fill the joints with mortar in laying the brick. Cracking may result from settlement, expansion and contraction, misalignment, or some other serious structural defect in the building. Spalling results if bricks are soft or tend to powder and crack, or when moisture freezes within the masonry. Porosity in brick walls is rare and can result only from the use of porous brick or mortar. Mortar containing large aggregate or poor quality brick may produce this porous condition. An engineering investigation of the causes of structural defects should govern the nature and extent of major repairs. Cracked and open joints should be repaired as recommended in the masonry repair section. Where masonry cracks appear at windows, doorways, offsets, and like places, an expansion joint or control joint should be provided at those locations to minimize the effect of foundation movement.

Major Masonry Problems

There are three major masonry problems which can be prevented. They are efflorescence, cracks, and parapet creep.

Efflorescence: Efflorescence usually appears as a light powder or crystallization caused by water-soluble salts deposited as water evaporates within the mortar on the masonry unit. Aside from detracting from the appearance of a wall, efflorescence may indicate moisture penetration into the wall. Efflorescence may be removed by vigorous and repeated scrubbing with a stiff fiber or wire brush and clean water. An inspection should be made, however, to determine the source of the stain. If efflorescence appears at the edges and not near the center of the masonry unit, the mortar is probably at fault; if it appears near the center of the unit only, the masonry unit is at fault. The most immediate remedy to prevent recurrence of efflorescence is to check causes of excessive moisture that contacts the wall, such as defective flashings, gutters, downspouts, copings, and mortar joints. See Figure 6.1 for methods of preventing efflorescence.

Cracks: Types of cracks encountered in brick, concrete block, and stone are described below. See Figure 6.2 for methods of preventing cracks.

- **Horizontal movement cracks** are usually long, wide cracks in the mortar joints that occur along the line of the floor or roof slab, or along the line of lintels over the window. Where these cracks turn the corner of a building, they frequently crack down.
- **Vertical and diagonal movement** cracks generally occur near the ends or offsets of buildings. They may also be found extending from a window sill to the lintel of a door or window on a lower floor. These may vary from 1/8 to 3/8 of an inch in width and follow the mortar joints, but in some instances they may break through the bricks or other masonry.
- **Shrinkage cracks** are the fine hairline cracks that are found in mortar, as well as in concrete walls. The most noticeable ones are those running vertically, but a close examination of a section of wall that leaks may also show them in the horizontal or bed joints of brick or block walls.

Preventing Efflorescence

Eliminate contaminating salts by:

1. Following ASTM specifications.
2. Using only washed sands.
3. Not using masonry units from a pile where efflorescence has been observed.
4. Using clean, potable water.
5. Using clean equipment, mortar boards and tools.
6. Using insulation fill that does not contain any harmful salts.

Eliminate moisture in walls by:

1. Providing well-tooled joints.
2. Installing vapor barriers to the inside of the walls.
3. Correctly installing flashings and copings.
4. Applying paint or other protective surface treatments to the exterior of the walls (but only after the units have thoroughly cured). Silicone coatings are effective.
5. Utilizing wide overlays.
6. Thorough planning so walls will not be exposed to frequent wetting from such things as lawn sprinklers.

Figure 6.1

Parapet Creep: Unequal expansion of roof slabs and brick parapets sometimes causes parapet corners to creep. When this happens, the horizontal mortar joint at the top of the slab or the nearest through-wall flashing usually shears open. The parapet then extends beyond the face of the wall below, at the corners. If the creep action continues to the extent that cracks extend through the parapet and concrete from the spandrel beam, and brick and mortar begin to break off, an engineering investigation and major repairs are necessary.

Maintenance

Routine maintenance procedures for masonry are listed below.

- Wash at regular intervals.
- Paint at infrequent intervals.
- Replace sealants in control joints.
- Tuck pointing.

Other items of maintenance, which are not performed as often as those listed above, include:

- Removal of efflorescence.
- Stain removal.
- Caulking.
- Repair of unit(s).

Cleaning Masonry Products

Prior to cleaning masonry products, test the compatibility of the cleaning agent and the masonry by applying it to a sample wall section and observing any reaction which takes place within a week's time.

Since this book deals with the maintenance and repair of existing structures, space will not be devoted to cleaning new masonry. This should be done at the time the building is built and in accordance with the contract documents.

There are five cleaning methods available for existing masonry walls. (Stain removal is explained in Chapter 3.) They are:

- Chemical and steam
- Hand washing
- High-pressure cold water blast
- High-pressure steam
- Sandblasting

Preventing Cracks

1. Use only masonry units that have been cured.
2. Install control joints at regular intervals.
3. Tie the cavity wall to the backing wall.
4. Minimize potential ground and/or foundation settlement through proper design procedures.
5. Utilize correct backfilling procedures.

Figure 6.2

Chemical and Steam: This method is used mainly to remove applied coatings to masonry, such as paint. This is a highly specialized field and, therefore, a contractor doing this type of work should be consulted.

Hand Washing: This method is accomplished by using soap or detergent with cold water. The type of soap should not be harmful to the masonry units. This method is costlier and must be repeated on a more frequent basis.

High-Pressure Cold Water Blasting: As the name implies, this method cleans by the blasting effect of the cold water, under pressure, hitting the wall. The major disadvantage is getting rid of the waste water. An experienced contractor should be employed to perform the work.

High-Pressure Steam: This method is by far the most used. As for the other processes, a specialized contractor should be consulted prior to cleaning the masonry units.

Sandblasting: This method should only be used when the units will not be damaged and/or when they cannot be cleaned with high-pressure steam. This method should also be done only by experienced professionals.

The material to be cleaned and the type of stains present determine the best cleaning method. Sandblasting and acid washing work well on some brick, concrete, granite, and other hard surfaces, but are harmful to soft stone and glazed finishes. Improper cleaning materials and methods may damage surfaces and mortar joints.

Concrete surfaces should not be cleaned with chemicals having an acid reaction because of the roughening effect on the surfaces. Stains of various kinds should be removed by the appropriate method recommended for concrete floors (see Chapter 3).

Removal of Efflorescence: The first step in the process of removing efflorescence is to determine the source of water in the wall and eliminate or control it.

To remove efflorescence, first attempt to scrub it off by applying water to it with a stiff scrubbing brush. If this doesn't work, the surface should be scrubbed with a solution of hydrochloric (muriatic) acid having a concentration not stronger than one part acid to nine parts water by volume.

Before applying the acid solution, dampen the wall surface with clean water. Apply acid solution to small areas of about four to five square feet at a time. Let stand five minutes and follow with a scouring of the treated surface with a stiff bristle brush. This should be immediately followed with a rinsing of the wall with clean water. Remember to protect all windows and trim from destruction by the acid.

Refer to Chapter 3 for methods of stain removal.

Mortar Repair

Weathering may cause spalling of mortar joints under the best conditions, but poor mortar mixes are usually at fault when the face of masonry walls is marred by stains and efflorescence. Impure water containing acids or other damaging matter and improper sand in the mortar mix not only cause stains through bleeding, but also allow wear in mortar joints in excess of that caused by normal weathering.

Repointing: Repointing with the best materials and skilled workmanship will correct defective mortar joints. The following procedure is recommended:

1. Cut out cracked and open mortar joints to a depth of at least one-half inch. Cutting can be done by hand, but if large areas are involved, it is usually cheaper to use power tools. Take care not to damage brickwork during the cutting process.
2. Remove all dust and loose material with brushes, compressed air, or a water jet. If water is used, no further wetting of joints may be needed unless the work is delayed.
3. Use mortar of about the same density as the original mortar, if this can be determined. Otherwise, use a prehydrated mortar mix in the following proportions by volume; one part portland cement, one part lime putty or hydrated lime, and six parts sand. Prehydrate the mortar by mixing it about two hours before use, adding only about one-half the mixing water, to eliminate excessive original shrinkage and volume change.
4. At the end of the two-hour curing period, rework the mortar, adding enough additional mortar to make the mixture plastic, but not enough to make it run.
5. Be sure the joints are damp, and then apply the mortar by packing it tightly into the joints in thin layers.
6. Tool the joints to smooth, compact, concave surfaces.

If openings in the mortar are small, cleaning out the joints is unnecessary. Use the following procedure:

1. Mix grout in the following proportions by volume: 1/2 sand, 3/8 portland cement, and 1/8 limestone flour, powdered flint, or fine hydrated lime.
2. Use a template or cover to keep the surfaces of bricks free of grout.
3. Wet the joints, and apply two coats of grout, brushing it vigorously into the joints.

Repair of Cracks: When cracks are observed in the wall, the first thing that should be ascertained is their cause and whether they are active or dormant. Refer to Chapter 3 for methods of determining their degree of dormancy. If the cause of the crack cannot be removed, accommodation must be made for future movements prior to repairing the crack. This can be done by installing new expansion joints.

Mortar Joint Cracks: These can be repaired by tuck pointing, using a mortar that will match the existing mortar. If this cannot be done, consideration should be given to painting the wall.

Unit Cracks (Wide): If the crack has crossed the unit, there are two methods of making the repair. The first is by breaking out the old unit, cleaning out the old mortar, remortaring, and installing a new unit of the same grade and color as the existing units.

The other method entails widening and undercutting the crack with a chisel so that the crack is about one-half inch wide at the surface and about three-quarters of an inch wide at the back and about one-half inch deep. Brush away all loose dust and fill the crack with an epoxy resin or a dry-pack mortar. If mortar is used, the adjacent unit should first be dampened. Use a mortar of one part portland cement to two parts masonry sand. If water is present (leaking through the crack) first attempt to eliminate the cause or bring it under control by installing small drains of plastic tubing. This is not the best method, since the tube can become clogged and thus become ineffective. Also, use a quick-setting portland cement in place of the normal cement. If epoxy resin is used, follow the manufacturer's recommendations.

Unit Cracks (Fine): Fine cracks should not be widened. Paint over them with a fine portland cement paint or scrub the surface with a lean mortar mix.

Racked-Down Corners: Racked-down corners occur where the horizontal crack not only continues around the corner, but forms part of a diagonal crack that takes a downward direction and meets a similar crack from the other side, forming a V. The bricks inside this V are loosened and must be reset.

1. First remove all the bricks inside the V, including any bricks that have been broken. This forms irregular sides and helps to hold or key the brick in place.
2. After the bricks are removed, clean the sound bricks and obtain as many new matching ones as are necessary to fill the opening. Relay the bricks in mortar up to and even with the horizontal lines running along the side and end of the building. If all joints are made the same width as the existing joints and the mortar tends to match the old mortar, a very presentable job will result.
3. Fill the top joint that is on line with the horizontal crack partially full with mortar. This can be done by pushing the mortar into the joint with a narrow pointing trowel. When about half the depth of the joint is filled, fill the remainder with sealing compound. In this system of mortaring, only half the joint supports the brick above, but forms a weak plane along the top of the racked-down areas. If movement takes place, the mortar joint breaks, but the relaid bricks remain in place. The sealing compound keeps the joint watertight.

Parapet Repair

In most cases, about 20 feet of the parapet, beginning at the corners, must be removed. A new section of parapet, doweled into the wall below, is then laid. A vertical expansion joint, one inch wide, is provided at the juncture with the existing parapet. Through-wall metal flashing and a continuous seal of bituminous plastic cement must be provided at the bottom of and in the expansion joint.

Glass Block Walls

Glass block panels require little maintenance other than occasional cleaning, by washing, and periodic inspection of joints. They can be damaged, however by superimposed loads transmitted through excessive deflection in beams and lintels, uneven settlement of foundations and footings, and by impact. The causes of serious damage must be corrected before repairs are made. To repair panels, use the following method:

1. Remove cracked or broken glass blocks.
2. Chip off fragments of broken glass adhering to undamaged blocks, taking care not to damage wall tiles and anchors.
3. Clean old mortar from exposed wall tiles and anchor, and from the mortar-bearing edges of adjoining blocks.
4. Replace panels, using materials and methods that match the existing work as closely as possible.

Mineral Surfaced Board

Recent developments in combining chemical and mineral elements have produced hard-surfaced, weather-resistant materials for exterior wall finishes. These materials require little maintenance except to keep fasteners secure and to remove stains from other sources. Painting should be accomplished according to the manufacturer's recommendations. Recent findings of the hazardous effects of asbestos upon humans will result in an eventual halt in the manufacture of this material.

Prevention of Deterioration

Mineral surfaced siding is a high-grade finish material and must be properly and carefully handled during transportation and application. It must be properly stored in the warehouse and on the job. While still in packages and until applied, this type of siding can be seriously harmed and discolored by moisture and dampness. Keep it dry. Pile on a solid, flat, unbending platform raised at least four inches above-ground. Keep it completely covered and protected from the weather by a watertight and waterproof cover. Do not apply siding over wet sheathing or underlayment, or during rain or snow storms. Do not permit water draining from uncovered sheathing or underlayment materials to drain over the siding, or staining may result. Fiberboard, gypsum board, and plywood sheathing can be seriously harmed by the absorption of moisture during careless storage. Wet sheathing can seriously and permanently harm any siding material applied over it. Keep these materials completely protected from the weather by a waterproof covering until installed.

Cleaning and Stain Removal

The areas to be cleaned should be dry-brushed thoroughly to remove the dust and loose material, and then scrubbed with a floor scrubbing brush, using one of the chemicals mentioned in the following paragraphs. It is very important that, immediately following the application of the cleaning compound, the entire area treated should be thoroughly rinsed with clean water. In the following treatments, the percentage of chemicals in solution is determined by mixing, in a porcelain or glass receptacle, the amount of chemical and clean water. For example, a 5 percent solution of phosphoric acid means 5 parts by volume of commercial phosphoric acid and 95 parts by volume of water. Where acids or strong chemicals are necessary, the cleaning solution should be handled carefully. It is recommended that rubber gloves be used. All painted trim and shrubbery should be protected, and any cleaning compound dropped on them should be immediately removed with clean water.

Window Screen Stains: Rain flowing from window screens is frequently the cause of discoloration of painted surfaces and mineral surfaced products. To remove stain, use a dilute solution (5 percent) of white vinegar. Screens should be periodically painted with dilute spar varnish to prevent further staining.

Copper Stains: Use paste consisting of one part ammonium hydroxide, ten parts water, 16 parts talc, and four parts ammonium chloride, then add the liquid in sufficient quantity to form a heavy paste. Carefully apply the paste with a trowel or scrubbing finish over the entire surface (dark rings may form if paste is applied only in spots). After 24 hours, thoroughly wash the dried paste off the stained surface.

Earth Stains at Grade Line: Dry-brushing and rinsing will generally remove ordinary earth stains.

Rust Stains from Nails, Hinges, Gutters, Leaders: Iron rust stains may be cleaned with a 2 percent solution of oxalic acid or a 5 percent solution of phosphoric acid. Wash the surface with the solution. Follow this with a thorough rinsing with clean water.

Stains from Unpainted Wood Trim: Stains caused by water running over unpainted wood trim may be removed by scrubbing them with common cleansers. If stains are deep, scrubbing with a strong solution of sodium hypochloride or a 2 percent solution of oxalic acid may be necessary.

Ordinary Dirt and Soot Stains: These stains may often be removed with a mild cleanser such as ordinary soap or a weak solution of trisodium phosphate or other commercial detergent. Use a stiff fiber brush.

Deeply Penetrated Dirt and Soot Stains: These stains may be removed by careful scrubbing with a common cleaner. Very severe dirt and soot stains may be removed with a bleaching agent, such as sodium hypochloride.

Discoloration from Oil-Stained Wood Shingles: In cases where surfaces are discolored from oil-stained wood shingles, it is generally necessary to use a two part solution treatment. First apply a 20 percent solution of sodium citrate, brushed on and allowed to dry. Then apply a 20 percent solution of phosphoric acid, applied with a scrubbing brush. This coat will readily remove the stains, but in some cases it will be found that it may leave a slight "bloom." Where this happens, it is necessary to wash the entire area with a dilute (5 percent) solution of phosphoric acid. It is most important with this method to scrub the wall surface as well as concrete foundation, sidewalk, etc., thoroughly with clean water to remove all traces of acid and bloom.

Paint Stains: Wipe off stains with a cloth soaked in turpentine or other paint solvent. If the paint is old and dried, repeated application may be necessary, after scraping with a knife blade.

Oil and Grease Stains: Use a paste consisting of 3-1/2 pounds of whiting mixed with one quart of cleaning solvent, benzine, or naphtha. Apply with a trowel and follow by wiping with a clean, dry cloth.

Repair and Replacement

Broken shingles and corrugated or flat sheets should be replaced promptly and secured by methods recommended by the manufacturer, using standard clips and fasteners wherever possible. For field repair involving the use of nails, bolts, or rivets, use the proper drilling tools and saws for appropriate cutting. Do not use fastening devices or common nails, bolts, and washers that are subject to weathering and corrosion. Where material is placed over wood sheeting, be sure the nailing base is in sound condition before adding the new materials.

Translucent Structural Panels

Siding produced from a combination of polyester resin, fiber glass, and plastics may be encountered in some newer constructions. As in the case of other flat and corrugated sidings, it is important to keep fastening devices tight and replace broken panels. Replacement fasteners should be similar to those recommended by the manufacturer. Siding material should match that existing in the building. Cleaning of translucent siding is paramount because it is used for lighting and decoration. Choice of detergents and chemical cleaning solutions must be within the limitations of the manufacturer's recommendations, or severe damage may result.

Stucco

The maintenance and repair of stucco surfaces is similar to other masonry materials. The reader is referred to the previous masonry sections.

Research has shown that most of the problems with stucco can be eliminated or prevented by good design and construction techniques (see Figure 6.3). The removal of stains and efflorescence would be identical to that of masonry materials.

Other Materials

New exterior wall finish materials are being introduced each year. Examples are vinyl siding and concrete-wood panels.

Each product requires its own unique cleaning and repair procedure. The reader should consult with the manufacturer of the siding for applicable maintenance procedure.

Weatherstripping

No building, old or new, can be made completely draft-free, but there is much that can be done to reduce drafts to a minimum. Heat loss through doors and windows (cracks) may account for as much as 25 percent of the fuel bill. Therefore, the application of weatherstripping can substantially reduce the heating bill. Other benefits of weatherstripping are: prevention of heat infiltration during the summer and shutting out of dust and small insects, such as ants. The main types of weatherstripping are:

- Flexible rolls or coils made of felt, plastic, or rubber.
- Rigid strips faced with foam, plastic, or rubber.
- All metal, spring bronze, or aluminum.

Problems with Stucco

Problem	Cause	Prevention
Color Variation	Nonuniform basecoat thickness.	Be sure backing is straight, well aligned and rigid.
	Overspray of extraneous material.	Adequate installation procedure.
	Poor materials or materials which have been inadequately prepared.	Follow good mixing procedure, utilizing the highest grade of materials.
	Failure to set-off joints in scratch and brown coat.	Be sure that the joints in the scratch coat do not occur at the same places as the joints for the brown coat.
	Uneven drying.	Brown coat should dry for five days.
	Placing stucco in inclement weather.	Be sure temperature and climate are suitable.
	Poor placing and finishing techniques.	Use trained personnel.
Coverage Too Thin	Too thin of a finish coat.	Increase coverage.
Effloresence	Materials.	Use highest quality of materials.
	Moisture action.	Insure that precautions have been taken not to allow water to continuously oversaturate the stucco.
Scaffold Marks	Improper scaffold or improperly erected.	Use correct type scaffold and erect properly.
Stains	Dirt, rust, wood sap, inadequate drainage.	Take care of these possibilities during the design stage.
Uneven Drying of Coats	Difference in thickness of coats.	Insure uniform thickness.
	Poor weather.	Do not install in bad weather.
Visible Joints	Cold joints.	Coverage operation should be continuous.
	Too much floating water.	Keep the quantity of water to a minimum.

Figure 6.3

The last type is the most permanent type. The metal stripping fits inside the frame so that it cannot be seen when the window or door is closed. The other types are applied to the surface, such as around doors and windows. These are easier to install and are more commonly used around the building.

The local hardware supply can aid in the selection of the correct type. Refer to the manufacturer's recommendations for specific installation directions.

Inspection

The inspection of weatherstripping should be included in any building inspection. If deteriorated or damaged weatherstripping is located, it is not economical to repair it. The only economical alternative is to replace it.

Weatherstripping is one of the details in a building that always seems to get in the way when attempting to clean and/or repair adjacent surfaces. Do not overlook this fact, and, when cleaning a wall, also clean the weatherstripping. Be sure that the cleaning method does not result in the failure of the weatherstripping.

Walks and Drives

Keeping walks and drives clean and in good repair is of utmost importance, since they serve as a means of getting to and from the building.

In order to keep repairs to a minimum, it is important to follow certain precautions relative to maintenance of the surface. The first of these is to refrain from applying harmful chemicals to the surface:

- Do not use salt to melt snow or ice on concrete.
- Do not spill gasoline, kerosene, or other such liquids on asphalt surfaces.

Secondly, whenever working on concrete and asphalt surfaces, place a protective covering over them to protect them from accidental spillage of chemicals that may be damaging to the surface. If an accidental spill occurs, immediately wash down the surface with water and a neutral detergent.

The section below addresses the maintenance and repair of asphalt surfaces. Refer to Chapter 3, "Concrete," for the maintenance and repair of concrete.

The major prerequisite for any effective maintenance program is knowing the material composition and the construction of the area.

Asphalt, as a material, is mostly found as surfaces (pavements) for parking lots, walks, and storage areas. If correctly placed, asphalt is a very flexible and durable surface. Like any other type of building material, it must be maintained. Outside forces that cause deterioration of asphalt are temperature changes, moisture, superimposed loads, and movement of underlying earth.

Inspection

Inspection should be made semi-annually, preferably in the spring and fall of the year. And just as for any other material, records should be kept detailing the maintenance history of the surface. The inspection procedure should include noting all cracks, holes, depressions and other types of visible distress.

Maintenance and Repair of Asphalt

Repairs need to be made as soon as possible after locating the defects. This is especially true with asphalt surfaces since, once deterioration begins, further deterioration progresses at a more rapid rate than for most materials due to the superimposed loadings and nature of the material.

One important precaution taken with asphalt surfaces is applying a sealer coat about six months after the surface has been constructed and once each year thereafter. Refer to Figure 6.4 for a summary of types of asphalt deterioration and maintenance.

Below is a general repair procedure for patching and sealing asphalt surfaces, utilizing a commercially available cold mix patching compound. For a more complete discussion of the various types of asphalt pavement deterioration and applicable repair procedures, refer to "Asphalt in Pavement Maintenance," published by The Asphalt Institute.

Equipment	Material
Shovel	Asphalt
Tamper	Patching material
Stiff bristle broom	Sealer compound
Water hose	Blacktop sealer
Squeegee	

1. Dig out all loose materials from the bottom of the hole.
2. If the hole is deep, gravel should be placed over the bottom and tamped firmly to provide a good foundation.
3. Fill the hole with the cold mixed patching compound to within one inch of the top.
4. Tamp the compound firmly with hand or mechanical tamper, depending upon the size of the repair.
5. Additional cold mix is then poured in the hole until it is about one-half inch higher than the surrounding surface.
6. Give the repair a final tamping.
7. For small cracks or other voids, spread dry sand over the surface. Sweep the sand off the smooth surface without removing it from the voids and crevices.

Asphalt Deterioration and Maintenance		
Deterioration	**Diagnosis**	**Maintenance**
Alligatored Cracks	Crack appearing like spider webs.	Firm and level area: seal coat.
		Loose or spongy area: Remove and patch.
Bleeding and Softening	Soft surfaces.	Refer to Asphalt Institute Booklets.
Chuckholes	Break in the surface extending into the sub-base.	Clean out chuckhole down to the base and patch.
Depressions	Holes.	Remove the defective area down to the soil, recompact the soil, provide drainage if a problem and patch.
Edge Breaking	Edge of pavement breaking off.	Provide lateral support by filling in other material. Follow by removing damaged area and patch.
Long Cracks	Long cracks.	Clean out crack (to 1 inch or wider) and fill with crack filler.
Numerous Wide Cracks	Cracking.	Clean out cracks (1 inch wide or more), fill with crack filler and resurface.
Raveled Surface	Loss of bitumin shows exposed aggregate overlay.	Clean surface and seal coat.
Softening	Surface is soft and spongy.	Clean and scrub well with strong detergent. If severe, remove area and patch.
Weathered Surface	Surface is gray in appearance and aggregate is bleached out.	Clean surface and seal coat.

Figure 6.4

Once all cracks and broken areas have been patched, the surface can be sealed. Do this on a dry sunny day.

1. Sweep the surface clean and remove all greasy areas.
2. Wet down the entire surface, making sure any puddles are swept away.
3. Apply the blacktop sealer and allow to dry overnight before using it.
4. A second coat may be needed, but don't apply it until the first coat has dried overnight.

Corrosion

The annual direct cost attributed to corrosion in the United States is estimated in millions of dollars. Such estimates include only the costs connected with corrosion control and the replacement of damaged equipment and structures; they do not include any of the various indirect losses resulting from corrosion. Losses resulting from corrosion are of several types, including those listed below:

- Direct loss or damage of metal structures as a result of corrosion.
- Direct maintenance costs due to corrosion.
- Costs attributed to the overdesign required to allow for corrosion.
- Direct losses resulting from corrosion.

The severity of corrosion varies mainly with the environment. Corrosion is defined as the destruction of a metal by chemical or electro-chemical reactions of the metal with its environment. This means that a flow of electric current is always associated with a corroding metal. In both chemical and electro-chemical reactions, water and oxygen must be present together in the environment in order for corrosion to take place. This can occur under any environmental conditions. The chemical and physical mechanics of corrosion are not treated here. This section contains discussions of methods for prevention and repair of corroded materials. Refer to Figure 6.5 for performances of the major metals in different environments relative to corrosion.

Metals in Atmosphere, Water and Soil-Corrosion Performances

	Atmospheres					Waters				
	Humid	Industrial	Marine	Rural	Salt Spray	Fresh	Distilled	River	Sea	Brine
Aluminum	1	1	1	1	1	2	1	2	2	2
Cast Iron	2	3	2	2	2	2	2	1	2	2
Copper	2	2	2	2	2	1	2	2	2	2
Iron	3	3	3	3	3	2	2	2	3	2
Magnesium	2	1	3	2	3	1	2	3	3	3
Nickel	1	1	2	1	1	1	1	2	2	2
Stainless Steel	1	1	2	1	2	1	1	2	2	2
Steel	2	2	2	2	3	2	2	2	2	3
Tin	2	2	2	2	2	2	1	2	2	2
Wrought Iron	2	2	2	1	2	2	2	2	2	1
Zinc	2	2	2	2	2	2	2	2	2	2

1 = Resistant;   2 = Attacked;   3 = Severe Attack

Corrosion increases with decreasing pH and soil resistivity. Will also vary with the amount of moisture, dissolved oxygen and organic materials.

Figure 6.5

Controlling Corrosion

The six methods presently used to control corrosion are listed below.

- Use of non-metallic materials
- Protection of the metal
- Passivity
- Change of the environment
- Cathodic protection
- Combination of these methods

Non-metallic Materials

The use of non-metallic materials in corrosive environments is often desirable. Although non-metallic materials will not fail as a result of corrosion, deterioration may result from other causes. The corrosion reaction of a metal is electro-chemical in character, whereas the deterioration of non-metallic materials is the result of physical and chemical processes. The failure of a non-metallic material may result from softening, bleaching out of the binder, mechanical factors, age hardening, or the equivalent of a stress corrosion mechanism, such as that occurring in metals. Consequently, care must be used in selecting a non-metallic material for eliminating corrosion to make sure that another equally costly type of deterioration will not occur. The non-metallic materials widely used in construction today may be broadly classified as follows:

- **Inorganic material:** cement, concrete, stone, tile, glass and ceramics.
- **Natural organic materials:** woods, fibers, oils and waxes.
- **Synthetic organic materials:** plastics.

Protection of Metal

When it is impractical to use a non-metallic material for a structure because of cost, deterioration, or structural design factors, it may be expedient to use a metal protected from corrosion by a non-metallic covering. Coatings or platings are intended to isolate the metal from an environment favorable to corrosion. In many cases, the use of coatings and treatments provides the most economical solutions to the corrosion problem. Such materials are often used in conjunction with cathodic protection, which prevents corrosion at faults or holidays in the material covering the metal.

Materials used for metal protection should have the following properties:

- Low moisture absorption.
- Inertness to soil chemicals.
- Resistance to stress and fracture.
- Ease of application.
- Adequate thickness and resistance to penetration by stones and sharp objects and other means of mechanical abrasion.
- Resistance to ozone and oxygen.

Materials used for isolating metals from their environment are used in the form of coating, tapes, and linings or combinations thereof. Some of the common types of materials are listed below:

- Bituminous coatings
- Plastic coatings
- Plastic tapes and coatings
- Concrete
- Ceramics
- Grease and waxes
- Paints, including enamel, varnishes, and lacquers
- Metallic protective coatings

Coatings used to protect metal surfaces may be applied in almost any thickness from a few thousandths of an inch to more than half an inch.

Bituminous Coatings:. Bituminous coatings are classified as coal tar coatings or asphalt coatings, depending on the base material from which they are made. Coal tar coatings are more effective than asphalt or fiber based coatings. Coal tar coatings are often applied to structures such as steel piles. However, they seldom provide adequate long term protection for the underwater portions of a structure. Coal tar coatings are effective for marine use above water, in the splash zone and in the portion of the final area where they can be repaired. Repair in the low tidal zone, however, where the time during which the coating can be repaired is limited by tidal and/or wave action, is difficult and usually ineffective. In such areas, it is very difficult to obtain a dry, clean surface for the application of even a thin, rapid-drying coating.

Coatings may be either hand- or machine-applied in the shop or in the field. Machine-applied coatings usually are more uniform than hand-applied coatings and therefore are preferable. On large projects, such as pipelines, it is advisable to use a traveling machine for field application of coatings. The coatings and wrapping are applied by the coating machine as it travels along the pipe after fabrication, and a very uniform coating is obtained. The use of such machines is economically restricted to sizable pipeline projects.

Asphalt coatings are used commercially for paints, coatings, and pavings. Application methods and techniques are similar to those used for coal tar coatings.

Bituminous coatings are often reinforced for special applications requiring resistance to stress. Fibrous compounds are used extensively for reinforcing bituminous coatings for underground use. Glass fibers are widely used for this purpose commercially, but are not recommended. The reinforcing makes the coating more resistant to soil stresses.

The best bituminous coating material is only as good as its application. The first step in good application is the surface preparation of the metal. Refer to a following section of this chapter for methods of surface preparation. After the surface preparation, the coating must be applied carefully in accordance with the manufacturer's specifications.

The selection of coating materials depends on the environmental conditions to which the material is exposed.

Mastic Coatings: Mastic coatings were developed for underground use, for tank linings; for abrasion-resistant coatings, such as those used for railroad cars; and for atmospheric exposures. Mastic coatings are comprised of rubber and asphaltic compounds as the base. Rubber-based mastic coatings do not require a primer and can be applied cold by either spraying or brushing. They are quick-drying for underground use. However, they are subject to deterioration when exposed to crude oil or its distillates.

The more effective a surface preparation, the longer the durability of the protective coating. Refer to a following section of this chapter on methods of preparation of metal surfaces. It is important to develop a strong bond so that the mastic coating penetrates the surface of the material.

Protective Jackets: Protective jackets are used to isolate metals, for example, to protect steel piling in the tidal zone. It is also used to isolate only a portion of a pipe or pile. The jacket is made of material that is non-corrosive in its environment. If a metal jacket is used, it must be of a material that does not cause galvanic corrosion (that is, corrosion between dissimilar materials) of the metal to which it is attached. For additional information, refer to Chapter 4.

The use of plastics for coatings and linings has become important in recent years. Pressure-sensitive organic plastic tapes, which are used extensively for pipe coverings, make excellent protective coatings. They may be applied directly to a clean pipe, either by machine or by hand. The economy of this method depends on the size of the job. Plastics are also used as electrical insulating tapes.

Concrete Coatings: Concrete coatings are used extensively to cover steel pipe on large aqueducts. Such coatings are very durable. They withstand rough handling and are not damaged by soil stresses. Concrete is also widely used as a pile capping material and for protection of steel pilings that have become corroded. On both steel and wood waterfront structures, concrete jackets have been used on pilings as a repair practice. It is of utmost importance that a continuous, homogenous, unbroken coating of concrete exists around the metal. If the concrete cover spalls off, cathodic currents can be established which will react with the materials in the concrete to initiate a corrosion process in the metal.

Concrete is used extensively for lining both steel and cast iron pipe. Concrete linings are comparable to other types of pipe linings. When properly applied, concrete makes an excellent lining which retards both rusting and pitting. Concrete lining is usually applied by a spinning method and tends to reduce metal losses. Concrete lining may also be used for reconditioning old installations. Old pipes that are badly pitted can be cleaned and cement-lined in place. Commercial techniques have been developed for such purposes.

Other Coatings: The application of ceramics to steel provides an excellent lining for storage tanks. Such linings are subject to chipping, however, and are thus very difficult to repair. Porcelain enamels are resistant to most mineral acids except hydrochloric acid.

The use of heavy greases for temporary protection of metal surfaces finds practical application in many cases. It should be emphasized that grease coatings are permissible only as temporary coatings.

Painting is the oldest, most widely used means of combating corrosion. Protective coatings must be selected to suit the particular corrosive conditions affecting the structure to be protected. The reader is referred to the section on painting techniques.

Metal coatings are applied to iron or steel surfaces to prevent rusting. They are also used for ornamentation. The various methods of metallic coatings are:

- Galvanizing
- Shredderizing
- Electro-plating
- Metalizing process

Passivity

The selection of a metal or alloy that is passive to its environment and does not corrode, or is so nearly passive that the rate of corrosion is less than the anticipated life of the structure, is a frequently used method of corrosion control. Cathodic protection and coatings may be used in conjunction with partially passive materials.

There are many available "corrosive-resistant" materials for substitution in different types of corrosive atmospheres. The most important point to consider is the type of atmosphere that will be attacking the material. The selection of a metal or alloy for use in a typical construction environment is based largely on economic factors closely related to local conditions. Corrosion prevention measures used in extremely cold climates, for example, are not necessarily used in a tropical climate. The compatibility of various materials with each other should be considered in relation to the economic factors of corrosion control. It would not be advisable to select a material for the compatibility without a thorough analysis of the total economic cost. Many alloys are too expensive for general use, and their applications are limited to environments that are extremely favorable to the formation of corrosion. The common types of steels and cast irons used in conjunction with the proper corrosion protection methods provide the most economical solution to the major portion of the corrosion problems in the construction industry. Some of the corrosion-resistant materials available are listed below.

- Copper
- Aluminum
- Lead
- Zirconium
- Alloys of steel
- Stainless steels
- Other alloys of metals

In the selection of combinations of metals that develop a minimum of corrosion in corrosive environments, the following factors should be considered:

- Minimum potential of corrosion between the two metals (galvanic potential).
- Cathodic protection of the metals with respect to each other.
- Relative areas of the two metals.
- Conductivity of the water or soils in which the metals are to be used.
- Geometric arrangements of the two metals.
- Feasibility of electrically isolating the metal.
- Possibility of applying an insulating coating to one or both metals.

In addition, consideration should be given to the practicability of substituting non-metallic materials. The selection of material combinations as specified above is typical of most installations in contact with a corrosive environment. There will always be cases that require special study of the corrosion problem. It is emphasized that, in addition to general corrosion, the coupling of dissimilar metals brought in contact with an electrolyte (a solution that acts as a catalyst for the corrosion) may cause accelerated local corrosion. See Figure 6.6 for a chart of performance of metals in different types of electrolytes. This type of bi-metallic corrosion is known as galvanic corrosion. If it is necessary to use two dissimilar metals, selection should be made with the idea of creating a minimum galvanic current. This problem becomes critical in connectors.

Figure 6.7 can be used in selecting fasteners or for ascertaining the compatibility of joining two members comprised of dissimilar materials. An alternative solution when using dissimilar metals is to isolate them by an insulating brushing or coupling.

Some methods of preventing or minimizing corrosion attack are as follows:

- Coating, painting, or insulating dissimilar metals.
- Avoiding irregular design stresses.
- Using fasteners made of an alloy or metal that can resist the corrosive environment.
- Not joining or bringing into contact dissimilar materials.
- Revising the design.

Changing the Environment

Sometimes it is economical to change the environment of a structure to prevent corrosion. Examples of controlled changes of environment are the use of selected types of backfills around a pipeline.

Cathodic Protection

Cathodic protection is an important means of preventing corrosion of new structures and halting it on existing structures. It is usually the only method economically feasible for halting corrosion on existing underground structures and marine structures that are permanently under water. Cathodic protection is defined as an electro-chemical method of reducing or preventing corrosion of a metal surface by making it cathodic to its environment. This is accomplished by forcing a direct electric current to flow in the metal surface. The current is induced from an external source.

The amount of current necessary for protection should be determined by a soil survey and by other electrical tests. After corrosion surveys have been made, the cost of cathodic protection systems should be evaluated against the cost of repair and replacement if protection were not applied. When cathodic protection systems for new structures are being designed, existing underground structures in the area must be considered.

Effect of Electrolytes on Major Metals

	Aluminum	Brass	Bronze	Cast Iron	Copper	Hastalloy	Lead	Magnesium	Nickel	Steel	Tin	Titanium	Zinc
Acetic Acid	R	R	R	D	R	R	U	U	U	U	D	R	U
Ammonia	R	D	D	R	D	R	R	D	R	R	D	R	U
Ammonium Nitrate	R	U	U	R	U	R	R	R	R	U	D	R	U
Calcium Chloride	D	R	R	R	R	R	D	U	R	D	U	R	R
Calcium Hydroxide	R	R	R	D	R	R	U	U	D	R	R	D	U
Chlorine	D	D	D	U	D	D	D	U	D	D	U	D	D
Chromic Acid	U	R	D	R	U	R	U	R	U	R	R	R	U
Citric Acid	R	R	D	U	R	R	U	U	U	U	D	R	U
Hydrochloric Acid	U	U	U	U	U	R	U	U	U	U	U	U	U
Hydrofluoric Acid	U	R	R	U	U	R	R	R	D	D	U	R	U
Hydrogen Peroxide	R	U	R	U	R	U	R	U	U	R	U	R	D
Hydrogen Sulphide	R	D	D	U	D	R	D	R	U	D	R	R	R
Magnesium Chloride	U	R	R	U	R	R	U	U	R	D	U	R	U
Nitric Acid	D	U	U	U	U	U	U	U	U	U	U	R	U
Phosphoric Acid	U	R	R	U	R	R	D	U	U	U	D	R	U
Potassium Carbonate	U	R	R	R	R	R	D	R	R	D	D	D	U
Potassium Hydroxide	U	R	R	R	R	R	U	D	R	D	U	R	U
Sodium Chloride	R	R	R	R	R	R	R	U	R	R	U	R	U
Sodium Hydroxide	U	R	R	R	R	R	R	R	R	U	U	D	U
Sodium Nitrate	R	R	R	R	R	R	U	R	R	D	D	R	R
Sulphur Dioxide	D	R	R	R	R	R	R	D	D	D	U	R	U
Sulphuric Acid	D	R	R	D	D	R	R	U	D	D	U	U	U
Zinc Chloride	U	U	U	U	U	R	U	U	R	R	U	R	U
Zinc Sulphate	U	R	R	R	R	R	R	U	U	D	U	D	U

R= Resistant

U= Unsuitable

D= Depends on grade of metal and state it is in (i.e., molten or solid).

Figure 6.6

Selection of Fasteners to Alleviate Corrosion Problems

Work	Connectors (in order of preference)
Cast Iron or Galvanized Steel	1. Ductile austenitic nickel-cast iron.
Painted Steel	2. Monel
Stainless Steel	3. Stainless Steel
	4. Galvanized Steel
Bronze	1. Silicon Bronze
Brass	
Copper	
Aluminum	1. Galvanized Steel
Aluminum Alloy	2. Plain Steel
	3. Aluminum Alloy

Figure 6.7

When an old structure has corrosion holes that have perforated the metal, but are temporarily sealed by corrosion products, cathodic protection will soften the corrosion products and cause the holes to re-open. On badly corroded structures, leaks will continue to appear at a decreasing rate after the application of cathodic protection.

Combination of Methods

Often a combination of the various methods is the most practical means of obtaining overall corrosion control. Although the choice of methods, or combinations of methods, in all instances depends to some extent on technical factors, it is usually determined by a proper economic balance.

Repair

If, on casual observation, corrosion is an apparent problem to the steel structure, a corrosion consulting firm should be called in. It is extremely important that prior to taking any kinds of precautionary methods or before making any repairs, a corrosion field survey is performed by experienced personnel. A thorough corrosion survey would include the following steps:

1. Study of maintenance records.
2. Examine and test the properties of the soil, water, and atmosphere.
3. Visual and physical inspection of the corrosion or the corroded area itself.

The reader is referred to Chapter 5 for further discussion of the subject of repair of corroded metals. Also refer to Figures 6.8, 6.9, 6.10, 6.11, 6.12 for methods of mitigating corrosion relative to specific items.

Painting

Painting is probably one of the most used and misused preventive maintenance procedures. Painting done at too frequent an interval is just as bad, if not worse, than painting at too infrequent an interval. It is often better to wash the surface than to repaint. If considering painting, one must remember that more than one painting system is well-suited for a combination of surfaces, environment, and service required. And, for surface preparation, the choice of equipment for and the method of application may vary according to the severity of the exposure, the coating used and the conditions under which it is applied. Each of these factors contributes to the effectiveness of the completed job. This section will focus on painting as a general maintenance item and the specifics that go along with exterior painting. Painting will also be mentioned in the balance of the chapters, pertinent to the specific items being covered.

Paint serves as a protective shield between the base construction materials and the elements which attack and deteriorate them. Painting, when regularly programmed, offers long-range protection that extends the useful life of the structure. Besides protection, some of the other purposes of painting are as follows:

- To maintain a sanitary and clean environment.
- To improve illumination and visibility.
- To insure a safe and efficient environment.
- To improve the building's appearance.
- Possible fire retardant.

Painting is primarily used for maintenance and to improve safety and efficiency. Decorative painting, for appearance, is of secondary importance and should be kept to a minimum.

A lasting appearance can be maintained at minimum cost by:

- Adequately protecting the surface.
- Using paints for finish coats that will retain, to a maximum, their color and gloss.

Structures Exposed to Atmospheric Corrosion

Structure	Major Causes of Corrosion	Materials of Construction	Corrosion-Mitigation Methods
Buildings, exterior: roof gutters, corrugated roofing, leader pipes, window frames, and vent flashing	Exposure to atmosphere (salt air or industrial vapors); trapped moisture.	Steel, aluminum, and copper.	Use galvanized steel, aluminum, or copper. Apply paint coatings. Eliminate moisture traps.
Buildings, interior: hardware	Exposure to atmosphere, (steam or industrial vapors); handling.	Steel, brass, bronze, and plastic.	Use brass, bronze, or plastic materials. Apply coatings (lacquer, paint, or wax).
Piping	Condensation on exterior of pipe (cold water).	Steel, galvanized steel, cast iron, and copper.	Wrap pipe or use copper pipe (with red brass fittings).
Structure members	Exposure to atmosphere (steam or other vapors).	Steel	Apply paint coatings; exhaust vapors to outside of building.
Pole lines & pole line hardware	Exposure to atmosphere (salt air or industrial vapors); or dissimilar-metal couples.	Steel, copper, aluminum, and brass.	Use galvanized steel, alclad aluminum copper-clad steel, or other protective coating. Avoid dissimilar metal couples.
Radio and electrical gear	Exposure to atmosphere, high relative humidity, flooding, or dissimilar-metal couples.	Steel, copper, brass, lead, aluminum, and silver.	Apply paint coatings to steel. Use heaters to keep dry. Drain manholes and vaults. Ventilate and seal enclosures. Avoid dissimilar-metal couples.
Towers and cranes	Exposure to atmosphere, trapped moisture, or electrolysis.	Steel.	Use galvanized steel. Apply and maintain paint coatings. Eliminate moisture traps. Ventilate properly.
Utility buildings	Exposure to atmosphere (steam or other vapors), trapped moisture, or inadequate ventilation.	Steel and aluminum.	Use galvanized iron or steel, or aluminum. Apply and maintain paint coatings. Eliminate moisture traps. Ventilate.

Figure 6.8

Power Plant Equipment — Corrosion

Structure	Major Causes of Corrosion	Materials of Construction	Corrosion-Mitigation Methods
Boilers	Oxygen, carbon dioxide, and high causticity	Steel.	Install and maintain deaerators. Establish correct chemical treatment for boiler water and maintain properly. Maintain alkalin pH of boiler water. Use welded tanks rather than riveted tanks. Repair minor leaks immediately to avoid caustic embrittlement. Install cathodic protection.
Condensers	Oxygen, carbon dioxide, excessive turbulence, high temperature, and stress.	Steel, Muntz metal, Admiralty metal, red brass, copper, aluminum-brass, and copper nickel.	Streamline water boxes, injection nozzles and piping. Avoid sharp angular changes in direction, low-pressure pockets, and obstructions. Use propeller-type circulating pumps. (Seal pump glands to prevent air intake.) Limit water velocity to 5–7 fps. Use lowest possible operating temperature. Remove dissolved air or gases from liquid. Use stress-reliever materials, and support adequately to prevent vibration and cyclic stresses. Use proper copper alloys for liquids involved. Install cathodic protection.
Electrical equipment	Exposure to atmosphere, high relative humidity, oxygen in cooling water, flooding in vaults, and electrolysis.	Steel and copper.	Apply and maintain paint coatings on exterior. Use inhibitors in cooling water. Drain vaults and manholes. Use heaters to keep dry. Ground equipment.
Intake flumes and screens	Oxygen, turbulence, high velocity, and marine organisms.	Steel, cast iron, brass, copper, and copper alloys.	Streamline flow characteristics. Limit velocity. Install cathodic protection. Use heavy galvanized steel or cast iron.
Pumps	Oxygen, carbon dioxide, and dissolved minerals; turbulence, cavitation, stress; high velocity; dissimilar-metal couples; high temperatures; foreign materials, and electrolysis.	Cast iron, cast steel, brass, bronze, and copper.	Deaerate fluid and streamline flow. Relieve stresses. Use lowest velocity and temperatures possible. Avoid dissimilar-metal couples. Use high-silicon cast iron.
Stacks	Sulfur in flue gases and traces of nitrogen oxides.	Steel and brick.	Use fuels low in sulfur content. Coat inside of stacks with asphalt or chromates. Use high-chromium or chromium nickel steel. Line brick stacks with refractory brick.

Figure 6.9

		Above-Ground Structures Containing Electrolytes	
Structure	Major Causes of Corrosion	Materials of Construction	Corrosion-Mitigation Methods
Hot water storage tanks	Oxygen and dissolved minerals, dissimilar-metal couples, excessive temperatures, and contamination of water by copper.	Steel and galvanized steel.	Avoid copper piping on inlet side of tanks. Avoid dissimilar metal couples. Use heavy-grade or galvanized tanks. Apply cathodic protection. Use glass or vitrified tanks.
Process tanks and vessels	Oxygen, temperature, turbulence, velocity, aeration, moisture contamination, acids, and dissimilar metals.	Steel, nickel alloys, nickel-chromium alloys, copper alloys, synthetic rubbers, plastics, ceramics, glass, lead, aluminum, tin, high silicon cast iron, and carbon.	Depending upon the process involved, use proper materials and coating system to resist corrosion. Closely control operation and maintenance to prevent changing conditions that could increase corrosion rate. Use ceramics, synthetics, glass, and plastics for electrical insulating properties. Apply cathodic protection.
Sewage disposal plants	Acid condition of sewage (low pH), exposure to atmosphere, dissimilar metals, temperature, and aeration.	Steel and concrete.	Treat sewage for alkaline pH. Apply paint coatings to metal above the sewage line. Apply cathodic protection.
Surface condensers	High temperatures, velocity, acidity, oxygen concentration, dissolved minerals, and dissimilar metals.	Steel, Muntz metal, Admiralty metal, red brass, copper, aluminum-brass, and copper-nickel.	Use lowest temperatures and velocity possible. Use inhibitors in cooling waters, deaerate, and add chemicals to make the water alkaline. Use electrically insulated parts. Apply cathodic protection.
Surface and elevated water storage tanks	Oxygen and dissolved minerals, exposure to atmosphere, dissimilar metals, galvanic cells, and corrosive water.	Steel, concrete, and wood.	Treat water. Apply paint coatings to interior and exterior. Apply cathodic protection to interior of all metal tanks and to bottom of surface metal tanks. Place surface tanks on pad of clean sand oiled with sulfur-free oil.
Water treatment plants, including flocculators and sedimentation basins	Dissolved minerals and gases, water treatment chemicals, dissimilar metals, galvanic cells, and concrete coated steel.	Steel, cast iron, concrete, copper, brass, bronze, babbitt, and galvanized steel.	Paint or coat metal parts. Avoid the use of dissimilar metals. Use cathodic protection when applicable. Treat water to remove minerals. Use insulating materials.

Figure 6.10

Underground Structures — Corrosion

Structure	Major Causes of Corrosion	Materials of Construction	Corrosion-Mitigation Methods
Buried power, communication, and fire alarm cables.	Corrosive soil and water; stray, long-line, and galvanic currents.	Lead sheath, neoprene or plastic jacket, and parkway cable.	Drain soil water when possible. Apply cathodic protection. Drain stray current, if present. (Supply negative return for ground currents.) Use insulating section in cable sheath. Use neoprene sheath over rubber-insulated cables. Use asphalt-impregnated jute coverings. Avoid dissimilar metal couples. Use clean sand backfill.
Building columns	Soil corrosion, dissimilar types of soil; and stray, long-line, and galvanic currents.	Steel and concrete.	Set column footings in concrete. Apply cathodic protection. Use a bituminous coating.
Compressed air distribution system	Corrosive soils, dissimilar metals, concentration cells, stray and long-line currents, and bacteria.	Steel, bronze, and brass valves, and copper.	Use coatings or galvanized steel. Select backfill with good drainage. Apply cathodic protection. Drain stray current, if present. Avoid dissimilar-metal couples. Install insulating joints to reduce galvanic currents.
Deep wells	Soil corrosion, bacteria, long-line currents from different soil strata, stray currents, dissimilar metals, and concentration cells.	Steel and steel alloys.	Isolate well from the surface piping by use of insulating joint. Use cathodic protection. Drain stray current, if present. Use corrosion-resistant steel alloys in very corrosive soils.
Domestic and fire protection water distribution systems	Corrosive soils, bacteria, dissimilar types of soil or electrolytes, concentration cells, dissimilar-metals and stray or long-line currents.	Steel, cast iron, asbestos-cement, copper, lead, brass, bronze, Monel metal, and stainless steel.	Use asbestos-cement pipe in very corrosive soils when pressure and surges do not exceed rating of pipe. Select backfill with proper drainage. Avoid the use of dissimilar metals. Use insulating joints between dissimilar metals and different types of soil. Use proper coatings. Drain stray current, if present. Apply cathodic protection when possible.
Electrical grounding systems	Soil corrosion; stray, long-line, and galvanic currents.	Copper, galvanized steel, and zinc.	Avoid dissimilar-metal couples. Use insulating joints and insulated wire. Apply cathodic protection. Drain stray current, if present.
Exterior steam lines and returns	Corrosive soils, dissimilar types of soil, concentration cells, dissimilar metals, stray or long-line currents, and bacteria.	Steel, wrought iron, cast iron or bronze valves, vitrified tile, concrete with vermiculite filler, asbestos-cement, natural asphalt, or resinous hydro-carbon.	Use proper coatings. Use sand backfill with good drainage. Apply cathodic protection. Drain stray current, if present. On metal-cased lines, coat seal and vent casings, and apply cathodic protection. Drain steam tunnels and conduit casings. Avoid dissimilar-metal couples. Install insulating joints.
Gas distribution systems	Corrosive soils, bacteria, dissimilar types of soil or electrolytes, concentration cells, dissimilar metals, and stray or long-line currents.	Cast iron, steel, bronze, and lead.	Use proper coatings. Select backfill with proper drainage. Apply cathodic protection. Drain stray current, if present. Avoid use of dissimilar metals. Use insulating joints between dissimilar metals, different types of soil, and other structures.
Hydraulic lines and tanks	Soil corrosion, dissimilar types of soil, differential moisture and oxygen in soil, dissimilar metals, and stray and long-line currents.	Steel, brass, bronze, copper, and cast iron fittings.	Use heavy bituminous coatings. Apply cathodic protection. Drain stray current, if present. When galvanized pipe is used, paint exposed threads. Use clean sand backfill. Avoid dissimilar metals. Install insulating fittings.
Lawn irrigation systems	Soil corrosion, dissimilar types of soil, dissimilar metals, stray and long-line currents, concentration cells, and bacteria.	Steel, iron, copper and brass.	Copper pipe is recommended for corrosion resistance. Zinc-coated (galvanized) iron pipe is adequate for moderate soils. Paint threads that are exposed to soil. Avoid dissimilar metals. Use red brass rather than galvanized iron pipe risers with copper piping. Install insulating couplings. Apply cathodic protection.
Metallic culverts	Soil corrosion, erosion, stray current, concentration cells, and bacteria.	Steel.	Use galvanized steel and/or a bituminous coating. Use clean sand backfill. Apply cathodic protection. Drain stray current, if present.
Metallic sewers	Soil corrosion, dissimilar types of soil, stray and long-line	Steel and cast iron.	Use heavy-grade cast iron pipe dip-coated with coal tar enamel. Bond across all joints in cast

Figure 6.11

Structure	Major Causes of Corrosion	Materials of Construction	Corrosion-Mitigation Methods
	currents, dissimilar metals, concentration cells, bacteria, and sewer gases inside pipe.		iron pipe with a bond wire where cathodic protection is to be applied. Use clean sand backfill in bad soils. Avoid using vitrified clay fitting with metallic piping. Apply bituminous coating both inside and outside of pipe. Avoid dissimilar metals. (Concrete-coated steel pipe can be used; however, it must not connect to bituminous-coated or bare pipe.) Prestressed concrete pipe may present special corrosion problems on the tension members. Concrete lining can be used if the upper portion is coated with chlorinated rubber paint to prevent attack by sewer gas. Drain stray current, if present.
Propane and butane lines and tanks	Soil corrosion, different types of soil, differential moisture and oxygen in the soil, dissimilar metals, bacteria, and stray and long-line currents.	Steel, galvanized iron pipe, copper, brass fittings.	Paint threads on galvanized iron and steel pipes. Use bituminous coating or tape covering on black iron pipe. Use a heavy bituminous coating on tanks. Apply cathodic protection to tanks and piping. Avoid dissimilar metals. Install insulating fittings. Drain stray current, if present. Use clean sand backfill.
Radiant heating systems in soil or concrete	Corrosive soils, differential environment, dissimilar metals, stray and long-line currents, elevated temperatures, and seepage of snow-melting chemicals through concrete.	Steel, copper, and wrought iron.	Drain stray current, if present. Avoid dissimilar metals. Avoid the use of copper plated steel tubing. Apply cathodic protection. Avoid contact between piping and reinforcing steel. Use concrete spacing blocks between piping and reinforcing iron. Copper pipe is recommended for use in soil. If galvanized pipe is used, it should be placed in clean sand backfill and exposed threads should be painted. In concrete, black iron or galvanized iron pipe can be used. Copper pipe is also recommended. All portions of the piping should be encased in concrete and should not be in contact with the soil at any point. Isolate any piping in the ground from piping encased in concrete by use of an insulating bushing or coupling.
Salt water lines	Soil corrosion, long-line currents from different soil strata, stray currents, dissimilar metals, and concentration cells.	Steel and cast iron.	Use coatings and clean sand backfill. Apply cathodic protection. (On cast iron, bond across pipe joints with bond wire.) Drain stray current, if present. Avoid dissimilar-metal couples.
Sheet piling, H-piling, and reinforced rods	Soil corrosion, different types of soil, differential moisture and oxygen in soil, stray currents, dissimilar metals.	Steel.	Use protective coatings and paints. Apply cathodic protection. Avoid the use of dissimilar metals. Use insulating sections. Drain stray current, if present.
Tower Footings	Soil corrosion, different types of soil, and stray and galvanic currents.	Galvanized steel.	Set tower footings in concrete. Tower footings may be given a bituminous coating before being set in concrete. The tower structure should then be grounded by the use of magnesium or zinc anodes. Anode grounds should be made where the soil resistivity is low. Apply cathodic protection to tower footings in soils of low resistivity. Avoid connecting towers together that are in different types of soil, by use of ground wire. The ground wire can be sectionalized at boundaries of different types of soil.
Underground fuel oil tanks and piping, other buried tanks, and avgas storage and distribution systems	Corrosive soils, concentration cells, stray and long-line current, bacteria, and dissimilar metals.	Steel, brass or bronze valves, and copper piping.	Select a backfill with good drainage, clean sand, if possible. Avoid dissimilar metal couples. Drain stray current, if present. Apply cathodic protection. The hazards of certain products such as avgas and other volatile or combustible liquids require that every precaution be taken to assure complete protection and to reduce hazards to personnel.

Figure 6.11 (continued)

125

Marine Structures — Corrosion

Structure	Major Causes of Corrosion	Materials of Construction	Corrosion-Mitigation Methods
Barges and other floating structures.	Atmospheric and sea water corrosion, splash zone, chemical pollution, differential oxygen content, and dissimilar metals.	Steel, bronze, and brass fittings.	Paint structures. Apply cathodic protection in conjunction with approved paint system to submerged portion of structure. Avoid dissimilar metals.
Drydocks, caisson gates, and lock gates.	Atmospheric corrosion, chemical pollution, splash and submerged zones, differential oxygen content, and dissimilar metals.	Steel, cast iron, and lead.	Paint structures. Apply cathodic protection to submerged portions of structures. Avoid the use of dissimilar metals without proper precautions.
Salt water intake lines, flumes, and intake screens.	Atmospheric corrosion, chemical pollution, splash and submerged zones, differential oxygen content and dissimilar metals.	Steel, cast iron, bronze.	Galvanize all steel and cast iron. Paint structures. Avoid dissimilar metals. Apply cathodic protection.
Seadrome lighting and harbor installations.	Atmospheric corrosion, chemical pollution, splash and submerged zones, differential oxygen content, and dissimilar metals.	Steel.	Paint structures. Avoid dissimilar-metal couples. Apply cathodic protection to submerged portions of structures.
Sheet-piling bulkheads.	Soil corrosion, different types of water in estuaries, chemical pollution, differential oxygen content, splash zone, atmospheric corrosion, and stray current.	Steel or steel with concrete capping.	Paint piling in the atmospheric and splash zones. Apply cathodic protection to portions of the structure below water surface. Use a bituminous coating on the portion of piling encased in concrete capping when capping is in contact with water or soil. Bond all sheet piling to negative bus of cathodic protection system. Drain current, if present.
Ships (active).	Atmospheric corrosion, splash and submerged zones, differential oxygen content, dissimilar metals and turbulence.	Steel, bronze, and brass.	Paint structures. Apply cathodic protection to submerged portion of ship. Avoid dissimilar-metal couples.
Ships (inactive).	Atmospheric corrosion, pollution, splash and submerged zones, differential oxygen content, dissimilar metals.	Steel, bronze, and brass.	Paint structures. Avoid use of dissimilar-metal couples. Apply cathodic protection to submerged portion of ship.
Steel piers (cellular).	Soil corrosion, different types of waters in estuaries, chemical pollution, differential oxygen content, splash zone, atmospheric corrosion, and stray current.	Steel or steel with concrete capping.	Paint piers in the splash and atmospheric zones. Apply cathodic protection to structures below water. Use a bituminous coating on portion of piling encased in concrete capping when capping is in contact with water or soil. Bond all sheet piling to the negative bus to assure complete drainage of cathodic protection currents. Drain stray current, if present.
Steel piers (H-piling).	Soil corrosion, different types of waters in estuaries, chemical pollution, differential oxygen content, splash zone, atmospheric corrosion, and stray current.	Steel or steel with concrete capping.	Paint piling in the atmospheric zone. Use a corrosion-resistant jacket in the splash zone that will not create any serious galvanic couple with the steel piling. Apply cathodic protection to the piling below water surface. Drain stray current, if present. Use a bituminous coating on the portion of piling encased in concrete capping when capping is in contact with water or soil. Bond all piling to the negative bus of the cathodic protection system.

Figure 6.12

Relative to building housekeeping, paint coatings provide smooth, non-absorptive surfaces that are easily washed and kept free of dirt. The coating of rough or porous areas seals out dust and grease that would otherwise be difficult to remove. Also, paint coatings will reveal built-up foreign substances, thereby indicating that better housekeeping practices are needed. Some paints delay the spread of fire in a system confining it to its origin. Use of fire retardant paints is restricted to appreciable areas of highly combustible surfaces and for other select uses. Do not consider fire retardant paints as substitutes for conventional paints. Include intumescent paint for fireproofing where appropriate.

Inspection and Testing

It is of utmost importance, when planning a major project involving painting, that the correct paint be purchased for the specific type of application. But the planning should not stop here. It is important that the paint be sampled and tested on the surface for which it is to be used along with planning an effective job inspection program. Painting requires continuous inspection for best results. The inspector should be trained in this area. The initial inspection should be to insure that all surfaces have been prepared correctly. This should be followed up by an inspection of the materials to see that they are those specified or selected for the job. The job should be inspected during application to determine that proper procedures are being used, that film thickness is as specified, and that the applied paint is of the correct color and appearance, and is uniform without sap or runs. Therefore, the following steps should be used as a checklist:

1. Inspect the surface to be painted.
2. Study the environment to which the coating will be exposed.
3. Select an appropriate coating.
4. Effective surface preparation.
5. Select a method of application.
6. Analyze coats to be incurred in the system selected.

	Marine Structures — Corrosion (continued)		
Structure	Major Causes of Corrosion	Materials of Construction	Corrosion-Mitigation Methods
Steel piling.	Soil corrosion, different types of soil, different types of waters in estuaries, chemical pollution, differential oxygen content, splash zone, atmospheric corrosion, and stray current.	Steel or steel with concrete capping.	Use a corrosion-resistant jacket in the splash zone that will not create any serious galvanic couple with the piling. Paint piling in the atmospheric zone. Use a bituminous coating on portion of piling encased in concrete capping when capping is in contact with water or soil. Apply cathodic protection under water and under ground portion of piling. Drain stray current, if present.
Texas towers and other stationary, cylindrical piling.	Atmospheric corrosion, pollution, splash and submerged zones, differential oxygen content, and dissimilar metals.	Steel.	Paint structures. Protect legs of towers in the splash and tidal zones with a corrosion-resistant protective jacket that will not create any serious galvanic couple with the structure. Avoid dissimilar metal couples. Apply cathodic protection to submerged portions of structures.

Figure 6.12 (continued)

Frequency of Painting

The principal objective of painting is to prevent deterioration of the material at a minimum cost per square foot per year. Of course, the real economic concern is the cost per square foot divided by the years served between repainting. One procedure frequently used for providing protection has been to completely repaint after the original has failed. This failure results in an unsightly surface, expensive preparation before repainting, and possible deterioration of structural members. Another procedure is to completely repaint by applying two and even three coats at arbitrary intervals. This may be too late in cases where deterioration has already taken place, and unnecessary in others. Extensive surface preparation will be required in the first case and film thickness will eventually become excessive in the latter case, leading to early failure by cracking and peeling. The most practical method of protection is a continuous program of inspection and painting as necessary.

Some suggested frequencies are as noted below for exterior painting:

- White paint: once every four years.
- Lightly tinted paints: once every four to five years.
- Deeply tinted paints: once every five to six years.
- Dark colors: once every five to eight years.

Note that in environments that are more destructive to the coatings, the frequencies noted above may have to be decreased. Therefore, another general rule would be to repaint at the first sign of chalking on the south side of a structure or general checking (50 percent or more of the area).

Effective Record Keeping

The main reason for gathering data and maintaining records is to use the information to arrive at meaningful evaluations of completed jobs and to determine the methods of improvement or cost reduction on future jobs. These records can be used for the purposes listed below.

- To determine the effectiveness of a particular paint system on different surfaces or in varying environments.
- To compare different paint systems under similar conditions.
- To compare the use of different equipment for surface preparation or application.
- To determine manpower efficiency under varying conditions.
- As a basis for the use of better or lower-cost painting systems on planned jobs.
- As a basis for more efficient use of manpower and equipment on planned jobs.
- To determine frequency of spot painting and repainting.

The proper use of records tightens guidelines and replaces haphazard action and guesswork with meaningful direction and planning. The ultimate result is systematic programs of preventive maintenance inspections and painting which provide an economical and efficient means of protecting facilities.

Deterioration

Paints are not indestructible. Even properly selected protective coatings, properly applied on well-prepared surfaces, will gradually deteriorate and eventually fail. The rate of deterioration under such conditions, however, is slower when proper painting operations are carried out. Painters and personnel responsible for maintenance painting must be familiar with the signs of various stages of deterioration in order to establish an effective and efficient system of inspection and programmed painting.

Exterior Deterioration: Paints which are exposed to the external environment normally proceed through two stages of deterioration: generally, a change in appearance followed by gradual degradation. If repainting is not done in time, disintegration of the paint takes place, ultimately followed by deterioration of the substrate (natural surface).

Change in Appearance: The first stage of deterioration shows up as a change in appearance of the coating, with no significant effect on its protective qualities. This change in appearance may result in any one or a combination of the following, depending on the type and color of the paint used and the conditions of exposure:

- **Soiling:** exterior coating normally gathers dirt and become increasingly soiled. Soiling increases as the paint becomes flat and somewhat rough. Dirt pickup is greater with softer paints, such as linseed oil paint, and is more visible on white or light colored paints. Soiling is less evident on paints which chalk rapidly since the dirt is readily washed off with the chalk during rainstorms.
- **Color changes:** many colors, especially the brighter ones, fade and turn duller with time. Tinted paints become paler. Enamel and latex paints fade less rapidly than the softer linseed oil paints. Whites, especially those based on linseed oil, will yellow in areas protected from the sunlight.
- **Flatting:** glossy paints lose their gloss and eventually turn flat with age. This is a sign of initial breakdown. Loss of gloss is soon followed by chalking. Enamels flatten (and chalk) less rapidly than the softer linseed oil paints.

Degradation: The second stage in normal deterioration occurs after continued exposure. The coating begins to break down, first at the surface, then, unless repainted, gradually through the coating and down through the substrate. There are two types of degradation that take place: chalking and checking or cracking. The degree of either depends upon the type of paint and the severity of exposure.

- **Chalking:** Chalking is the result of weathering of the paint at the surface of the coating. The paint is broken down by sunlight and other destructive influences, leaving behind loose powdery pigment, which can easily be rubbed off with the finger. Chalking takes place more rapidly with softer paints and is most rapid in areas exposed to large amounts of sunshine. Chalking can be an asset, especially in white paints, since it is a self-cleaning process. Chalked paints are also generally easier to repaint, since the underlying paint is in good condition and generally little surface preparation is needed.
- **Checking or Cracking:** Checking or cracking are breaks in the paint film, which are formed as the paint becomes hard and brittle. Temperature changes cause the overlying paint to expand and contract. As the paint becomes hard, it gradually loses the ability to expand without breaking to some extent. Checking is characterized by tiny breaks which appear only in the upper coat or coats of the paint film without penetrating to the substrate. The pattern usually resembles a crow's foot. Cracking appears in the form of larger and longer breaks which extend through to the substrate. Cracking will generally take place to a greater extent on wood than on other substrates because of its grain. Checking and cracking are aggravated by excessively thick coatings because of their reduced elasticity.

Disintegration: As the coating degrades, it finally reaches the point of disintegration. The following types of disintegration are possible:

- **Erosion:** as chalking continues, the entire coating wears away or erodes and becomes thinner. Eventually, it becomes too thin to hide the substrate.
- **Crumbling:** if the cracks are relatively small, the moisture penetrating through the coating will cause small pieces of the coating to lose adhesion and fall off the substrate.
- **Flaking and Peeling:** moisture goes into the cracks and loosens relatively large areas of the coating. The paint then curls slightly, exposing more of the substrate and finally flakes off. Peeling is an aggravated form of flaking, in which large strips of paint can be easily removed.

Complete Deterioration: When large areas of substrate become exposed, the coating has reached the point of complete deterioration and is in a state of neglect. Such surfaces require extensive and difficult preparation before repainting. All the old coating may have to be removed to be sure that it does not create problems by continuing to lose adhesion, taking the new coating with it. Furthermore, complete priming of the exposed substrate will be required, thus adding to the cost and time. Continued neglect may also lead to deterioration of the structure, resulting in expensive repairs in addition to painting costs.

Abnormal Deterioration: When coatings deteriorate sooner than anticipated or in an abnormal manner, the cause of such premature failure must be found and corrected before repainting. The cause may be due to the substrate, the structure, the environment, or the paint.

Substrate Peculiarities

Many substrates have individual characteristics which can present abnormal problems if not corrected or eliminated before or during painting operations. The following is a discussion of a few of these for the major types of materials:

Wood: Wood is a natural product which can vary in its chemical and physical properties in a number of respects. The major parameters which will affect or cause a premature deterioration of the paint are:

- Color.
- Uniformity of grain.
- Inherent soluble dyes in the wood fiber.
- Difference in the softness of the spring and summer wood in any one tree.
- Types of grain.
- Knots.
- Resinous materials inherent in the fibers of the wood.
- Whether or not the wood has been treated.

Metal: All metals are much more uniform than wood. They expand uniformly in all directions so that adhesion loss because of uneven stresses is much less of a problem than with wood. Some types of materials, listed below, do present certain problems which can cause abnormal deterioration.

- **Iron and Steel:** Both iron and steel rust when they are exposed or unprotected. If moisture penetrates through breaks in the film, rust is formed. This rust will increase in area, lifting the edge of the film around the break, then creep underneath the film and continue to progress. This will result in an earlier deterioration of the paint.
- **Galvanized Steel:** The inherent chemical properties in the galvanizing process often create problems of adhesion of applied coating after exposure. If the incorrect paint system is used, extreme flaking and peeling may take place after a year or so of exposure. Allow galvanized steel to weather.
- **Non-ferrous Metals:** Aluminum and copper. Although both of these metals corrode, their corrosion products do not tend to expand as rapidly as do iron and steel. They should be cleaned thoroughly to obtain optimum adhesion.

Concrete, Stucco, Masonry, Plaster: All of these materials are hard, they all contain lime and other soluble salts, and they are relatively porous.

Abnormal Environment: Unusual conditions of exposure are a major cause of abnormal deterioration.

Effects of Humidity or Moisture: Moisture may cause abnormal deterioration two ways: it may cause flatting or formation of mildew (fungi).

- **Flatting:** If moisture in the form of fog, rain, or dew lies on the surface of newly applied paint before it is thoroughly dry, it may cause a spotty appearance or complete loss of gloss of the paint. This is primarily an appearance problem which makes a new paint job look inferior.
- **Mildew:** Paint coating exposed to humid climates or in warm, damp rooms may be attacked by fungi. Mildew will grow and become quite unsightly, and eventually will accelerate degradation of the coating. In its early stages, it looks like dirt, but it cannot be washed off as easily. The presence of mildew may be determined by using household bleach; it will bleach mildew, whereas it has no effect on dirt. Hard-drying paint, such as enamels or paints containing zinc oxide, are more resistant to mildew. Use specially formulated moisture-resistant and mildew-resistant paint for these exposures.
- **Atmospheric Contamination:** Smoke, fumes and sulfur-containing gases can adversely affect paint coating, causing discoloration and rapid failure. They will also accelerate chalking and erosion.
- **Rapid Temperature Changes**
- **Wind Velocities:** Excessive wind velocities may cause the paint to dry too rapidly, make application difficult, and thus lead to a poor painting procedure. Do not paint when the wind velocity is above 15 miles per hour.

Incompatibility of Paints: The entire coating system must be compatible through each layer, from the surface of the raw material to the surface of the last coat of paint, to achieve optimum durability. Any incompatibility between substrate and paint systems, and between coats, will reduce adhesion and accelerate deterioration associated with the loss of adhesion (lifting, peeling, etc.). Some of the results of using incompatible paints are listed below:

- **Lifting:** This is an effect produced by the solvent in the applied paint acting as a paint remover on the coating underneath. Lifting is more likely to occur when a second or third coat is applied over an undercoat which has not dried hard enough.
- **Alligatoring:** Alligatoring describes the pattern in a coating which looks like the hide of an alligator. It is caused by uneven expansion and contraction of a relatively hard top coat over a relatively soft or slippery undercoat. Alligatoring may be caused by applying an enamel over an oil primer, painting over an oil primer, painting over bituminous paint, or painting over grease or wax.
- **Crawling:** Crawling occurs when the new coating fails to wet and forms a continuous film over the preceding coat. Examples are applying latex paints over high-gloss enamel or applying paints on concrete or masonry treated with a silicone water repellent.
- **Intercoat Peeling:** The loss of adhesion caused by the use of incompatible paint may not be obvious until after a period of time may elapsed. Then the stresses in the hardening film will cause the two coatings to separate and the top coat will then flake and peel.

Along with the items described above, improper painting operations may also cause premature failure of the paint. To minimize this problem, consider the following points in planning the painting operation:

- Always use recommended coating systems.
- Be sure the surface is properly prepared and the painting conditions are within specified limits (temperature, humidity, etc.).
- Follow application directions exactly.

Stains on Paint

Refer to Figure 6.13 for the removal of common stains on painted surfaces.

Surface Preparation

The first, and often most important, activity in the painting process is surface preparation. The best quality paint will not perform effectively if applied to a poorly prepared surface. The selection of surface preparation methods is dependent upon the following factors:

- Nature of substrate.
- Condition of surface to be painted.
- Type of exposure.

Removal of Stains on Paint		
Type of Stain	Symptoms	Cure
Black or brown stains	Sudden appearance.	Will bleach out when the cause is eliminated.
	Occur in industrial environments from hydrogen sulfide or similar gases.	Wash with hydrogen peroxide or commercial bleach.
		Repaint with fume-proof paint.
Black or gray fibrous stains (mildew).	Mold or discoloration of the surface.	Scrub surface with a strong detergent in warm water, then treat surface with a household bleach.
		Apply paints containing chemicals which kill the mold.
Black or gray round or irregular stains.	Similar to mildew.	Wash with a warm detergent solution, then repaint the surface.
Brown stains over knots.	Brown staining over knots followed by exudation of resinous materials.	Remove any resin and paint from the surface.
		Seal the knot.
		Repaint.
Brown or green stains adjacent to metals.	Stain beginning below a metal due to rust runoff from water.	Finish and seal metals so that they are not exposed to further weathering.
		Clean or paint stained areas.
Irregular brownish stain.	Not associated with metallic objects.	Eliminate the source of moisture, bleach out the dye from the wood, and then repaint the surface.
	Appear on white or light-colored paints applied over red cedar or redwood.	
	Usually appears after heavy rains.	
Premature disappearance of film.	Rapid weathering without visible cracking or peeling.	Wire brush surface, followed by application of paint.
Uneven changes in color or gloss.	Color or gloss will be different from that of the rest of the surface.	Add another coat of paint.
	On masonry surfaces, the pattern will follow the mortar lines.	
	On cinder or concrete blocks, individual blocks may appear different.	

Figure 6.13

- Limitations, such as time, location, space, and the availability of equipment.
- Economic considerations.
- Type of paint to be applied.
- Safety factors.

Many surface contaminants reduce adhesion and cause blistering, peeling, flaking, and under-film rusting. Among these contaminants are dirt, grease, rust, rust scale, mill scale, chemicals, moisture, and effloresence. In addition, the following surface defects will affect adhesion adversely:

- Irregular weld areas.
- Metal burrs.
- Crevices
- Sharp edges.
- Irregular areas.
- Weld splatter.
- Weld flux.
- Knots.
- Splinters.
- Nail holes.
- Loose aggregates.
- Old paints.

Surface Cleaning

The various methods of mechanical cleaning are hand cleaning, power tool cleaning, blast cleaning, and flame cleaning. Each of these methods is described in the following sections.

Hand Cleaning: Hand cleaning will remove only loose or loosely adhering surface contaminants. These include rust scale, loose rust, mill scale, and loosely adhering paint. Hand cleaning is not to be considered an appropriate procedure for removing tight mill scale and all traces of rust. Hand cleaning cannot be expected to do more than remove major surface contamination. It is recommended for spot cleaning in areas where corrosion is not a serious factor and/or areas which are not accessible to power tools. For hand cleaning, the surface must be free of oil, grease, dirt, and chemicals. This can be accomplished best with the use of solvent cleaners. Then remove rust scale and heavy build-up of old coatings with impact tools such as chipping hammers, chisels, and scalers. Remove loose mill scale and non-adhering paint with wire brushes and scrapers. Finish up by sanding, especially on woodwork. All work must be done to avoid deep marking or scratches on the surface by the tools used. Start painting as soon as possible after cleaning.

Power Tool Cleaning: This method provides faster and more adequate surface preparation than hand tool methods. Power tools are used for removing small amounts of tightly adhering contaminants which hand tools cannot remove, but they remain uneconomical and time consuming as compared with blasting for large area removal of tight mill scale, rust, or old coatings. Chipping hammers are used for removing tight rust, mill scale, and heavy paint coats. Rotary and needle scalers are used for removing rust, mill scale, and old paint from large metallic and masonry areas. Wire brushes are used for removing loose mill scale, old paint, weld flux, slag, and dirt deposits. Grinders and sanders are used for complete removal of old paint, rust, or mill scale on small surfaces and for smoothing rough surfaces. It is important that care be taken not to cut too deeply into the surface, since this may result in burrs that are difficult to protect satisfactorily. Care must be taken when using wire brushes to avoid polishing metal surfaces and thus prevent adequate adhesion of the subsequent coatings. Power tool cleaning is to be preceded by solvent or chemical treatment and painting must be started and completed as soon as possible after power cleaning.

Blast Cleaning: Blast cleaning abrades and cleans through the high velocity impact of sand, metal shot, metal grit or other abrasive particles on the surface. Blast cleaning is most often used on metal structures in the field, but may also be used, with caution, on masonry surfaces. It is the most thorough of all mechanical operations. However, blast cleaning requires trained personnel often not available in one's own maintenance organization. Therefore, outside consultants must be called in to perform the work.

The abrasive discharged may be either wet or dry. The wet system differs from the dry in that water or a solution of water and a rust inhibitor are incorporated with the blast abrasive. All blasted metal surfaces require the prime coat to be started and completed on the same day to prevent new rust from forming, since such blast-cleaned surfaces are subject to rapid rusting if not immediately coated. The grit used in the blasting process must be of a size sufficient to remove contamination without working the surface in excess. Over-working creates extreme peaks and valleys on the surface which require an additional build-up of the paint film for adequate protection. There are two general methods of dry blasting: *conventional* and *vacuum*.

Conventional blast cleaning is a term used to designate the usual method of field blasting in which no effort is made to alleviate the dust hazard or reclaim the blast abrasive. The procedure precludes the need for special rinsing as required for wet blasting, but requires that health precautions be taken to protect the operator and other personnel in the area. After blasting, the surface must be brushed, vacuumed or air cleaned to remove residues or trapped grit. Vacuum blasting, also known as dry honing, allows practically no dust to escape and contaminate the atmosphere. The type of blasting is less efficient than conventional blasting methods on highly irregular surfaces because of the poor vacuum on such surfaces. However, vacuum blasting is very efficient and economical for cleaning repetitive, small scale surfaces in a shop. The process results in considerable savings in abrasive costs and also reduces the dust and health hazard.

Wet blasting reduces to a minimum the dust associated with blasting, but it is not suitable for all types of work. Steel structures, containing a large number of ledges formed by upturned angles and horizontal girders, present a large amount of troublesome cleanup work if the wet method of blasting is used. Wet sand and other blast residues trapped on these edges are more difficult to remove than dry materials. Also, a sufficient amount of sludge adheres to wet blasted surfaces necessitating removal by rinsing, brushing, or compressed air. Moreover, there is a tendency for wet blasted surfaces to rust even though an inhibitor is present in the mixing and rinsing water. The blasted surface must be thoroughly dry before coatings are applied. Refer to Figure 6.14 for rates of blast cleaning.

Flame Cleaning: Flame cleaning, to be used on metal surfaces only, is a method of passing high-velocity flames over a metal surface. Oil and grease must be removed prior to flame cleaning, both for safety and adequacy of preparation. Wire brushing normally follows flame cleaning to remove loose matter. Extreme safety precautions must be taken when utilizing this method of cleaning.

Rates of Blast Cleaning*

Method	S.F. Per Hour
White-Metal	100
Near-White	175
Commercial	370
Brush-Off	870

*Approximate cleaning rates using 100 PSI.

Figure 6.14

Chemical and Solvent Treatment

Solvent cleaning is a procedure used for removing oil, grease, dirt, paint, stripper residues, and other foreign matter from the surfaces prior to painting or mechanical treatment. The simplest procedure is to first remove the soil, cement splatter, and other dry materials with a wire brush. The surface is then scrubbed with brushes or rags saturated with solvents. Clean rags are used for rinsing and wiping dry. More effective methods include immersing the work in a solvent or spraying a solvent over the surface. It is essential that several clean solvent rinses be applied to the surface.

Alkali Cleaning: Alkali cleaning is more efficient, less costly, and less hazardous than solvent cleaning, but is more difficult to carry out. The alkali solution can be applied by brushing, scrubbing, or spraying, or by immersion of the surface into soak tanks. Thorough water rinses are absolutely necessary to remove the soapy residue as well as all traces of alkali to avoid chemical reaction with the applied paint. Otherwise, cleaning may do more harm than good. The rinse water should be hot and preferably applied under pressure. Do not use alkali cleaners on aluminum or stainless steel.

Steam or Hot Water Under Pressure: Steam cleaning is another method of cleaning. A detergent should be included for added effectiveness. The steam or hot water removes oil and grease by liquifying them under the high temperature, then emulsifying them and diluting them with water. Wire brushing may be necessary to augment the cleaning process by removing any residues.

Acid Cleaning: Acid cleaning is used for cleaning iron, steel, concrete, and masonry by treating the surface with an acid solution. Iron and steel surfaces are treated with solutions of phosphoric acid containing small amounts of solvent, detergent, and wetting agent. This chemical should not be used on aluminum or stainless steel. Such cleaning effectively removes oil, grease, dirt, and other foreign materials. This method also removes tight rust and faintly etches the surface to assure better adhesion of applied coatings. There are many commercial acids available and the manufacturer's recommendations should be followed in their application. There are basically four methods of application: wash off, wipe off, hot dip, and spray.

The wash-off method involves, first, an application of a cleaner, followed by an interval of time to let it set. Then thoroughly rinse and allow surface to dry before painting. The wipe-off process is used when rinsing is impractical and involves the application of the cleaner, a time allowed for it to act, wiping off the surface with clean, damp cloths and, finally, drying with clean, dry cloths prior to painting. The hot-dip method involves immersion of the work in a hot cleaner, a rinse in hot or cold water after the surface is sufficiently cleaned, a second rinse in a weak cleaner solution, and drying time before painting. The last method (spray) involves the same steps as the wash-off method, but requires pressurized spraying.

Concrete and masonry surfaces are washed with a 5–10 percent muriatic acid to remove efflorescence and laitance, to clean the surface, to remove any glaze and to etch the surface. To acid clean these surfaces, thoroughly wet the surface with clean water, then scrub it with a 5 percent solution of muriatic acid using a stiff fiber brush.

In extreme cases, up to 10 percent muriatic acid solution may be used, and may be allowed to remain on the surface for up to five minutes before scrubbing. Work should be done on small areas not greater than four square feet in size. Immediately after the surface is scrubbed, wash the acid solution completely from the surface by thoroughly sponging or rinsing with clean water.

Alternative Methods: Another method is of cleaning is to use a paint and varnish remover. Paint and varnish removers generally are used for small areas only. Solvent-type removers or solvent mixtures are selected according to the type and condition of the old finish as well as the nature of the materials. Removers are available in many types and consistencies and should be selected according to the need. It is absolutely essential that any residue left be removed from the surface prior to painting to prevent loss of adhesion of the applied coating. In such instances, follow the manufacturer's directions or use mineral spirits to remove any wax residue.

Chemical methods of surface cleaning are usually more suited to paint shop application while mechanical methods are generally more practical in field work. On the basis of overall effectiveness and efficiency, chemical cleaning is superior to mechanical methods with the exception of blast cleaning. The coating and environment determine the degree of surface cleaning required. Refer to Figures 6.15 and 6.16 for a list of materials and cleaning procedures for iron, steel, and non-ferrous metals.

Pretreatments

Pretreatments are applied on metal surfaces after cleaning to improve the adhesion and to improve the effectiveness of the applied paint. The major types of pretreatment solutions are as follows:

- Hot phosphate treatments.
- Cold phosphate treatments.
- Wash primers.
- Chemical conversion coatings.

Hot and Cold Phosphate Treatments: Hot phosphate treatments greatly increase the bond and adhesion of applied paints while reducing underfilm corrosion. These types of solutions require careful controlled conditions and cleaned surfaces. When painting, it usually is necessary to apply thicker paint coats over the heavier phosphate coatings, if a gloss finish is desired, because the heavier phosphate coatings absorb considerably more paint. Hot phosphate treatments are excellent procedures to insure tight bonds between the surface and applied paint. The mechanics of

Methods for Surface Preparation of Ferrous Metals

Types of Cleaning	Methods of Cleaning
Flame cleaning	Alkyd and phenolic vehicle paints, baked enamels.
Iron phosphate	Baked enamels.
Pickling (phosphoric acid)	Natural-drying-oil and resin vehicle paints.
Pickling (sulfuric acid)	Vinyl, alkyd and phenolic vehicle paints, baked enamels.
Rust removers	All types of paint used for maintenance and on-site painting.
Sand blasting and grit blasting	Coal-tar enamels and vinyl vehicle paints; baked enamels; also used for alkyd and phenolic vehicle paints.
Solvent cleaning	Oil-base paints.
Wire brushing	Natural-drying-oil and resin vehicle paints.

Figure 6.15

hot phosphate treatment limit its use to the paint shop. There are many types of hot phosphates available and manufacturers should be consulted for the types to be used relative to the exposure or environmental conditions. Cold phosphate treatments are similar to hot phosphate treatments, but are composed of different chemical bases. Adhesion properties are not quite as good as for hot phosphate treatments. The procedures used for cold phosphating are adaptable to field use in large or small structures.

Wash Treatment: Wash primers are actually a form of cold phosphating. They perform more efficiently than the standard cold phosphating treatments and are generally replacing them for field use. Wash primers develop to extremely good adhesion to blast clean or pickled steel and provide a sound base for top coating. They are also used to promote adhesion of coatings to surfaces generally considered difficult to paint, such as galvanized or stainless steel or aluminum.

Conditioners
Latex paints do not adhere well to chalky masonry surfaces. To overcome this problem, an oil-base conditioner is applied to the chalky surface before the latex paint is applied. The entire surface should be wire brushed by hand or power tool, then dusted to remove all loose particles and chalk residue. The conditioner is then finished on freely to assure effective penetration and allowed to dry. This surface conditioner is not intended for use as a finished coat.

Sealers
Sealers are used on bare wood to prevent resin bleeding through applied paint coatings. Since the sealer is not intended for use as a priming coat, it should be used only when necessary and applied only over the affected area.

Methods of Cleaning Nonferrous Metals for Painting		
Metal	Method of Cleaning	Type of Paints Used With This Method of Cleaning
Aluminum	Let weather for a month; wipe with turpentine or mineral spirits.	Never use a lead base paint directly on an aluminum surface.
Copper, bronze, and their alloys	Remove loose corrosion by sanding, and wipe with turpentine or mineral spirits.	Use paints recommended by the manufacturers of copper, bronze, and their alloys.
Galvanized iron	Three methods: 1) wipe surface with turpentine or mineral spirits; 2) let weather for six months or until it turns dull; 3) apply dilute solution of hydrochloric, phosphoric, or acetic acids, and then rinse.	Zinc-dust types of priming paints.
Terneplate	Wipe with turpentine or mineral spirits; do not allow to rust before painting.	Use paints recommended by the manufacturers of terneplate.

Figure 6.16

Fillers

Fillers are used on porous wood, concrete, and masonry to fill the pores to provide a smoother finished coat. Wood fillers are used on open grain hardwoods. See Figure 6.17 for a summary of when fillers are needed. If filling is necessary, it is done after any staining operations. Stain should be allowed to dry for 24 hours before filler is applied. If stain is not used, the filler is applied directly to the bare wood. To apply, first thin the filler with mineral spirits to a creamy consistency, then liberally brush it across the grain, followed by light brushing along the grain. Allow to stand five to ten minutes until most of the thinner has evaporated, at which time the finish will have lost its glossy appearance. Before it has a chance to set and harden, wipe the filler off across the grain using burlap or other coarse cloth, rubbing the filler into the pores of the wood while removing the excess. Finish by stroking along the grain with clean rags. It is essential that all filler be removed. Allow the filler to dry for 24 hours before applying finished coats.

Characteristics of Wood				
	Soft Wood	Hard Wood		
Wood Species	Closed Grain	Open Grain	Closed Grain	Notes on Finishing
Ash		X		Requires filler.
Alder	X			Stains well.
Aspen			X	Paints well.
Basswood			X	Paints well.
Beech			X	Paints poorly; varnishes well.
Cedar	X			Paints and varnishes well.
Cherry	X			Paints and varnishes well.
Chestnut		X		Requires filler; paints poorly.
Cottonwood			X	Paints well.
Cypress			X	Paints and varnishes well.
Elm		X		Requires filler; paints poorly.
Fir	X			Paints poorly.
Gum			X	Varnishes well.
Hemlock	X			Paints fairly well.
Hickory		X		Requires filler.
Mahogany		X		Requires filler.
Maple		X		Varnishes well.
Oak		X		Requires filler.
Pine	X			Variable depending on grain.
Redwood	X			Paints well.
Teak		X		Requires filler.
Walnut		X		Requires filler.

Figure 6.17

Masonry fillers are intended for use on coarse surfaces, such as rough concrete, concrete block, stucco, and other masonry surfaces. There are two types of filler coatings; a solvent thin material, which cannot be applied to damp masonry, and a water thin cementitious or latex emulsion coating, which can be applied to damp masonry. Prior to the use of fillers, the surfaces must be cleaned, whether they are old, new, or have been previously painted. On previously painted surfaces, only the solvent-type filler should be used. Residual form oil or other organic material on the surface should be removed by sandblasting, or strong detergent treatment, including proper rinsing or, if time permits, allowing natural weathering to remove the oils. On previously painted surfaces, all loose, powdery or flaky material, dirt, and old paint may be removed effectively by controlled sandblasting. Allow the filler to dry for 24 hours before painting. When applying a cement water or emulsion-type filler, a brush with relatively short bristles is needed to work the filler into the voids. Solvent thin filler may be applied by brush, roller, or spray; however, a brush is preferred to most effectively work the material into the pores. Before this filler becomes tacky, excess material is removed with a rubber squeegee.

Repair of Surfaces

All surfaces must be in good condition before painting. Repair or replace degraded wood, concrete, masonry, stucco, metal, plaster, and wallboard. Remove and replace all loose mortar and brickwork. Replace broken windows and loose putty or glazing compound. Securely fasten or replace loose gutter hangers and downspout bands. Fill all cracks, crevices, and joints with caulking compound or sealants. Drive all exposed nailheads below the surface. Patch all cracks or holes in wood, masonry, and plaster. The final surface should be smooth, with no openings or defects of any kind. These preparatory procedures eliminate the major areas for the entrance of moisture, which leads to blistering and peeling of the paint. If peeling is a problem, refer to Figure 6.18 for an inspection checklist.

Caulking Compounds: Caulking compounds are oil-resin or latex-based. They are used to fill joints of wood, metal, or masonry, and joints with very limited movement. Sealants, on the other hand, are elastomeric, rubber-like compounds. They are intended for use in expansion or other movable joints. Caulking compounds are available in two grades: a gun grade and a knife grade. The gun grade is most popular since it is easier to use and faster in application because it employs the use of a caulking gun, whereas the knife grade must be applied by hand using a putty knife. The gun grade is applied in two forms; that is, in bulk and factory-filled cartridges. The cartridge type fits directly into a caulking gun and is preferred for convenience of use. Caulking compounds tend to dry on the surface, but remain soft and tacky within the crevice. The applied caulking should be painted each time the surrounding area is painted to help extend its life. The life expectancy of caulking compounds is from 5 to 15 years.

Sealants: A more durable type of calking compound. They have better adhesion to the walls of the crevices, have better extensibility so they do not pull away when the walls contract in cold weather and they remain flexible for much longer periods of time. Although they are considerably more expensive than calking compounds, their longer life (15 to 30 years) is often well worth the difference in cost.

Putty: Used to fill nail holes, cracks, and imperfections in wood surfaces. It is supplied in bulk form and is applied with a putty knife. Putty is not flexible and should not be used for joints or crevices. It dries to a harder surface than caulking compound.

Glazing Compounds: Used on both interior and exterior wood and metal window sash, either as beading or bedding or face glazing. They are used to cushion glass in metal or wood frames and are not intended to hold or keep the glass in position. Glazing compounds are set firmly, and have some limited flexibility. They are more flexible than putty. They tend to harden upon exposure, with life expectancy estimated to be about ten years if they are properly applied. Painting over glazing compounds will extend their useful life.

Application of Caulking, Glazing, Putty and Sealants: All surfaces must be cleaned and dried to obtain good adhesion. Remove all oil, grease, soot, dirt or loose paint. Also remove old materials, especially when they show signs of wear. Be sure the crevice openings are large enough to allow an adequate amount of materials to be inserted. Prime the surface, when recommended by the manufacturer, in accordance with his directions. If the opening is deep, first insert backup materials, such as oakum, foamed plastic or rubber, fiberglass, or fiberboard. Follow manufacturer's instructions for application of the caulking, glazing, putty, and sealant type compounds.

Checklist for Causes of Paint Peeling

Below Grade Area
—— Water coming through floors or walls.

Exterior Drainage
—— Downspout transmit water away from the building.
—— Water does not stand around building after a rain.
—— Drains in area ways are adequate and operating.

Flashing
—— Around windows, doors, re-entrant angles, and all openings.

Gutters
—— Broken, rusty, or stopped.
—— Holes or breaks in down spouts.
—— Formation of ice barriers in winter.

Roof
—— Broken or loose shingles.
—— Holes or worn places.

Siding
—— Split, broken or warped.
—— Unprotected ends.

Ventilation
—— If moisture condenses on windows in winter, make sure:
    1. There is an adequate vapor barrier in the wall.
    2. Sources of moisture are ventilated.
—— Attic ventilation is adequate and operating.
—— Crawl spaces are adequately ventilated or covered.

Figure 6.18

Other types of patching materials available for cracks, holes and crevices in masonry, plaster, wallboard, and wood are as follows:

- **Patching plaster:** Use for repairing large area in plaster.
- **Spackle:** Use to fill cracks and small holes in plaster and wallboard.
- **Joint cement:** Use to seal the joints between wallboards and repair large cracks.
- **Portland cement grout:** Use to repair cracks in concrete masonry.
- **Plastic wood:** Use to fill gouges and nail holes in wood.

The reader is referred to Figure 6.19 for a summary of the various types of treatments available for surface preparation before painting.

Preparation Procedures

One of the most essential parts of any paint job is proper surface preparation. Paint will not adhere well, provide the required protection, or have the desired appearance, unless the surface has been properly prepared. The following sections describe the preparation procedures for the various exterior materials.

Wood:

1. Sand all new surfaces smooth with # 1/0 or # 2/0 sandpaper; use # 1 sandpaper first on rough spots. Use fine steel wool on rounded or irregular surfaces.
2. Remove loose paint by scraping, sanding carefully with coarse sandpaper, such as # 2, or by use of paint and varnish remover. Then sand as for new surfaces.

Treatment of Various Substrates					
		Metal		Concrete	Plaster
	Wood	Steel	Other	Masonry	Wallboard
Mechanical					
Hand Cleaning	S	S	S	S	S
Power Tool Cleaning	S*	S		S	
Flame Cleaning		S			
Blast Cleaning					
Brush-Off		S	S	S	
All Other		S			
Chemical and Solvent					
Solvent Cleaning	S	S	S		
Alkali Cleaning		S		S	
Steam Cleaning		S		S	
Acid Cleaning		S		S	
Pickling		S			
Pretreatments					
Hot Phosphate		S			
Cold Phosphate		S			
Wash Primers		S	S		
Conditioners, Sealers, and Fillers					
Conditioners				S	
Sealers	S				
Fillers	S			S	

S = Satisfactory for use as indicated.
* = Sanding only.

Figure 6.19

3. Thoroughly clean all surfaces.
4. Seal all knots or resin areas with knot sealers and allow to dry at least two hours before priming.
5. Fill all gouges, dents, and small openings with plastic wood.
6. Use putty for larger openings (after priming). Allow to dry hard, then sand lightly with # 2/0 sandpaper before painting.
7. Wood must be dry before any painting is done.

Concrete and Masonry: New concrete may be rough or very smooth, depending upon the finishing method used; concrete blocks are usually rough.

1. If a smooth finish is desired, fill rough surfaces with a masonry filler.
2. Treat smooth trowelled surfaces with a solution of 3 percent zinc chloride plus 2 percent phosphoric acid to etch the surface and allow to dry.
3. Remove old paint by mechanical treatments.
4. Patch all cracks, openings and broken areas.
5. Thoroughly clean all surfaces.
6. If cement water paint is to be used, dampen surfaces within one hour of painting. Use a garden hose or portable pressure-tank sprayer, adjusted to a fine spray. Do not use a brush. The surface must be moist, not dripping wet, when the paint is applied.

Metal: Unpainted iron and steel may have loose rust or mill scale, both of which will affect paint adhesion. Rust may be present under loose paint on old work.

1. Remove all loose and scaling paint, rust, mill scale, dirt, oil, and grease.
2. Sand edges of painted surfaces surrounding areas cleaned to substrate.
3. Wire brush or sand metal and spot prime with an appropriate primer.

Paint Materials

A knowledge of the types of materials used for painting is useful in determining their capabilities and limitations. There are sound reasons for the existence of each coating specification and these become more apparent with some insight into the composition of the various types specified. It is not the function of this book to delve into the chemistry of paints. Information presented in this section is to aid the reader in determining which product should be used for the job.

When trying to determine which type of coating to use, one must consider the following parameters:

- Area to be painted.
- Surface to be painted.
- Substrate to be painted.
- Reason why painting is to be done.
- Conditions under which painting is to be done.

Relative to any one type of coating material, the maintenance manager must evaluate the following characteristics:

- Adhesion ability.
- Abrasion resistance.
- Workability.
- Stain resistance.
- Chemical resistance.
- Fire resistance.

The manager must also obtain the following information about each product being considered in order to make a thorough economic analysis:

- Recommended surface preparation.
- Coverage per gallon.
- Film thickness per coat.
- Total cost of applying the coating.
- Ease of application.
- Drying and curing time.
- Ease of repairing damaged areas.
- Estimates of life.

The first step in selecting a coating is to eliminate those that require most surface preparation. Choosing the correct coating is hard since there are literally hundreds of manufacturers, each one producing many types and grades of paint. To present every type of paint available would take up a volume in itself. Therefore, tables are presented in Figures 6.20, 6.21, 6.22, 6.23, 6.24, and 6.25 to help the reader select a general category of paint for specific purposes.

Mixing: The paint should be mixed properly and completely prior to beginning the painting process. See Figure 6.26 for proper mixing procedures.

Painting of Nonferrous Building Materials

Material	Preparation of Surface	Pretreatment	Priming Paints	Finish Paints
Aluminum	Solvent cleaner, inhibited caustic, acid pickle, steam clean; or non-metallic abrasive cleaning.	Anodic (chemical or electrolytic), or wash primer or phosphate or proprietary chemical treatment.	Zinc chromate priming paint.	Any paints compatible with service environment and priming paint.
Asbestos board	Clean grease and dirt with solvents, detergents, and soap.	Seal surface with one coat of boiled linseed oil; allow to dry.	Exterior house paint.	Exterior house paint.
Asphalt	Brush or wipe off loose materials.	None required.	Asphalt sealer such as shellac, resins soluble only in alcohol, styrene butadiene emulsion paint, polyvinyl acetate emulsion.	As desired, but should not dissolve sealer.
Brass and bronze	Solvent clean, sand or wirebrush; pickle or sandblast.	None required.	Oil or alkyd paint.	Any paints compatible with service environment and priming paint.
Cement, concrete, masonry	Surface should be roughened by weather or abrasion. Remove grease or oil. Surface must be dry.	Water solution of 3% phosphoric acid and 2% zinc chloride; allow to dry completely. Brush off any loose or powdery deposits. Omit if aged.	Oil paints, styrene-butadiene copolymer paints, polyvinyl acetate emulsion paints, resin-emulsion paints, chlorinated rubber paints.	Same as priming paint.
Cement, concrete masonry	Clean	None required.	Cement-water paints.	Cement-water paints.
Coal tar	Clean to remove grease, oil, dirt, loose material.	Same as asphalt.	Same as asphalt.	Same as asphalt.

Figure 6.20

Painting of Nonferrous Building Materials (continued)

Material	Preparation of Surface	Pretreatment	Priming Paints	Finish Paints
Copper	Solvent clean, sand or wire brush; pickle or sandblast.	None required.	Aluminum paint often used or use any oil or alkyd paint.	Any paints compatible with service environment and priming paint.
Lead	Solvent clean; light wire brushing or sanding.	None required.	Use any oil or alkyd paint.	Any paints compatible with service environment and priming paint.
Masonry	Same as cement.	Same as cement.	Same as cement.	Same as cement.
Plaster	Allow to age; emulsion paints or resin emulsion paints only should be used on new or damp plaster.	None required.	Use primer sealer, styrene butadiene emulsion paint, or polyvinyl acetate emulsion paint, or non-penetrating oil or alkyd paints.	Same as for priming paints except only one coat of primer-sealer is used with semi-gloss or gloss olioresinous finish paints.
Terneplate	Solvent clean to remove oil film; wire brush rusty spots.	None required.	Any good oil or alkyd paint.	Any paints compatible with service environment and priming paint.
Tin, tin plate	Solvent clean to remove oil film.	Wash primer advantageous for adhesion.	Use any good oil or alkyd paint.	Any paints compatible with service environment and priming paint.
Wood, exterior	Scrape, sand or brush; remove old thick layers of cracked paint.	None required.	New or exposed wood: prime with wood primer or add up to one quart of raw linseed oil per gallon of house paint. Previously painted: use finish paint as primer.	White or tinted, chalking or non-chalking, oil paints, Titanium dioxide, white lead, zinc oxide, linseed oil paint, colored oil paints.

Figure 6.20 (continued)

Types of Paint Used for Painting Metal

Type of Paint	Surface Preparation and Treatment	Priming Coat	Intermediate Coat	Finish Coat
Alkyd vehicle	"Commercial" blast cleaning, pickling flame cleaning. No pretreatment needed.	Red-lead alkyd varnish primer.	Same as priming coat.	Aluminum alkyd, black alkyd, white or tinted alkyd paint.
Coal-tar	Blast cleaning to white metal; surface to be cleaned and prime coat immediately applied.	Coal-tar enamel primer applied hot.	None.	Coal-tar enamel applied hot.
Oil-base vehicle	Solvent cleaning, wire brushing; no pretreatment necessary.	Red-lead oil-base primer.	Same as priming coat, except tinted.	Aluminum varnish or black, white or tinted oil-base paint.
Phenolic vehicle	"Commercial" blast cleaning, pickling, flame cleaning; no pretreatment necessary.	Red-lead mixed pigment phenolic varnish primer.	None.	Aluminum phenolic, white or tinted phenolic paint.
Vinyl vehicle	"Commercial" blast cleaning, pickling; after cleaning surface to be penetrated with basic zinc chromate vinyl butyral washcoat.	Vinyl red-lead primer.	Same as priming coat except tinted.	Aluminum vinyl, black vinyl, or vinyl-alkyd paint in white black, red, yellow or orange.

Figure 6.21

Types of Paint Used on Wood

Type of Paint	Surface Preparation	Priming Coat	Intermediate Coat	Finish Coat
Alkyd vehicle	Sandpapering, puttying, filler.	Alkyd primer for three-coat work can be white; for two-coat work, tint to same color as finish coat.	Same as finish coat.	Alkyd vehicle, glossy, semi-glossy or matt; white or tinted.
Lacquers	Sanding, filler, puttying and staining if necessary.	Lacquer primer, clear or tinted.	Same as finish coat.	Lacquer, clear or tinted.
Oil-base vehicle	Puttying, filler for exterior and interior work; sandpapering for interior only.	Oil-base primer for three-coat work can be white; for two-coat work, tint to same color as finish coat.	Same as oil-base coat.	Oil-base vehicle for exterior, white or tinted.
Shingle stains	Puttying for exterior.			Oil-base stains.
Varnish	Puttying, filler, sanding and staining, if necessary.	Clear varnish primer.	Same as finish coat.	Clear varnish; glossy, semi-glossy or matt for interior; high-grade spar varnish for exterior.
Varnish for floors	Sanding, filler.	Lacquer floor sealer for floors that have been oil treated.	Varnish sealer.	Varnish sealer.

Figure 6.22

145

Paints Used for Concrete, Plaster, Masonry, and Other Materials

Type of Paint	Surface Preparation	Priming Coat	Intermediate Coat	Finish Coat
Asphalt-base paint (emulsion type)	Surface free of foreign material and not dripping wet.	Asphalt-base paint (emulsion type).		Asphalt-base paint, emulsion type.
Cement-base paint	Surface clear and fairly smooth; efflorescence removed and surface thoroughly wetted.	Same as finish coat; slightly moisten priming coat before applying finish coat.		Cement-base paint, white or tinted.
Coal-tar paint	Surface free of foreign matter and thoroughly dry.	Coal-tar paint.	Coal-tar paint.	Coal-tar paint.
Concrete-floor paint (varnish base or rubber base)	Clean smooth surface etched with 10% to 15% muriatic acid solution.	Varnish or rubber base primer.		Varnish-base or rubber-base cement floor paint.
Oil-vehicle paint	Thoroughly clean, dry, smooth surface; efflorescence removed & pretreatment with chemicals to inhibit alkaline reaction, using 2% zinc chloride and 3% phosphoric acid water solution.	Oil-base primer coat tinted same color as finish coat.		Oil-base paint, white or tinted.
Varnish-base paint	Surface clean, smooth and dry.	Varnish-base primer sealer.		Varnish-base paint, white or tinted.
Water-base paint (latex casein and resin-emulsion)	Surface clean, smooth and dry.	Same as finish coat.		Water-base paint, white or tinted.

Figure 6.23

Comparison of Paint (Binders) Principal Properties

	Alkyd	Bituminous	Cement	Epoxy	Latex	Oil	Phenolic	Rubber	Moisture Curing Urethane	Vinyl	Zinc
Ready for Use	Yes	Yes	No	No	Yes	Yes	Yes	Yes	Yes	Yes	Yes
Brushability	A	A	A	A	+	+	A	A	A	–	A
Odor	+	–	+	–	+	A	A	A	–	–	–
Cure — Normal Temp.	A	A	A	A	+	–	A	+	+	+	A
Low Temp.	A	A	A	–	–	–	A	+	+	+	A
Film Build/Coat	A	A	+	+	A	+	A	A	+	–	A
Safety	A	A	+	–	+	A	A	A	–	–	A
Use on Wood	A	+	–	A	A	A	A	–	A	–	–
Use on Fresh Concrete	–	A	+	+	+	–	–	+	A	+	–
Use on Metal	+	A	–	+	–	+	+	A	A	+	+
Corrosive Service	A	A	–	+	–	–	A	A	A	+	+
Gloss — Choice	+	A	–	+	–	A	+	+	A	A	A
Retention	+	A	X	–	X	–	+	A	A	+	A
Color — Initial	+	A	A	A	+	A	–	+	+	+	A
Retention	+	A	–	A	+	A	–	A	–	+	A
Hardness	A	A	+	+	A	–	+	+	+	A	A
Adhesion	A	A	–	+	A	+	A	A	+	–	A
Flexibility	A	A	–	+	+	+	A	A	+	+	A
Resistance to: Abrasion	A	A	A	+	A	–	+	A	+	+	+
Water	A	+	A	A	A	A	+	+	+	+	A
Acid	A	A	–	A	A	–	+	+	+	+	–
Alkali	A	A	+	+	A	–	A	+	+	+	–
Strong Solvent	–	A	+	+	A	–	A	–	+	A	–
Heat	A	–	A	A	A	A	A	A	A	–	A
Moisture Permeability	Mod	–	High	Low	High	Mod	Low	Low	Low	Low	A
Min. Application Temp.											
Degree F.	40	40	50	50	40	40	50	40	40	40	40
Drying Time											
To Touch (Hours)	2–12	1–12	4–8	½–3	¼–2	8	½–2	¼–4	½–12	¼–1	¼–4
To Recoat (Hours)	3–48	12–72	12–48	3–96	1–12	8–48	8–48	¼–24	3–48	½–8	¼–24
Final Coat (Days)	4–14	3–15	2–4	1–10	5–30	12–14	7–12	7	2–14	3–7	1–7

+ = Among the best for this property

– = Among the poorest for this property

A = Average

X = Not applicable

Figure 6.24

Paints Available for Typical Exterior Building Exposure

	Oil or Oil-Alkyd	Cement Powder Paint	Exterior Clear Finish	Aluminum Paint	Wood Stain	Roof Coating	Trim Paint	Porch and Deck Paint	Primer or Undercoater	Metal Primer	Latex	Water Repellent Preservative
Masonry												
Asbestos Cement	X*								X		X	
Brick	X*	X		X					X		X	X
Cement and Cinder Block	X*	X		X					X		X	
Concrete/Masonry Porches and Floors								X			X	
Coal Tar Felt Roof						X						
Stucco	X*	X		X					X		X	
Metal												
Aluminum Windows	X*			X			X*			X	X*	
Copper Surfaces			X									
Galvanized Surfaces	X*			X*			X*			X	X*	
Iron Surfaces	X*			X*			X*			X	X*	
Metal Roof	X*									X	X*	
Metal Siding	X*			X*			X*			X	X*	
Steel Windows	X*			X*			X*			X	X*	
Wood												
Clapboard	X*			X					X		X*	
Natural Wood Siding and Trim			X		X							
Shutters and Other Trim	X*						X*		X		X*	
Wood Frame Windows	X*			X			X*		X		X*	
Wood Porch Floor								X				
Wood Shingle Roof					X							X

X = Applicable.

* = A primer, sealer, or fill coat is necessary before the finishing coat is applied.

Figure 6.25

It is important that manufacturer's recommendations be followed relative to:
- Temperature of the surface to be painted.
- The temperature of the paint at the time of application.
- Wind velocity and humidity of the environment.
- Recommended spreading rate.
- Drying time between coats.
- Surface preparation prior to and after painting.

Other Types of Coatings

Besides paints, there are other types of coatings available for special uses. Surface preparation, type of material, exposure conditions, method of applying the coating and follow-up inspections are just as critical as for painting. Some of the other major types of coatings are listed below:

- Abrasion resistant finishes.
- Anti-sweat coatings designed to prevent condensation of water on cold surfaces, such as water pipes.
- Bituminous finishes.
- Fire-retardant paints.
- Heat-resistant coatings.
- Mildew-resistant (fungi-resistant) paints.
- Non-slip coatings.
- Odorless paints.
- Textured finishes.

The reader should note that it is important, when using these special coatings, to follow the manufacturer's recommendations explicitly.

Painting Variables

Parameters which will affect the painting operation are as follows:
- Supervision.
- The condition of the surface.
- Weather.
- Environmental conditions.
- Painting materials.

Only persons having experience and training in painting should be utilized as supervision. All jobs must be checked frequently to be sure that personnel are preparing surfaces properly: are using paints properly with regard to mixing and conditioning; that application is uniform and at proper film thickness: and that proper drying time is allowed between coats.

Recommended Mixing Procedures for Different Paints	
Type of Paint	**Mixing Equipment**
Enamel, semigloss or flat oil paints	Manual, propeller or shaker
Extremely viscous finishes, such a coal tar paints	Propeller
Clean finishes, such a varnishes and lacquers	Manual
Two-component systems	Manual, propeller or shaker
Two-package metallic paints, such as aluminum paint	Propeller
Water-based latex paints	Manual, propeller or shaker

Figure 6.26

Conditioning Surfaces: The conditioning of the surface to be painted is of utmost importance, particularly in maintenance painting where protective requirements are most stringent and premature film failure is costly. Refer to the previous section on surface preparation for more information.

Environmental Conditions and Weather: Surface, ambient, and material temperatures, moisture conditions, and wind velocity affect the application, drying time, and adhesion of paints. Low temperatures thicken paints, making them difficult to apply, prevent smooth leveling and retard drying. High temperatures cause the opposite to take place, that is the viscosity is low so that the paint spreads too far, resulting in inadequate film thickness. Also, paints tend to sag and set up too rapidly. This results in lap marks and may even result in wrinkling and loss of adhesion under extreme circumstances. Humidity, dampness, and frost retard drying time and are common causes of blistering and poor adhesion.

Paint Application

The most common methods of applying paint are by brush, roller, and spray. Brushing is the slowest method, rolling is much faster, and spraying is usually the fastest of all. The choice of method is based on many additional factors such as environment, type of surface, type of coating to be applied, appearance of finish desired, and skill of personnel involved in the operation.

General surroundings may prohibit the use of spray application. Typical of these are parking lots and open storage areas. Adjacent areas not to be coated must be masked when spraying is performed. This results in loss of time and, if extensive, may offset the advantage of the rapidity of spraying operations.

Roller coating is most efficient on large flat surfaces. Corners, edges, and odd shapes, however, must be brushed. Spraying is most suitable for large surfaces, but can also be used for round or irregular shapes. Brushing is ideal for small surfaces or for cutting in corners and edges.

Rapid drying, lacquer type products (vinyls), should be sprayed. Application of such products by brush or roller is extremely difficult, especially in warm weather or outdoors on breezy days.

Coatings applied by brush may leave marks in the dried film; rolling leaves a stippled effect, while spraying yields the smoothest finish, if done properly.

Personnel require the least amount of training to use rollers and the most training to use spray equipment. The degree of training and experience of personnel will influence the selection of the application method.

To obtain optimum performance from a coating, there are certain basic application procedures which must be followed, regardless of the type of equipment selected for applying the paint:

1. Cleaned, pretreated surfaces must be first coated within the specific time limits established. It is essential that surface and ambient temperatures are between 50 and 90 degrees Fahrenheit for water-thin coatings and 45 to 95 degrees Fahrenheit for other coatings, unless the manufacturer specifies otherwise. The paint material should be maintained at a temperature of between 65 and 85 degrees Fahrenheit at all times. Paint is not to be applied when the temperature is expected to drop to freezing before the paint has dried.
2. Wind velocity should be below 15 miles per hour.
3. The relative humidity should be below 80 percent.
4. Masonry surfaces that are damp (not wet) may be painted with latex or cementitious paints. Otherwise, the surface must be completely dried before painting.
5. Paints should be applied at recommended spreading rates.
6. When successive coats of the same paint are used, each coat should be tinted differently to aid in determining proper application and to assure complete coverage.

7. Sufficient time must be allowed for each coat to dry thoroughly before top coating. Allow the final coat to dry for as long as is practicable before service is resumed. Refer to Figure 6.27 for a summary of application methods for specific types of surfaces.

Brushes: Brushes, like any other tools, must be of the best quality and must be maintained in perfect working condition at all times. Brushes are identified, first, by the type of bristle used. Bristles are either natural, synthetic, or mixed. The various types available and the characteristics of each are listed below.

- **Chinese Hog Bristle:** The finest of the natural bristles because of their length, durability, and resiliency.
- **Hog Bristles:** The unique character of this bristle is that it has end forks at the end of the bristle which reach out like tree branches. This permits more paint to be carried on the brush and leaves finer brush marks on the applied coating.
- **Horsehair Bristles:** Used in cheap brushes and are very unsatisfactory.
- **Mixture of Bristles:** The quality of this type of brush depends upon the percentage of each type of bristle used.
- **Nylon:** Readily available and replacing many of the natural-type bristles. Superior to horsehair. Suitable for applying lacquer, shellac, and some other coatings that would soften or dissolve the natural-type bristles. It is especially recommended for use with latex paints.

Brushes are further identified by types; that is, the variety of shapes and sizes that are required for specific painting jobs. The following are the various types available:

- **Wall brushes:** Flat, square-edged brushes ranging in widths from three to six inches and used for painting large, continuous surfaces, either interior or exterior.

Types of Paint and Methods of Application

Type of Paint	How Applied
Cement paints	Brush
Caulking compounds	Caulking gun
Cement floor paint	Brush, spraying
Clear finishes	Brush, spraying
Clear finishes for floors	Brush
Dampproofing	Brush, spraying, trowel, mopping
Enamels	Brush
Filler	Brush, hand or small trowel
Flat paints	Brush, spraying
House paints	Brush, spraying
Masonry paints	Brush, spraying
Metallic paints	Brush, spraying
Multicolored paints	Spraying, brush
Porch and trim paints	Brush
Putty	Hand or small trowel
Rust-inhibiting paints	Brush
Shingle stains	Brush, cloth, sponge, spraying
Stains	Brush, cloth, sponge
Waterproofing	Brush, spraying, trowel, mopping
Water repellants	Spraying, brush

Figure 6.27

- **Sash and trim brushes:** Available in four shapes: flat square-spaced edged, flat angle-spaced edged, round, and oval. Range in width is from one and one-half inches to three inches or a diameter of from one-half to two inches. Use for painting window frames, narrow boards, and interior and exterior trim surfaces.
- **Enameling and varnish brushes:** Flat, square-space edged or chisel-edged brushes available in widths from two to three inches.

Avoid a brush that is too small or too large. The latter is particularly important. A large area job does not necessarily go faster with an over-sized brush. If the brush size is out of balance for the type of painting being done, the user tends to apply the coating at an uneven rate, general workmanship tends to decline, and the applicator actually tires faster because of the extra output required per stroke. Synthetic (man-made) brushes are ready to use when received. The performance of natural bristle brushes is very much improved by a previous 48-hour soak in linseed oil, followed by thorough cleaning in mineral spirits.

Application: Dip the brush into the paint up to one-half of the bristle length, then withdraw and tap against the inside of the bucket to remove excess paint. Hold the brush at an angle of 45 degrees to the work. Make several light strokes in the area to be painted; this will transfer much of the paint to the surface. Spread the paint evenly and uniformly. Do not bear down on the brush. When one section of surface is painted, adjacent areas should be painted so that the brush strokes are completed by sweeping the brush into the wet edge of the paint previously applied to the first section. This eliminates lap marks and provides a more even coating. Finally, cross-brush lightly to smooth the painted surface and eliminate brush or sag marks. When using fast-drying finishes, the paint should be applied, spread rapidly, and then allowed to dry undisturbed. To go back over such paint will only cause a piling up of the coating.

Start major work on the topmost area first, such as the ceiling of a room, then work downward, painting walls down to the floor. Begin painting at a corner or other logical vertical division. Cover only that area which can easily be reached without moving the ladder. Work downward, painting progressive sections to the floor or ground level, then start at the top of the adjacent area and work down again. Paint trim, doors, windows, or similar areas after walls or ceilings or other major surfaces are completed. A major possible exception would be painting jobs where scaffolding is required. In such cases, paint both major surfaces and any trim in the section as the scaffolding is moved along from area to area. When painting clapboards, moldings, or other surfaces with narrow or indented edges, and other similar areas, paint these first and then paint the surrounding continuous surfaces. Corners and edges are always painted so that the stroke is completed by sweeping off the corner or edge. Avoid poking the brush in the corners or crevices. Instead, use the edge of the brush and twist it slightly, if necessary, to cover the rough surfaces.

Roller Application

A paint roller consists of a cylindrical sleeve or cover which slips on a rotatable cage to which a handle is attached. The cover may be one and one-half to two and one-fourth inches inside diameter and usually three inches, four inches, seven inches, and nine inches in length. Special rollers are available in lengths from one and one-half inches to 18 inches. Proper roller application depends upon the selection of the specific fabric and the thickness of the fabric used based on the type of paint used and the smoothness or roughness of the surface to be painted. The fabrics for rollers generally used are as listed below:

- **Lambs wool:** Available in nap lengths up to one and one-fourth inches. Recommended for synthetic finishes for application on semi-smooth and rough surfaces. Mats badly in water and is not recommended for water paints.

- **Mohair:** Solvent resistant and is supplied in 3/16 and 1/4-inch nap lengths. Recommended for synthetic enamels and use on smooth surfaces. Can be used with water paints.
- **Dynel:** Has excellent resistance to water. Best for application of conventional water paints and solvent paints, except those which contain strong solvents, such as ketones. Available in all nap lengths from 1/4 inch to 1-1/4 inches.
- **Dacron:** Best suited for exterior oil or latex paints. Available in nap lengths from 5/16 inch to 1/2 inch.
- **Rayon:** Not recommended, mats badly in water.

Refer to Figure 6.28 for a roller selection guide.

Application Process: Pour the premixed paint into the tray to about half the depth of the tray. Immerse the roller completely, then roll it back and forth along the ramp to fill the cover completely and remove any excess paint. The first load of paint on a roller should be worked out on newspaper to remove entrapped air from the roller cover. Always trim around ceilings, mouldings, etc., before rolling the major wall or ceiling surfaces. Then roll as close as possible to maintain the same texture. Always roll paint onto the surface, working from the dry area into the just painted area. Never roll completely in the same or one direction. Do not roll too fast and avoid spinning the roller at the ends of the stroke. Always feather out final strokes to pick up excess paint on the surface. This is accomplished by rolling the final stroke out with minimal pressure.

Roller Selection Guide

Type of Paint	Smooth (1)	Semi-smooth (2)	Rough (3)
Aluminum	C	A	A
Enamel or Semigloss (Alkyd)	A or B	A	
Enamel Undercoat	A or B	A	
Epoxy Coatings	B or D	D	D
Exterior House Paint:			
Latex for Wood	C	A	
Latex for Masonry	A	A	A
Oil or Alkyd — Wood	C	A	
Oil or Alkyd — Masonry	A	A	A
Floor Enamel — All Types	A or B	A	
Interior Wall Paint:			
Alkyd of Oil	A	A or D	A
Latex	A	A	A
Masonry Sealer	B	A or D	A or D
Metal Primers	A	A or D	
Varnish — All Types	A or B		

Roller Cover Key	Nap Length (Inches)		
A — Dynel (Modified Acrylic)	1/4–3/8	3/8–3/4	1–1¼
B — Mohair	3/16–1/4		
C — Dacron Polyester	1/4–3/8	1/2	
D — Lambswool Pelt	1/4–3/8	1/2–3/4	1–1¼

(1) Smooth Surface: Hardwood, smooth metal, smooth plaster, drywall, etc.
(2) Semi-smooth Surface: Sand finished plaster and drywall, light stucco, blasted metal, semi-smooth masonry.
(3) Rough Surface: Concrete or cinder block, brick, heavy stucco, wire fence.

Figure 6.28

153

Spray

Most materials that can be brushed or rolled can be sprayed. Exceptions include very thick or stringy materials, some textured materials, and some rubber-based coatings. Control over the spray operation requires control over the following variables:

- **Viscosity of the paint:** Must be low enough to permit proper atomization, but high enough to apply without running or sagging. Generally, the trial-and-error approach is required.
- **Pot pressure:** Determines the amount of material forced through the nozzle. It is controlled at the air regulator or at the gun.
- **Atomizing pressures:** This is the air pressure supplied to the gun to atomize the material and produce a uniform wet film. Too much pressure will cause excessive overspray or a dry spray. Too little pressure produces a speckled or dimpled effect.
- **Air cap on spray gun:** This controls the amount and distribution of air mixed with the coating at the gun. The amount of air, and air pressure, controls atomization while distribution of the air determines the shape of the spray pattern.
- **Material nozzle:** The size of this opening controls the amount of material that can be passed through the gun.
- **Air and material controls on the gun:** These are for rough adjustment of amount delivered.

The viscosity of the paints should be according to manufacturer's instructions. Excessive thinning results in needless overspray, excessive runs and sags, and poor hiding of inferior surface protection. Use the lowest material and air pressures that result in a quality finish with good flow out. Material pressure is best adjusted starting at the point where a solid stream of paint will flow out about 24 inches from the gun with the atomization air turned off. When the material is heavy or viscous, when the fluid hose is extra long, or when a more rapid rate of application is required, it will be necessary to increase the material hose pressure. In this case, it will be necessary to also increase the air pressure, since it is important to maintain proper ratio of material pressure to atomizing air pressure. Refer to manufacturer's recommendations for air pressure. Use the minimum pressure necessary to reduce overspraying. As the width of the pattern is enlarged, increase the flow of the paint to maintain the same coverage over the wider area. Keep the spray gun six to ten inches from the surfaces being coated. Holding the gun too far away causes "dusting," in which the paint solvent evaporates in mid-air and the coating hits the surface in a nearly dry state. Tilting the gun causes the paint to be more heavily applied in one area than another in the spray pattern. Use a free-arm motion and feather out at the end of the stroke by pulling the gun trigger after beginning the stroke and releasing it before the stroke is completed. When spraying corners, stop one or two inches short of the corner. Then hold the gun so as to sweep up and down along the edges of the corner and hit both sides at the same time.

Paint Mitt Application

The paint mitt is a mitten made of lambskin with the wool exposed and lined to prevent paint leaking through to the user's hand. It is excellent for painting small pipes, railings, wrought iron, and similar surfaces.

Other Equipment

Besides the basic painting equipment discussed above, it is mandatory that the following items also be available.

- Paint brushes
- Dust brush
- Stiff bristle brush
- Wire brush
- Caulking gun
- Cans (for cleaning brushes with solvents)
- Drop cloths
- Emery cloth (for cleaning and polishing metal surfaces)
- Hammer

- Ladder
- Masking tape
- Mixing paddle
- Paint
- Paint bucket
- Paint scraper
- Paint strainer, wire mesh, or cheesecloth
- Paint tray (for painting with rollers)
- Patching plaster
- Putty or glazing compound
- Putty knife
- Rags
- Rollers
- Roller extension poles
- Sandpaper or production paper
- Spackling compound
- Steel wool
- Turpentine or other solvents
- Wire comb

Cleanup

It is absolutely essential that all the tools and equipment be cleaned thoroughly and immediately after use before the paint materials have a chance to set up. Remove as much paint as possible, then clean thoroughly with a compatible solvent. Clean two or three times in fresh solvent until no paint is noticeable. Then wipe clean and dry. After cleaning, wash all brushes with mild detergent and warm water; rinse in clean water, then twirl to get rid of excess water, comb bristles straight with a metal comb, and place in brush keepers or wrapping paper and allow to dry flat. Also wash cleaned rollers in a mild detergent and water. Rinse in clear water and twirl to get rid of excess. Then stand on end to dry. When dry, cover to keep clean. Spray equipment should be cleaned thoroughly. When clean, empty, wipe clean, and dry.

Labor Productivity

Figure 6.29 is an excerpt from a page from *Means Building Construction Cost Data*, 1988, showing average labor productivity. Expected average labor productivity, which includes preparation, would be as follows:

- Brushing: 1,000 square feet/8 hour/man
- Rolling: 1,000 square feet/8 hour/man
- Spraying: 4,000 square feet/8 hour/man

The reader should note that the productivity will vary, depending upon such items as:

- Weather.
- Capability of workers.
- Height off ground level.
- Type of surface and amount of prior preparation.
- Type of paint and method being used in application.

Coverage

See Figure 6.30 for information on paint thickness versus coverage. The reader should note that instrumentation is available which can measure the thickness of the paint. When the thickness is critical (such as when a minimum is required for protection) it should be specified in the technical specifications and measured using the applicable instrumentation.

Safety

Every painting activity exposes maintenance personnel to conditions and situations that represent potential danger to themselves and to others in the area. Not only are the personnel working with toxic and flammable materials, but they are also working in precarious places involving ladders, scaffolding, and rigging. All these present a potential hazard to the painters and thus all painting operations must be planned for and carried out safely.

Painting (Div. 099)

Item	Coat	One Gallon Covers			In 8 Hrs. Man Covers			Man Hours per 100 S.F.		
		Brush	Roller	Spray	Brush	Roller	Spray	Brush	Roller	Spray
Paint wood siding	prime	275 S.F.	250 S.F.	325 S.F.	1150 S.F.	1400 S.F.	4000 S.F.	.695	.571	.200
	others	300	275	325	1600	2200	4000	.500	.364	.200
Paint exterior trim	prime	450	—	—	650	—	—	1.230	—	—
	1st	525	—	—	700	—	—	1.143	—	—
	2nd	575	—	—	750	—	—	1.067	—	—
Paint shingle siding	prime	300	285	335	1050	1700	2800	.763	.470	.286
	others	400	375	425	1200	2000	3200	.667	.400	.250
Stain shingle siding	1st	200	190	220	1200	1400	3200	.667	.571	.250
	2nd	300	275	325	1300	1700	4000	.615	.471	.200
Paint brick masonry	prime	200	150	175	850	1700	4000	.941	.471	.200
	1st	300	250	320	1200	2200	4400	.364	.364	.182
	2nd	375	340	400	1300	2400	4400	.615	.333	.182
Paint interior plaster or drywall	prime	450	425	550	1600	2500	4000	.500	.320	.200
	others	500	475	550	1400	3000	4000	.571	.267	.200
Paint interior doors and windows	prime	450	—	—	1300	—	—	.333	—	—
	1st	475	—	—	1150	—	—	.696	—	—
	2nd	500	—	—	1000	—	—	.800	—	—

Figure 6.29

Expected Dry Film Thickness of Applied Paint

Coverage Rate (Square Foot Per Gallon)	Approximate Dry Film Thickness Finish Coats (Mils)
450	2.25
500	2.00
550	1.75
600	1.50
650	1.30
700	1.10
750	1.00

Figure 6.30

Moisture Control

The backfill around a foundation is usually less dense than the surrounding natural, undisturbed earth. Thus, the foundation area has a tendency to be a reservoir of excess surface and underground seepage water unless it is properly drained. Confined water builds up hydrostatic pressure against foundation surfaces and, given cracks, porosity, or voids in joints of foundation walls, will seep into the structure and cause wet walls and floors. If drainage is neglected for a long time, over-irrigation of the bearing soil may reduce its stability and lead to major dislocation of the foundation. Where the nature of subgrade soil resists, deters, or stops free drainage of subgrade water, waterproofing of usable areas below grade (such as basements, cellars, pools, pits, vaults, igloos, tunnels, utility trenches, and manholes) is usually necessary.

Causes and Control of Ground Water

The most common solution to ground water problems is improved drainage. The following are typical causes of ground water problems with suggested solutions.

High Water Table: Moisture in structures caused by a high water table can be drained away from a foundation by the installation of tile, porous wall, or perforated pipe drains surrounded by loose gravel fill. The drains should be laid so as to drain the water away from the footings and wall into a sump with a float-controlled electric pump.

Roof Drainage: Where roof drainage causes a foundation water problem, gutters and downspouts should be installed, preferably connected to a storm sewer.

Gutters that are improperly hung or allowed to become clogged will overflow and lose their effectiveness. Leaks in gutters should be repaired promptly. Splash blocks or tile drain should be installed in the absence of storm sewer connections to prevent pooling of water below downspouts.

Surface Drainage Toward Building: The drainage of surface water toward a building can be reversed by sloping the ground surface away from the foundation wall. Where that is not practical, ditching or installing tile drains will serve the same purpose. The general grade of crawl spaces should not be lower than the surrounding area, which should be graded to drain away from the building.

Utility Leaks: Breaks in water, sewage, heating, and drain pipes should be repaired as soon as discovered.

Moisture Problems

Moisture is the most prevalent cause of failure of exterior walls. Stains, paint deterioration, and rot are usual signs of moisture damage. Weathering leaks and cracks allow moisture to enter and collect behind exterior walls. Condensation within and behind walls is a less obvious, but equally damaging factor.

Condensation: The air contained in all buildings holds some moisture in vapor form. The vapor condenses on any surface that resists its flow and that is colder than the dewpoint temperature. Generally, condensation occurs when interior warm air, which holds more moisture than cool air, strikes the inner surface of an outside wall. Three means of condensation prevention in outer walls are available: venting the outer surface so the vapor passes through without condensing, reducing indoor humidity, and using insulation to keep vapor from reaching a cold surface.

Ventilation: Venting of attic and crawl spaces to allow escape of vapor to the outside air can be best accomplished by installing insect screen at the eave, louvered vents in gable or fabricated ridge vents. For attic areas it is desirable to provide free vent area of 1/300 of the ceiling area to be vented with 50 percent located at the eave. The ratio of free ventilation for crawl spaces is covered later in this chapter. Vents should be located and baffled to prevent the entrance of snow or rain; they should be screened against insects. These vents may be placed in loft spaces below flat roof decks, in gable walls, roofs, or eaves.

Vapor Barriers: A vapor barrier is the best means of preventing condensation. When existing buildings have not been provided with a vapor barrier and condensation problems exist, some relief may be obtained by use of vapor-resistant paint on interior wall faces; however, application of a vapor barrier on the inside of the warm face of the outside wall is preferable. There are a variety of vapor barrier materials such as impregnated paper, plastic, and metallic sheet. In most instances where a vapor barrier is to be applied within a wall, it is best to use an insulation board, batts, or similar material with an integral factory-applied vapor barrier. This accomplishes the purpose and incorporates insulation at the same time. The reader is referred to Figure 6.31 for methods of curing condensation problems.

Condensation Cures

Type and Location	Methods of Curing
Concealed Condensation	Reduce the humidity in the building.
	Add a vapor resistant paint coating to the interior walls and ceilings.
	Add a vapor barrier between ceiling and joists.
	Improve attic and crawl space ventilation.
Visible Condensation	
Attic Area	Install vapor barriers between joists under the insulation.
	Install or enlarge attic inlet and outlet ventilators.
	Decrease humidity in areas adjacent to the attic (dehumidifier).
Concrete Slabs	Keep windows closed.
	Use dehumidifier.
	Raise indoor temperature.
Crawl Spaces	Install vapor barrier over soil.
	Install vents.
Glass Surfaces	Use of storm windows.
	Replacing single glass with insulating glass.
	Use dehumidifier.

Figure 6.31

Insulation

Exterior wall insulation is usually installed at the time a building or structure is built, according to area climatic conditions and building occupancy. The economic feature of saving and equalizing heat and/or air conditioning is of primary importance. Insulation also serves a valuable purpose in moisture control, which prevents rot and fungus growth. In buildings and structures without adequate insulation, or where insulation has deteriorated or been displaced, it should be installed or replaced after a study of the best type and method of application for prevailing conditions. The fireproofing and vermin-proofing qualities of insulation should also be considered.

Types: Roll or batt blankets may be used where access to the space between studs allows their placement and fastening. Loose material (pellets or wool) may be used where areas are accessible from a limited opening only, such as around windows and doors and in wall utility compartments. Other types of insulation include rigid and semirigid composition board, which is generally used around concrete slabs and as sheathing under the siding. Utility batts may be used where no vapor barrier has been provided.

Replacement: Roll or batt insulation is most satisfactorily replaced when either the inside or outside surface of a building or structure is uncovered. Replacement of insulation should be considered when the outer face of a wall is resheathed or when an inner surface is replaced. When both wall faces are covered, it is necessary to pour loose insulation from the top or force it in by compressed air from some opening in the wall. In any case, care should be exercised to fill all small crevices and to place material into confined spaces and around piping and wiring. The vapor barrier side of insulation should face toward the warm side of the wall. In placing any type or kind of insulation, follow the manufacturer's recommendations for proper thickness, form, and fastening.

Crawl Spaces

Considerable deterioration extending from foundation to building superstructure can be caused by neglect of crawl spaces, especially in climates where it is necessary to enclose the space to maintain comfortable floor temperatures. Improperly ventilated crawl spaces contribute materially to rapid absorption of moisture into structural wood and other materials, and the spaces soon become a natural habitat for fungus growth and termites. Sills, joints, and subflooring may be affected by wood decay. Condensation may occur in the studding spaces above the floor level and cause paint failures.

Maintenance: Routine good housekeeping requires that crawl spaces be kept clean, clear, and accessible. An accumulation of rubbish in the space may provide a natural harbor for insects and rodents, as well as impede access and possibly interfere with drainage. Scrap wood is a clear invitation to termites. Crawl spaces should be checked periodically, and an adequate program of pest control carried out. Disorganized storing of any materials in crawl spaces should be prohibited. In some instances, 12 to 18 inches of water in crawl spaces have been reported. Children (also dogs, cats, and rats) have drowned in these spaces. Such areas breed mosquitoes, cause fungus growth, and weaken soil bearing under footings. All ventilation openings should be covered with suitable hardware cloth or copper screening to prevent entry of birds and rodents. Access doors to crawl spaces should be provided with a suitable padlock and kept closed.

Ventilation: Adequate ventilation of enclosed crawl spaces is necessary to prevent decay resulting from condensation. Standard guides for ventilation requirements are listed below:

1. For buildings of 5,000 square feet, or less, provide two square feet of air per 100 linear feet of building perimeter plus 1/3 of one percent of total crawl space ground area.
2. For buildings of more than 5,000 square feet, provide two square feet per 100 linear feet of building perimeter plus 1/4 of one percent of total crawl space ground area.
3. For interior foundation walls, provide one square foot per 25 feet of foundation, with cross-ventilation for all spaces. (The above requirements may be reduced in arid or semi-arid climates. In severely cold temperature zones, operable louvered vents should be closed during the cold season. Vents should always remain open in temperate and tropical zones with high humidity.

Crawl spaces may be ventilated by the installation of gratings or louvers of adequate size in the foundation walls. Small round louvers that can be installed by drilling with an expansion bit and tapping into place are commercially available. Vents through interior walls are equally important. Both types of vents should be installed to assure free air circulation through all parts of the underfloor space.

Below Grade Moisture in Concrete Walls

Moisture in a below-grade building area appears in three forms, leakage, seepage, and condensation. Refer to the chart in Figure 6.32 for the causes of moisture.

Use the following test to determine whether the moisture is caused from condensation or seepage: Paste a small pocket mirror to the foundation wall in the middle of one of the damp areas using a small dab of mastic adhesive. Allow to remain overnight. If the surface of the mirror is fogged over (the next day) or covered with dampness, condensation is the problem. If the surface of the mirror is clear, seepage is the problem. Water drops forming on cold water pipes is another sign of condensation problems.

Causes of Moisture in Concrete Walls

Form	Cause
Leakage: Actual flow of water that can be seen pouring in, usually through openings such as cracks in the foundation wall (especially during spring thaws, after heavy rains, or after long rain spells).	Excess amount of ground water pressure in the soil outside the wall.
Seepage: General dampness in certain sections, rather than trickle or flow; usually more pronounced along bottom of the wall or along joints where the floor and wall meet.	Excess amount of water pressure on the outside or capillary action drawing water through pores in the wall.
Condensation: General damp spots (like seepage); but differs in that the moisture is not coming through the wall.	Moist warm air strikes cool concrete wall, leaving dampness.

Figure 6.32

Moisture transferred through the floor by capillary action may result in a high humidity condition in a basement, or the accumulation of moisture or efflorescent salts behind surface finishes applied to walls or floors. To determine if water is moving by capillary action, place a rubber mat on a floor for a period of several days. If moisture collects under the mat, this indicates that water is saturating the concrete under the mat. The presence of salt deposits on floors or walls may indicate slow seepage or capillary movement of water.

Leakage and Seepage: Listed in Figure 6.33 are the possible causes and solutions of leakage and seepage.

Capillary Action: There is no simple method of preventing capillary moisture movement, short of a major effort to reduce the moisture content of the soil adjacent to the wall or floor. Deep drainage might be effective in granular soils, but would probably be ineffective in clay soils.

Causes and Solutions for Leakage in Concrete Walls

Cause	Solutions
Plugged-up perimeter drainage (sum well not filling during rain)	Rod out drainage pipe (contact specialist or rent a rodding tool)
Collapsed drainage pipe (sump well may not be filling up during rain dependent on location of break)	Locate break, dig out covering dirt and replace pipe and dirt.
Sump pump not operating (sump well filling during rain)	Repair or replace pump (see later unit)
Wall not waterproofed (wall leaking is more than one place)	Excavate around walls and apply asphalt or tar based waterproofing compound
Space between top of foundation and bottom of sill open (leak begins at top of foundation)	Fill in with mortar.
Inoperative roof drainage system	Clean out gutters of leaves and ice blocks and insure downspouts extend a minimum of ten feet from foundation wall and empty onto splash block.
Improper grading of soil (ponding of surface water near foundation)	Reshape ground (with caution)
Crack in wall	Repair crack. Refer to Chapter 3.

Figure 6.33

Waterproofing

To prepare the wall surface, excavate a trench, wide enough for working space, around the outside of foundation walls and to the bottom of the footings. Thoroughly clean the exposed surfaces with water, detergents, live steam or other available agents. Carefully examine the exposed surface for cracks, holes, fractures, and other damage that would permit moisture to enter or penetrate the walls. Repair concrete and masonry walls as outlined earlier in this chapter and in Chapter 3.

Bituminous Mopped-On Membrane: Bituminous mopped-on multiple membrane is one of the most positive methods of waterproofing foundation walls subject to considerable hydrostatic pressure. Membranes are of bituminous-saturated felts or fabrics. Roofing-grade hot coal-tar pitch or asphalt is used to mop on the membranes. Five piles of membrane are recommended for hydrostatic pressures to 12 feet high. All component materials used in any one application should be the products of one manufacturer, and his application instructions should be followed carefully. Membranes should be applied free from wrinkles and buckles, with each ply coated completely so as to separate one from another. After the last ply has been placed, the entire surface should be mopped with not less than 70 pounds of pitch or 60 pounds of asphalt, according to the membrane type, per 100 square feet of surface. The final coat should be protected from backfill by a layer of fiberboard embedded in the hot coat. Care should be taken not to rip, scar, tear, or cut the finished membrane during backfilling.

Nonbituminous Sheets: Membranes of pure plastics, thermoplastics, metal (usually copper or aluminum), and sheathed fibrous building papers are usually applied in more than one layer. They are embedded in portland cement, mortar, mastics, plastic cements, or bituminous coatings. All component materials should be products of one manufacturer and used in strict accordance with his instructions.

Cold Asphalt Application: A primer coat of thin cutback asphalt should be applied to porous masonry. A heavy coat of asphalt mastic should be troweled on the primer coat at least 1/8-inch thick at the rate of one gallon per 12 square feet. Where trowel application is not feasible, two coats of asphalt mastic should be brushed over the primer coat at the rate of one gallon per 80 square per coat. On concrete, a nonfabricated asphalt emulsion primer diluted with 15 percent cool water should be applied by brush or spray. At least two heavy coats of nonfabricated asphalt emulsion should be brushed or sprayed over the primer coat at the minimum rate of one to one and one-half gallons per 100 square feet per coat.

Hot Asphalt Application: A coat of thin penetrating asphalt primer should be mopped or sprayed over the surface to be treated. Coverage should be 200 to 500 square feet per gallon, depending on the porosity of the surface. Two coats of hot roofing-grade asphalt should be mopped or sprayed evenly over the entire surface at the rate of 25 pounds per 100 square feet per coat. The finished coat should be bright, glossy black; any dull areas should be recoated.

Cold Coal-Tar Pitch Application: A penetrating creosote-oil-base bitumen should be brushed or sprayed evenly over the surface at the approximate rate of two gallons per 100 square feet. Application should be repeated up to four coats until all pores and voids are filled. Each additional coat should be applied at cross angles to previous coats to provide full coverage. After the last primer coat has been absorbed and the surface is dry, a complete brush coat of cold bitumen (80 to 100 degrees Fahrenheit) should be applied at the approximate rate of one-half gallon per 100 square feet. Areas that are not bright, glossy black should be recoated. The finished coat should be hard and dry before backfilling is begun.

Hot Coal-Tar Pitch Application: A coat of creosote oil primer should be brushed evenly over the surface at the approximate rate of one gallon per 100 square feet. Two coats of hot coal-tar pitch should be mopped over the primer coat at the approximate rate of 25 pounds per 100 square feet per coat. Pitch should be heated until completely liquid, but not over 375 degrees Fahrenheit. Backfill should be placed and tamped immediately after the last coat has been applied.

Paints and Water Repellents for Dampproofing: Where porosity of masonry is the only cause of water entry, paints and water repellents may be used for dampproofing. For more serious moisture infiltration, an acrylic resin emulsion paint is recommended. This paint may be applied over previously unpainted concrete and masonry surfaces or surfaces covered with the same type of paint. For effective waterproofing, the paint must be applied with stiff brushes. Oil-type concrete and masonry paint may be applied to previously unpainted surfaces, without the use of a primer, and to surfaces coated with cement-water paint. Silicone water repellents having a minimum nonvolatile content of 5 percent silicon resin dispersed in cylene or toluene provide a "colorless" transparent finish and permit breathing. In recent years, many new resin preparations with excellent waterproofing qualities have become commercially available. Selection of the type to be used should be based on an investigation of the nature of the problem and actual conditions. Many of the preparations are difficult to apply because of temperature and humidity controls required during the setting and curing period. The manufacturer's recommended application procedures should be followed.

Metallic-Type Waterproofing: Where exterior walls are inaccessible and waterproofing must be applied to the interior face, the metallic method is recommended. The wall should first be exposed, cleaned, and roughened to provide a key for the waterproofing materials. Holes, cracks, and other soft or porous places should be cut back to solid material, cleaned, and pointed with mortar. All pipes, bolts, and similar construction should be caulked with lead wool and waterproof cement and made watertight. Surfaces should then be dampened with clean water and given a bonding coat composed of one part cement and one part metallic material mixed to a creamy consistency and applied with bristle brushes. The surface should be thoroughly finished to seal all pores rather than merely provide a surface veneer. This is followed by application of two coats of mortar composed of one part portland cement, three parts sand, and 25 pounds of metallic materials to each bag of cement. This first coat should be trowelled on and scratched when partially dry. The second coat, mixed to a heavy brushing consistency, should be brushed on carefully and floated with wood floats. Total thickness of all coats should be approximately 5/16 to 3/8 inch. After each coat has set, but not dried out, surfaces should be wetted down frequently over a period of at least 72 hours. Sufficient time should be allowed between coats to permit thorough oxidation of the material.

Masonry Joint Fillers

Leaks occurring in the joints between precast concrete panels or slabs are usually caused by expansion and contraction of the panels or slabs. Joints should be filled with okum or ethafoam bacher rod to within about 3/4 inch of the exterior. The remaining depth of each joint should be caulked.

Foundations

The foundation of a building or structure transfers the dead and live loads of the superstructure to soil that has enough bearing capacity to support the structure in a permanent, stable position. Footings are used under foundation components, such as columns and piers, to spread concentrated loads over enough soil area to bring unit pressures within allowable limits. Foundation design is determined not only by the weight of the superstructure, the occupancy or use of the building or structure, the load-bearing capacity of the soil, and the location of the ground water table at the site. The latter conditions may change and introduce maintenance and repair problems even in initially well designed foundations.

Inspection

Where accessible, foundations should be inspected at least annually and more often where climate, soil conditions, or changes in building occupancy or structural use present special problems. Evidence of incipient foundation failure may be found during routine inspection of other parts of the structure.

Foundation Displacement

Foundations should be checked regularly for proper elevation and alignment. Complete failure in foundations is rare; however, some settling or horizontal displacement may occur. Common causes of foundation movement include: inadequate size footings; overloading the structure; excessive ground water which reduces the bearing capacity of soil; inadequate soil cover, which fails to protect against frost heaving; and adjacent excavations that allow unprotected bearing soil to shift from under foundations to the excavated area. Severe localized foundation displacement may show up in cracked walls, damaged framing connections, sloping floors, sticking doors, and even leakage through a displaced roof.

Material Deterioration

Foundations are subject to deterioration, whether from material or construction deficiencies or from environmental conditions. In order to determine the cause(s), the deterioration of foundation materials must be observed directly whenever possible. Excessive moisture from surface or subsurface sources is a major cause of timber deterioration, providing the necessary condition for wood decay and encouraging insect infestation. Improperly seasoned wood is subject to cracking, splitting, and deflection. Refer to Chapter 5 for wood maintenance. Concrete and masonry are subject to cracking, splitting, and setting, particularly under adverse ground and climatic conditions. Refer to Chapter 3 for methods of concrete maintenance. Steel and other ferrous metals are subject to corrosion in the presence of moisture and sometimes by contact with acid-bearing soils. Signs of corrosion are darkening of the metal, rusting, and pitting. Refer to Chapter 4 for maintenance of steel.

Soil Investigation

Soil can be generally classified into four types depending on size. The largest is gravel, the next sand, then silt, and finally, the smallest, clay. All gravels, sands, and larger size silts are known as course-grain soils. The remaining silts and all clays are known as fine-grain soils.

Coarse-grain soils are more permeable than fine-grain ones and thus are good for draining water. This is why they are placed over drain tile or as fill material beneath slabs or against foundation walls. Coarse-grain soils have a tendency to draw water through capillary action. For this reason, it is important that all foundations be placed below the frost line or be protected from the freeze-thaw action. Relative to load carrying capacity, coarse-grain soils are ineffective unless contained by such methods as sheet piling or adjacent stabilizing soil medium or chemicals such as grout. The strength of this type of soil is measured in terms of relative density.

Fine-grain soils are non-permeable and thus water will not drain through them as well as coarse-grain soils, if at all. Fine-grain soils derive their load carrying capacity from their shearing strength (or resistance), which may change with ground conditions at the site. Clay bearing values, for example, vary with moisture content. A foundation laid on the basis of local standards for bearing value for the site may settle because the clay loses optimum moisture content (and therefore loses shearing strength) when pumping operations are undertaken in the area, or when an adjacent deeper excavation drains moisture from the foundation-bearing clay. Clay also loses moisture when molded or worked; frost action is similar in effect to working the clay.

Before making any repairs to foundations, refer to the soil report on which the original foundation design was based. If a report cannot be found, or there appears to be some question as to whether or not the same subsurface conditions exist, have one prepared. Consult with a registered professional soils engineer to obtain this report. Furthermore, any new construction should be preceeded by a soils report. This is a preventative measure to minimize future maintenance problems.

Maintenance

Excess moisture is generally a direct cause of, or contributing factor to, foundation and crawl space damage. Moisture control is treated in detail in the previous section. Damage from fungus attack and termite infestation and appropriate controls are discussed in Chapter 5.

Foundation Stabilization:

1. Replace immediately any missing or dislodged part of the foundation; repair cracks or open joints in concrete or masonry foundation walls; replace defective wood members.
2. Provide proper drainage away from buildings and structures.
3. Replace unstable fill around the foundation with clean, properly compacted fill.
4. Remove growing roots of trees or shrubs that may dislodge footing or foundations.
5. Increase bearing area of inadequate footings.
6. Maintain enough soil cover to keep the base of footing below the freezing zone.
7. Prohibit loads exceeding the design loading of building and structures; isolate foundations from heavy machine operations by providing independent footing and foundations for the machines. Air conditioning equipment, such as cooling towers, and compressors should be provided with cork or rubber isolation mounts to prevent transmission of vibrations to the structural frame of the building.

Material Protection:

1. **Wood Piling:** Wood piles are customarily treated with a preservative and placed totally and permanently below the low-water table. Treated piling when exposed along wharfs, piers, etc., should be inspected periodically for the presence of fungi and termites. Infested wood should be replaced with sound treated timber.
2. **Steel Piling:** Steel piling is protected against corrosion by the application of coatings. Pilings should be inspected for failure of coating and recoated if necessary. If concrete capping has not been used or has failed, piles should be properly capped.
3. **Concrete:** Concrete should be inspected for soundness and continuity of surface. Concrete is subject to failure through corrosion of reinforcing bars. Concrete should be chipped out and spot-grouted to cover any reinforcing steel that shows on the surface. Minimum cover of concrete over steel reinforcement should be in accordance with American Concrete Institute requirements. In maintenance patching, it is desirable to bring the repair work to the plane of the original finish surface.

Fences

Fences are of many types and are used for a variety of purposes. Chain link fences are usually used as security fences. This type of fence may or may not be protected on top by barbed wire strung on angle brackets. Other fences are made of stranded wire, wood rail or picket, poles, chicken wire, and cables.

Routine Maintenance

Vines, weeds, high grass and shrubs should not be permitted to grow up and around fences. Moisture retained by the vegetation accelerates rust or rotting of fencing materials, and vegetation, when dry, becomes a fire hazard. Fire causes considerable damage to, or loss of, wood fences, and fence posts. Trash must not be permitted to accumulate against fences for several reasons. Combustible trash is a fire hazard. Metal trash of a different metal from the fence may cause corrosion by galvanic action. Piles of trash may also be used as a stain to permit access over the fence.

Erosion around fences often defeats the purpose for which the fence was intended. Earth washed out under the fence leaves holes for persons or animals to crawl through. In other cases, earth washes down against the fence, piles up against it, and finally damages the fencing from the pressure. The earth pile also makes it easier for intruders to climb over the fence.

Metal Fences

Metal fences are either aluminum or steel. Maintenance of steel fences involves protection against rust or other damage. Except in highly corrosive climates, such as seaside locations or certain industrial areas, maintenance, as far as rusting is concerned, is not a major problem. Aluminum fencing usually requires little attention. However, repairs to fencing damaged by accident or vandalism are often necessary. Generally, maintenance of metal fences involves painting to prevent rust or corrosion, keeping the fence material, ties and braces in place, fence posts straight and firmly anchored in the ground, gates hung properly without sagging, holes under fences filled, and the fence line free of vegetation or trash. Replace damaged mesh, supporting wires, barbed wire, and other members with materials to match the existing fencing. Care must be taken in handling fencing material, particularly barbed wire, to prevent injury. Use proper tools, such as wire or fence stretchers, cable cutters, block and tackle, and other tools.

Patching: Small holes in wire fabric fencing can usually be patched. Fasten the patch on one side and stretch it across the hole, then fasten the other side, the top and the bottom, using heavy tie wires twisted inside the fence. The cut ends of the wires of the patch are generally inside the fence to complete the job.

Replacing Chain Link Fencing: Chain link fencing is generally fastened to or through terminal posts by tension bars, approximately 3/16 by 3/4 inch, threaded through the links and clamped to the terminal posts. When a section of fencing is replaced, the intermediate posts on each side of the damaged section become terminal posts while the work is being done, and the tension bars should be used while the damaged section is being replaced. Tie wires should be used, or temporary braces should be installed, to support these posts against the pull of the adjacent fencing when the damaged section is removed, unless the remaining fence is adequately braced by a top rail. After removing the damaged section, cut off the proper length of fencing to span the gap, and fasten one end to one of the temporary terminal posts, using a tension bar and clamps. Stretch the new section across the gap and pull it tight, then secure the free end to the other temporary terminal post with a tension bar and clamp. Cut the excess material in a neat manner and remove the temporary tie wires and braces. Install the tension wire at the top of the new section or tie the top of the new section to the top rail to match the type of support used in the fence being repaired. Cut ends of steel wire should be touched up with aluminum paint to avoid rusting.

Wood Fences

Wood fences are costly to maintain and require considerable attention. Wood fences must be maintained by keeping pickets, rails, planks, braces and other members securely nailed. Damaged members must be replaced. This type of fence should be painted as often as necessary to prevent decay. However, painting is required less often when the fence is constructed of treated lumber. Repairs to wood fences are fairly simple, and are generally performed by carpenters. When damaged members are removed, however, they should be replaced with treated lumber and all cut edges and holes brushed with preservative treatment before installation. The treatment used depends on whether the fences are to be painted.

Posts

Wood posts, pressure treated with preservatives, should be used for fencing. If pressure-treated posts are not available, dry wood posts should be submerged in preservative solution for 24 hours. Posts not to be painted should be treated with creosote. Wood posts damaged beyond repair, should, of course, be replaced.

Posts used to support metal fencing are of several types. Chain link is usually secured to pipe posts, but wood posts are often used. Garden wire, chicken wire, and other types of woven wire fences are erected using angle-iron posts or steel posts of special cross section. This type of post sometimes has a spur or clip punched from the post which is used to secure the wire to the posts. Wood posts are generally used for stranded wire or barbed wire installation. In some instances, concrete posts have been used for fencing. Metal or concrete posts are usually set in concrete. When posts become loose, the earth must be compacted around the base or a new concrete base poured around the posts. Angle-iron posts are usually driven into the ground and also may be set in concrete. If the driven post becomes loose, the earth should be retamped around it or the earth excavated and the hole filled with concrete. Concrete bases should be brought above grade and crowned to shed water. Metal posts that have been bent should be straightened or replaced. All braces and hardware must be tightened and the missing parts replaced. The same applies to framing where a framework is used to support the fencing, such as gates and playground backdrops. Wood posts should not be set in concrete unless the soil cannot be adequately packed around the posts.

Gates

The care of gates is the same whether they are metal or wood. Gates should be maintained so that they will operate easily and close properly without binding. Hardware should be maintained in working condition and missing parts replaced as soon as possible. Sagging gates should be puffed back into shape and braced by wires or rods with turnbuckles so that adjustments can be made as necessary. Hinges, brackets, etc., should be maintained tight to prevent wobbling of the gate or interference with its operation. Wheels and pulleys of sliding gates must be lubricated as needed. Tracks must be kept straight and level.

Electric Fences

Electric fences are rarely used. This type of fence is usually a wire suspended from insulators and is hooked up to a power source or shocking machine. The machine consists of induction coils energized by a battery or power supply. High voltage at low amperage is transmitted over the wire in pulses giving a harmless, but unpleasant shock to any person or animal touching it. The wire must be firmly attached to the insulation which in turn are firmly attached to the wood posts. The batteries and power supply must be checked from time to time to see if proper voltage and current are being maintained. Broken wires and insulation must be replaced in order the keep the fence operable.

Snow or Drift Fences

This type of fence is used in areas where drifting snow or sand is a problem. They are usually placed where drifts endanger roads or runways and are installed for the period of time they are needed. Fences used to retain drifting sand remain in place as long as the sand must be controlled and are often left permanently in place.

Snow fences are only installed during the winter when snow is involved. The snow or drift fence is constructed of thin slats held together with wire and supported on driven metal stakes in temporary locations or wood posts in more permanent areas. The slats are usually treated with a preservative stain. Care must be taken in erecting, removing, or storing the fence, since the slats are easily broken.

Repair of snow or drift fences involves replacing broken slats and wires and straightening bent posts. Maintenance involves carefully erecting and removing the fencing, rolling it into bundles and storing the bundles on sills or rails in areas where they will not hinder operations, be damaged or be exposed to fire. Fence pickets or slats should be replaced before the fence is stored and, if necessary, dipped or sprayed with a preservative treatment.

Painting Fences

Fence painting is expensive and, in certain cases, cannot be economically justified. Repeated painting of fences can soon equal or exceed the cost of replacing the fencing. If fences are properly inspected and maintained, they will last for years without more than occasional spot painting. Heavy chain link fencing will often last for many years without attention. Wood fences will require painting more frequently than metal fences, particularly when they are constructed of untreated wood. Due to the high replacement rate, untreated and/or unpainted posts should be used to a minimum. Painting for maintenance should be done as necessary, but before a paint job is scheduled, the remaining useful life of the fence should be considered and the cost of painting it compared with replacing the portions of fence that may fail at the end of the estimated useful life period. Unless complete painting will extend the life of the fence to a considerable degree, only spot painting of the more seriously deteriorated portions is justified, even though a poor appearance may result. However, fencing exposed to the public should be maintained so that it does not create an unsightly appearance or become offensive to the adjacent property owners. Barbed wire, strand wire, or frame type fences are usually not painted. Barbed wire that is severely twisted should be replaced. Wood fences with a stained finish should be restained as needed.

Methods of Painting Fences: General information on the methods of painting and the materials to be used is covered in the section of this chapter devoted to painting. Before any painting is done, the surfaces must be properly cleaned by being scraped or wire brushed. Remove rust or loosen scale and then prime as necessary after any damaged portions of the fence have been repaired and replaced. Surfaces to be painted must be dry. Priming should be done as soon after surface preparation as possible.

Except in special cases, the use of rollers is considered to be the best and most economical method of painting chain link fences. This type of painting should be done only with deep nap rollers that are designed specifically for this purpose. To insure complete coverage (and to prevent holidays), two painters, one on each side of the fence, working simultaneously, are recommended. Touch-up, especially around the posts, should be done with a brush. Unless unusual circumstances dictate otherwise, chain link fences are normally painted with aluminum paint. Wood fence structures that are badly weathered, cracked, or peeled must be scraped to remove loose or scaled paint, and a prime coat brushed in thoroughly.

Spraying is one of the quickest and most economical methods to use for painting fences, posts, and railings. However, damage to adjacent property from over spray may occur unless proper protection methods are used.

Brushing may be used for touch-up work painted by other methods, but it is not normally recommended for painting fences because of the large areas involved and the amount of detail work required. Alternately, paint may be economically and satisfactorily applied by deep pile mitts. This method could be as economical as roller methods and more economical than other methods.

Retaining Walls

Although primarily designed to hold earth or other materials in place and to prevent slides and displacements, retaining walls serve a dual function in that they may also assist in the prevention of erosion and the channeling of storm water runoff. Types of retaining walls are:

- Gravity.
- Cantilever.

Retaining walls may be constructed of concrete, rubble, brick masonry, steel sheet piling, and combinations of the above materials.

Because erosion at the toe of a retaining wall decreases the stability of the wall, it must be prevented. Divert the flow of water causing erosion, and fill in any eroded places with crushed rock or other suitable material, well tamped into place. Keep weep holes properly cleaned out so that standing water will not build up excessive pressure behind the wall. Where inspection indicates necessity for repair, replacement, or addition to retaining walls, corrective action must be taken promptly in order that additional damage may be prevented.

Concrete, Rubble or Brick Masonry

Where scouring or undermining has occurred, secure the wall by blocking. Excavate in sections to two feet or more below probable future scour lines and fill sections with concrete. Follow this procedure until all sections of the footing are completed. Backfill to the original earth line and rip-rap thoroughly. In extreme cases, it may be necessary to make the new footing wider than existing ones and to extend it to a point well above the top of the original footing in order to hold the latter in place. Repair or replace loose masonry, fill all joints with mortar or crack sealer, correct the drainage at the back of the wall, and keep weep holes open. Tie new work thoroughly to the original construction. For concrete construction, use steel dowels spaced at suitable intervals and sealed into holes drilled into the old wall. Use the same type of construction materials as the original wall. Evidence showing that scour has occurred on top of the structure is evidence of a possible structural failure. A structural engineer should be consulted. The design and construction of a higher wall to replace the affected one may be needed.

Timber or Metal Sheet Piling

When either timber or metal sheet piling is displaced by overload and the affected section is threatened with failure, remove part or all of the overload, replace members in the original position and backfill. Reinforce the section by adding waling strips as needed. In unsuitable earth, use longer piles. If needed, install deadmen of dimension lumber or heavy logs with cable attached to outside strips. A professional registered structural engineer should be consulted to make the structural analysis prior to making any type of repairs.

Bulkhead Retaining Walls

Bulkhead retaining walls fail due to the overloading of horizontal members. This occurs when the pile spacings are too great, horizontal members are too light, or both. Place intermediate piles, if necessary, and substitute heavier horizontal members. Where piles have been forced out of line by walls, remove part of the load against the wall, pull the piles back into line with ratchet jacks, and secure piles with cables attached to deadmen on about the same elevation as the top of the slope and far enough behind the wall to secure good anchorage.

Sacked-Concrete Retaining Walls

Sacked-concrete walls fail because of scouring and undermining. Tear out old damaged work and excavate to two or more feet below the probable future scour line. When scour occurs on top of the original wall, a professional structural engineer should be consulted. Replacement of the section of wall may be needed, in which case a new design is indicated.

Material Maintenance

Wood is subject to decay and steel to corrosion, so they must be properly protected when used in permanent structures. Use wood that has been pressure-treated with preservatives to prevent rot and decay. If raw wood is exposed, paint the exposed areas. Paint steel before installation and touch up any spots where paint is knocked off during installation. Subsequent painting is seldom justified because the critical areas are not available for painting. When members decay or corrode to the point of incipient failure, replace them.

Athletic and Other Recreational Facilities

Recreational facilities that are constructed of concrete or steel should be maintained and repaired according to previous and future discussion in this book. Some types of athletic and other recreational facilities are:

- Spectator stands
- Playing fields and courts
- Golf courses
- Big beaches
- Guard towers
- Picnic grounds
- Archery ranges

It is not the purpose of this text to describe the maintenance and repair procedures for playing fields, courts and golf courses. However, many of the maintenance and repair procedural discussions in this book, used for pavements and flatwork, would also pertain to those items.

CHAPTER 7

INTERIOR CEILING AND WALL MATERIALS

Interior walls are either plastered (wet walls) or covered with materials such as gypsum wallboard, plywood, tongue-and-groove wood paneling, tile or glazed-face masonry, and other new wall finishing materials as they become available. Partitions may be of plywood, plasterboard, hard-pressed fiberboard, structural clay tile, gypsum block, metal, and glass. Ceilings are usually made of gypsum board, plastered, paneled with wood, or covered with sheet metal or acoustical materials. In this chapter, the repair procedures for walls and suspended ceilings comprised of the same materials are identical, except for the amount of time and the necessary equipment.

When maintaining and repairing interior ceilings and walls, take care to maintain the effectiveness of special interior finishes (i.e., fire resistance, insulation, appearance, and acoustical functions). Figures 7.1 and 7.2 contain guidelines for the selection of interior wall and ceiling finishes.

Plaster

The most frequent maintenance problems with plastered and gypsum board surfaces are cracks, holes, loose segments, and water stains. Cracks, holes, and looseness in plastered surfaces are signs of excessive internal or external stresses. Defects may be caused by poor workmanship, such as improper proportions or application of the plaster, imperfect lathing, and poor atmospheric conditions during plastering; by moisture infiltration or an excess of moist air generated inside a building; or by the settling or other movement of some part of the building frame. Moisture causes spalling, efflorescence, and stains. External stresses that cause plaster damage should be investigated and corrected before repairs are made to the plastered surfaces themselves.

Other causes for defects are listed below.

- Settlement of the structure
- Defective materials
- Operational abuse
- Inadequate protection from environmental attack
- Inadequate maintenance procedures (incorrect for material)
- Incomplete engineering or design
- Vibrations
- Structure expansion and shrinkage
- Wind pressures
- Lack of proper inspection and supervision

Refer to Figure 7.3 for the various types of plaster deterioration and corresponding symptoms and causes.

Plaster Maintenance

To effectively maintain plaster surfaces, first make sure that they have been sealed with an acceptable high-quality sealer or paint that is commercially available. This should have been done at the time of the original plastering. Follow the routine maintenance procedures listed below.

Interior Finishes — Wall Materials							
Wall Materials	Insulating Values	Resistance to Dampness	Resistance to Grease and Dirt	Durability	Ease of Application	Sound Deadening	Ease of Cleaning
Brick or Block	4	3	4	1	3	2	4
Corrugated Plastic Sheeting	2	1	1	1	3	3	3
Glass Blocks	2	1	1	1	3	3	1
Plaster — Bare	3	4	4	3	4	3	4
Plastic Blocks	2	1	1	2	3	3	1
Stone and Artificial Stone	4	2	4	1	3	2	4
Wallboards and Plasterboards Asbestos Board	2	1	4	1	2	2	1
Fiberboard Unfinished	1	3	4	3	2	1	4
Hardboards Prefinished	3	1	1	1	1	3	1
Insulating Wallboards	1	3	4	3	2	1	4
Perforated Hardboard	4	2	2	1	1	4	3
Plasterboard Unfinished	3	4	4	3	2	3	4
Wood Plywood Prefinished	3	3	3	2	2	3	2
Plywood Unfinished	3	3	4	2	1	3	4
Solid Hardwood Unfinished	2	2	4	2	3	3	4
Solid Softwood Unfinished	2	3	4	2	3	3	4
	1 = Excellent	2 = Good	3 = Fair	4 = Poor			

Figure 7.1

Dusting: The following equipment and materials should be used for dusting.

Equipment

Dust mop (yarn duster) or vacuum cleaner (and extensions)
Clean Cloths
Ladder

Materials

Dusting Compound (may be used)

1. Cover or remove all furnishing and/or equipment that should not be exposed to the dust.
2. In dusting walls and ceilings, start with the ceiling and follow by dusting the walls from the top to the floor. It is imperative that the dusting tools (and components) do not contain abrasive materials. Do not use brushes or feather dusters.

Interior Finishes — Wall Coverings					
Wall Covering	Resistance to Dampness	Resistance to Grease and Dirt	Durability	Ease of Application	Ease of Cleaning
Cork	4	4	3	2	1
Enamel-on-felt	1	1	3	2	1
Glass Sheets	1	1	3	3	1
Leather	2	2	3	2	1
Linoleum — Inlaid	1	1	2	2	1
Paint					
Clear Finishes (Varnish, Shellac, Plastic Paints, Lacquer)	2	2	2	2	1
Linseed Oil	2	3	2	1	2
Paints Mixed with Water	4	4	3	1	4
Paints with Oil, Rubber Alkyd, Emulsion Bases	3	2	2	1	1
Tiles					
Asphalt, Rubber	1	2	2	2	1
Glazed Ceramic	1	1	1	3	1
Metal	1	1	1	2	1
Plastic	1	1	1	2	1
Unglazed Ceramic	1	2	1	3	2
Wallpaper					
Coated Fabrics (Oil, Pigment, Vinyl, Pyroxylin)	2	2	2	2	1
Non-washable	4	4	4	2	4
Scrubbable, Fabric Backed	1	1	2	2	1
Thermosetting Plastic Sheets (Micarta, Formica, etc.)	1	1	2	2	1
Uncoated Fabrics (Burlap, Canvas)	3	3	4	2	4
Washable	3	4	3	2	2
1 = Excellent 2 = Good 3 = Fair 4 = Poor					

Figure 7.2

3. During the dusting operation, give special attention to corners, window
sills, doors, and other trim work.
4. Upon completion of the above, clean all furnishings and equipment prior to returning them to their proper location.

Labor Production	Frequency
Wall Dusting—2–3 sec./S.F.	Depends on location, room utilization, and contact with foreign matter.
Vacuuming—4–5 sec./S.F.	At least once every six months.

Washing: Periodic washing initiates good housekeeping. Do not use abrasive or caustic cleaners, since they can scratch the surface. If anything stronger than mild soap is required to remove dust, then the housekeeping frequency is incorrect and wall surface treatment inadequate. To eliminate water splatter and slopping conditions, keep excess water from surface being cleaned. In cases where the ceiling requires cleaning, wash the ceiling prior to washing the walls. The same equipment and materials are used for washing both the ceiling and the walls.

Deterioration of Plaster — Symptoms and Causes

Type Deterioration	Symptom	Cause
Loose Plaster	Bulging and cracking of large areas of the surface.	Excess moisture in the surface. Deficient materials.
Map Cracks	Penetrate through the plaster, but do not extend entirely across the surface.	Improper bonding between the plaster and lath or masonry base.
	Generally occur as a series of irregular cracks running at various angles.	Deficient materials. Inferior workmanship.
Shrinkage Cracks	Resemble map cracks.	Inferior workmanship.
	Confined to the finish coat.	Too rapid drying of the surface.
	Do not extend entirely through the plaster surface.	Insufficient troweling.
	Covers a much smaller portion of the wall or ceiling surface.	
Structural Cracks	Large, well defined cracks.	Foundation settlement.
	Extend across the surface and entirely through the plaster.	Failure of masonry wall section by shrinkage or cracking.
	Develop during the first year in the life of a building.	Sagging, warping or shrinkage of wood members.
	May extend diagonally from the corners of door and window openings, run vertically in corners where walls join run horizontally along the junction of walls and ceilings, or occur in walls where two unlike materials join.	

Figure 7.3

Equipment	Materials
Buckets	Steel wool # 00
Clean cloths	Mild detergent suitable for
Drop cloths	type of wall surface
Rubber gloves	
Ladders of scaffolding	
Sponges	
Wall washing machine with attachments (for optional method)	

The recommended procedure for washing plaster is listed below.

1. Prepare the equipment and materials needed for washing the walls and prepare the washing detergent.
2. Move any furniture away from the walls.
3. Remove pictures, drapes, or other furnishings from the walls.
4. Spread drop cloths over the furniture and floors.
5. Dust walls as described in the previous section.
6. Set ladders and scaffolding in place as needed.
7. When manually washing walls, dip the sponge or clean cloth into the solution or squeeze out excess detergent.
8. Apply the cleaning detergent to the wall, beginning at one of the ceiling corners. Wash an area of approximately ten square feet, wetting the surface without rubbing to remove the dirt.
9. Repeat the washing operation to remove the dirt.
10. A straight rubbing motion should be used either side to side or up and down.
11. Lap over into the clean area when washing, far enough to prevent border marks.
12. Continue washing the balance of the areas. By employing a scaffold, the whole top section of the wall can be washed before completing the bottom section. In using a ladder, a quarter of the top section may be washed, followed by washing the bottom half before proceeding to the next top section of the wall.
13. Wash the woodwork and doors as reached.
14. If using a ladder and the solution spills on the wall below, wipe off immediately with a clean, damp cloth.
15. When washing by machine, fill the machine compartment with detergent. Start washing the wall at one of the ceiling corners.
16. Never apply heavy pressure which may cause surface damage. It is recommended to wash the walls twice if excessive foreign material is embedded. Hold the trowel flat against the wall and move it with light pressure in long straight strokes, from side to side when the surface permits.
17. Overlap area just washed in order to avoid border marks.
18. Remove streaks or marks with the heel of the trowel.
19. Wash an area of approximately five square feet at a time.
20. For manual and machine washing, use fine steel wool (with care) on spots.
21. Do not use soiled towel when washing the walls.
22. Change the water and cleansing solution frequently to avoid redistributing the dirt to the walls.
23. Wipe area just washed with a rinsing towel to remove soiled cleaning solution.
24. Use a drying towel to pick up moisture still on the surface.
25. Remove drop cloths, replace furnishings, clean, and return equipment to its proper storage.

Labor Production	Frequency
Spot washing—125–175 S.F./hour	As required.
Thorough cleaning of walls and ceilings—275–500 S.F./hour	

See Figure 7.4 for a summary of wall finishes and recommended cleaning procedures.

Plaster Repair

The appropriate repair procedure will depend upon the type of plaster being repaired. The various types of plaster are similar in composition and application, but are designed for specialized uses.

Gypsum Plaster: Gypsum plaster is most generally used in ordinary construction because it can be readily applied to furred and lathed surfaces of exterior masonry and directly to interior masonry. It can be applied over metal lath and gypsum products.

Lime Plaster: Lime plaster, found in many older buildings, is similar in application to gypsum plaster, but should not be used in conjunction with gypsum products. Lime plaster should be repaired with like material.

Wall Cleaning Procedures

Finish Wall Material	Cleaning Procedures
Acoustically Treated Finishes	Use vacuum cleaner to remove loose dirt. Clean with a machine or by hand (using a stiff palmetto brush). Rinse and wipe dry.
Fiberboard Finishes	Cannot be washed unless the surface is painted. Can be dusted.
Glass, Vitrous-China and Glazed Tile Solution	Wash with a neutral cleaning solution. Do not use scouring powders or abrasives. Can be dusted.
Marble, Granite, Onyx and Other Neutral Stone Surfaces	Wash with synthetic detergent, rinse with clear water and dry. Do not use an oily duster or mop.
Oil or Enameled Painted Walls	Wash with an all-purpose synthetic detergent. Can be dusted.
Oil and Latex Base Painted Gypsum Board	Wash with an all-purpose synthetic detergent.
Unpainted Plaster	Do not wash. May be dusted with a soft bristle brush or vacuum. Do not dust in damp weather.
Walls Painted with Water-Based Paints	Can be washed, except surfaces that have been white washed. Can be dusted.

Figure 7.4

Keene's Cement Plaster: Keene's-Cement plaster produces a hard, moisture resistant surface suitable for spaces given hard use, particularly wainscots, and in areas subject to continued moisture, such as baths, kitchens, and certain hospital areas. Keene's-Cement plaster should be applied only over a gypsum-plaster base.

Portland Cement Plaster: Portland cement plaster may be applied directly to interior and exterior masonry walls and over metal lath; it should never be applied over gypsum products. It is recommended for use in plastering walls and ceilings of large, walk-in refrigerators and other cold-storage spaces, basement spaces, rest rooms, janitors' closets, and similar areas.

Insulating Plaster: Insulating plaster differs from other plasters in that light-weight vermiculite and perlite are used as an aggregate instead of sand. It is light-weight, provides some thermal insulation, and gives a fire-retardant surface to a degree determined by its composition and method of application.

Accessory Materials: The materials used in conjunction with plastered finishes are listed below.

- Lath
- Furring
- Grounds
- Corner Beads
- Corner Lath
- Strip Lath

Gypsum and Lime Plaster Repair

The following equipment and materials are necessary for the repair of structural cracks.

Equipment

Spatula or putty knife
Small diamond-shaped pointing trowel
Sharp chisel
Linoleum knife
Mixing pan and tools

Materials

1. Generally, two coats of patching material are required to repair wide structural cracks. The first coat may be a mixture of one part fibered gypsum plaster and two and one-half parts plastering sand, by volume, mixed with clean water to a uniform color and workable consistency.
2. Material for the second coat may be either a neat gypsum plaster or a mixture of one part hydrated lime and one-half part calcined gypsum mixed with water to a suitable consistency.
3. A small amount of casein glue added to the above mixes insures easier application because it tends to retard the setting time of the mix. The addition of glue also prevents shrinkage of the mix and helps form a better bond with the original plaster.
4. Instead of the mixes described in (1) and (2) above, a commercial patching plaster may be used. This material, known as spackling compound, is a mixture of plaster of Paris and powdered glue, which is mixed with clean water to the consistency of soft putty.

Follow the procedure outlined below:

1. Cut out and remove any loose material with a linoleum knife or chisel.
2. Form the crack to a V-shape. Do not over-widen the crack.
3. Clean all expanded metal or wire lath and be sure the mesh is opened.
4. If a wood lath is used, wet it prior to applying patching material.
5. Brush out all loose material, remove all grease or dirt from the surrounding surface areas.
6. Wet the edges of the grooves.
7. Prepare the plaster. The quantity of any patching material must not exceed what can be applied within 30 minutes after mixing. Plaster material should not be retempered (by adding water) once it has begun to dry out and harden. Retempered plaster placed on the wall will dry and become soft and crumbly. Press the first coat of patching plaster firmly into place, filling the groove nearly to the surface.
8. Allow it to set until nearly dry, but not hard.
9. Complete the patch by applying a coat of finished plaster, strike off flush, and trowel smooth.
10. If the edges of the original plaster and the wood lath are not thoroughly wetted, they serve as a wick to draw the water from the fresh plaster, causing it to dry out, remain chalky, and crack aroused the edges of the patch.
11. Clean up the working area.
12. Clean and return the equipment to proper storage.

Labor Production	Frequency
5–10 S.Y./hour/worker	As needed.

Map Cracks: The following equipment, material, and procedure should be used in the repair of map cracks.

Equipment	Material
Same as for structural crack repair.	Spackling compound

Map cracks are repaired in much the same manner as described for shrinkage cracks. For extreme cracking, use the same method as for repairing loose plaster, as discussed later in this chapter.

Labor Production	Frequency
See applicable method used.	As needed.

Shrinkage Cracks: The following equipment, material, and procedure should be followed for repairing shrinkage cracks.

Equipment	Material
Same as for structural crack repair.	Mixture of white lead and turpentine.

1. For light surface cracking that does not penetrate the base, scrub the mixture of white lead and turpentine into the cracks.
2. Where shrinkage cracks penetrate the base course, they should be repaired as a structural crack.

Labor Production	Frequency
See structural cracks (deduct an average of two square yards per hour).	As needed.

Holes: Holes are repaired using the same procedure as for structural cracks. Same labor production as for shrinkage cracks can be applied.

Bulges: Create a hole where the bulge appears and repair as a hole defect. See shrinkage cracks for labor production figures.

Loose Plaster: For temporary repair to prevent loose plaster from falling until permanent repair can be accomplished, secure the loose plaster with a section of wallboard nailed securely to the wall or ceiling over the area affected. Nails should be of sufficient length to penetrate through the plaster and obtain a firm bearing in the studs or joists.

Repairs of a permanent nature should be made as soon as practicable. Remove all loose plaster around the break, working well back in the surrounding area to a point where solid plaster (well keyed to the lath, which in turn is solidly secured to the structural frames) is obtained. Remove defective lath and replace with suitable plaster backing, such as metal lath or plasterboard, and securely refasten all lath that has become loosened. Repairs are made similar to structural cracks.

Old, worn, and crumbling plaster is repaired in the same manner as loose plaster. However, in the attempt to repair old plastered surfaces, the operation of removing the affected areas sometimes causes adjoining areas of apparently sound plaster to fall. This is evidence that the entire plastered surface has deteriorated to the extent that replastering of the whole area is necessary.

Portland Cement Plaster

The types of failures and methods of repairing portland cement plaster are basically the same as described above for gypsum plaster. In patching small areas, the edges of the surface surrounding the defective area must be thoroughly and continuously wetted for at least one hour before application of the patching material. Just prior to applying the patch, dust the entire edge of the exposed plaster with a light coat of portland cement. Press the first coat of portland cement plaster compound, composed of one part portland cement, three parts plastering sand, and one-quarter part lime putty, firmly into the groove or hole nearly to the surface, using particular care to insure that no voids are left around the perimeter of the patch. Scratch or roughen the surface of the patch with a wire brush or nail to make a base for receiving the finish coat. Thoroughly cure the patch by keeping it moist for at least 72 hours, then let it dry thoroughly, for not less than seven days. Just prior to application of the finish coat, which should be of the same composition as the first coat, moisten the patch thoroughly and firmly press the plaster into the remaining cavity, float to a smooth even surface, then trowel to the same texture as the surrounding surface. Keep the patch moist for at least three days.

Drywall

Drywall needs periodic maintenance similar to that of plaster. Repair of interior wallboard materials may be accomplished in the same manner as for plaster surfaces. Also, it is important that nails, screws and other fasteners be secured.

Joints in drywall construction that fail must be recemented and taped. Broken sections of interior wallboard are generally best corrected by replacement of an entire panel. Walls and ceilings should be inspected periodically for marks, dents, scratches, cracks or other surface blemishes.

Ceramic Tile

Newly tiled surfaces should be cleaned to remove job marks and dirt. Cleaning should be done according to the tile manufacturer's recommendations to avoid damage to the glazed surfaces.

Depending on the location of the tile, daily dusting may be needed. Follow the same basic procedure previously described for dusting walls.

Ceramic tile should be periodically washed with a non-caustic, non-abrasive agent. Care must be taken not to harm the mortar in which the tile is set. Follow the same procedures as those recommended for washing plaster surfaces.

A sealant can be applied to the wall to prevent future deterioration of the mortar. If such a sealant is not provided, periodic cleaning and minor tuck pointing may be needed at the joints. An abrasive cleaner, such as scouring powder, can be used to clean out the joints, using a small stiff brush.

Ceramic Tile Repair

Cracked and broken tile should be replaced promptly, to protect the edges of adjacent tile and to maintain waterproofing and appearance. Timely pointing of displaced joint material and spalled areas in joints is necessary to keep tiles in place. The equipment, materials, and methods needed to repair cracked and broken tile are listed below.

The following equipment, materials, and procedure should be followed to repair ceramic tile.

Equipment	Materials
Cold chisel	Setting mortar mix
Hammer	
Caulking compound	
Brush	
Protecting finish compound	
Trowel	
Cloths	
Bucket	

1. Obtain all equipment and materials.
2. Cut around the damaged tile with a cold chisel to break the cement which holds it. Tap gently to avoid breaking adjoining tiles or chipping their edges.
3. Use the chisel to get behind the tile and pry it out. If not successful, break the tile.
4. Clean the mortar from the edges of the surrounding tile.
5. Roughen the concrete under the bed to provide a good bond for the new setting concrete.
6. Dampen the underbed and the surrounding tiles with water and place the setting mortar mix in proportion of one part cement to three parts sand, or as per specific instructions for special compounds required for special tile types.
7. Wall tiles should be thoroughly soaked in clean water before they are set.
8. Set the tile, tapping it into place gently.
9. Fill the joints with grout or pointing mortar. The joints should be tooled slightly concave and the excess mortar cut off and wiped from the face of the tile.
10. Any interstices or depressions in the mortar joints after the grout has been cleaned from the surface should be roughened at once and filled to the line of the cushion-edge before the mortar begins to harden.
11. All joints between wall tile and plumbing or other built-in fixtures should be made with a light-colored caulking compound.
12. Immediately after the grout has initially set, tile wall surfaces should be given a protective coat of non-corrosive soap or other approved protection.

Labor Production	Frequency
7–10 S.F./hour/worker	As needed.

Adhesive Tile Setting: Use an adhesive to install wall tile over existing, patched, or new plaster surfaces.

Equipment	Materials
Trowel-notched	Adhesive
Level	
Calking compound	
Clean cloths	
Bucket	
Hammer	

1. Where wall tile is to be installed in areas subject to intermittent or continual wetting, the wall areas should be primed as recommended by the manufacturer of the adhesive.
2. Wall tile may be installed either by the floating method or by the buttering method. In the floating method, apply the adhesive uniformly over the prepared wall surface, using quantities recommended by the adhesive manufacturer. Use a notched trowel held at the proper angle to insure a uniformly spread coating of the proper thickness. Touch up thin or bare spots by an additional coating of adhesive. The area coated at one time should not be any larger than what is recommended by the manufacturer of the adhesive. In the buttering method, daub the adhesive on the back of each tile in such amount that the adhesive, when compressed, will form a coating not less than 1/16-inch thick over 60 percent of the back of each tile.
3. Joints must be straight, level, plumb, and of even width not exceeding 1/16 inch. When the floating method is used, one edge of the tile is pressed firmly into the wet adhesive, the tile snapped into place in a manner to force out all air, then aligned by using a slight twisting movement. Tile should not be shoved into place. Joints must be cleaned of any excess adhesive to provide for a satisfactory grouting job. When the buttering method is used, tile is pressed firmly into place, using a "squeeze" motion to spread the daubs of adhesive. After the adhesive partially sets, but before it is completely dry, all tiles must be realigned so that faces are in same plane and joints are of proper width, with vertical joints plumb and horizontal joints level.
4. If the joints are to be grouted, allow adhesive to set for 24 hours. Joints must be cleaned of dust, dirt, and excessive adhesive, and should be thoroughly soaked with clean water before grouting. A grout consisting of portland cement, lime, and sand, or an approved ready mix grout may be used, but the grout must be water-resistant and non-staining.
5. Non-staining caulking compound should be used at all joints between built-in fixtures and tilework, and at the top of ceramic tile bases, to insure complete waterproofing. Internal corners should be caulked before corner bead is applied.

Labor Production	Frequency
9–12 S.F./hour/worker	As needed.

Pointing of Tile Joints

Pointing of Tile Joints: If the mortar in existing joints has cracked or crumbled, follow the procedure below to repoint them.

Equipment	Material
Trowel	Mortar mix (pointing mortar)
Bucket	
Cloth	

1. Thoroughly clean the joints of all loose mortar.
2. Grout the joints using one of the following methods, depending on the width of the joints:
   A. Grout joints an inch or less in width with portland cement grout of the consistency of thick cream.

B. Point joints 1/8 to 1/4 inch wide with a pointing mortar consisting of one part portland cement to one part pointing sand.
C. Point joints wider than 1/4 inch with pointing mortar consisting of one part portland cement to two parts pointing sand.
D. In locations where the wall is directly exposed to the effects of corrosive agents, use acid-resistant joint material to fill joints. The acid-resistant mortars should be mixed in accordance with the manufacturer's recommendations.

3. Clean adjacent tiles, allow mortar to cure and seal.

Labor Production	Frequency
10–15 S.F./hour/worker	As needed.

Acoustical Tile

The major causes of defective acoustical tile are listed below.
- Moisture
- Broken or loose tiles
- Inferior material
- Operational abuse
- Inadequate protection from environmental attack
- Inadequate maintenance procedures

Maintenance

The basic maintenance program is similar to that of gypsum board. Remove loose dirt and dust with a soft brush or cloth. The most effective method would be to use a vacuum cleaner.

In most cases, spots and streaks can be removed using an art gum eraser. Use wallpaper cleaner for larger marks.

As for plaster finishes, acoustical tile can be washed with a mild detergent using reasonable care. Commonly available cleaners especially designed for acoustical tile can be obtained from a local maintenance material supplier.

Equipment	Material
Buckets	Cleaning agent (mild soap)
Sponges	

1. Get equipment and materials ready.
2. Provide protection for the furniture by covering it with drop cloths.
3. Using either a cloth or sponge, wash ceiling, using long, sweeping, gentle strokes. If the tile is striated, clean in the same direction as the striations.
4. Do not saturate the tiles with water.
5. Rinse the tiles, using a damp cloth and warm or cold water.
6. If the tiles are removable, they may be taken down to clean.

Labor Production	Frequency
Same as for plaster.	At least annually.

Repair

In most cases, it is recommended that a broken piece of acoustical tile be replaced instead of repaired. If repair is deemed necessary, an acceptable "glueing process" should be used which is consistent with the type of material the tile is made of (consult with the tile manufacturer).

Wood

Generally, wood used for walls and ceilings is obtained in the form of panels, such as 4' x 8' sheets of paneling. Refer to Chapter 6, Timber, for further information on maintenance problems associated with wood.

Maintenance

1. At regular intervals, wood walls and ceilings should be dusted with a soft cloth or vacuum cleaner (see section on plaster walls).
2. If wall and ceiling material is not prefinished, a sealant coat that is absorbed into the pores of the wood should be applied.
3. Avoid using water on wood, finished or unfinished. Use only recommended wood cleaner.
4. Periodic waxing with a high quality wax helps to preserve the wood.

Also refer to the section in Chapter 9 on maintenance and repair of wood floors for additional information.

Repair

Minor repairs can be made to the wood as long as the appearance is not ruined. These repairs may include:

1. Filling an indentation (nail, hole, etc.) with a wood·base filler, then sanding followed by painting or staining.
2. Filling an indentation with a panel filler which matches the wood.
3. Repairing localized splitting or checking by covering the damage up with a filler type material, followed by a protective sealer, stain, or paint.
4. Refinishing, due to water staining or stain from substances such as atmospheric corrosion, by sanding or fine steel wool, followed by refinishing, resealing, and buffing, as for wood floors.

If the extent of the damage is widespread or the appearance is marred by a localized repair, the wood member must be replaced. In this case, remove the portion of the member which has been marred back to the supporting element of the member (such as removing a portion of a wood panel back to the supporting framework). Replace with a new member of the same grade, quality, and appearance along with needed finishing work such as filling, staining or painting. For additional information on the repair of wood members, refer to Chapter 5.

Additional Wall and Ceiling Finishes

The maintenance and repair procedures for wall and ceiling finishes made of the materials listed below are the same as or similar to others already covered in other sections of the book.

- Aluminum, Perforated: Refer to Chapter 6, "Exterior Materials," for maintenance and repair procedures.
- Ceramic Tile: Refer to Chapter 8, "Floors."
- Concrete and Cement Block: Refer to Chapters 4 and 7, "Reinforced Concrete" and "Exterior Materials," respectively.
- Cork: Refer to Chapter 8, "Floors."
- Glass: Refer to Chapter 10, "Windows and Doors."
- Plaster: Refer to Chapter 8, "Floors."
- Stainless Steel: Maintenance and repair is similar to aluminum (see above).
- Stone or Other Masonry Materials: Refer to Chapters 6 and 8, "Exterior Materials" and "Floors," respectively.
- Fiberboard and Hardboard: Maintenance and repair is similar to wood panels (see previous section of this chapter).

Trim and Wainscot

Trim and wainscot material vary according to design requirements for different interiors. Wood trim and wainscot are secured by nails, screws, or other fastening devices. When broken areas occur, they should be repaired or replaced with similar material and fastening devices. Ceramic tile may occasionally be patched with plaster or plastic materials, but it is usually more suitable to extract the broken piece and replace it with a new one. Joints should be well maintained; place new joining material where severe splitting occurs. Keep synthetic molding securely fastened to the wall so that moisture and dirt do not collect between them. Use mastic of the type used in the original construction for repair or replacement of moldings and wainscots of this type.

Wall Coverings

Wallpaper, paint, and other wall coverings are usually placed over plaster, gypsum board, and wood. Thus, any maintenance problems are linked to the wall itself. The only direct problem may be with the wall covering material, its application, and/or its maintenance.

Wallpaper should be dusted at the same time that adjoining surfaces are cleaned. At less frequent intervals, if the wallpaper is washable, clean with a mild, neutral detergent. Do not use abrasive cleaning agents on the paper.

It is almost impossible to repair damaged wallpaper. If pieces of the wallpaper were saved at the time of the original installment, an attempt could be made to replace the damaged portion. On a commercial basis, this is not an economical job. The complete wall should be repapered, which is why wallpaper is often considered an uneconomical wallcovering.

Painting of Interiors

Refer to Chapter 6 for complete details on painting. However, there are a few items that are unique to interior painting, which are reviewed in this section.

Coatings used for interior painting generally are not the same as those used for exterior painting, since the latter are primarily designed to withstand exterior environments. Availability of interior finishes is comparatively broad, allowing considerable choice of products to meet the properties desired.

Interior coatings do not generally require repainting as a result of normal deterioration. The most common reasons for painting are cleanliness, illumination, and general appearance. The areas where the environment can be detrimental are limited to specific areas in the building.

Interior Deterioration

With time, interior coatings generally change slowly in appearance, but usually do not degrade to any significant extent. Interior finishes do change in appearance with aging, but not as rapidly as exterior finishes. The changes are somewhat similar, but for different reasons.

Degradation is a relatively minor problem with interior coatings. Furthermore, it is generally confined to relatively small areas. Enamels on woodwork may become brittle with age and crack. Cracking may also show up on wall paints when the building settles slightly. The cracks usually are quite fine and may be easily repaired and touched up. Because areas around switches or door handles may be cleaned often during the life of the paint in order to remove smudges, eventually the paint will be removed by the abrasion of the cleaner.

Surface Preparation

1. Make a list of materials and the quantity of each needed. The amount of paint will depend on the type of paint and the condition of the surface. Also list all repairs to be made.
2. Nail all trim down with finishing nails and countersink below the surface. Patch holes with spackling compound, then allow to dry.
3. If a gap exists between the baseboard molding and the floor, remove the baseboard and replace as flush to the floor as possible.
4. Fill all cracks between pieces of trim with spackling compound and sand with fine sandpaper.
5. Scrape off all loose or peeling paint with a putty knife. Follow this by sanding with fine sandpaper and wiping clean with a dry cloth.
6. Repair any punctures in the wall.
7. All nails that have "popped" should be countersunk with a nailset, filled with spackling compound and sanded.
8. All other holes and cracks must be filled with spackling compound, and sanded.
9. Wash the wall and allow to dry.
10. Seal all spackled spots by dabbing a bit of the finish paint over the sanded spackle and allow to dry.

11. Remove all switch cover plates, door handles, curtain rods, and other fixtures which could be splattered.
12. Remove lightweight furniture from the room.
13. Spread paper or dropcloths over the floor. Cover any furniture with dropcloths especially if the ceiling is being painted (including all mechanical equipment).
14. Use masking tape to cover areas that do not require paint or where the paint does not have to blend with the original surfaces.
15. Paint materials and paints are reviewed in Chapter 6.

Refer to Figures 7.5 and 7.6 for assistance in the selection of interior coatings.

Application of Wood Finishes		
Type	Initial Steps	Final Steps
Blond Finish	On light woods brush on a coat of pigmented white (or pastel-colored) stain, then wipe off. On dark woods, bleaching with a prepared wood bleach is required first.	Allow the stain to dry, then apply two coats of water-white, satin-finish varnish or clear plastic coating in satin finish. Let the first coat dry hard before applying a second one.
Clear or Natural (using penetrating sealer)	Brush on first coat of penetrating sealer. Wipe off excess with cheese-cloth after allowing it to set for five or ten minutes.	Let first coat dry for 24 hours. Then brush on a second coat and rub off. Allow to dry hard, then rub on a thin coat of paste wax.
Clear or Natural (using surface-type coating)	Brush on the first coat of varnish, shellac or lacquer.	Let first coat dry overnight, then sand lightly. Apply a second coat.
Stained Finish	For a uniform effect on soft woods, apply a thin coat of clear resin sealer before staining. Allow to dry, then brush on oil stain. Wipe off after five or ten minutes.	After the stain has dried, brush on two coats of varnish. Two coats of shellac can also be used.

Figure 7.5

Interior Coatings

	Flat Paint	Semi-Gloss Paint	Enamel	Rubber Base Paint (Not Latex)	Emulsion Paint (Including Latex)	Casein	Interior Varnish	Shellac	Wax (Liquid or Paste)	Wax (Emulsion)	Stain	Wood Sealer	Floor Varnish	Floor Paint or Enamel	Cement Base Paint	Aluminum Paint	Sealer or Undercoater	Metal Primer
Plaster Walls and Ceiling	X*	X*		X	X	X												X
Wall Board	X*	X*		X	X	X												X
Wood Paneling	X*	X*		X	X*		X	X	X		X	X						
Kitchen and Bathroom Walls		X*	X*	X	X													X
Wood Floors								X	X	X*	X*	X	X*	X*				
Concrete Floors									X*	X*	X			X				
Vinyl and Rubber Tile Floors									X	X								
Asphalt Tile Floors										X								
Linoleum								X	X	X			X	X				
Stair Treads								X			X	X	X	X				
Stair Risers	X*	X*	X*	X			X	X			X	X						
Wood Trim	X*	X*	X*	X	X*		X	X	X		X						X	
Steel Windows	X*	X*	X*	X												X		X
Aluminum Windows	X*	X*	X*	X												X		X
Window Sills			X*				X											
Steel Cabinets	X*	X*	X*	X														X
Heating Ducts	X*	X*	X*	X												X		X
Radiators and Heating Pipes	X*	X*	X*	X												X		X
Old Masonry	X	X	X	X	X										X	X	X	
New Masonry	X*	X*	X*	X	X										X		X	

Asterisks indicate that a primer or sealer may be necessary before the finishing coat (unless surface has been previously finished).

Figure 7.6

CHAPTER 8 FLOORS

Repair and maintenance of floor coverings are among the most important aspects of building operations, since they are the most used and abused portions of a structure. From a statistical standpoint, floors probably receive about 90 percent more wear than any other part of a building and account for about 50 percent of the overall cost of the building operations. Therefore, it is of utmost importance that the proper flooring material is carefully selected and maintained. It is not enough to inspect floors on a yearly or semi-annual inspection tour of the facility. Floors should be continually inspected and minor defects corrected before they become major problems.

Figure 8.1 contains a comparison of some of the major types of flooring materials available, including their advantages and disadvantages. Figure 8.2 is a comparison based on cost. The major types of flooring materials (i.e., resilient, wood, etc.) are presented in order of increasing costs.

General Floor Repair

It is of no use to implement and proceed with a repair procedure until the cause of the deterioration has been found. The major causes of deterioration of flooring materials are as listed below:

- Moisture
- Settlement
- Shrinkage
- Structural defects
- Inadequate materials
- Inadequate installation
- Chemical attack—environment exposure
- Operational abuse
- Poor construction details

Once the deterioration has been exposed and the cause found, the facility manager must decide how to proceed. The options available are listed below:

- Repair
- Abandonment
- Isolation of the floor from the cause
- Living with the cause
- Replacing the floor system with a different material or supporting medium

Selection Criteria for Flooring Material

Type of Flooring	Location	Grease	Alkalis	Abrasion	Stains	Cigarette Burns	Indentation	Resilience	Quietness	Ease of Maintenance	Durability	Under-foot Comfort	Warmth to Touch
Aluminum	BOS	A	B	A	A	A	A	D	E	B	C	E	E
Asphalt	BOS	D	B	D	D	D	D	E	E	C	B	C	C
Brick	BOS	B	D	C	D	A	A	E	D	D	D	E	E
Carpet (General)	BOS	E	B	B	B	C	B	B	A	B	B	A	A
Clay Tile (Ceramic Tile)	BOS	C	D	C	C	C	A	E	D	A	A	D	D
Concrete	BOS	C	C	B	C	A	A	E	D	B	A	E	E
Cork	S	C	C	C	D	C	C	A	A	B	E	A	A
Linoleum	S	A	E	C	D	B	E	D	D	C	C	B	B
Rubber	BOS	B	B	C	B	A	A	B	B	E	B	A	B
Steel	BOS	A	B	A	A	A	A	E	D	A	A	E	E
Stone	BOS	B	B	B	C	A	A	E	C	A	A	D	D
Terrazo (& Oxychloride)	BOS	B	B	B	C	A	A	E	C	A	A	C	C
Vinyl Solid	BOS	A	A	C	B	B	B	C	C	B	A	A	B
Vinyl & Rubber Backing	BOS	A	A	C	B	C	B	C	C	B	A	B	B
Rag Felt Backing	S	A	B	C	B	D	C	C	C	B	A	B	B
Vinyl Composition	BOS	B	B	D	D	B	D	E	E	B	B	B	B
Vinyl Cork	S	A	B	C	B	E	B	B	B	B	C		
Wood	BOS	D	D	C	D	D	B	C	C	B	B		

1. Alphabetical rating of each product compared to other products with A being the highest rating.
2. Location — B = Below grade;   O = On Grade;   S = Suspension
3. See Figure 8.14 for a more thorough examination of various carpets.

Figure 8.1

Comparison of Average Costs of Flooring Material

Flooring Group	Group Items	Relative Costs* as compared to flooring material within this flooring group	Relative Costs* as compared to all flooring material
Resilient:	Asphalt Tile	A	A
	Linoleum	B	A
	Vinyl Composition Tile	B	A
	Rubber Tile	C	A
	Homogeneous Vinyl Tile	C	B
	Cork Tile	D	B
	Backed Vinyl Tile	B	A
	Vinyl Cork Tile	E	B
Wood:	Oak Strip	A	B
	Maple Strip	A	B
	Block	B	B
	Parquet Block	C	C
Carpet:	Polyropylene	A	A
	Nylon	B	B
	Antron Nylon	C	C
	Acrylic	D	C
	Wool	E	D
Formed in Place:	Mastic	A	A
	Polychloroprene	B	B
	Epoxy	C	B
	Polyurethane	C	B
	Polyester	D	D
	Silicone	E	E
Masonry:	Concrete	A	A
	Magnesite	B	C
	Brick	C	E
	Ceramic Tile	C	D
	Terrazzo	D	E
	Slate	D	E
	Marble	E	E

*Ratings used are based on per square foot installed price for material and labor.

A – Very Inexpensive
B – Inexpensive
C – Moderate or average
D – Expensive
E – Very expensive

Figure 8.2

Whichever choice is made, one must consider the cost of the decision relative to the functional role that the flooring material plays within the facility. Therefore, the scope of any floor repair should be dictated by the present and future use of the building, its projected lifetime, and the relative cost of replacement or repairing of flooring materials.

Housekeeping Procedures

The daily maintenance of floors must be protective and preventive to the finishing materials. For this reason, floor maintenance requires much consideration and effort, more so than for any other type of interior material. The following factors should be considered when establishing a floor maintenance program:

- Frequency of cleaning (brushing, sweeping, washing, etc.).
- The pH and concentration of cleaning solutions.
- Weight of fixed and moving objects.
- Amount and frequency of traffic (abrasion on floors).
- Quantity of airborne dust settling out on floor.
- Grinding in of settled dust.
- Formation of oily films.
- Oxidizing effects of atmosphere.
- Exposure to color-fading light rays.
- Attack by industrial gases and chemicals (possible stain producers).
- Extreme changes in temperature and humidity.
- Attack by fungi, insects, and animals.

As noted in Chapter 1, the correct installation of the flooring materials alleviates many potential maintenance problems. Manufacturers' specifications and recommendations for installation should be closely followed. Once the flooring material has been installed, the initial treatment of them is just as important as, if not more important than, the daily routine maintenance program. The initial treatment should be performed as part of the original building construction, and observed by the personnel who will be involved in the maintenance of material once the facility is occupied.

Cleaning Products

The facility manager should take care to select appropriate detergents, waxes, sealers, and equipment. These products should not cause further deterioration. The reader is referred to the charts in this chapter for help in the selection process. The manufacturer or supplier should also be consulted to insure that the correct materials are being utilized for the specific flooring material. There is no one always appropriate cleaning and/or finishing compound for any one floor material. It depends on such parameters as the exposure of the floor, the amount of frequency of traffic, the location of the flooring material, the material of the finished flooring material, etc.

Equipment Selection

There are numerous types, styles, and models of automated cleaning equipment available. Before purchasing any equipment, an evaluation must be made relative to the needs of the maintenance staff and budget available. Equipment manufacturers or dealers should be consulted in the evaluation process. The company's historical records or past experiences should also play a role in determining the equipment types and needs for a specific function. Some general rules that can be followed in the selection process are as follows:

1. When selecting the type of equipment to be utilized in cleaning flooring materials, parameters such as the nature of flooring materials, degree of congestion, type of training and labor available, minimum hygiene standards, and costs should be considered.
2. Use automated machines, whenever possible.
3. Any machines that are to be used should require a minimum amount of labor time to set up and operate.

4. Purchase and use the largest machine which can be accommodated in the building or in the space where it is to be used.
5. Machines should be as portable or as light as possible; they should not be bulky and thus a hindrance to the operator.
6. A floor machine should be flexible, have variable speed, pressure, and brush diameters, and should be able to not only scrub, but also polish.

Parameters for Product Evaluation

Time, manpower, and resources should be made available to ascertain the most effective and efficient cleaning products, equipment, and method for the specific type and location of flooring materials. The parameters that should be utilized in formally evaluating the effectiveness of any cleaning product, equipment, and procedures are:

- Shiny and lustrous glow
- Scuffing
- Scratching
- Water spot resistancy
- Other chemical (used in the floor location) resistancy
- Black marks
- Powdering
- Wearing ability versus amount of traffic
- Discoloration
- Soil resistancy
- Slip resistancy
- Workability of cleaning or finishing product in terms of application, reconditioning, and stripping.

Frequency and Productivity Rates

Many of the finished coverings (surface) for floors are the same as for walls and ceilings as covered in Chapter 7. The main difference, as stated above, is that the floor surface will be exposed to more wear. Therefore, the frequency of preventive maintenance and of possible repair will increase. Supplementary tables have been included in this section to aid in the planning, maintenance, and repair of flooring materials.

Not only will this chapter include step-by-step procedures for repair of most flooring materials, but it also will present the different types of preventive type housekeeping practices, including stain removal, that should be implemented to increase the lifetime of the flooring material. The reader will note that with each maintenance procedure is given the equipment, materials, labor productivity, and frequency. The labor productivity rates are given as ranges due to such parameters as location of material relative to work area, number and frequency of obstructions, and degree of finished appearance desired. The given frequencies are a beginning toward the development of a company's own production records. It must be kept in mind that, due to the wide variety of flooring materials available and soil conditions possible, maintenance practices will vary in the method, equipment, and materials utilized to keep the floor in an operational or functional condition. Floors can be ruined by implementing improper maintenance practices. Once an effective combination of materials, equipment and methods is found, document and continue to use it. Refer to Figures 8.3 through 8.7 for aid in developing a thorough floor maintenance program.

Equipment Commonly Used in Housekeeping and Repair of Floors

Broom, Upright, 32 lb. per doz.
Broom. Push, 18"
Broom, Push, 24"
Brush, Floor Sweeping 18", 100% Horsehair
Brush, Floor Sweeping 24", 10OZ Horsehair
Brush, Floor Sweeping 30", 100% Horsehair
Brush, Floor, Sweeping Sidewalk 24"
Brush, Scrub, Hand 8"
Brush, Scrub, 10" for Handle
Brush, Dusting, Bench
Cleaner, Vacuum, Household, Tank Type
Cleaner, Vacuum, Industrial
Cloth, Polishing (Oil Treated)
Cloth, Polishing (Untreated)
Chamois, Leather, Sheepskin, 16" x 21"
Gloves, Gauntlet, Work
Gloves, Rubber
Gloves, Short, Work
Handle, Wood. Threaded End 72" Brush Broom
Handle, Wet Hop, 60", Screw Type
Handle, Dust Mop, 72"
Knife, Putty, 1 1/4"
Machine, Scrubbing 26" Twin Brush
Bucket, 14 qt., Galv.
Mophead, Dusting, Cotton
Mop Bucket, 26 qt.
Mop Wringer, Squeeze Type
Mop Truck, Two-Tank, 28 gal. Capacity per Tank, 3" Diameter Wringer Rolls
Hop, Dusting, Cotton, 17", Straight
Mop, Dusting, Cotton, 27", Straight
Mop, Cellulose. Sponge, Yarn, 24 oz.
Mop, Cellulose, Sponge, Yarn, 12 oz.
Mophead, Wet, Cotton, 1 1/2 lb.
Mophead, Dusting, Cotton
Pad Steel Wool Floor Polishing
Polisher, Floor, Electric, 11", Disk Type
Polisher, Floor, Electric, 15", Disk Type
Polisher, Floor, Electric, 18", Disk Type
Polisher-Scrubber, Vacuum, Floor Electric
Scrubbing Machine, Floor Electric, Power Batter
Sponge, Cellulose, 3-13/16" x 6: Laminated
Sponge, Natural
Squeegee 18", Floor, with Handle
Truck, 2 Wheel, Hand, General Utility Type
Truck, Platform, Hand, Non-Tilt, Type Wood, 60" Long
Wax Applicator
Wax, Applicator Pad
Wringer and Bucket, Mop
Pad, Steel Wool, Grade D
Pad, Steal Wool, Grade 1
Pad, Steel Wool, Grade 2
Pad, Steel Wool, Grade 3
Pad, Lambswool Polishing
Pad, Synthetic, Fine, Polishing
Pad, Synthetic, Medium, Scrubbing
Pad, Synthetic, Coarse, Stripping
Pads, floor polishing machine, nonwoven nylon web, for maintenance and cleaning of resilient tile, terrazzo, marble, ceramic and wooden floors
Pads, floor, curled hair, extra thick durable, washable and reversible
Pad, thick nylon for rotary polisher—Coarse 15"
Pad, thick nylon for rotary polisher—Medium "
Pad, thick nylon for rotary polisher—Fine "
Baseboard pad-holder for floor machine brush

Figure 8.3

Summary of Basic Cleaning Equipment

Item	Use	Advantages	Disadvantages	Comments
Brushes and Brooms	Removal of fine and coarse particles or dirt, especially on rough surfaces.	Good for small areas. Will not remove wax.	Relatively inexpensive method of cleaning large areas. Will not remove soil stuck to floor.	Bristle should be firm, but flexible. Use method on rough concrete, unpainted or unsealed wood.
Mops: Dust Mops	Removal of dust, dirt and litter on smooth surfaces.	Quick. Good for smaller areas. Will not remove wax.	Will not remove embedded dirt or dirt stuck to floor.	Use untreated mop on smooth concrete, linoleum, asphalt, vinyl composition, vinyl, terrazzo, mosaic, rubber tile and other smooth surfaces. Use treated mop on smooth and sealed or painted wood.
Damp Mopping	Removal of water soluble soil deposits and stains.	Reduces the amount of time between waxings.	May remove wax finishes.	Part of soil removed by mopping action and part by detergents.
Wet mopping	Removal of surface and embedded dirt; done along with stripping.	Will remove coated-on dirt.	More costly than damp mopping. Water can ruin floor, furniture and molding.	Do not use on wood.
Automatic Scrubbers, Polishers (with or without controlled liquid feed). Scarifiers, Shampooers, Auto-Scrubbers	More effectively and efficiently cleans, polishes, and buffs floors. Removes embedded dirt. Scarifiers are used when dirt is deeply embedded in a coarse surface.	Faster action. Will remove stubborn dirt. Good for small areas.	Equipment must be maintained. Operators must be trained. Expensive method of cleaning large areas. Equipment must be maintained. Will not pick up embedded dirt.	Good on all surfaces.

Figure 8.4

Summary of Basic Cleaning Equipment (continued)				
Item	Use	Advantages	Disadvantages	Comments
Carpet Sweeper	Use on carpets to remove small size surface litter.			
Vacuum Cleaner (Portable, canister, etc.)	Picks up litter and loose dirt on rug backing.	Large area cleaning. Faster and more area covered. Can get at hard to reach places.	Equipment to maintain. Will not get out dirt stuck to carpet. Will not clean heavily soiled carpet.	Can also get wet-vacuum which will pick up water faster than with cloth. Dirt (water) picked up must be disposed of properly. Use on all surfaces.
Vacuum Sweeper	Picks up embedded dirt in carpet.	Does more thorough job on carpet.	Will not clean a soiled carpet. Equipment to maintain.	
Miscellaneous Equipment				
Squeegee	Aids in picking up water.	Control water.	Can only be used on flat surfaces.	
Sponges	Aid in picking up water.	Rapid pickup of water in small area.	Sponges need to be maintained so they don't rot, etc.	
Cloth	Varied cleaning uses.			

Figure 8.4 (continued)

Kind of Floor Surface	Bristle Floor Brush	Fiber Floor Brush	Corn Broom	Sweeping Mop Treated	Sweeping Mop Untreated	Powered Equipment
Rough, unpainted, or unsealed open-grained wood floor		XX	X			
Smooth, unpainted, or unsealed wood floor	X	XX				
Smooth, sealed, or painted wood floor — not waxed	X			XX		
Smooth, sealed and waxed wood floor	X			XX	XX	XX
Linoleum — waxed				XX	XX	XX
Asphalt tile — not waxed	X			XX	XX	XX
Asphalt tile — waxed	X			XX	XX	XX
Vinyl — not waxed	X			XX	XX	XX
Vinyl — waxed	X			XX	XX	XX
Terrazzo					XX	XX
Mosaic Tile				XX	XX	XX
Quarry Tile				XX	XX	XX
Rubber Tile	X			XX	XX	XX
Rubber Tile — waxed	X			XX	XX	XX
Rough concrete		XX	X			
Smooth concrete — not treated to eliminate dustiness	X	XX				
Smooth, treated or painted concrete	X	X		XX	XX	XX
Oxychloride cement	X				XX	XX

A Summary of Sweeping Tools For Various Types of Floors

1. XX means that the tool is to be used if equipment is available. X means that the tool is to be used only if tool marked XX is not available. Where two tools are checked with the same symbol, either may be used.

2. Use a liquid emulsion mop-treating compound. Do not use an oil emulsion or an oil base.

Figure 8.5

A Summary of Basic Cleaning Solutions

Solution	Summary of Properties
Soap-based compounds (alkali salts, fatty acids)	Solid variety: —Not as easily rinsed —Less soluble in cold water Liquid variety: —More soluble in cold water —More thorough for deep-cleaning action
Alkaline-salt compounds (alkaline salts, no pH higher than 10.0)	—Less expensive —Heavy duty cleaner —May be used with soap-based detergents —Potentially damaging to certain types of floors: linoleum, asphalt tile vinyl composition, vinyl or rubber —Effective in removing stains, oil and stubborn soil
Synthetic detergents (alkaline salts)	—Available in solid, powered, granulated or liquid form
Types: Anionic	—Can be used on all types of flooring
Cationic	—Liquid variety is less likely to cause any floor damage
Nonionic	
Abrasive compounds	Small areas on small areas where detergents won't do the job Properties vary according to the type of abrasive Cannot be used on all surfaces
Special-purpose compounds	Available as additive to above types, such as germicidal solution
Solvents	May have a dissolving effect on paints and plastics Are fire and toxic hazards Safety precautions must be taken
Alcohol	Dissolves resins, waterproof inks, iodine, etc. No effect on starches or gums Little use in dissolving waxes, rubber, tar or plastic adhesives Can be used on rubber or plastic material

Figure 8.6

A Summary of Basic Cleaning Solutions (continued)	
Solution	**Summary of Properties**
Acetone	Dissolves resins, paint, nail varnish and plastics Use only on inorganic surfaces such as concrete Use only for emergency stain removal
Gasoline	Very volatile Dissolves tar, rubber, wax and pitch Little effect on paints or resins
Kerosene	Same as gasoline
Trichlorethylene	Dissolves fats, waxes, oils, tars, pitch, paint, resin, rubber and many plastics Use only on concrete, magnesite and inorganic materials
Carbon Tetrachloride	Very toxic
Tetrachloroethylene	Similar to Trichlorethylene
Methylene Chloride	Dissolves fats and waxes Good for paint stripping
Sweeping Compounds	Not a good idea to use on resilient floors Use to obtain dust-free floor Takes more time to sweep a floor

Figure 8.6 (continued)

A Summary of Waxes and Polishes

Type	Properties
Buffable Types:	
Liquid Emulsion	Produces a gloss upon drying. gloss can be highlighted by buffing.
	Susceptible to scuffing, marring and heel marking
	More resistant to water spotting, if buffed.
	Excellent slip resistance.
	Do not use on cork.
Solvent Base Liquid or Paste	Do not use on asphalt, rubber or vinyl.
	Can be used on asbestos and linoleum.
	Should be used on wood and cork.
	Easier to apply liquid form.
	Paste type results in thicker, more wearable coat.
	May be buffed to achieve a high gloss.
	Must be polished to last.
Paste-Emulsion	Use on asphalt, vinyl, asbestos and rubber.
	Can also act as a cleaning agent.
	Less durable surface.
	Do not use on cork.
Non-Buffable Types:	Appearance level drops off faster than buffable.
Water Base Emulsion	Time period between stripping and refinishing is shorter than buffable.
	Produces harder and less tacky surface.
	Becomes scratched instead of showing scuff marks.
	Much greater care must be taken in applying.
	Will break down faster than buffable waxes under low temperature and humidity conditions.
	More difficult to strip.
	Do not use on cork. May use on all other surfaces.
	Do not apply over waxed surfaces.
	Do not use in area exposed to sand or gritty soil.
	Excellent resistance to oil and grease.
Water Based Polymer	Highly resistant to alkaline cleaning compounds.
	Special stripping methods needed.
	May damage floor.
Wash-wax Products (Combination of polymer and synthetic detergents)	Can clean and wax the floor in one step

Figure 8.7

Manual Sweeping and Dust Mopping

The following equipment, materials, and procedure are recommended for manual sweeping and dust mopping.

Equipment	Materials
Broom	Sweeping compound
Dust Pan	(Do not use on all surfaces)
	Brush or straight broom

1. Lay out sweeping compound in line with path intended to be swept.
2. Begin at one end of the sweeping compound line and sweep it forward as far as you can reach conveniently without taking a forward step.
3. Use one hand to push and the other to guide the broom. Use a smooth and uniform rhythm.
4. Don't pile up too large a pile of dirt.
5. Use a counter brush or straight broom to sweep out corners.
6. Pick up dirt from piles using a dustpan and disposing into a movable waste container.

Labor Production (per worker)	Average Frequency
Dust mopping 5–20 minutes/1000 S.F.	Daily (or as needed)
Sweeping 10–25 minutes/1000 S.F.	

Use higher figures when more than the average number of obstructions are encountered.

Manual Floor Mopping

The following equipment and materials are used for all forms of manual floor mopping. Procedures for damp mopping, wet mopping, and scrubbing are outlined below.

Equipment	Materials
Mops	Detergent suitable for
24–36 oz. for men	mopping
12–16 oz. for women	
Dust mop	
Putty knife	
Safety signs	
Counter brush	
Abrasive mop pad	
Bucket and wringer	
Dust pan	
Clean cloth	

Damp Mopping: Damp mopping, or buffing, is usually done between waxings. A suggested procedure for damp mopping is described below.

1. Fill bucket with clear, fresh, warm water.
2. Remove any obstructions.
3. Sweep area clean (see previous instructions).
4. Place warning signs where they can be seen.
5. Using putty knife, remove tar, gum, and other substances from the floor.
6. Proceed to mop the floor. Be sure to wring out the mop as dry as possible.
7. First mop along the baseboards by drawing the mop parallel to them.
8. If you need to mop under furniture, use a push-pull technique.
9. On the rest of the floor use a side-to-side mopping stroke which results in the greatest coverage and speed with the least amount of fatigue. Turn mop from time to time.
10. Rinse and wring mop and repeat. Keep mopping equipment ahead of the work area next to be mopped.
11. Attach an abrasive pad to the mop for getting out stubborn soil.
12. Using a dry, clean cloth, wipe furniture legs and baseboards dry.
13. When mop appears to be getting dirty, change the water.

14. Do not allow the water to stand on the floor and do not drop on the baseboards or on the furniture legs.
15. When finished, clean all equipment and return them to proper storage. Hang the mop to dry.
16. Remove warning signs.
17. Replace any objects moved.

Labor Production (per worker)	Average Frequency
Buffing: 15–40 minutes/1000 S.F.	As needed
Spray Buffing: 20–50 minutes/1000 S.F.	

Use higher figures when more than the average number of obstructions are encountered.

Wet Mopping: A suggested procedure for wet mopping is described below.

1. Prepare detergent solution in the bucket.
2. Place rinse water in another bucket or in the second compartment of the main bucket.
3. Repeat steps 2 through 5, inclusive, of the Damp Mopping section.
4. First apply detergent solution along the baseboards, same as for damp mopping, and under any furniture. Apply the solution to an area of about 8 by 20 feet. The quantity applied should be such as not to allow wetted area to dry out before rinsing. Be sure to get into the corners.
5. For embedded dirt use more mop pressure or an abrasive pad or abrasive cleaner (do not use abrasives on all types of surfaces).
6. Pick up any detergent and wring out the mop in a separate bucket compartment or in another bucket. Other methods, such as a water vacuum, can be used. Do not allow the mop used for placing detergent to remain in contact with the floor for any length of time; staining could result.
7. Using another clean mop, apply the rinse water to the floor.
8. Pick up the rinse water and wring out.
9. Repeat steps 12 through 16 of the Damp Mopping section.

Labor Production (per worker)	Average Frequency
Damp Mopping: 15–30 minutes/1000 S.F.	Heavily used area—daily
Wet Mopping: 30–50 minutes/1000 S.F.	Secondary areas—Weekly
Scrubbing: 50–140 minutes/1000 S.F.	

Scrubbing

The following procedure is recommended for scrubbing using an electric floor machine.

Equipment	Materials
Power scrubbing machine	Detergent for specific
Pads	surface
Buckets, wringer	
Mops	
Putty knife	
Clean dry cloth	
Warning signs	

1. Prepare equipment and materials.
2. Sweep area and dispose of dirt properly.
3. Remove any items, such as furniture, and set up warning signs.
4. Wet the mop with the detergent solution and spread the solution over the floor using side strokes. Along baseboards, use parallel strokes. Avoid splashing the furniture. Follow those applicable items under the Damp Mopping job description for the balance of this activity.
5. Using the power scrubbing machine, proceed to scrub, covering the maximum area without creating a fatigue condition.
6. Use the mop, scrubbing to get to those areas which are inaccessable when using the power scrubber.

7. After scrubbing, use a well wrung-out mop to pick up the detergent solution and wring the mop out in another bucket. Pick up using a side-to-side, front-to-back method, turning the mop regularly.
8. Rinse floor, if required.
9. Wipe the baseboards and furniture legs clean.
10. Clean and store equipment properly.
11. Return the furniture to their original locations and remove the warning signs after the floor has dried to the touch.

Labor Production (per worker)	Average Frequency
15–30 minutes/1000 S.F. Other areas—weekly	Heavily used areas—daily

Size Machine Suggested	Size Area Scrubbed in Square Feet
19″	up to 10,000
21″	10,000 to 20,000
24–26″	20,000 to 50,000
30–35″	50,000 to 100,000
60″	100,000 to 200,000
72″	200,000 +

Stripping

The following equipment, materials, and procedure are recommended for stripping floors.

Equipment	Materials
Power floor machine	Stripping compound or
Machine pads	detergent solution
Vacuum	
Clean cloth	
Sponges	
Utility brush	
Mops	
Bucket and wringer	

1. Dust mop or sweep floor, picking up the dirt and disposing of it correctly.
2. Obtain and make ready all equipment and the solution.
3. Remove the furniture and erect warning signs. Turn back any rugs so that they don't get soaked.
4. Remove any tar, gum, etc., with your putty knife.
5. Lay down the stripping detergent as outlined in the Mopping job description.
6. Using power floor machine, work in solution to strip wax (sealer).
7. Pick up the resulting slurry either with water vacuum or as described in the Wet Mopping job description.
8. Rinse the floor twice, using the mop, and change the rinse water whenever necessary.
9. Hand-strip corners and baseboard regions.
10. Proceed with the waxing process (see Waxing job description).
11. Clean and store equipment.

Labor Production (per worker)	Average Frequency
100–200 minutes/1000 S.F.	As needed for type of flooring and wax

Waxing and Buffing

The following equipment, materials, and procedure are recommended for waxing and buffing.

Equipment	Materials
Power Machine	Wax suitable for flooring
Pads for buffing	material
Clean mop or sponge mop	
applicator	
Buckets and wringer	
Dry cloths	

1. Obtain the equipment and materials and proceed to the job site. If working on stripping first, be sure the floor is dry before waxing begins.
2. Saturate the lower two-thirds of the mop with wax. If using water emulsion wax, first wet the mop with water and wring out.
3. Apply wax to the floor, using long sweeping strokes, with slight overlap. Do not apply too thick a coat. It is better to put down thin coats. Do not begin at regions adjacent to baseboards. Stay a tile-distance away.
4. When applying thin coats, apply second coat at right angles to the first coat. Let the wax dry thoroughly.
5. When using water-emulsion wax, rinse the mop periodically.
6. If using paste wax, apply to a small area with a cloth applicator.
7. If using buffable wax, proceed to buff.
8. Begin to buff where you first started to wax.
9. Using a side-to-side stroke, polish the floor to a sheen. Do not over buff and do not nick baseboards or furniture legs.
10. Clean all equipment (mops, pads, etc.) and return to the proper storage area.
11. Store materials properly. Do not pour any leftover wax into the original container. Either dispose of it properly or pour into another container.

Labor Production (per worker)	Average Frequency
Rewaxing (one coat) 15–30	
minutes/1000 S.F.	60 days
Stripping and Rewaxing (two coats)	
100–300 minutes/1000 S.F.	
Waxing and Buffing (one coat)	
30–70 minutes/1000 S.F.	

Power Sweeping (Vacuuming)

The following equipment and procedure are recommended for power sweeping.

Equipment

Carpet vacuum sweeper
Vacuum bags
Attachments

1. Police the area by picking up surface litter.
2. Be sure the vacuum bag is empty.
3. Adjust the brush height according to type of carpeting. See manufacturer's specifications.
4. Push back and forth in a sawtooth motion, paying particular attention to the surfaces along the baseboards and not forgetting to vacuum behind the doors and under the furniture.
5. The most heavily traveled areas should get more attention. Vacuum over those areas more than once.
6. Use attachments to get in corners, etc.
7. Replace any items previously removed from the floor area.
8. Dispose of bag and return the vacuum to storage.

Labor Production (per worker)	Average Frequency
Dry vacuuming—15–40 minutes/1000 S.F.	Daily
Wet vacuuming—30–50 minutes/1000 S.F.	As needed

Use higher figures when more than the average number of obstructions are encountered.

Carpet Mopping

The following equipment, materials, and procedure are recommended for carpet mopping.

Equipment

Bucket and wringer
Clean dust mop-with locking head
Warning signs

Materials

Mopping solution for specific type carpet

1. Prepare the equipment and material for use.
2. Remove any obstacles and place warning signs.
3. Dip the dust mop into the solution so that only a small amount of the solution gets on the mop top.
4. Using a push-pull stroke, mop the carpet.
5. Re-dip the mop as needed and repeat the procedure.
6. Do not soak the carpet.
7. Let the carpet dry.
8. Vacuum the carpet.
9. Return the furniture, etc., to their original locations and remove the warning signs.
10. Clean the equipment and store properly.

Labor Production (per worker)

20–40 minutes/1000 S.F.

Average Frequency

As needed

Shampooing

The following equipment, materials, and procedure are recommended for shampooing carpets.

Equipment

Power scrubber
Brushes
Vacuum cleaner

Materials

Rug Shampoo

1. Remove all loose items, such as furniture.
2. Vacuum the rug thoroughly.
3. Protect drapes, curtains, woodwork, and baseboards.
4. Remove any stains before proceeding.
5. Using the wet shampooing method, spread a small amount of shampoo and let the scrubber work up a lather.
6. Shampoo a small area in about eight to ten foot squares. Move from left to right releasing a small amount of shampoo, as you progress, to insure that a uniform layer of suds is spread over the carpet. Overlap so streaking doesn't occur.
7. Scrub heavily soiled areas twice.
8. Do not saturate the carpet.
9. Use manual technique with a hand brush to get into corners and along baseboards.
10. Wipe dry the baseboards and furniture legs.
11. Set the pile by using a pile brush.
12. Allow the carpet to dry thoroughly (12–24 hours).
13. Vacuum the carpet.
14. Replace the furniture and take down the warning signs.
15. Clean all equipment and store properly.

Labor Production (per worker)

175–250 minutes/1000 S.F.

Average Frequency

Seasonal or as needed
Minimum-two times per year

Resilient Flooring

The first series of floor coverings to be discussed are termed resilient. They are a series of floor tiles which share a number of similar characteristics such as their manufacture, installation, and general physical properties. Those which will be dealt with in this text are linoleum, rubber, cork, asphalt, and a group of smooth-surfaced floor coverings whose chemical base is a vinyl-chloride resin.

Resilient floor coverings are available in several forms. The first form is a sheet which can be bought in several widths and delivered in rolls. Nowadays, however, resilient flooring is more often obtained in tile form in either nine or twelve-inch squares.

General Maintenance Guidelines

Basically, maintenance procedures are the same for all types of resilient flooring. However, the types of materials used in the cleaning and repairing process will differ depending on the composition of the floor. Resilient floors are somewhat porous and tend to be softer than other flooring materials; therefore, dirt is more easily ground into this type of flooring by foot traffic. This foot traffic causes an abrasive action that scratches the surface and fills the pores with dirt. This dirt is not easily removed by the everyday sweeping process. Wet cleaning with a cleaning compound that is suitable for that type of covering is needed. But even this type of maintenance procedure will not remove all the scratches and surface irregularities.

The floors must be cleaned and then protected from their exposure conditions. It is important that once the floor is cleaned, it is finished to fill the irregularities and seal the pores. This will result in easier future cleaning, will improve the appearance, and prevent any further damage from abrasive dirt. The protective finish itself will be affected by the same abrasive dirt. Therefore, it is important that the type of protective finish that is placed on the floor surface is easily removable. A good grade of wax should be applied to the surface in thin coats.

Each type of resilient flooring is covered independently in this section. General items that should be considered for all the different types are listed below:

1. Always use a cleaning detergent or product specifically designed and recommended for the type of resilient floor. Many times the facility will have more than one type of resilient floor and therefore, this point should be noted at the time of construction. Manufacturers of the flooring product can recommend specific types of cleaning products. There are also manufacturers associations, listed in the Appendix, that can help you in choosing the right cleaning product.
2. Excessive use of water or cleaning solutions should be avoided.
3. Do not use abrasive agents on the floor surface.
4. Avoid using harsh alkaline solutions.

For the type of floor wax to use, refer back to Figure 8.7. Waxing should be held to a minimum, since wax build-up can cause discoloration. The following precautions should be taken when waxing.

1. Always clean the floor before waxing. If the floor is not clean, an unsightly appearance will result, preventing wax from taking hold or bonding to the floor. A milky or flaky appearance can result if soil is left on the floor or other impurities are mixed with the wax.
2. Always be sure that the floor is dry before waxing. If you wax over a wet floor, the wax will not bond to it and a milky or cloudy appearance will result.
3. Always apply wax in thin coats.
4. Do not use wax as a preventive maintenance item just because the floor looks dull. Dry-buff the floor before attempting to wax again.
5. Be sure the wax is stored properly. Containers should not be left to stand open since the vehicles of the wax will evaporate and increase solid content, thus causing a heavier film to be applied.

6. It is advisable to apply wax to floor surfaces when the wax is at a temperature of about 60 to 75 degrees Fahrenheit. Cold wax on a cold floor will cause the wax to become heavier in viscosity and result in a thicker film on the floor.

Some of the major problems encountered with resilient floors are listed in Figure 8.8, along with a listing of the probable causes and solutions.

Asphalt Tile

Asphalt tile is considered to be moisture-proof and decay-proof. However, it is easily indented by heavy or pointed objects. The following substances are harmful to asphalt tile: oils, grease, gasoline, naphtha, turpentine, carbon tetrachloride, and kerosene. Oil, grease, or organic solvents will soften and dissolve the tile. Alkaline or caustic cleaners will dry out the tile and cause possible cracking, curling, and bleeding. The use of varnishes, lacquers, and shellacs can permanently damage an asphalt floor. Asphalt tile will fade from direct sun exposure and soften due to high temperatures. Too much exposure to low temperatures will result in brittleness and cracking, especially if the subfloor is not smooth and rigid. Under normal conditions and proper design installations, a reasonable life expectancy is about 15 years.

For the periodic maintenance (housekeeping) of an asphalt floor, refer to the earlier job descriptions of sweeping, mopping, etc. In addition, the following specifics should be incorporated into the housekeeping program for asphalt floors:

1. Use a soft bristle broom or a dust mop to clean the floor. If a sweeping compound is to be used, be sure it is water, rather than oil-based.
2. Apply a mild, neutral chemical cleaner and allow it to penetrate. Do not use an alkaline cleaner with a pH greater than 11. Follow by a good rinsing with warm water and a dry mopping. Do not allow the water to stand on the floor.
3. Do not use scouring powder or solvent type cleaners.
4. Sealing is mandatory for asphalt tile. Consult the asphalt tile manufacturer for recommended washing compounds and recommendations for their application and maintenance.
5. The sealing should be followed with the application of a water-based emulsion wax or a water-based emulsion resin floor polish. Apply two thin coats rather than one thick coat. Buff the dried floor.

Cork Tile

Cork tile is the most porous and resilient of all of the floor coverings. Cork flooring is not slippery as many of the other tiles are, especially when waxed. Cork has a natural tendency to fade and discolor. Cork is good insulation, practically odorless, and has no tendency to splinter or warp. Substances such as naphtha, gasoline, or similar organic solvents, oil mops, or sprays are harmful to cork. Cork can be divided into the following categories:

- Natural
- Factory-waxed
- Resin reinforced
- Vinyl

Natural cork tile must be sanded, sealed with an alcoholic shellac or other sealer, given several coatings of a solvent-base paste or liquid wax, and buffed. This may be required every 12 to 18 months if a heavy traffic condition exists. Other housekeeping aspects that should be made part of a total maintenance program for natural cork tile are:

1. Sweep with a brush or chemically treated mop. Do not use oil-based compounds.
2. The floor can be damp mopped if needed, but water should be avoided, if possible. Use a neutral cleaner.
3. If the tiles are badly soiled, use a No. 00 steel wool pad with the cleaning solution.

Checklist of Maintenance Problems for Resilient Floors

Problem	Possible Causes	Solution(s)
Black Marks	Exposure to environmental abuse by shoes, furniture, etc.	Replace wheels, etc., with non-marking material.
		Scrub with No. 0 steel wool wetted with cleaning solution for specific material.
	Wrong polish.	Strip, clean and rewax.
Bleeding of Asphalt and Rubber Tiles	Oils, grease and/or other harmful chemicals.	Scrub with an abrasive floor pad followed with application of floor polish. If extensive, replace tiles.
	Inadequate floor material for exposure conditions.	Replace tile.
	Incomplete wax removal or excessive build-up.	Completely strip the floor, rewax and buff.
Broken Tiles	Inadequate subfloor	If bouncy: Need to replace subfloor and/or use more flexible filing.
		If unlevel: Need to relevel subfloor or use underlayment.
	Brittleness: Due to inadequate maintenance procedures and/or cleaning solutions.	If extensive, replace tiles and revise maintenance program.
Brittle and Porous Cork or Linoleum	Exposure to harmful cleaning agents.	Strip wax, apply sealer and rewax.
	Washing too often.	Change washing schedule.
Cuts	Heavy loads and/or moving loads.	Replace tiles. Remove causes.
Darkening of Floor	Incomplete wax removal or excessive buildup.	Completely strip floor and rewax and buff.
Discoloration of Vinyl Floors Due to Rubber Marking	Chemical reaction between materials.	Remove with a fine steel wool and rewax.
Edges Curling	Exposed to an excessive amount of moisture.	Remove source of water, remove tiles, and replace after allowing the area to dry completely.
Fading	Exposure to direct sunlight.	Utilize harsh floor pads or hone surface. Possible tile replacement.
	Harsh chemical exposure.	Utilize harsh floor pads or hone surface. Possible tile replacement.
Fine Scratch Marks	Soil and traffic abrasion.	Strip floor, apply sealer and remove causes of everyday soiling.
Holes	See cuts.	
Loose Tiles	Tiles unsuitable for exposure conditions and/or location.	Replace tiles.
	Inadequate installation.	Replace tiles.
	Powdering (whitish) from efflorescence.	Remove tiles, treat floors with a mild acid and replace tiles.
	Exposed to excessive water.	Replace tiles after removing cause and allowing area to dry.
Pitting	See Bleeding.	
Pock Marks	Indentation by heavy painted objects.	Hone surface or replace tiles.
Poor Gloss	Wax coating too thick.	Strip and rewax with thin coats.
	Porous surface.	Strip wax, apply sealer and rewax with thin coats.
	Residual film due to improper cleaning.	Strip and rewax with thin coats.
	Improper waxing technique and/or material.	Strip and rewax with thin coats.

Figure 8.8

Checklist of Maintenance Problems for Resilient Floors (continued)

Problem	Possible Causes	Solution(s)
Poor Leveling (spreading) of Wax	Improper application and/or materials for waxing, washing and rinsing.	Strip wax, wash and rinse thoroughly. Apply new wax in thin coats.
Powdering	Inadequate wax, improper cleaning and waxing procedures.	Strip, rinse, dry and rewax correctly.
Rapid Soiling	Improper type wax for exposure conditions.	Completely strip old wax, rinse and rewax using thin coats of proper wax.
Roughness of Asphalt and Rubber Tiles	See Pitting.	
Scratches	See Cuts.	
Shrinkage of Solid Vinyl	Using improper cleaning solutions.	Replace tiles and revise maintenance materials.
Softening of Asphalt and Rubber Tile	See Pitting.	
Stains	Refer to Figure 8.15	
Sticky Floors	Improper cleaning and/or waxing procedures. Inadequate waxes.	Completely strip, rinse and rewax floor.
Streaking of Polish	Improper cleaning and/or waxing procedure. Use of inadequate cleaners and/or waxes Porous floor surface.	Completely strip, rinse and rewax floor using thin coats. Seal porous floor.
Swelling of Rubber Tile	Utilization of oily base soapy cleaner.	Completely strip, rinse and rewax. Revise cleaning solution being used.
Tears	See Cuts.	
Uneven Appearance (dull and glossy areas occurring on same surface)	Improperly installed subfloors. Incomplete or uneven wax stripping.	Hone surface or install new subfloor. Completely strip, rinse and rewax floor. Check equipment.
Water Spotting	Improper cleaning procedures. Improper cleaners and/or waxes. Moisture, environmental.	Strip completely, rinse and rewax. Revise procedures and/or products. Remove source and follow up with procedures above.
Whitening	Incorrect tile. See Water Spotting.	Replace tile.

Figure 8.8 (continued)

Factory-waxed cork tile is impregnated with a wax composition at the factory and does not require a finishing treatment at the time of installation, where many of the others do. This flooring must be stripped, sanded, sealed, re-waxed and buffed periodically (not as often as natural cork). For the most part, the other maintenance procedures are the same as for natural cork tile.

Resin-reinforced cork tile is saturated with a resin-reinforced wax at the factory. It is somewhat less porous, and therefore, smoother than the first two varieties; therefore, there is less of a tendency for dirt to become impregnated in this kind of cork tile. There is no need for sanding, sealing, waxing, and buffing. It can be damp mopped or scrubbed with a detergent solution. Daily maintenance procedures are the same as for vinyl cork tile.

Vinyl cork tile is a denser product that can stand up to substantially more wear. It is the one that is most impervious to straining and accumulation and ingraining of dirt. Sanding, sealing, polishing, and buffing are usually unnecessary; in fact, sanding and sealing should not be done at all on vinyl cork tile.

The same housekeeping procedures used for the other resilient floors can be used for cork, keeping in mind the following:

1. Daily sweeping with a chemically treated mop or damp mopping.
2. Scrubbing, at more frequent intervals, with a water solution. Do not allow water to stand on the floor.
3. Application of a coat of solvent-base paste or liquid wax, followed by buffing.
4. Occasionally, buffing may be necessary; use a very fine steel wool pad to remove embedded dirt.

Linoleum Tile

Linoleum tile has a load limit of about twice as much as resilient tiles, such as asphalt tile. All forms of alkali, along with grease and oils, are injurious to linoleum. Highly alkaline materials will deteriorate linoleum, causing brittleness. Also color fading, whitening and eventually disintegration will occur due to contact with highly alkaline materials. As is true for other types of resilient flooring, excess amounts of water accumulation will damage the tile.

Abrasives should not be used except on isolated stains. Non-alkali cleaners should be used. Either solvent-type water wax or emulsified resin finishes may be used on linoleum tile.

The important facts to keep in mind when planning the cleaning program are as follows:

1. Sweep or dust mop, using a non-oily floor dressing on a daily basis.
2. For more stubborn dirt, an occasional damp mopping may be needed, followed by a buffing, using a No. 1 steel wool pad.
3. Re-buff with a nylon fiber pad, as needed.
4. If a more thorough cleaning is needed, scrub the floor using a mild cleaner (soapless detergent).
5. Apply two thin coats of wax. Avoid the use of a wood sealer, varnish, lacquer, or other hard finishes.

Rubber Tile

Rubber tiles are somewhat resistant to alkali, acids, and stains. They are also soundproof and slip resistant. The tendency for them to crack is minimum and they can withstand large amounts of pressure. Rubber tile should not be cleaned with regular soap. Special synthetic detergents made for rubber tile should be used according to the manufacturer's recommendations. Do not use substances such as kerosine, gasoline, naphtha, benzine, turpentine, mineral solvents, or harsh alkali. Avoid the use of sweeping compounds which contain oil, abrasives, or chemicals. Never apply any varnish, lacquer or paint to a rubber tile. Sunlight tends to deteriorate the tile.

For stubborn accumulations, use mild abrasive cleaners or steel wool with a synthetic detergent. Steel wool pads and a good wax stripper are effective in removing old wax coatings. Use only a water emulsion type wax. Other types of waxes will soften the tile. Some other things to keep in mind when maintaining a rubber floor are as follows:

1. On a daily basis, remove any dirt with a broom or a clean dust mop.
2. Damp mop the floor occasionally, to remove stubborn dirt.
3. To remove embedded dirt, wash the floor using a soapless detergent. Do not use a high alkaline cleaner.
4. Buff the waxed floor at regular intervals, using a fine steel wool pad.
5. At more infrequent iervals, strip the old wax (using an approved stripper) and re-wax with two thin coats of an approved wax.

Vinyl Composition

Vinyl composition tile is a type of vinyl tile that has many advantages over asphalt tile. It is immune to mineral solvents, oils, and greases along with alkalis and acids. It has a higher indentation resistance than asphalt tile, but it will still indent under heavy static loads. It is easier to maintain since it is a less porous type tile. It is decay- and mildew-proof.

The following considerations should be a part of the routine maintenance procedures for the vinyl composition flooring:

1. Dust mop or sweep on a daily basis.
2. Damp mop as needed. Buff, using a nylon buffing pad or a No. 00 steel wool pad.
3. Thoroughly wash as needed, using any one of the commercially used soaps and detergents. Avoid using a strong alkali cleaner.
4. Strip old wax when the floor becomes worn. Use a water-base wax. Follow with buffing. Solvent-type waxes, including paste wax, can also be used.
5. Do not use abrasive cleaners on unwaxed floors.
6. This type of floor should be sealed. It should be resealed at every sixth waxing thereafter.

Vinyl Plastic Tile

Vinyl plastic tile has all the qualities of vinyl composition tile in addition to having a higher indentation resistance. All the types of vinyl tiles may be maintained in the same manner as asphalt tile except that they can tolerate solvent type cleaners, waxes, and sweeping compounds.

Vinyl tiles are the most durable and have excellent resistance to grease and are seemingly unaffected by high alkaline cleaners. The biggest disadvantage of vinyl tiles is that they will show permanently many types of burn marks, such as those left by a lighted cigarette. It is impossible to remove this type of strain or mar. Depending upon the wear of the tile, it may or may not need to be sealed. If it is poorly maintained, pitted, and is rough-surfaced, some type of rejuvenation, such as sealing, is in order. Otherwise, it is not needed, due to the denseness of its composition. Sometimes it needs to be sealed if specific types of waxes are used. There are specific types of vinyl available on the market today advertised as not needing waxing; however, an application of wax will insure a longer lifetime of the tile. Some of the things that should be considered when planning a maintenance program are the same as for vinyl composition tile.

Resilient Floor Repair

The repairs which follow can be used for both tile and sheet forms of flooring, but repairs are easier on a tile-type floor.

The biggest problem in patching a tile floor is obtaining new tiles or sheet material which will match the color and pattern of the existing flooring material. There are two solutions to this problem. The first is, when putting down a new floor, always purchase extra tiles for future use. Unfortunately, due to sun fading and wear, the existing tile may fade and the new tile will not match the old. Therefore, the solution to this problem is to take up additional tiles to form a pattern with the new tiles and thus result in a more eye-appealing floor.

Once a floor has become badly marred, scuffed, or pockmarked, there is little that can be done, except either to replace the floor or to grind it down below the indentation mark.

Grinding: The following materials, equipment, and procedure are recommended for grinding resilient floors.

Equipment	Materials
Floor machine	100 grit abrasive discs
Industrial vacuum	150 grit abrasive discs
Clean cloth	

1. Make ready the equipment and material.
2. Sand down the area with the 100 grit disc and continually pick up dust with the vacuum.
3. Follow with sanding using the 150 grit disc.
4. Clean area thoroughly of all dirt and wipe clean all surfaces adjacent to the repair area.
5. Apply two coats of wax, for the specific type of tile, and buff.
6. Clean and return equipment to proper storage.

Labor Production (per worker)	Average Frequency
50–80 S.F./hour	As needed

Replacement

The following materials, equipment, and procedures are recommended for replacement of resilient tile.

Equipment	Materials
Open-ended wood frame	Dry ice
Wide-blade chisel	New tile
	Burlap

Alternate Methods	Alternate Materials
Blowtorch or electric iron	Hot water
Trowel	Butcher paper
	Adhesive

1. Make ready the equipment and material.
2. If using dry ice as the method of removal, place the wood frame over the tile, fill with dry ice, and cover with burlap. Leave this in place for at least five minutes (longer for rubber and solid vinyl tile).
3. Remove the frame, the dry ice, and the cover.
4. Lift off the old tile with the chisel.
5. If using a heat source (blowtorch, electric iron, or hot water) apply the heat for from three to five minutes, being careful not to damage adjacent tiles. If using an iron, be sure you place a piece of paper or wet cloth over the tile prior to the application of heat. As heat is applied, pry up the tile carefully with a chisel.
6. Scrape off old adhesive and be sure the surface is clean.
7. Spread new adhesive with a trowel or brush keeping new adhesive away from the edges of existing tile (see Figure 8.9 for types of adhesives).
8. Fit new tile into place. Press firmly down around all edges. Asphalt and vinyl composition tile can be flattened if a slight amount of heat is briefly applied, while other tile can be smoothed down by applying pressure with a small hand roller.
9. Clean up adjacent surfaces with a solvent that won't deteriorate the tile.
10. Clean equipment and store properly.

Labor Production (per worker)

Removal: Tiles—100–130 S.F./hour
(by hand) Sheet—120–160 S.F./hour
Hardwood—40–60 S F./hour

Replacement

Ceramic—10–20 S.F./hour
Resilient—40–70 S.F./hour
Hardwood—25–35 S.F./hour

Add for related items: Replace wood subfloor—80–100 S.F./hour
Replace underlayment—75–90 S.F./hour
Replace floor moulding—10–30 S.F./hour

Replacement of Sheet Tile: Lay out the area to be repaired. Place an oversized section of the new matching sheet tile over the damaged area. Cut through the two areas of tile simultaneously so as to give a tight fit. Remove the damaged section and clean the exposed underfloor. Replace damaged felt lining. Follow the same steps as for tile floor in installing the new sheet flooring.

Masonry Floors

Masonry floors are composed of such materials as natural stone, brick, and ceramic tile. They also include poured masonry floors, such as concrete and terrazzo. Floors made of such materials are hard and durable and available in a wide range of colors, patterns, textures and designs. They are especially suited, when installed properly, for use in areas intended for heavy traffic. These floors are especially resistant to staining and to deterioration by solvents, acids, and alkalis.

Adhesives for Resilient Flooring			
Finish Flooring	Wood Subfloor	Concrete Subfloor (suspended)	Concrete Subfloor (on and below grade)
Asphalt	Asphalt Emulsion	Asphalt Emulsion	Asphalt Emulsion
	Asphalt Cutback	Asphalt Cutback	Asphalt Cutback
	Asphalt Rubber	Asphalt Rubber	Asphalt Rubber
Cork	Linoleum Paste	Linoleum Paste	Do not use
	Waterproof Resin	Waterproof Resin	
Linoleum	Linoleum Paste	Linoleum Paste	Do not use
	Waterproof Resin	Waterproof Resin	
Rubber	Linoleum Paste	Waterproof Resin	Latex
	Waterproof Resin	Linoleum Paste	Epoxy
Vinyl (composition backing)	Latex	Latex	Latex
	Linoleum Paste	Linoleum Paste	
Vinyl (rag felt backing)	Linoleum Paste	Linoleum Paste	Do not use
	Waterproof Resin	Waterproof Resin	
Vinyl (solid)	Waterproof Resin	Waterproof Resin	Latex
	Latex	Latex	Epoxy
Vinyl Composition	Asphalt Emulsion	Asphalt Emulsion	Asphalt Emulsion
	Asphalt Cutback	Asphalt Cutback	Asphalt Cutback
	Asphalt Rubber	Asphalt Rubber	Asphalt Rubber

Figure 8.9

The same basic periodic housekeeping operations should be used on this type of floor as those described under resilient flooring. Refer to the earlier section for procedures. The following information will relate specifically to the type of flooring material and repair procedures to be used for many of them.

Ceramic Tile

Ceramic tile is available in either a glazed or unglazed surface. The unglazed variety is used more widely for floors. Both types are very dense, hard, and non-porous.

Ceramic tile can be used both indoors and outdoors. The three varieties that are used for floors are: 1) ceramic mosaic tile; 2) quarry tile; and 3) pavers. The tile is available in many shapes, sizes, and patterns.

The following procedures should be part of the maintenance program:
1. Sweep the floors daily with a wax dressing. Do not use oily dressings.
2. Clean the floor with a non-alkali synthetic detergent. Do not use a soap-based detergent since it tends to build up film which catches dust and causes slipperiness of the surface.
3. Wax floor, only when needed, with a water-emulsion wax.
4. The floor should be sealed with a penetrating (terrazzo type) sealer. This is needed to protect the masonry joints.
5. Do not use grit or abrasive powder cleaner, which may cause scratching.
6. Do not use steel wool since rust stains could result.
7. Keep stray acids off the floor when the mortar is composed of an acid-resisting base.

The reader is referred to Chapter 6 for a discussion of the repair of ceramic floor tile.

Concrete Floors

A concrete floor is smooth, one-piece, fire-resistant, and low in maintenance cost. It also stands up better to heavy floor traffic than most floorings. However, concrete lacks the color and general appearance that many finishes have. This is of utmost importance when appearance is a major consideration.

A concrete floor should be sealed whether or not a finish flooring is to be installed on top of it. Sealing will: 1) lengthen the floor life; 2) prevent dusting; and 3) make maintenance easier and more economical. Two coats of sealant are recommended to produce maximum wear resistance and a smooth, easily maintained surface. The type of seal will depend on the exposure condition of the floor.

Exposure Condition	Type of Sealant
Heavy traffic	Modified phenolic or epoxy resin
Mineral oils, greases, mild acids and alkalines.	Chlorinated rubber base
Concrete has been saturated with oil.	Resin modified nitrocellulose base.

For an unsealed concrete floor:
1. Remove all soil and previous treatments by machine scrubbing with a concrete cleaner. Do not use a soap solution as it will leave a scum. Allow cleaning solution to penetrate into the floor. Pre-moisten with warm water.
2. If the floor is heavily greased, it is best to use a mild alkaline cleaner. The use of a scarifying machine may be needed if the grease and dirt accumulation is deeply set into the concrete.
3. If a concrete floor is old, and severe cracking and grazing are evident, grinding the top surface may be needed. Once this is done, the surface is replaced with a new topping of similar concrete. However, this is seldom necessary if good patching is done. See Chapter 3 for further details on this point.

4. Do not use solvents such a kerosine, gasoline, etc., since they will cause the soil to penetrate deeper into the concrete.
5. Remove any dirty solution, rinse, and allow to dry.
6. Apply two coats of concrete sealer and allow to dry.
7. Apply two thin coats of wax. The type of wax depends on the existing surface and type of surface desired.
8. Dirt and dust should be removed daily with a treated mop or vacuum.
9. When stubborn soil can't be swept up, a wet mop or a floor machine and a neutral cleaner should be used. For extra stubborn dirt, use a slightly alkaline solution and remember to pre-wet the surface.
10. If the floor becomes exceptionally soiled, consider normal stripping. Waxing and buffing procedures are the same as those outlined for resilient flooring.

For a discussion on the repair of concrete floors, refer to Chapter 3.

Granolithic Floors

Granolithic floors are similar to terrazzo. The difference is that granite chips an used in place of marble chips. This type of floor usually has no metal dividers. See the section of this chapter devoted to terrazzo floors.

Magnesite (Oxychloride)

Magnesite floors, sometimes referred to as oxychloride, are very durable surfaces. The surface is extremely high in resistance to grease and oil, but it has a poor resistance to water. Besides being durable, magnesite floors are strong, non-combustible, economical, dimensionally stable, and flexible. A variety of coloring and surface textures are possible with this type of flooring. Another type of magnesite is known as oxychloride terrazzo and should be maintained in the same way as terrazzo. This type of flooring should be sealed with a penetrating, alkali-resistant floor sealer. Maintenance procedures are described below.

1. On a daily basis, vacuum or sweep the floor. Mop sprays and sweeping compounds may be used.
2. Periodically, (as needed) scrub the floor using a mild, neutral synthetic cleaner. Degreasers can be used to remove grease and hardened soil. Do not use trisodium phosphate, silicate soda ash, or any type acid cleaner. Do not allow the solution or any water to stand on the surface.
3. After washing the surface, seal with a solvent type or water wax.
4. After the wax has dried, buff. Buffing should also be done on a regular basis.
5. About once every six months, apply penetrating dressing. Be sure the floor has been cleaned and is dry before the application. The dressing can be placed on a waxed floor.

Repair of this type of flooring material is similar to terrazzo flooring.

Terrazzo

Terrazzo has a high strength, is termite-proof, is cool in the summer, warm in the winter, easy to maintain, and long lasting. It is also noted for its beauty.

There are basically two types of terrazzo flooring. The first type is a mixture of Portland cement and marble chips. The second type is sometimes called thin terrazzo and is a mixture of an epoxy-type resin and marble chips. Many times it is hard to tell the difference between the two on sight. It is generally safe to assume that if there are metal divider strips, the terrazzo is probably the Portland cement type. The other type does not need metal strips, which provide expansion and contraction leeway for the cement mixture.

The surface should be treated with a penetrating type sealer, for reasons of protection, attractiveness, safety, and ease of maintenance. This will help prevent dusting and staining. The sealer is mainly to protect the grout material. The floor should be periodically re-sealed, to eliminate the need for waxing. Refer to Figure 8.10 for a summary of types of deterioration affecting terrazzo floors.

The following steps should be incorporated into an effective maintenance program for terrazzo flooring:

1. On a daily basis, sweep or mop the floor. If a sweeping compound is to be used, be sure it is non-oily.
2. To wash the floor, use a neutral liquid cleaner having a pH of 10 or less. Soaps and scrubbing powder containing water-soluble organic salt should not be used. Place the cleaning solution on the floor and allow to set a few minutes. Do not allow the solution or water to stand on the floor for any length of time.
3. Mop up the dirty solution, rinse, and dry the floor. It is suggested that the new terrazzo floors be washed two times a week and damp mopped on alternate days.
4. Do not use steel wool on the floor.
5. Buffing will restore the luster to the surface once washed.
6. Do not expose the floor to acid, alkalis, and/or oils.
7. Wax is not needed, but will enhance the beauty. If used, a slip-resistant type should be used.

Terrazzo floors can be repaired in accordance with the specifications for putting in the new original floor. Only floor specialists who are capable of this type of workmanship should be entrusted to repair a terrazzo floor.

Marble

The utilization of marble results in a surface that is beautiful, brilliant, and permanent. Marble flooring should be treated with a penetrating sealer such as that used for terrazzo flooring.

A recommended maintenance program for marble floors is described below.

1. Vacuum and sweep the floor on a daily basis. Never use an oily sweeping compound.
2. Periodically wash the floor using a neutral synthetic detergent. Do not use soap, acid, or alkaline cleaners, since they will have a dull surface and cause disintegration. Do not use a water-ammonia solution.

Terrazzo Floor Deterioration		
Problem	Cause	Solution
Acid Damage	Acid exposure	Wash acid out of topping using a neutralizing agent; follow by grinding or honing.
Cracks	Settlement of floor foundation	Hydraulically injecting cement under slab and raising the tipping (floor) to its original state. This is followed by caulking cracks with cement or a special caulking compound.
	Overload	Use same solution as above but without the hydraulic injecting of cement.
		Replace flooring.
Dusting	Improper installation	Thoroughly clean floor and reseal with a penetrating seal.
	Worn surface seal	
Powdering	See dusting	

Figure 8.10

3. For stubborn dirt, a mild alkaline abrasive cleaner that contains no caustic or harsh filler can be used. Pre-wet the floor with hot water before applying the cleaner. Do not use any coarse abrasives, such as scouring powder or steel wool.
4. Rinse and wipe the floor dry. Follow this with buffing.
5. Polishes are not recommended.

Minor repairs can be made using commercially available fillers. If major or extensive damage occurs, replacement of the tile is almost a necessity. Consult with a marble contractor to develop specifications for the repair procedure.

Slate Tile

Slate tile floors are very suitable for areas subjected to heavy traffic because of their durability and ease of maintenance. However, slate tile tends to be slippery when wet. If the area is often subjected to wetness, either provide a covering surface such as a rubber mat, or do not use this type of tile. It is suggested that when this tile is used indoors, the surface should be sealed with a penetrating type of sealer as a base coat and then covered with a surface type sealer to prevent dusting. The maintenance and repair procedures used for slate tile would be the same as that used for ceramic tile.

Brick Flooring

Using brick for floors is relatively economical. Brick is resistant to abrasion and has a relatively low maintenance cost. The best type of brick to use is a hard, dense, vitrified brick. Porous brick can become a maintenance headache in any company, since it has a greater tendency for discoloration and staining. Once the brick is in place, and cured, it is best to seal it with two coats of commercially available sealer. Once this is done, the floor can be polished for longer wear. The sealer will inhibit the possibility of discoloration. Make sure a brick polisher is used. Do not use varnish or shellac as finishes for brick floors because these products do not stay in place and after a short time will present a poor appearance.

The maintenance and repair procedures for brick tile are essentially the same as for ceramic tile.

Metal Floors

Not too many metal floors are found in manufacturing plants today. If they are found, they are usually composed of sheet, plate, and grating.

The following steps should be incorporated into the maintenance program:
1. On a daily basis, sweep the floor surface.
2. For more stubborn dirt, scrub the floor using an electric floor machine. Steam cleaning is also very effective in cleaning the floor.
3. Sanding and sandblasting are two techniques which can be used to clean deeply set-in dirt and stain.
4. It is not advisable or beneficial to wax metal flooring.
5. Sealers are also no aid to a metal floor.

Repair is usually accomplished by replacement or plating (covering) the damaged floor or grating.

Seamless Floors

Seamless floors are one-piece synthetic resins poured-in-place plastic floors that are tough, colorful, durable, and nearly maintenance-free. There are many different types of seamless flooring, depending upon the basics of the sealer primers. The greatest asset of seamless floors is their durability. Scratches in the top surface will eventually occur, especially in heavy traffic areas which, in turn, will dull the finish. The floor can be restored to its original condition by adding an additional clear coat of plastic. Also, cigarette burns become a problem since they will mar the plastic coating.

A typical maintenance procedure for a small surface area would be as follows:

1. Dust mop daily and damp mop when needed.
2. For more stubborn dirt, use a mop or machine to scrub the floor using a neutral detergent.
3. Do not use nylon abrasive pads, steel wool, or scouring powder.
4. Buffing will bring out the highlights of the floor.
5. Sealing and waxing are not needed for this type of floor, from the standpoint of wear. However, a thin coat of wax will enhance the beauty of the floor, and make a somewhat easier removal of spills and stains.
6. If organic solvents are spilled onto the floor, allow them to dry out and then remove by damp mopping.
7. Do not use nitric acid and/or acetic acid around the floor.
8. Do not allow grease, oil, or water to stand on the floor for any length of time.

Repairs of extensive areas must be accomplished using the same method that was utilized in installing the original floor.

For repair of small areas use the following procedure:

Equipment	Materials
Metallic scraper	Mastic paste (plastic)
Mixing container	Plastic chips
Clean cloth	Solvent cleaner
Applicator	

1. Scrape the area clean.
2. Apply more plastic chips and plastic paste (in essence, a small section of the floor is actually rebuilt on the spot). At times it is difficult to get a good match between the original floor and the patch.
3. Finish off with a sealer.

Labor Production (per worker)   Average Frequency

5–15 S.F./hour As needed

Conductive Floors

Conductive floors are used in areas where explosive vapors, chemicals, and materials are present. Sparks resulting from the accumulation of static electricity in the floor or from the floor create a very real hazard in these areas. Conductive floors can be purchased in ceramic tiles, terrazzo, oxy-chloride, rubber, linoleum, and various types of vinyl. It is of extreme importance to consult with an expert in this area to insure that the floor is functioning as designed.

Special care should be taken in setting up a maintenance program for this type of flooring, due to its importance in the facility. In setting up a maintenance program for conductive floors, refer to the specific section in this text for the type of material in addition to incorporating the following:

1. Mop the conductive floor after any spillage and also on a daily basis.
2. Do not use any soap on the floor. Use a synthetic detergent especially made for conductive floors.
3. Conductive floors should not be sealed or waxed. Even though conductive waxes and sealers are available, the risk involved in the possibility of the floor losing its conductive properties is neither worth the savings in floor wear, nor the improved appearance.
4. It is of utmost importance that these floors be frequently checked to see that they are performing within the range of the subscribed conductive factor.

Repairs should be performed in accordance with procedures presented in this text for the specific material, taking all precautions to maintain the conductive effectiveness of the floor.

Pedestal Floors

Pedestal floors refer to those floors that are elevated up above the main structural floor. These floors are found in computer rooms and often in hospital facilities. Electronic and special ventilating systems are contained in the space between the structural floor and the pedestal floor. The type of finished flooring usually used is a non-carpet surface such as vinyl-asbestos and at least one special flooring made exclusively for computer rooms. The maintenance and repair procedures utilized should be consistent with the type of finished flooring material being used. The reader is referred to those respective sections in this chapter.

Bituminous Floors

Since most types of bituminous floors are found on the exterior of the building, the reader is referred to Chapter 6 for information on maintenance and repair of bituminous floors.

Wood Floors

Wood flooring is durable, decorative, warm, versatile, and fairly easy to maintain and restore to its original appearance. Wood floors come in many forms. The principal forms are strips, planks, block, and parquet. The two basic types of species available in wood flooring are the soft woods (such as pine) and the hardwoods, the hardwoods being more common. The common types of hardwoods are oak, beech, birch and maple; however, there are many other types of hardwoods used for flooring today. Refer to the latter part of this section for more discussion of wood-block type floors.

The maintenance program for wood flooring should include the following points:

1. Sweep daily with an untreated dry mop or bristle brush. Do not use an oil-based sweeping compound.
2. The surface should periodically be dry buffed using a fine steel wool No. 0 pad on a rotary type electric buffing machine. Vacuum after buffing.
3. Wipe up any spills immediately.
4. All marks and discoloration can be removed with a finish-type renewer. Never use water on a wood floor or use a scrubbing type machine.
5. Strip and re-wax floor periodically. Use a solvent to remove the wax with a No. 0 steel wool pad or soft cloth. Use a special solvent or naphtha for this procedure.
6. Wipe the surface down and apply a thin coat of paste wax or liquid wax (spirit-base type). Follow by buffing the surface.
7. Never apply soaps, strong detergents, or emulsion-type waxes.
8. It is not a good idea to apply wax to an unsealed floor surface.

Floors that have been worn, cut, gouged, indented, stained, or that have splintered, loose, or warped strips or planks should be refinished or covered. Pine or hardwood floors can be sanded at least 3/16-inch before floor covering is needed.

The process of refinishing can include all or any of the following activities: removing old finish; refinishing surface; or covering floors. The preparation of the surface must be carefully done prior to performing any of the above. Refer to Figure 8.11 for a summary of wood finishing products. Also refer to Figure 8.12 for a listing of special wood maintenance problems and the solutions to those problems.

Finishes Used to Refinish Wood Floors

Finishes	Advantages	Disadvantages
Epoxy (one-package)	Durable finish. Easy to apply.	
Epoxy (two-package)	Durable finish.	Difficult to use and apply.
Lacquer	About the same durability as varnish. Invisible spot retouching. Dries rapidly. Resistant to water. Seldom darkens with age.	
Paint (enamel)	Easy application.	Not used much because qualities of paints have not been able to withstand abrasive effects that floors undergo.
Penetrating wood sealers	Wear resistant surface is produced.	Requires more attention than other finishes.
Shellac	Easy to spot repair. Applies fast. Quick drying.	Water spots. Should not be used where durability is required. Scratches easily.
Urethane (polyurethane)	High abrasion resistance. Many of the same advantages and disadvantages as varnish.	
Varnish	Many different characteristics depending on the type of varnish. Lustrous appearance. Scratch and water resistant.	

Figure 8.11

Special Wood Maintenance Problems

Problem	Cause	Solutions
Dark spots	Environmental exposure or chemical stains.	Clean spot with a #1 steel wool and solvent base polish or mineral spirits. If not effective, try bleaching spot out. Remove or replace tile.
Heel and caster marks	Heel markings, caster markings.	Rub with fine steel wool and solvent-base polish or mineral spirits. If not effective, try bleaching spot out. Remove or replace tile.
Ink stains		Use same procedures as for dark spots.
Light spots	Excessive exposure to cleaning solutions, milk, dried food or standing water.	Sand with fine sandpaper, follow with a cleaning with #1 steel wool and mineral spirits or wax. Apply a finishing material which will match the adjacent floor.

Figure 8.12

Special Wood Maintenance Problems (continued)

Raised grain	Use of water on improperly sealed floors.	Buff with #2 steel wool pad. In severe cases, resand. Apply a matching finish, then follow by polishing.
Scratches	Environmental exposure conditions.	Use commercially available finish restoring compound sold for furniture.
Squeaks	Movement of the flooring system.	First, determine the cause of the movements: deflection of supporting floor joists; subflooring not adequately supporting the structural flooring; poorly manufactured finished flooring resulting in a loose fit; warped flooring or subflooring; inadequate nails. First, eliminate cause. Regardless of cause, one way is to lubricate the tongue of the tongue in groove with mineral oil introduced into the opening between adjacent boards. Do not use too much oil, in that stains could occur. Another method is to drive a nail through the face of the flooring to the subfloor, preferably also into the joints. The nail should be set and the hole filled with an acceptable filler. Where flooring is warped, and the undersurface of the floor is exposed, screws through the subfloor will be effective. This procedure is less objectional than face nailing, relative to appearance.
Stains	See section on Stain Removal	
Streaks and smears	Too heavy application of polish. Incorrect or insufficient buffing.	Polish. If this does not work, one would have to follow the directions for refinishing a wood floor, as presented earlier.
Uneven at joints	Result of inadequate nailing or loosening of nails under heavy traffic, during the shrinkage period of the floor material.	Renail flooring. Replace flooring.
Warped or buckled floors	Water Not providing for expansion at room perimeter.	First determine the cause and eliminate it. Allow for the floor to return to a flat condition, which may take some time, followed by sanding and refinishing. Screwing the subfloor into the finished floor from below may be required.
Wood decay	Moisture or condensation under the floor.	Replace the flooring.

Figure 8.12 (continued)

Wood Floor Preparation

The following preparations should be accomplished prior to sanding and varnishing a wood floor.

Equipment		Materials
Hammer	Power sander	Nails-finishing
Nailset	Scraper	Crack filler
		Degreasing cleaning solution
		Sanding paper or discs

1. Make ready the equipment and materials.
2. Re-nail all loose and warping boards (see suggestions in floor repair section for wood floors).
3. Remove all tacks and set nail heads well below the floor surface.
4. Scrape or machine-sand all high joints to make them level with the adjacent floor boards.
5. Remove and replace boards that are damaged beyond reconditioning (see repair of wood floors).
6. Remove loose splinters and fill all holes and large cracks with crack filler.
7. Remove grease, oil, and other foreign matter with a cleaning solution.

Labor Production (per worker)	Frequency
See productivity for Sanding.	As needed.

Sanding

The following equipment, materials, and procedure are recommended for sanding wood floors.

Equipment	Materials
Power-operated sanding machine	Sandpaper discs (See Figure 8.13 for type and grade)

1. Old varnish, shellac, and wax material should be removed from the floor by roughing off with the proper type of abrasive paper. Go over the floor twice diagonally. Don't cut into the wood.
2. Finish with a fine sandpaper or steel wool parallel with the grain.
3. Edges that cannot be reached by machine should be hand sanded.
4. Floor should be sealed and waxed as described in the following repair section.

Labor Production (per worker)	Average Frequency
For preparation and sanding: 40–60 S.F./hour	As needed.

Old Varnish and Paint Finishes

The following is a recommended procedure for removing old varnish and paint finishes.

1. Apply an organic solvent remover by brushing it on and allowing it to stand until the varnish or paint softens.
2. The old finish is then scraped off with a steel blade scraper or rubbed off with a steel wool pad.

Oil Finishes

The procedure below is recommended for removing oil finishes.

1. Apply a water solution of trisodium phosphate, washing soda, or commercial solution.
2. Apply to a small area at a time and allow to stand for a few minutes.
3. Scrub off with a stiff brush.
4. Scrub and mop the floor to a dry or slightly damp surface. Old shellac can be removed with steel wool and turpentine.

Sealing and Waxing Wood Floors

The equipment, materials, and procedure below are recommended for sealing and waxing wood floors.

Equipment

Vacuum cleaner
Clean cloths
Power polishing machine

Materials

Wood floor sealer

1. Remove all dirt and dust from the sanding or from other methods of preparation used.
2. Apply liberally a sealer of your choosing. Spread or spray it along the grain of the wood.
3. After the sealer has dried completely, buff the floor with a floor polishing machine. If portions of the floor look lusterless, dry, or dead after the buffing, continue sealing and polishing until floor surface has a uniform appearance.
4. Apply two thin coats of wax. Buff the wax after each application.

Labor Production (per worker)

For sealing only:
200–300 S.F./hour
Add waxing time as per waxing section.

Average Frequency

As needed.

Sandpaper Selection Chart		
Grade	**Type**	**Use**
3-1/2	Open	Preliminary roughing off of stubborn varnish, shellac, floor oil, wax and deep penetrating filler compounds. Not to be used for cutting into wood surfaces.
3	Open	Used in place of No. 3-1/2 for surfaces of less resistance. Preferred if it does the required work.
2-1/2	Open	Preliminary roughing off of floor finishes such as shellac, wax, floor oils, alcohol stain and lacquered surfaces. Use as a follow up paper for floors roughed off with No. 3-1/2.
2	Close	Use instead of No. 2 and 2-1/2. Use open coat where surface permits cutting without gumming. Close coat should be used in preference to open coat, whenever possible.
1-1/2	Close	Use as a first paper on all new floors.
1	Open	Use as a follow up for No. 2 and No. 2-1/2 in all cases.
1	Close	Use the same as No. 1 open coat to provide a smooth floor finish.
1/2	Close	Use as a final finish on most floor work.
1/0 & 2/0	Close	Use as a final finish on best hardwood floor work.
3/0 & 4/0	Close	Use for finishing fine woodwork, such as furniture, and for rubbing down paint and varnished finishes.

Figure 8.13

Repair

Wood floors should be inspected for loose nails, loose boards, raised ends, slivers, cracks, loose knots, raised nails, and damage from improper cleaning, condensation and wood decay. Procedures for repair of specific problems associated with wood floors are described in the following sections.

Loose Wood Boards or Tiles: Loose boards and tiles should be immediately repaired. A recommended repair procedure for this problem is outlined below.

Equipment	Materials
Hand tools	Nails
	Colored wood filler

1. Locate the loose boards or tiles.
2. Drive two 3″ finishing nails in at an angle so that they form a "V".
3. Use a nail set to drive the heads below the surface so you don't damage the floor.
4. The remaining holes can be filled with colored wood filler.
5. Wood tiles (parquet) can also be repaired similar to resilient floor tile if mastic was used originally. Try nailing first.

Labor Production (per worker)	Average Frequency
Dependent on extent of work. Average: 50–250 S.F./hour	As needed.

Wood Strip Floor Replacement: If wood damage requires replacement of strips or planks, use the following method.

Equipment	Materials
Hand tools	Replacement strips or planks
Sanding machine	Nails
	Sandpaper

1. Make two longitudinal cuts in the damaged strip or plank.
2. Remove the section between the two cuts by cutting the strip with a chisel at midpoint.
3. Remove the remainder of the damaged strip taking care not to damage the tongues and grooves of the adjoining board.
4. Remove the lower part of the groove of the new enclosure or plank.
5. Insert tongue of the closure into the groove of the adjoining board and nail with two six-penny angle-ring finishing nails to the top surface.
6. Set the exposed nails.
7. Bring the new portion to the level of the adjacent floor by sanding both areas to a continuous smooth plane.
8. Dry-sweep the area to remove all particles of dust, seal, wax, and then buff.

Labor Production	Average Frequency
30–60 S.F./hour	As needed.

Wood Block Floors

Wood block flooring is manufactured using the following species of wood: Southern Yellow Pine, Port Oxford Cedar, Douglas Fir, and California Redwood. The wearing surface is the end grain of the wood. Sizes available range from 1-1/2″ to 4″ wide and from 3-1/2″ to 8-1/2″ long.

When planning a maintenance program keep the following points in mind:
1. Sweep with a stiff bristle brush or a power sweeper.
2. Areas of hard-to-remove dirt and oil should be scraped with a metal scraper or cleaned with a power buffing machine equipped with a wire bristle brush or coarse steel wool pad.

3. For spillage of oil or water, apply a soaking compound and follow this by sweeping it up.
4. Follow the other steps outlined for wood floors in this section.

Wood block floors are repaired by the replacement of tiles (blocks) or replacement with another material such as concrete (see Chapter 3 for concrete replacement techniques).

Equipment	Materials
Wood chisel	Surface cleaner and primer
Hammer	Sandpaper
Crowbar	Coal-tar pitch or plastic bituminous cement

1. Make ready the equipment and materials.
2. Remove all deteriorated and loose blocks.
3. Clean the subfloor, removing all loose particles, dirt, oil, and grease.
4. Once the subfloor and surrounding blocks are completely dry, apply one coat of a primer designated for this purpose to the subfloor and sides of the surrounding wood blocks.
5. After the primer has thoroughly dried, apply a coat of coal-tar pitch or plastic bituminous cement to the subfloor in the quantities needed for permanent adhesion between the subfloor and wood blocks.
6. Fit the individual blocks or strips of blocks into place, matching the existing floor lines as closely as possible. Fill any voids of less than one full block with a section of a block cut to fit. Retain existing expansion joints.
7. Fill the joints and finish or seal the floor, as needed.

Labor Production (per worker)	Average Frequency
20–60 S.F./hour	As needed

Carpets

Rugs or carpets are installed in a building for floor covering to provide a softness under foot, warmth, quiet, and beauty. They also offer other functional, decorative, and psychological values. The selections of colors, styles, patterns, sizes, and fibers are too numerous to mention. The reader is referred to Figure 8.14 for a comparison of the basic types of carpeting.

Carpet Maintenance

There are four major methods of cleaning carpets for the removal of stubborn dirt. These are:

- **Dry-foam or granular**—Good surface job and easy to use.
- **Wet shampoo**—Better and deeper cleaning job. Requires greater skill and care.
- **Steam cleaning**—Most thorough on-site cleaning. Consult with a specialist.
- **In-plant cleaning**—Most efficient yet most impractical for all except area rugs.

A regular routine maintenance program is essential to the upkeep of a carpet and should include the following:

1. Vacuuming should be done on a daily basis in high traffic areas, at least twice weekly in a medium area, and at least once a week in light traffic areas. Using a vacuum sweeper will result in a more effective job.
2. Spot clean and de-stain as needed. Refer to Figure 8.15 for stain removal methods.
3. Absorbent powder cleaners can be used between wet cleanings to prolong the period between wet cleaning and to enhance the beauty of the rug. Select a powder suitable for the type of rug and follow the manufacturer's suggestions for application.

4. When using the wet type of cleaner, use a neutral detergent sold especially for carpet cleaning and follow the directions for dilution and application. Do not use soap, ammonia, washing soda, or any of the strong household cleaning agents intended for use on hard surfaces, such as linoleum or tile. Don't get the carpet too wet. It is advisable to remove all room furniture when using a wet method. If this isn't practical, place thick plastic film or aluminum foil under and around the legs of the chairs and tables and other furniture to prevent rust or furniture stains. Wet clean at least twice a year.

5. After the carpet is cleaned, brush the wet pile all in one direction with a soft brush or special tool for this use. Dry the carpet as quickly as possible and follow with a thorough vacuuming. Aerosol foam sprays are also available for carpet cleaning. This procedure is to spray a thin layer on the carpet, work it in with a wet sponge mop and vacuum when dry. This method, however, does not do as thorough a job as a powered machine.

Selection Criteria for Carpeting

Criteria	Wool	Nylon	Acrylic	Modacrylic	Polypropylene	Rayon	Polyester
Resistance To:							
Abrasion	B	A	B	B	A	A	A
Alkalis	C	B	B	B	B	B	B
Acids	C	B	B	B	B	B	B
Insects and Fungi	A	A	A	A	A	A	A
Burns	B	C	C	C	C	B	C
Compression	B	B	B	B	C	C	B
Crushing	A	B	B	B	C	C	B
Staining	B	B	B	B	A	C	A
Soiling	B	C	B	B	A	C	B
Static Buildup	C	C	B	B	A	B	B
Texture Retention	A	B	B	B	C	C	B
Wet Cleanability	C	B	B	B	A	C	B
Durability	B	A	B	B	B	C	B
Appearance Retention	A	B	B	B	B	A	B
Ease of Maintenance	B	B	B	B	A	B	B
No problem with fuzzing, pilling	B	C	B	A	A	B	B
Dyeing, Colors	B	B	B	A	C	A	B

Rating on A–C scale — A being the highest rating (Excellent)

Note: The ideal commercial carpet is one that is light beige and grey tweeds (in muted middle tones), having a very tight level loop pile.

Figure 8.14

Carpet Repair

Extensive repair of carpeting is often uneconomical, particularly if reweaving is needed. As a general rule, repair work should be deferred when:

- More than 10 percent of the carpeting requires skilled repair work.
- The carpeting is approaching the end of its useful life.
- Areas with worn spots, holes, tears and frayed edges can be covered by furniture or by the use of scatter rugs.
- The carpeting may be cut to useful dimensions for other applications.

Replacement of carpeting is the only effective repair procedure to use for areas of extensive deterioration. Minor, temporary repairs can be made by sewing and/or taping.

Some of the more common problems with carpets are discussed below with suggested ways of correcting them. No attempt is made to provide information such as labor production figures since this will depend on the scope of the job.

Replacing Tuft (Furling): The following procedure is recommended for replacing tuft.

1. Obtain matching yarns.
2. Remove damaged tufts with scissors.
3. Using a curved needle, insert it under the shot and sew in loops. If the carpet is cut pile, construct the loops higher than the adjacent tufts and then trim the loops flush with the surface of the pile. If repairing loop pile, use the same technique except for the fact that the loops are made the same height as the carpet pile.

Holes in Carpet: The procedure outlined below is recommended for repairing holes in carpet.

1. Using a knee-kicker, kick the carpet toward the damaged area and stay-tack from all sides.
2. Outline a rectangle after the damaged area is blocked in and the tension removed.
3. Open up the nap of the carpet, using a screwdriver along the outline.
4. Using a sharp knife, cut out the damaged section, cutting along the outline. Never cut across the face yarn, only with it.
5. Place face-up tape under all four sides of the cut-out.
6. Apply seam cement on the entire tape and carpet edges.
7. Insert the patch and push down on edges to bring it into contact with the tape and cement.
8. Place a weight over the patch and allow to dry.

Rough Seams: To repair rough seams, cut the protruding end of the cut loops with nap scissors to a height a little less than the surrounding pile.

Damaged Seams: The procedure outlined below is recommended for repair of damaged seams.

1. Stretch the carpet toward the seam and tack. Once tacked, the carpet should appear to be bulging between track rows with the seam in-between.
2. Cut the tape at the seam and remove it.
3. Place the edges of the carpet, one on top of the other. Determine the quantity to be cutoff.
4. Make the cut down the entire length of the seam. Take off an equal amount from both edges.
5. Lay enough face-up tape on the floor for the total length of the seam and secure it at both ends.
6. Place the carpet with the cut edges on the center of the tape and check the fit.
7. Lift the edges of the carpet and apply cement to the tape and carpet edges.
8. Set the carpet on the tape and butt together using a knee-kicker. Roll the seam to insure good contact.
9. Tack in place to hold the repair until the cement dries.

Figure 8.15

Summary of Stain Removal for Major Types of Stains

Stain	Carpets	Wood	Linoleum Cork	Asphalt Tile	Vinyl	Marble, Terrazzo or Oxychloride Cement	Concrete & Glazed Tile
Blood	Allow household ammonia to stay on only a moment. Rinse using a cloth and cold water. Brush nap.	Rub with cloth dampened in clear, cold water. When stain persists, dampen cloth with ammonia.	Same as for wood.	Same as for wood.	Same as for wood.	Rub with cloth dampened in clear cold water. Bleach with peroxide when stain persists.	Same as for marble.
Chewing gum	Scrape off with dull knife. Sponge off with clear water.	Remove gum with putty knife. Apply alcohol.	Same as for wood.	Remove gum with putty knife. Do not use alcohol on asphalt tile.	Remove as much as possible with putty knife. Rub with #0 steel wool pad dipped in all-purpose synthetic detergent solution.	Same as for wood.	Same as for wood.
Grease or oil	Apply kerosene to the spot using circular motion, working from edge to center to avoid a ring. Shampoo, dry and brush nap.	Pour kerosene on spot. Permit to soak for short time. Wipe dry with a clean cloth. Wash with all-purpose synthetic detergent solution, rinse, then dry.	Scrub with warm all-purpose synthetic detergent solution. Rinse with clear water.	Same as for linoleum.	Same as for wood.	Pour solvent on spot, cover with Fuller's earth and let stand for several hours. Repeat, if necessary. Scrub with a mild, soapy water and rinse thoroughly with clear water.	Pour alcohol on spot. Rub with clean cloth, or spray stain with commercial degreaser product, and flush with clear water according to manufacturer's instructions.
Ink	Remove immediately. Same as for vinyl.	Apply solution of 1 part oxalic acid to 9 parts warm water. Permit to stand until dry. Mop with clear water.	Use warm all-purpose synthetic detergent solution. If stain persists, mix 2 TBSP sodium perborate in pint of hot water. Mix whiting to form paste. Apply to spot and let dry.	Same as for linoleum.	Wash with all-purpose synthetic detergent, rinse, dry. If stain persists, rub with cloth dampened with ammonia.	Same as for linoleum.	Same as for linoleum.
Iodine or mercurochrome	Blot immediately. Wash with all-purpose synthetic detergent, rinse, then dry.	Apply alcohol and rub with clean cloth.	Same as for wood.	Warm neutral soap solution.	Wash with all-purpose synthetic detergent, rinse, then dry. If stain persists, scrub with scrubbing powder and warm water.	Apply alcohol and cover with Fuller's earth.	Apply alcohol and rub with clean cloth.

Summary of Stain Removal for Major Types of Stains (continued)

Stain	Carpets	Wood	Linoleum Cork	Asphalt Tile	Vinyl	Marble, Terrazzo or Oxychloride Cement	Concrete & Glazed Tile
Paint	Apply turpentine or mineral spirits. Wash with strong luke-warm soap suds. Rinse with clear water, wipe with clean cloth and let dry.	Use oxalic acid solution, or 1 lb. tri-sodium phosphate in 1 gal. warm water.	Rub with #0 steel wool pad, dipped in turpentine. Wash with all-purpose synthetic detergent solution and rinse with clear water.	Rub with steel wool pad and all-purpose synthetic detergent solution. If area is large, use steel wool on buffing machine.	Rub with #0 steel wool pad dipped in kerosene.	Rub with #00 steel wool pad dipped in turpentine.	Scrub with 1 lb trisodium phosphate in 1 gal. hot water. Rinse with clear water.
Rust	Wash with all-purpose synthetic detergent. Rinse using a cloth and cold water. Brush nap.	Wash with all-purpose synthetic detergent. Rub with #0 steel wool pad if necessary.	Apply solution of 1 part oxalic acid to 9 parts warm water. Let dry. Rinse thoroughly with clear water.	Rub with #0 steel wool and all-purpose synthetic detergent solution.	Same as for linoleum.	Horizontal Surfaces: To 3/4 gal. water add 1.9 sodium citrate and hydrosulfite. Add enough water to make a gallon solution. Cover stain with solution and let stand 1/2 hour. Absorb with cloth by rubbing. Rinse with clear water. Vertical Surfaces: Make paste with whiting and 3 oz. sodium citrate and 3 oz. sodium hydrosulfite. Apply	with putty knife and allow to remain 1 hr. Wash with sodium citrate. Dissolve 1 part sodium citrate in 6 parts water. Mix with equal parts of gylcerine. Make paste with whiting and apply to stain. For bad stains wash with sodium citrate solution 1-6 parts, add pad soaked in sodium hydrosulfite for 10-15 minutes. Wash thoroughly with water.
Sole and heel marks. Crayon marks.	Shampoo lightly with all-purpose synthetic solution and rinse with clean water. Wipe off and vacuum. Brush nap.	Rub with #0 steel wool or wash with all-purpose synthetic detergent solution.	Same as for wood.	Rub with #0 steel wool dipped in all-purpose synthetic detergent solution.	Wash with all-purpose synthetic detergent solution and rinse.	Same as for marble.	
Tar	Same as for grease or oil.	Remove tar with putty knife. Soak with kerosene. Rub with clean cloth. Wash with all-purpose synthetic detergent solution.	Same as for wood.	Remove surplus with putty knife. Do not put kerosene on asphalt tile. Wash with warm all-purpose synthetic detergent solution.	Same as for wood.	Remove surplus with putty knife. Soak with alcohol and cover with Fuller's earth.	Same as for wood.

Figure 8.15 (continued)

Other Carpet Repairs: Miscellaneous carpet repair procedures are described in this section.

For *loose carpet* and *buckling*, re-stretch the carpet.

If *crushing* occurs, follow the procedure below:

1. Tease up by brushing or applying steam using a hot electric iron. Be sure to protect the carpet from damage by wetting the area and covering with two layers of dry cloths.
2. Brush the hot, dampened tufts erect with a dull knife.

For *shading*, follow the procedure below:

1. Minimize by a good daily vacuuming, to lift the fibers and lay them in one direction, since shading is nothing more than the fibers laying in different directions.
2. Wet shampooing is also effective.

For *shedding*, allow the carpet to shed. Do not pull at the shedding fiber. Shedding should cease after a sufficient length of time.

For *sprouting*, clip the protruding end even with the pile surface. Do not try to puff the pile.

For carpet *discoloration*, a thorough dry cleaning or wet shampooing should be performed.

For *pilling*, or the formation of fuzzy balls, shear or clip the balls from the carpet.

For *bearding*, remove only loose fibers or shear off at height of adjacent piling.

If *mildew* occurs, follow the procedure outlined below:

1. Remove any visible mold using a brush or vacuum.
2. Dry the area completely.
3. Shampoo the rug and treat with a fungicide especially made for rugs.

For *static* problems:

1. Increase the humidity in the room.
2. Apply an anti-static agent to the carpet.

For *cigarette burns*, re-weave or re-stuff the carpet.

Stain Removal— All Surfaces

There are many workable and effective means of removing stains from finish, wall, ceiling, and floor surfaces. Presented in Figure 8.15 are some suggested means of stain removal for some of the major stain producers. If using any commercially available agent, follow manufacturer's recommendations. If specific methods don't work, or stains are produced by agents not covered in Figure 8.15, consult with the supplier or manufacturer of the material.

The two most important factors in preventing permanent stains are:

1. Give the matter your immediate attention.
2. Use the correct stain removal agent and/or procedure.

The following general rules should be followed when removing stains:

1. Wipe up spills with an absorbent cloth, sponge, or toweling as soon as possible.
2. Identify the stain-producing agent or at least narrow the cause down to as few reasons as possible.
3. Select the removal method and perform the operation slowly to insure that the method neither produces a stain itself nor causes additional deterioration.

CHAPTER 9 ROOFS

The roofing system is one of the most important parts of any building. A roofing system includes the structural roof deck and a field-fabricated weathering surface, such as shingles or metal, built-up or single membranes. Some of the considerations in the design stage that influence roof selection are the type of structure, the configuration of the roof deck surface, the forces the roof must resist, and the size of the area. A roofing system requires several components to provide resistance to environmental elements and forces that are detrimental to the building and its contents. In this chapter, the maintenance and repair procedures for roof decks and coverings of various materials and designs are discussed in detail.

Roof Decks

Commonly used roof deck materials are listed below.
- Wood
- Metal
- Cementitious wood fiber
- Structural concrete
- Lightweight insulating concrete
- Gypsum

Metal, structural concrete, and lightweight insulating concrete are most commonly used. Metal and structural concrete normally require the addition of insulation to the deck. Lightweight insulating concrete is self-insulating, and may not require any supplementary insulation, depending on the prevailing building codes.

Gypsum is used mainly for its fireproofing qualities and relatively low resistance to heat transmission. However, since gypsum is slow to dry during the curing process, it tends to extend the construction period. Thus, its use is being minimized.

Roof Coverings

The various types of roof covering available are listed below.
- **Shingles:** Composition asphalt, fiberglass, tile, slate or metal.
- **Sheet Metal:** Copper, zinc-coated galvanized steel, or various alloys and commercial coating.
- **Membranes:** Elastomer sheeting, liquid silicone rubber, mineral surface composition rolled roofing, field-fabricated (built-up) membrane consisting of reinforcing plies of felt or fiberglass felt embedded in bitumen and covered with a protective surface.

Refer to Figure 9.1 for a summary of the available types of roofing and roofing sandwich components.

Available Roofing and Roofing Sandwich Components

Roofing Type	Description
Flat Roofs:	
Built-up (Membrane) Asphalt	Asphalt reinforced with cellulose felt plies and covered with embedded gravel or slag for protection wearing surface.
Coal Tar Pitch	Coal tar pitch reinforced with cellulose plies and covered with embedded gravel or slag.
Single-ply Sheeting	Thermosetting or thermoplane the material placed over insulated roof surface.
Fiberglass Reinforced	Asphalt reinforced with fiberglass mat (felt) and covered with embedded gravel or slag.
Sheet Metal	Flat seam. Similar application to standing seam (see Sheet Metal listed under Sloping Roofs) except all joints are flat and soldered.
Modified Bitumen	Composite sheet of two or more materials and additives. Similar to built-up roofing except the individual plies are factory laminated and the modified bitumen is applied in one layer.
Spray-on Foam and Coating	Special formulated coating(s) applied to urethane foam.
Single Ply Membrane	
Sloping Roofs:	
Composition Shingles	245 to 300 pounds per square (100 S.F.). Asphalt composition shingles, strip or individual over asphalt saturated felt underlay.
Exotic Shingle Materials	Wood, tile, slate, metal. All applied over asphaltic felts.
Mineral Surface Composition Roll Roofing	Applied over asphaltic felt underlay.
Sheet Metal	Standing seam, lock seams and cleats, installed over rosin paper. Cleats nailed to wood deck or screwed through 1/2 inch rigid insulation on metal deck.

Figure 9.1

Roof Slope

Flat roofs are the most difficult to make watertight and to maintain. Sloped roofs, even of a slight amount, are used with a greater frequency.

The three major categories of roof slope are listed below.
- Flat roof—roof slope of 1 inch in 12 inches and less
- Sloping roof—roof slope in excess of 1 inch in 12 inches and up to but not including 3 inches in 12 inches
- Steep sloping roof—roof slope of 3 inches in 12 inches or more

Materials for Specific Slopes

Do not use slag or gravelled-surfaced built-up type roofing on steep roofs. Mineral-surfaced, wide-type roofing and smooth-surfaced built-up roofing requiring nailing, asphalt shingles, cement composition shingles, corrugated protected metal, corrugated aluminum, corrugated vinyl-coated steel, corrugated ceramic-coated steel, or any other suitable type of roofing can be used for steep roofs.

On roof slopes of 1/2 inch to 12 inches or less, low-slope asphalt or coal-tar pitch can be used.

On shingle type roofs, asphalt, asphalt-cement, slate, and wood may be used on slopes of five inches or more per foot but are not advised for roofs of lesser slope.

Built-up and single membrane roofs are best suited to roof decks that range from flat to a slope of two inches per foot.

Metal roofs in sheet and corrugated form should not be used on slopes of less than three inches per foot. These include special shapes of galvanized and aluminum sheet, protected metal sheets, and all batten and standing seamed metal roofings.

Soldered seamed metal roofs (copper and terne) may be used on flat or nearly flat decks.

Inspection

The first step in establishing a roof maintenance program is the development of an inspection schedule to anticipate trouble areas in the early stages in the life of the roof. This procedure eliminates serious and costly damage, such as the buckling of wood decks, deterioration of roof mats, and interior water damage. Early discovery and correction of minor defects forestalls major repairs and extends the date when reroofing will become necessary. Since a large proportion of early roof failures are due to flashing failures, the regular inspection of flashing is of vital importance. Inspection should be made even though a roof has been in service for less than one year. For roofs in northern hemispheres, one of the inspections should always be scheduled in the spring, because it follows a severe condition and is the period best suited for roofing work. The first yearly inspection is of great importance because it frequently discloses minor defects that were not apparent when the roofing or reroofing job was completed.

The roof inspection record should contain a surveyed classification of all roof areas, with entries including age and condition. The record should be made a part of the history file maintained for each structure.

Deterioration

It is safe to assume that most failures are the direct or indirect result of faulty design, workmanship or simple maintenance. Most failures are premature. Few roof installations, especially field-fabricated membrane types, last the expected life of between 15 to 20 years without some failure requiring repair or replacement. There may appear to be an implied warranty when a 20 year roof is purchased; however, no roofer will guarantee such a roof installation against defects of material or workmanship for a period exceeding two years. Manufacturers' long-term bonds cover only limited responsibility. Both exclude resulting damage.

The main problem with workmanship is that manufacturer's specifications are not adhered to. This book will not concern itself with the design application except as related to failure. Some problems can be corrected by repair; some problems cannot be corrected by repair and may require chronic maintenance or complete replacement.

Causes of Roofing Failures

In order to inspect a roofing system and select the correct repair procedures, the major types of and reasons for roof failure and the most effective maintenance and repair procedures must be identified and understood. The most common manifestations of roof failure are splits in the roofing, moisture, and improper drainage.

Some of the reasons for the above roof failures are listed below.

- Temperature and moisture interaction
- Improperly mounted objects on or through roof
- Incompatibility of construction materials and thermal effects
- Improper patching and reroofing
- Inadequate inspection
- Inadequate testing of roofing materials
- Inadequate or improper specifications
- Ignorance of specifications
- Reverse shingling of flashing material
- Improper selection or mixture of materials in a system
- Use of asphalt for flat roofs
- Use of tar and tar-based cements for steep roofs
- Bleeding of pitch at eaves and into roof drains and at projections through the roof
- Inadequate storage of materials
- Installation of wet components
- Speed of application
- Weather
- Overheated bitumen
- Cold bitumen
- Long exposure of felts before application of gravel
- Too little gravel
- Use of vapor barrier or base sheet as a temporary roof
- Water behind flashings
- No ventilation of insulation
- Improper use of vapor barrier
- High roof drains
- Lack of maintenance
- Use of unseasoned lumber

Refer to Figure 9.2 for additional information on causes of failures for specific roofing types.

Reroofing

To prepare a deck for reroofing, follow the initial procedures outlined below.

1. Remove the old membrane entirely.
2. Restore the deck to as nearly new a condition as possible.
3. On wood decks, remove all rolled boards, replace with sound boards, and renail loose boards.
4. On cracked concrete or gypsum decks, apply at least 1/2 inch of insulation before applying a new membrane.

Once the deck has been prepared, the reroofing operation proceeds as for any new work. A specification for reroofing is similar to that used for a new roof.

Refer to other sections in this chapter for information on roof deck preparation and reroofing procedures specific to the types of materials for the design involved.

Common Roofing Application Faults

1. Built-up Roofs
   A. Improper cleaning of surfaces to be bonded, particularly in patching leaks.
   B. Improper temperatures of asphalt or pitch bonding coats.
   C. Failure to provide smooth flat fit of roofing felt against cant strips and parapet walls and similar areas of flashing, causing air pockets and nonsupport of felts, resulting in holes being kicked in the roofing.
   D. Use of untried or unproven new types of products.
   E. Failure to apply a layer of building paper over wood decks or other decks with open joints.
   F. Failure to cover knotholes and cracks in wood decks with metal.
   G. Application of built-up roofs in cold or wet weather.
   H. Inadequate nailing of felts in built-up roofs resulting in the blowing off or slipping of bits on steeper slope.
   I. Inadequate moppings of hot bitumen in the construction of slag or gravel surfaced roofs.
   J. Inadequate moppings of hot bitumen between the plies of felt.
   K. The application of too much asphalt on the weather surfaces of smooth-surfaced built-up roofs.
   L. Failure to broom felt smoothly behind the mop, resulting in poor adhesion, wrinkles and buckles.

2. Single Membrane Roofs
   A. Not following manufacturer's instructions.
   B. Using improper installation techniques and equipment.

3. Asphalt Rolled Roofing
   A. Failure to cement the seams of rolled roofing or to use the proper kind of cement.
   B. The application of rolled roofing with exposed nails.
   C. The application of rolled roofing with the nails too close to the edge of the sheet.
   D. Failure to cut rolled roofing into short lengths and failure to permit it to lie flat.
   E. Failure to cover resinous knots, knotholes or wide cracks in the roof deck with sheet metal.

4. Asphalt Shingle Roofs
   A. The application of asphalt shingles with nails too high.
   B. Failure to cement down the tabs of asphalt shingles with quick-drying cement in windy areas.
   C. Failure to use rolled roofing underlay over eaves in areas subjected to freezing and thawing

5. Metal roofs
   A. Failure to provide adequate side and end laps with corrupted sheet roofing.
   B. Failure to paint tin roofing.
   C. Failure to provide adequately for expansion and contraction with changes in temperature.
   D. Failure to fasten sheets adequately.

6. Rigid Roofing Materials (Slate and Tile)
   A. Nailing too tightly.
   B. Using improper nails.

Figure 9.2

Built-Up Roofs

This section presents maintenance problems and solutions particular to built-up roofs. In general, for mineral surfaces, asphalt and asphalt-saturated felt should always be used in the maintenance of asphalt built-up roofs and coal-tar pitch and coal-tar-saturated felt in the maintenance of coal-tar pitch built-up roofs. Asphalt and coal-tar pitch are not compatible; contact between the two should be avoided.

In the following paragraphs, the term bitumen and bituminous are used to indicate, respectively, asphalt, when asphalt roof maintenance is discussed, and coal-tar pitch, when coal-tar pitch roofs are discussed.

Bare Areas (Small Areas)

For bare areas where the bituminous coating is exposed, first brush loose gravel or slag from bare areas. Then cover the bare area with hot bitumen poured on at a rate of about 70 pounds per square and embed fresh gravel or slag. Old gravel or slag may be reapplied when the dirt and dust have been screened from it. Do not attempt to apply hot bitumen over slag or graveled surface because it will not adhere.

If felts are exposed and appear to be weathered, brush all dust and dirt from the exposed area and in the case of asphalt roofs, apply one thin coat of asphalt primer. When primer is dry, treat as described for bare areas with the coating exposed. Coal-tar pitch roofs are treated similarly, except that no primer is required before the coal-tar pitch is applied.

If bare, *small blisters* and *buckles* should be treated as bare areas; otherwise, they should be disregarded.

Weathering

When the bituminous coating is weathered severely over the entire roof area, remove coating and mineral surfacing. Removal is accomplished best in cool weather.

When large roof areas are involved, the use of mechanical equipment to remove the bituminous coating and surfacing material may prove to be more economical than removal by hand. However, the purchase of such mechanical equipment is usually not justified. Repair disintegrated felt, damaged areas, blisters and buckles, as described below. Sweep off loose material. Recoat with hot bitumen poured on at a rate of about 70 pounds per square, and embed 400 pounds of gravel or 300 pounds of slag in the hot bitumen.

Felts Exposed

For small areas where felts are exposed and partially disintegrated, scrape off all surfacing materials to at least 2-1/2 feet beyond the area of disintegrated felts. Remove disintegrated felt layers and replace them with new 15-pound bituminous-saturated felts of approximately the same size, mopped in place with hot bitumen. Apply at least two additional layers of 15-pound saturated felt, mopped on with hot bitumen and extending at least 12 inches beyond the area covered by the replacement felts. Apply hot bitumen to the repaired area at a rate of 70 pounds per square and into it, while hot, embed fresh gravel or slag. Old gravel or slag may be reapplied if dirt and dust are screened from it.

Roof Membrane Cracked

Treat as described for disintegrated felts, except that it is usually necessary only to mop on at least two plies of 15-pound saturated felt, followed by heavy pouring of bitumen, with slag or gravel surfacing.

Large Blisters or Buckles

Blisters or buckles, though large, will not leak when intact. However, if they allow water to penetrate they should be repaired. To treat, scrape off all surface material to a dry felt surface at least 2-1/2 feet beyond the edge of the blister or buckle. Make two cuts at right angles to each other extending 12 inches beyond the edge of the blister or buckle. Fold back the four corners of the membrane and allow to stand until thoroughly dry. When dry, apply a liberal mopping of hot bitumen, fold down the four corners of the membrane and press them firmly into the hot bitumen to produce a flat area. Apply at least two additional layers of 15-pound saturated felt, mopped down with hot bitumen and extend at least 18 inches beyond the edges of the cuts. Apply a pouring of hot bitumen at a rate of 70 pounds per square and while hot, embed gravel or slag.

When felts have been exposed and are disintegrated in numerous areas, no definite criteria can be established to determine whether the existing membrane should be repaired by adding plies of felt or whether the old membranes should be removed entirely and a new one applied.

Fishmouths or Buckled Open End Laps
Repair by scraping off all surfacing material to a distance of at least 12 inches beyond the affected area. Cut the fishmouths or buckled end lap and cement down the loose felts with hot bitumen. Apply at least two layers of 15-pound saturated felt, mopped on with hot bitumen and extending at least eight inches beyond the end of the cut felt and eight inches below the lap edge. Apply a heavy pouring of hot bitumen and into it, while hot, embed gravel or slag.

Small-Surfaced Built-up Roofs
This section presents maintenance problems and solutions particular to small surfaced built-up roofs.

Felts Exposed (Small Areas): Repairs to felts exposed in small areas should be made in a manner similar to those for mineral-surfaced built-up roofs, except that 20 to 25 pounds of asphalt should be mopped per square and the mineral surfacing applied.

Small Blisters and Buckles: When felts are exposed, treat as described for exposed felts of mineral-surfaced built-up roofs, applying 25 pounds of asphalt per square and omitting mineral surfacing. If felts are not exposed, disregard small blisters and buckles.

Severely Weathered Asphalt Coating: Smooth-surfaced, asphalt built-up roofs on which the surface mopping is relatively thin usually show definite alligatoring of the surface coating within three to five years. Alligatoring is always most severe where the asphalt coating is the thickest. If allowed to proceed, alligatoring will develop into cracking. Once the surface coating is cracked, water enters the membrane, leaks may appear, and the roof deteriorates rapidly. The kind of maintenance chosen for this problem should depend upon the future use of the structure as indicated below.

Maintenance of Roof to Be Used More Than Four Years
To treat, remove all loose dirt by sweeping, vacuuming or air blasting. Apply a thin coat of asphalt primer. Preferably apply it by brushing to avoid excessive primer. After the primer is dry, apply one of the following coatings:

- Asphalt emulsion by brush or spray at a rate of three gallons per square.
- Asphalt-based roof coating by brush or spray at a rate of three gallons per square.
- Fatty acid pitch-based roof coating applied at a rate of approximately three gallons per square.
- Hot mopping asphalt applied at a rate of 20 to 25 pounds per square.

If the asphalt coating is alligatored, but not cracked, and the felts are not exposed, the primer may be omitted with the asphalt and fatty acid-based coatings. If emulsion coating is to be applied to such surfaces, dust and dirt may be washed off with a stream of water from a hose. The emulsion can be applied to a damp, but not a wet, surface.

Maintenance for Prolonged Use
Clean off all loose dirt or dust by sweeping, vacuuming, or air blasting. Apply one coat of asphalt primer by brushing, avoiding excess primer. After the primer is dry, apply an asphalt emulsion at a rate of two gallons per square. Immediately after applying the emulsion, while it is still wet, embed

strips of fibrous glass mesh in the emulsion, lapping the strips two inches and using a moistened brush to force the mesh into the emulsion. Clay-type emulsion should be used with a glass membrane. Over the fibrous glass strips, while the first coat of emulsion is still wet, apply a second coat of emulsion at a rate of one gallon per square. Brush the second coat into the mesh with a fiber brush. Allow the second coat of emulsion to set firmly at least 12 hours in good drying weather, longer in damp humid weather. Apply a final coat of emulsion at a rate of two gallons per square. If the asphalt surface has alligatored, but not cracked, and the felts are not exposed, treat as described in the previous section, omitting the asphalt primer.

Felt Built-up Roofs

Following are maintenance problems and solutions associated with felt built-up roofs.

Felts Exposed (Small Areas): For maintenance of roofs to be used four years or less, smooth-surfaced, felt built-up roofs may be surfaced originally with hot asphalt or with a cold applied asphalt-based coating. After four or five years of exposure (sometimes earlier with a cold-applied coating) light gray or even white areas appear, indicating that the felts are partly exposed. For an expected use of not more than four years, no treatment is needed. Disregard small blisters and buckles.

For maintenance of roofs for indefinite use, recoat with asphalt emulsion at a rate of three gallons per square at intervals of four to five years. To be most effective for long periods, the initial coating should preferably be an asphalt emulsion or thin asphalt emulsion or thin asphalt cutback, rather than a hot applied or conventional asphalt fibrous roof coating. The use of fibrous glass mesh is not recommended in the maintenance of felt roofs.

Cold-Processed Roofs: Cold-processed roofs are constructed with coated organic felts cemented to each other and to the roof deck with special quick-setting adhesive and surfaced with asphalt emulsion or with a cold-applied coating.

If felts are exposed on cold-processed roofs, recoat with an asphalt emulsion at a rate of three gallons per square or with the material with which they were coated originally. Disregard small blisters or buckles. Large blisters and buckles should be treated as described under repair procedures.

Major Repairs for Built-up Roofs

When deciding whether to repair or replace a roof, the following criteria should be met if a major repair is to be made:

1. The roof has not reached its expected life, which is approximately: for five-ply roofs, 20 years; a four-ply roof, 15 to 20 years; and three-ply roof, 10 years.
2. Base felts are in sound condition and are not water-logged.
3. Installation, if present, is dry.
4. Leaks that have developed are few in number and are not serious.

Reroofing is mandatory when:

1. The roof has exceeded its expected life with little or no maintenance.
2. Felts have disintegrated and desponded and the entire membrane is water-logged.
3. Installation is wet and/or disintegrated.
4. Numerous leaks of a serious nature covering more than 25 percent of the roof surface have developed in the membrane.

Repairing by Applying Additional Plies of Felt: In applying additional plies of felt to an existing built-up roof, the additional felts must never be mopped solidly to an existing membrane. The repair membranes must be isolated from the old membrane by a dry felt layer, spot or strip mopped, or nailed, or by 1/2 inch of roof insulation strip mopped and, if possible, nailed through the old membrane to the roof deck. When the slope of the roof is greater than one inch per foot, some provision must be made for fastening the felts and insulation. The new membrane should contain not less than three plies of felt.

Surface Preparation: Remove mineral surfacing, bituminous coating, and all loose and disintegrated felts. Repair blisters, buckles, cracks, and fishmouths, as indicated earlier in this section, but omit final pouring of bitumen in mineral surfacing material. When new flashing is to be installed, cut off existing membrane flush with the base of fire and parapet walls.

Application of New Membrane

Following are methods for application of a new membrane to an existing membrane on a concrete insulated deck.

Saturated Organic Felt: To repair saturated organic felt with insulation, mop solidly with hot bitumen, 1/2 inch of roof insulation to the existing membranes. Follow with at least three plies of 15-pound saturated felt mopped solidly to the insulation and to each other with hot bitumen at a rate of 25 to 30 pounds per square. Then pour hot bitumen at a rate of 70 pounds per square, and embed gravel or flag at a rate of 400 pounds gravel or 300 pounds slag per square.

To repair saturated organic felt without insulation, spot or strip mop with hot bitumen, one ply of 15-pound saturated felt to the existing membrane. Follow with at least two plies of 15-pound saturated felt mopped solidly and finished as described in the preceding paragraph.

Asphalt and Asphalt Saturated Felt: To repair asphalt and asphalt saturated felt, with and without insulation, proceed as described for saturated organic felts, using asphalt-saturated felts. In this case, however, the finished coat should consist of an asphalt emulsion and should be applied at a rate of three gallons per square.

Wood Decks: To apply a new membrane to an existing membrane on a wood deck, proceed as described for concrete decks, except that insulation felt should always be nailed over wood decks in addition to the mopping recommended. Spot or strip mopping may be omitted when the new membrane is applied directly to the old, the first ply being simply nailed.

Gypsum Decks: To apply a new membrane to a membrane of a gypsum deck, proceed as described for wood decks. If the roof deck shows cracks on the underside, the use of insulation under the new membrane is recommended to provide protection against cracking and misalignment of the deck.

Insulated Roofs: In applying a new membrane to an existing membrane over an insulated roof, first make sure that the old insulation is dry and not disintegrated. If the insulation is not thoroughly dry or if it shows any evidence of disintegration, the existing roofing and insulation should be removed down to the roof deck and new insulation and roof applied. When the insulation is in good condition, apply an additional layer of 1/2 inch insulation to provide a smooth surface, followed by three layers of 15-pound saturated felt, mopped on and finished as described for concrete decks.

Asphalt Shingle Roofing

Asphalt shingle roofs that are properly applied usually require no special maintenance or repair treatments. Shingles normally last for a number of years with little change in appearance. The first indication of normal weathering is the loss of mineral surface granules, slight at first, but accelerating as the loss of granules exposes more of the asphalt coating to the weather.

No definite periods can be ascribed to the various phases of weathering because they will vary with the direction of exposure, the climate, and the slope of the roof. Weathering is more rapid in hot, humid climates, on southern and western exposures, and on low pitched roofs.

Asphalt shingle roofs with at least double coverage usually will not leak, even if most of the granular surface has disappeared. However, without the surfacing, weathering proceeds rapidly, and the shingles become brittle and vulnerable to wind damage with large bare areas indicating that the roof will soon need attention.

Apply asphalt shingles in accordance with new specifications for construction. Consult with a roofing contractor for further information.

Recoating
The recoating of weathered asphalt shingles is not recommended. Shingles that have weathered to the stage that recoating would seem to be indicated are probably so brittle that they are likely to be damaged severely by the recoating operation. Experience has shown that the cost of recoating and sealing the edges of the tabs in most cases approaches so nearly that of reroofing that reroofing is the most economical procedure.

Improperly Nailed Shingles
The defect that is found most frequently in the application of square-tab strip shingles is the improper placement of nails, nailing too near the top of the shingles, rather than 1/2 to 3/4 inch above a line drawn through the top of the cut out portions. Too high nailing of shingles should be corrected by placing a spot of quick-setting asphalt plastic cement under the center of each tab and pressing the tab down firmly. The spot of cement should be approximately 1-1/2 inches in diameter when pressed flat. Special quick-setting cements are obtainable from most manufacturers of asphalt roofing products. Approximately 1/3 gallon of cement is required per square of shingles. The shingle tabs should not be bent up further than needed to place the cement. No attempt should be made to renail shingles in the proper location since the bending required may damage the shingle tabs.

Areas of Strong Wind
The aforementioned cement treatment should also be used to prevent wind damage to shingles that are nailed correctly but are located in areas where strong winds are prevalent.

Hail Damage
Severe hail storms may damage asphalt shingle roofs beyond repair, particularly when the shingles have been exposed for a number of years. With such damage, both layers of shingles are broken, the roof will leak severely, and reroofing is needed. If the shingles are not broken, the removal of the asphalt coating and surfacing granules from numerous small areas will not cause the roof to leak. However, the life of a roof, so damaged, will be shortened. Maintenance is not practical because the cost would probably equal that of a new roof.

Minor hail damage, that is, when the coating and surfacing granules have been loosened from only occasional areas, may be repaired by covering the bare areas with asphalt-based roof coating or plastic cement, asphalt emulsion or cottonseed pitch based roof coating.

Reroofing with Asphalt Shingles
Asphalt shingles may be used for reroofing over wood and asphalt shingle roofs and over smooth and mineral surfaced asphalt roll roofing. The best practice, however, is to remove the existing roof covering. By removal, an opportunity is given to correct deficiencies in the roof deck, such as warped or rotted framing and sheathing; better nailing is provided, since shingles applied on a smooth surface render better surface than on an uneven surface; and the appearance of the reroofing job is better. When asphalt shingles have been applied over an existing roof and reroofing is needed, the two roofs should be removed without question. The application of asphalt shingle roofing over existing metal, slate, or asbestos cement roofs should not be attempted because of the difficulties of nailing.

Deck Preparation: When existing roofing is removed, to restore the roof deck to a nearly new condition as possible:

1. Remove all protruding nails and renail sound sheathing where necessary.
2. Remove rotted or warped sheathing boards or delaminated plywood and install new decking.
3. Cover all large cracks, knotholes, and resinous areas with sheet metal.
4. Repaint ferrous metal drip edges and other flashings that are in good condition or remove badly corroded metal flashing and install new ones of non-ferrous metal.
5. Remove old rolled roofing flashing and install new base flashing of non-ferrous metal or rolled roofing, and new non-ferrous metal counter flashings.
6. Use woodstrips of the same thickness as the existing sheathing to fill in between the spaced sheathing to which a wood shingle, slate, or tile roof was previously applied.
7. Sweep all loose debris from the roof deck.

When existing roofing remains, follow the appropriate procedures outlined below:

1. When reroofing over asphalt roofing:
   A. Remove all loose protruding nails.
   B. Cut all wrinkles and buckles and nail cut edges securely to the roof deck.
   C. Repaint ferrous metal drip eaves and counter flashings that are in good condition. Install non-ferrous metal drip edges at eaves and rake where none exist.
   D. Install new non-ferrous metal or rolled roofing base flashing and new metal counter flashing in accordance with new construction specifications where needed.
   E. Install new non-ferrous metal or roll roofing flashings.
   F. Sweep all loose debris from the roof deck.
2. When reroofing over asphalt shingles:
   A. Remove all loose or protruding nails.
   B. Nail down or preferably cut away the butts of all curled or lifted shingles.
   C. Treat flashings as described above.
   D. Sweep all loose debris from the roof deck.
3. When reroofing over wood shingles:
   A. Remove all loose and protruding nails.
   B. Nail down or cut off corners of warped shingles.
   C. Replace the decayed or missing shingles with new ones.
   D. Cut back shingles at eaves and rake far enough to apply one-inch by four-inch wood strips securely nailed.
   E. Apply beveled wood strips approximately four inches wide over each course of wood shingles. The thick side of the strips should be thick at the butts of the wood shingles, with the other side feathered to negligible thickness.
   F. Treat flashings as described above.
   G. Sweep all loose debris from the roof deck.

Asphalt Rolled Roofing

Leaky Seams

Leaks at the seams of both smooth and mineral-surfaced rolled roofings applied with two-inch laps and exposed nails are the most common rolled roofing failures. These leaks are caused principally by inadequate lapping, cementing, or nailing of the roofing, buckling of the roofing at the seams, and loose nails.

Smooth-Surfaced: Smooth-surfaced asphalt rolled roofings, used normally on temporary structures, should receive only the minimum amount of maintenance or repair that will keep them leakproof. The first effect of normal weathering on smooth surfaced rolled roofing is the loss of the fine mineral matter used to prevent sticking in the rolls. The coating asphalt, being exposed directly to the weather, deteriorates more rapidly than that of mineral, or mineral-surfaced roofing. Recoating is usually required within three to five years. Recoating will be required earliest in hot, humid areas and on buildings such as kitchens and washrooms where excessive moisture conditions prevail.

Mineral-Surfaced: Weathering of mineral-surfaced rolled roofing is similar to that of asphalt shingles, that is, normal weathering proceeds slowly and is first evidenced by the loss of granular surfacing. Recoating of mineral-surfaced rolled roofing is not generally recommended, although it may be used to extend the life of a roof if re-roofing does not appear to be justified.

Recoating Smooth-Surfaced Rolled Roofing

Smooth-surfaced rolled roofing that has been exposed for three to five years will usually show one or more of the following conditions that indicate a need for recoating:

- Asphalt coating alligatored or cracked, small coating blisters cracked or opened and other small breaks in the coating that expose the felt.
- Minor hail damage, where the asphalt coating is damaged, but the felt remains intact, also indicates a need for recoating.

To treat smooth-surfaced rolled roofing, remove all loose dirt and dust by sweeping, vacuuming, or air blasting. Apply asphalt emulsion, asphalt-based coating, or fatty acid pitch-based roof coating. The coating of asphalt primer should be omitted for all but the most severely weathered roofs. For methods of making minor repairs before the recoating operation, see further sections.

Recoating Mineral-Surfaced Rolled Roofings

Because of the protection afforded by the mineral surfacing granules, mineral-surfaced rolled roofing weathers much more slowly than a smooth-surfaced type and, consequently, requires less maintenance. As with asphalt shingled roofing, by the time deterioration through the loss of granules becomes serious, the roofing will be so brittle and generally deteriorated that recoating will not be economically feasible. However, when the future use of the building is in doubt, or when funds for reroofing are not available, mineral-surfaced roofs may be recoated as described for smooth-surfaced roofs. Mineral-surfaced rolled roofing that has been in place not more than half the period of its expected lifetime, but has been damaged by hail, should be recoated if the damage consists of loss of mineral surfacing without breaks in the felt.

Repair Methods for Asphalt Rolled Roofings

Since repair methods for smooth- and mineral-surfaced rolled roofing are identical, they will be treated together in the following section.

Small Breaks: When small breaks are limited in number, nail holes and other small breaks should be repaired by applying asphalt plastic cement.

Large Breaks: Large breaks should be repaired by opening the horizontal seam below the break and inserting through it a strip of roofing of the type used originally. Extend this strip at least six inches beyond the edges of the break, with the lower edge flush with the horizontal exposed edge of the covering sheet. Before inserting the strip, coat it liberally with a lap cement or it will come in contact with the covering sheet. After inserting the strip, press down the edges of the roofing firmly and nail securely with nails spaced approximately two inches apart and about 3/4 inch from the edges. Apply lap cement at the horizontal seam, press down firmly and renail if the original seam was nailed.

Large Areas: When a considerable area of a roof has been damaged, but the main area remains intact, repairs should be made by removing the roofing from the damaged area and applying new roofing of the same type. Full-width sheets should be applied in the same manner as the original roofing.

Leaky Seams: After all other necessary repairs to the roof have been completed, repair of leaky seams should be made. Such leaks occur most frequently as a result of loose nails, fishmouths, and inadequate lapping, nailing, or cementing. Follow the methods below..

For expected use of not more than one year, follow the procedure below. In making repairs, sweep the seams to remove accumulated dust and dirt, cut out all buckles (fishmouths) that terminate at the seams and insert a strip of roofing. Renail where necessary. Apply an asphalt plastic cement to the seams by using a trowel to feather the edge of the cement at the top of the strip. Approximately six pounds of cement is required per square of roofing.

For expected use of more than one year, follow these procedures. Permanent repairs to leaky seams of rolled roofing are best done by using a membrane such as cotton fabric or light-weight, smooth-surfaced rolled roofing, cemented over the seam and coated with a bituminous compound as described in the following paragraphs.

Repairs must be made in clear, mild weather when the outside temperature is 50 degrees Fahrenheit or higher. Sweep the seams to remove accumulated dust or dirt, cut all buckles that terminate at the seams mid insert a strip of roofing, as described in the previous paragraph. Replace all loose and missing nails. Apply the roof coating to the seams and strips approximately six inches wide. Use approximately one gallon of coating to 80 linear feet of seam and work with a three-knot roofing brush or other satisfactory brush. Embed the four-inch strip of saturated fabric in this coating, pressing it firmly into the coating until it lies flat without wrinkles or buckles. The center of the fabric must be directly over the exposed edge of the roofing. Then apply another coating directly over the strip of saturated fabric so that the fabric is completely covered and the first and second coatings are continuous. The strips of fabric can be handled most easily when cut into approximately 12-foot lengths.

Treat the vertical seams of the roofing next to the eaves first, extending the cement and fabric two inches above the exposed edge of the horizontal seam. Next, treat the horizontal seam next to the eaves, placing the fabric directly over the top of the fabric on the vertical seams. Finally, treat the vertical seams of the second sheet of roofing, extending the fabric and coating over the fabric on the first horizontal seam to its lower edge. Continue this alternate treatment of vertical and horizontal seams until all seams are repaired.

Seams treated by this method should be maintained by recoating them periodically with an asphalt coating of the type used in the original treatment.

Reroofing
To prepare the deck for reroofing with rolled roofing; (a) when the existing roofing is removed or (b) over existing rolled roofing, asphalt shingle or wood shingle roofs, proceed in the manner described in the section on reroofing with asphalt shingles. Rolled roofing should be applied in accordance with current specifications for new construction, except when applying rolled roofing over existing asphalt shingles, rolled roofing, or wood shingles, longer nails should be used.

Asphalt Cement Roofs

Investigation has shown that failure of fasteners and the mechanical damage that results from hail, traffic, contact with tree limbs, and the warping of roof decks are the principal reasons why maintenance and repair work are needed on asphalt-cement roofs.

A rigid frame is required for this type of roofing because of its tendency to crack at the fasteners when small movements occur. A double overlap of the joints and careful sealing during installation are required to reduce moisture penetration during windblown rains. Roofs should have a pitch of not less than four inches per foot.

When only a few shingles or corrugated sheets are broken, they should be removed and new ones applied. When a large percentage (25 percent or more) is broken, they should all be removed and a new roof applied. The present condition of undamaged units should determine whether those salvaged from the old roof should be reused with the new units.

When an asphalt-cement roof fails because of the deterioration of the fasteners, the failure is usually a general one and piecemeal repair is futile. When such failure occurs, usually on a very old roof, it is best to remove the entire roof. Whether the old roofing should be reapplied must be determined by its age and condition. Normally, if the roofing can be removed without damage, it may be reapplied safely.

Asphalt-Cement Shingles

To replace a broken asphalt-cement shingle that has been applied by the American method, follow the procedure described in the section on asphalt shingles. The same procedure can be followed with multiple unit shingles.

To replace a broken asphalt-cement shingle that has been applied by the hexagonal method, straighten the anchors, shatter the shingles and remove the broken pieces. Use a nail ripper to cut or draw the nail. Punch a hole in a small piece of copper, galvanized iron or painted tin. Place over bottom anchor and nail firmly to the roof deck. Notch a new shingle to allow the side anchors to pass. Slide the new shingle into place over the bottom anchor and bend down anchors to hold it in place.

To replace a broken asphalt-cement shingle that has been applied by the Dutch lap method, remove the metal anchor and nails, then remove the broken pieces of shingles and insert a new shingle with a new anchor. If the nails cannot be withdrawn, notch the new shingle to avoid them.

Asphalt-Cement Corrugated Sheets

Broken asphalt-cement corrugated sheets should be replaced with new ones fastened in the same manner as the original sheets. When this is not practical, toggle bolts with lead or plastic washers may be used. These bolts pass through the holes somewhat larger than the bolt, and when drawn tight, the washer forms a waterproof seal.

Reroofing

To prepare the deck for reroofing with asphalt cement shingles when (a) the existing roofing is removed or (b) the shingles are to be applied over an existing asphalt shingle, rolled roofing, or wood shingle roof, proceed in the manner described for asphalt shingles. However, a 15-pound asphalt-saturated felt should be applied over the wood shingle deck. Though it is entirely possible to apply asphalt cement shingles over the roofings mentioned, the long service that is normally expected of asphalt cement shingles indicates that the better practice is to remove the existing roofing, make repairs to the deck to bring it to as nearly a new condition as possible, and cover the deck with a 15-pound asphalt-saturated felt, laid horizontally with a four-inch head lap. End lap should be a minimum of six inches except at the hips, ridges, and valleys, where 12-inch laps are recommended. Nail the felt with sufficient large-headed roofing nails to hold it in place during application of shingles.

Asphalt shingles should be applied according to a complete set of specifications. Consult with a roofing contractor for more information.

Metal Roofing

Metal roofings may be of copper, tin (terne), galvanized steel, aluminum, or so-called protected metal. Because the different metal roofing materials normally require treatments, they will be considered separately; however, copper, tin, and aluminum roofs have one thing in common—when they have been well applied and adequately maintained, reroofing is seldom needed.

Copper Roofing

Copper is one of the least chemically active metals used for roofing. Consequently, copper roofing of adequate weight, applied properly, renders long service. When copper roofing is exposed to the atmosphere, a green coating forms on the exposed surface. This coating aids in protecting the metal from further corrosion.

The most common cause of failure in copper roofs is the failure to provide adequately for expansion and contraction, particularly in flat seamed roofing. Broken soldered seams and banks in the metal at points other than seams indicate inadequate provision for expansion and contraction. Copper roofs do not normally require maintenance. Any repair needed, following failure, should be done immediately.

Repair: The following repair methods are suggetsed for repair to provide adequate expansion and contraction.

For flat seamed roofs, when broken soldered seams indicate inadequate provision for expansion and contraction, new expansion joints sufficient to provide a joint at intervals of not more than ten feet in each direction should be installed.

For batten and standing seam roofs, if soldered horizontal seams are broken, loose lock seams that will permit movement of the sheet should be installed.

Small holes in copper roofing can be repaired with a drop of solder. In soldering copper, scrape the metal with a sharp instrument or emery cloth until bright metal shows on any surface that is in contact with the solder. Then apply zinc chloride or resin as a flux and tin the surface with a thin coating of solder. It is a poor practice to coat copper roofing, flashing, and gutters with an asphaltic mastic coating for the repair of small holes and breaks.

Larger breaks, not caused by inadequate provisions for expansion and contraction, may be repaired by soldering a piece of copper over the hole.

Reroofing with Copper: The deck should be restored to as nearly new condition as possible by doing the following:

1. Remove all protruding nails and renail sound sheathing where necessary.
2. Remove rotted or warped sheathing boards and install new decking.
3. Cover all large cracks, knotholes and resinous areas with sheet metal.
4. Replace or repair copper flashings.

The copper roof should then be installed in accordance with current specifications for new construction.

Tin Roofing

Tin roofs must be maintained by periodic painting. The frequency of painting will vary with different conditions of exposure, but painting should never be put off until rust appears nor should thick coatings of paint be built up by too frequent painting.

Repair:

1. For broken soldered seams, resolder in accordance with the instructions for soldering copper.
2. For leaky formed seams, reform or caulk with a plastic caulking material.
3. For small breaks, repair with a drop of solder.
4. For larger breaks, solder a piece of tin roofing over the break.

Reroofing: The roof deck should be prepared in a way similar to that of copper. Apply the tin roof in accordance with current specifications for new construction.

Galvanized Steel Roofing

Galvanized steel roofs need not be painted immediately upon exposure. In fact, without special treatment or the use of special paints, it is better to postpone the painting of galvanized steel for several months, at least, to assure adhesion of the paint. Painting may be postponed until the appearance of a bright yellow corrosion product that indicates that the zinc coating is no longer serving its protecting function.

Repair: Leaks at seams and fasteners in galvanized steel roofings may be repaired by caulking the seams or, in severe cases where caulking is impractical, by stripping the laps as described in the previous sections.

The best method of repairing breaks in galvanized steel roofing is to replace the defective sheet of roofing with a new one.

Reroofing: Prepare the deck in a way similar to that for preparing for new copper roofing. Apply galvanized steel roof in accordance with current specifications for new construction.

Aluminum Roofing

Aluminum roofing, properly applied, does not normally require maintenance. However, if evidence of severe atmospheric corrosion occurs, the roofing may be preserved by regular maintenance painting.

Failures in aluminum roofing that result from improper application are essentially the same as with galvanized steel roofing, and are similarly repaired.

Protected Metal Roofing

Protected metal roofing takes its name from the fact that it consists of steel-base sheet that is protected from the weather by a factory-applied coating. Since these are such special items, the recommendations of the manufacturer should be followed with regard to maintenance and repair. With materials of this kind, it is practically impossible to avoid some breaks in the protective covering during application. To cover such breaks and to renew the protective coating when exposure makes a renewal necessary, only materials furnished by the manufacturer should be used to insure that the new material is compatible with the old.

Slate Roofing

As with asphalt-cement roofing, mechanical damage, such as that sustained from wind, traffic, limbs of trees, and warping of the roof deck, and the failure of fasteners constitute the principal reasons why maintenance and repair work are needed. Actual failure of the slate as a result of weathering will occur eventually. With slate of poor quality this may happen in less than 25 years. With slate of good quality, it may be after more than 100 years of exposure and usually after the dates have been relayed because of the failure of the original fasteners.

If only a few pieces of slate are broken, they should be removed and new ones applied. If a large percentage (25 percent or more) are broken, they should all be removed and a new roof applied. The age and condition of the undamaged slates should be determined to decide whether they may be reused. No definite criteria can be given to determine whether a slate should be reused. However, if the part that has been exposed is not faded appreciably and shows no evidence of disintegration, and if the slate gives a sound ring when it is held as lightly as possible between the thumb and forefinger by one corner and struck a sharp blow with the knuckles, it may be used safely.

When the failure of a slate roof results from failure of the fasteners, the failure is usually a general one; as with asphalt-cement roofing, piecemeal repair is futile and it is preferable to remove the whole roof. Whether the old slate should be re-applied must be determined by its age and condition.

Maintenance and Repair

In replacing a broken slate, all pieces should first be removed and the nails should be cut with a ripper. Insert a new slate of the same color and size as the broken one, and nail it through the vertical joint of the next course above, driving the nail about two inches below the butt of the slate in the second course above the nail, and bend the strip slightly concave to hold it in place. The strip will usually extend about two inches under the course and will cover the nail and extend two inches below it.

Reroofing with Slate

To prepare the deck for reroofing with slate when (a) the existing roofing is removed or (b) the slate is to be applied over an existing asphalt shingle, rolled roofing, or wood shingle roof, proceed in the manner described for asphalt shingle roofs. However, a 30-pound asphalt-saturated felt should be laid over a wood shingle deck before the slate is applied. When slate is used to replace a tighter material or when it is applied over another material, the roof framing should be checked to determine whether it has adequate strength. As with asphalt-cement shingles, the long service that is normally expected from a slate roof indicates that the best practice is to remove the existing roofing, make the repairs to the roof deck to as nearly a new condition as possible and cover with a 30-pound asphalt-saturated felt laid horizontally with a four-inch head lap and six-inch side lap. Secure the felt with large-headed roofing nails as needed to hold it in place until the slate is applied. Apply the roofing in accordance with current specifications for new construction.

Sloping Roofs

Failure of fasteners and the mechanical damage that results from hail, traffic, and contact with tree limbs are the principal reasons why maintenance and repair work are necessary on tile roofs. The principles for sloping tile roofs are essentially the same as those that determine the treatment of slate roofs, except that occasionally after long periods of service, tile roofs in otherwise satisfactory condition may leak because of disintegration of the felt under layer.

Flat Roofs with Promenade Tiles

The frequent maintenance or repair work on a promenade tile roof is usually caused by using too few expansion joints between the promenade tile or by permitting expansion joints to become filled with non-resilient material. Maintenance and repair methods for the built-up roofing membrane are discussed under built-up roofs.

Tile Roofs

For sloping tile roofs, broken shingle tiles should be replaced with new ones by the methods described under maintenance and repair methods for slate roofs. Spanish tiles should be replaced by troweling Portland cement mortar on the new tile surface that will be lapped by the tile in the course above and on the surface that will lap the tile in the course below. Fasten the new tile in place with copper wire. Interlocking tiles use special fastenings and are replaced easily. It is sometimes impossible to match the exact shape or color of old tiles. When a number of buildings are roofed with the same kind of old tiles, it may be necessary to reproof the first with new tiles, then match the old as nearly as possible, keeping sound tiles salvaged from the first roof to patch the other roofs and to replace broken tiles when reroofing of other roofs is needed.

When insufficient joints cause raising of the promenade tiles, new joints should be installed. Expansion joints on ten-foot centers and at skylights, curbs and walls are considered adequate. If expansion joints have become filled with non-resilient material or if the expansion joint material has deteriorated, it should be raked out and replaced with new material.

Reroofing

To reroof sloping roofs with tile, existing roofing should be removed and the roof tile restored to as nearly new condition as possible by removing rotted or warped sheathing boards and replacing them with new ones and applying a 30-pound felt horizontally, with a four-inch head lap and a six-inch side lap. Secure the felt laps and expose sheets with large-headed roofing nails as needed to hold it in place until the tile is laid. Apply the roofing in accordance with specifications for new construction.

Wood Shingle Roofing

Roofs of good quality wood shingles correctly applied normally render long and satisfactory service. Failures in wood shingle roofs usually occur because of warping or splitting of the shingles, decay, normal weathering, or failure of the shingle nails.

Warping and splitting are found most frequently with flat grain shingles. They occur usually on the portion of the shingle that is exposed to the weather. Neither warping nor splitting is likely to affect the waterproofness of the roof, because a wood shingle roof should have three layers of shingles throughout. Splits in shingles, however, may shorten the life of the roof by permitting water to reach the nails in the shingles underneath and hasten decay in moist wood. Therefore, decay is found most frequently in low-pitched and shaded roofs that remain moist for long periods of time.

Normal weathering is slow in wood shingles. Factors in normal weathering are wind, driven rain, snow, hail, or sand, and alternate freezing and thawing in winter. Very old wood shingle roofs usually show the exposed shingle butt just below the area protected by the overlaying shingle to be much thinner than the protected area. Failure of shingle nails is most frequently caused by the splitting of shingles or the improper placement of nails.

The effects of all the deteriorating factors mentioned are lessened appreciably by impregnating the shingles with creosote oil. Impregnation by dipping or by pressure process before the shingles are applied is much more effective than treatment after application.

Maintenance and Repair

Warped shingles do not usually cause leaks and, except for appearance, are not immediately objectionable. Warped shingles will probably crack eventually, in which case they should be removed. Warped shingles should never be face-nailed except in preparation for reroofing. The nailing is likely to crack the shingles and the nails will work loose, permitting the roof to leak.

The following method is suggested for removing cracked or rotted shingles. Broken wood shingles can be removed by the methods earlier recommended for removing broken slates, but at least four nails must be cut. After the broken shingle is removed, insert a new one of the same size and nail it through the exposed butt, preferably with thin copper nails. Also nail the shingle immediately above through the exposed butt.

Reroofing with Wood Shingles

Wood shingles may be used for reroofing over wood and asphalt roofs and over smooth and mineral surface asphalt roofing. However, as with other materials, the best practice is to remove existing roof cover. Reroofing of wood shingle roofs is usually required when:

- Evidence of leaks shows in more than a single localized area after the roof has been exposed to more than the usual local rainfall.
- It has been determined that the leaks result from shingle failure and not from defective flashings.

Because of difficulties in nailing, no attempt should be made to apply wood shingles over metal, slate, or asphalt cement roofs.

Wood shingles should be applied in accordance with new construction specifications for this type of roofing.

Preparing the Deck: When preparing the deck, it is desirable to remove the existing roofing. Proceed by the method described for asphalt shingle roofs.

When existing roofing remains, it is desirable to lay wood shingles over an old roof of wood shingles, remove all loose and protruding nails and replace decayed or missing shingles with new ones. Nail down or cut off corners of curled and warped shingles and renail loose shingles.

Cut back shingles at eaves and rakes far enough to apply one-to four-inch strips securely nailed. Remove weathered shingles at the ridge and replace them with a strip of beveled siding, thin edge down, to provide a solid base for nailing the ridge shingles. Treat hips the same as ridges. Fill open valleys with wooden strips to make them level with the shingled surface. Install new non-ferrous metal flashing. Where necessary, install a new non-ferrous metal base and counter flashing in accordance with new construction specifications. Sweep all loose debris from the roof deck.

Reroofing over Asphalt Rolled Roofing: Inspect carefully all sheathing boards at the eaves, and replace rotted boards with new ones.

Remove all loose or protruding nails. Cut all blisters and buckles and nail cut edges to the roof deck. Repaint ferrous metal flashings that are in good condition. Where needed, install new non-ferrous metal base flashing and new metal counter flashings in accordance with new construction specifications. Install new non-ferrous metal valley flashings. Sweep all loose debris from the roof.

Reroofing over Asphalt Shingles: When it is desirable to reroof over asphalt shingles, proceed by the method described above and, in addition, nail down or cut away the butts of all curled or lifted shingles.

Single Membrane Roofs

This type of roof finish is relatively new. There are many varieties and thus manufacturers of this type of roofing membrane. Because of these factors, very little is known about proven maintenance procedures which can be effectively used on this type of finish.

Until experience and further research has resulted in maintenance procedures which are applicable to these types of roofs, the reader is referred to the Single Ply Roofing Institute, Glenview, Illinois, for specific information about maintenance and/or repair procedures.

Flashing

The function of a flashing is to provide a watertight junction between the roofing material and the other parts of the structure, and between roof sections. Flashings are the most vulnerable part of any roof because the majority of leaks result from failures of these vital areas. There are numerous causes of flashing failures and, most often, these failures result from inadequate or faulty construction. Many roof and flashing failures can be eliminated by constant, painstaking inspection by inspectors during installation. Some common causes of flashing failures are:

- Weathering resulting from insufficient or absence of protected coatings.
- Punctures usually resulting from the omission of a cant strip.
- Open laps or seams.
- Separation of flashings from vertical surfaces.
- No allowance made for expansion and contraction of metal flashings.

In many instances, leaks have been attributed to flashing failures when no such failures were evident. The actual cause may result from open joints in a masonry wall or chimney into which the water enters and works its way down behind the flashing and into the roofing. In masonry walls, this condition may be eliminated by through-the-wall flashings.

Flashings may be divided into two main classes: base and cap or counter flashings. The base flashing is the actual junction between the roofing material and the vertical wall or projection and should be considered as a component part of the roof construction. The base flashing may be either of plastic or metallic construction. In this text, the term plastic flashing includes all types other than metal.

The cap or counter flashing is usually constructed of metal and serves as a protective cover for the base flashing. This flashing should extend a minimum of eight inches and a maximum of 12 inches above the roof line and should be set into a cut extending into the wall at least four inches, through-wall flashings being preferable. The function of the cap or counter flashing is to overlap the base flashing so that all flashing strips and nails are completely covered. In some instances, a cap flashing of felt fabric is employed. This system usually consists of a four-to-six-inch strip of saturated felt or fabric embedded in plastic cement and placed two to three inches above and two to three inches below the top edge of the base flashing so that all nail heads are completely covered. A uniform coating of plastic flashing cement is then trowelled to a feather edge over the felt or fabric strip.

Flashing Failures

The following section presents maintenance problems associated with flashing failures.

Because numerous failures attributed to the roofing material are frequently flashing failures, areas with *plastic based flashings* should be one of the first to be inspected when leaks in a structure are reported. A good procedure to follow is to first make a careful inspection of the roofing material near the flashing for signs of moisture. In built-up roofs, blisters in this area are a sure indication that moisture has found its way beneath the membrane. When blisters are evident and the membrane seems intact, a flashing failure, however small, is indicated. Punctures, broken laps or seams, separation of flashings from vertical surfaces, and deterioration from weather are causes of failure. If a cant strip is present, it can usually be detected by gently tapping the flashing with a solid object in the area midway between the roof and the vertical surface. The area in question should be clearly marked for future maintenance and repair.

The common failures that occur in *metal based flashings* are cracks, broken joints and deterioration of ferrous metal flashing caused by lack of paint. Exterior and interior corners are most vulnerable areas. Usually no cant strip is employed when the base flashing is of metal.

The most likely causes of failure in *plastic cap* or *counter* flashings are:
- Separation of flashing from the vertical surface and/or from base flashings.
- Deterioration resulting from the lack of a protective coating.

The common failures of *metal cap* or *counter* flashings are:
- Location of the flashing too high or too low above the roof deck.
- Deterioration of ferrous base flashings resulting from the lack of paint.
- Cracks and broken joints from expansion and contraction.
- Separation of the flashing from the vertical surface.

Plastic Based Flashings

Punctures in plastic based flashing are usually caused by traffic or falling objects striking the base flashing or where a cant strip has been omitted. For temporary repair, make the puncture watertight by coating with plastic flashing cement. For permanent repair, remove the broken flashing, install a cant strip and reflash in accordance with standard specifications for new roof construction.

To make repairs when vertical laps of flashing are open, smooth laps back in place and cement with plastic flashing cement and coat entire lap with plastic flashing cement.

If a separation or sagging of base flashing from a wall, chimney, or monitor occurs, refasten the base flashing to the vertical surface by nailing or cementing. Recoat with a plastic flashing cement and replace appropriate counter flashing.

For disintegration of the surface coating of plastic base flashing, brush off all loosely adhering coating and apply a trowel coating of asphalt or coal tar plastic cement as used for the original coating.

Metal Based Flashings

To replace protective coating of paint on metal that is not severely deteriorated, remove all rust, moisture, loose scale, grease, and dirt and apply a fresh coating of paint.

For a lack of protective coating of paint on metal which is severely deteriorated, remove and discard the deteriorated area of flashing and reflash in accordance with standard specifications for new roof construction.

When vertical joints of base flashing are open, straighten the metal flashing and put it in place. Resolder open joints and install additional expansion joints where needed.

For a separation of metal base flashing from a wall, chimney, or monitor, make repairs in accordance with methods described above for plastic flashings.

Plastic Cap or Counter Flashing

For separation of plastic cap flashing from a vertical wall and/or base flashing, recement the plastic cap flashing and apply a trowel coat of flashing cement.

For deterioration caused by a lack of protective coating, brush off all loosely adhering coating and apply a trowel coating of plastic flashing cement.

Metal Cap Flashings

For a lack of protective coating of paint on metal which is not severely deteriorated, repaint according to the method described for metal base flashing.

For a lack of protective coating of paint on metal which is severely deteriorated, make repairs according to metal base flashings.

If the metal cap flashing is located so that it will not function properly, remove the loose flashing and trim off flush with the wall any flashing metal left in the joint. Make needed repairs to the vertical surface and reflash in accordance with standard specifications for new roof construction into a new reglet at least two inches above the old one.

Repairing Flashing When New Membrane is Applied

When it is necessary to apply a new membrane, remove the flashing and cut the old membrane flush with the base of the vertical surface. When the metal counter flashing is in good condition, it need not be replaced. Install a new base flashing after repair membrane is applied in accordance with standard specifications for new roof construction.

Parapet Walls

It is generally poor practice to coat parapet or fire walls with an impervious coating. To repair deteriorated mortar joints, rake out all loose mortar and repoint with a 1:1:6 Portland cement hydrated lime and cement mortar proportioned by volume. To repair open joints and coping, rake out all loose material and repoint with one part Portland cement and three parts sand or caulking compound.

Vent Flashings

When roof leaks in a structure are reported, all vents and vent flashings should be inspected, including the underside of the roof. Damp areas or stains near the vent indicate a flashing failure. Possible common causes are:

- Broken seams caused by expansion and contraction.
- Exposed nails that work loose, causing separation of metal flashing flange from the roof.
- Omission of felt stripping over the edge of the flange.
- Standing water around the vent or stains on adjacent roofing.
- Deterioration of metal caused by lack of protective coating of paint.
- Clogged drains.

When broken seams caused by expansion and contraction occur, resolder and install additional expansion joints, if needed.

If the flashing frame is separated from the roof, a result of exposed walls working loose, raise the flashing flange high enough to force plastic cement beneath it and redrive loose nails. Apply two plies of felt fabric cemented to each other and to the flange with asphalt, pitch, or plastic cement. The outer edge of the felt or fabric should extend not less than three inches beyond the flange and that of the second ply of fabric or felt not less than six inches. Apply finish surfacings similar to roofing surfacing.

For omission of felt stripping over flashing flange, treat edges by applying two layers of felt or fabric as described in the preceeding section.

If standing water exists around vent or water stains appear on adjacent roofing, remove old flashing and reflash. When severe leakage occurs, the reinstallation of a vent may be needed.

If there is a lack of protective coating of paint (with or without metal deterioration), remove all rust, moisture, loose scale, grease and dirt and apply a protective coating. When metal is seriously deteriorated, the installation of a new vent may be needed.

Repair During Reroofing: Before reroofing, metal vents and flashing flanges should be carefully inspected for signs of deterioration. When deterioration is serious, remove the old vent and install a new vent in accordance with standard specifications for new roof construction. When the old vent is deemed serviceable, proceed as follows:

- Flange Vents—If the flashing flange is securely fastened to the old roof, the new membrane should be cut to fit around the vent and applied over the flashing flange.
- Curb Vents—Raise the vent and reflash so that the new flashing can be brought up and over the curb and replace and refasten the vent to the curb.

Valley Flashings

Valleys must be kept clean in order to function properly. Roof dams and debris-blocking valleys may cause water to back up beneath the roofing or laps of the valley. The valley incline should be smooth and uniform to assure a rapid runoff of water.

Copper valley flashings usually require little maintenance when properly installed. However, tin valley flashings require protective paint coating and galvanized valley flashing should be painted when first signs of rust appear. Separations at the end laps and openings in the metal resulting from corrosion are common failures.

Metal valleys must be kept clean to function properly. When leaks occur at the laps, do not attempt to solder. Treat the seam or lap with a white lead paste which consists of basic carbonate and 8 percent boiled linseed oil. When small holes are found in copper flashings, clean the surface around the hole with emery cloth, apply a flux of zinc chloride or resin, and repair with a drop of solder. Where large holes occur in copper flashings, prepare the surface in the manner previously mentioned and solder a piece of copper over the holes. When holes occur in galvanized metal flashing, replace the smallest unit of the flashing.

Mineral-Surfaced Rolled Roofing Valley Flashings

Signs of normal weathering will most likely appear in valleys because water concentrates there. Weathering, which is first indicated by the loss of granules, is slight in the beginning and then accelerates as the asphalt is exposed to the elements. Separation of end laps and separation of the roof proper from the flashing often result in leaks.

Mineral-surfaced roofing valley flashing must be kept clean in order to function properly. When granules have been lost as a result of normal weathering and water concentrating in a mineral surfaced rolled roofing valley, remove all loose granules, dust, and dirt by sweeping, vacuuming, or air blasting, and apply one thin coat of asphalt primer by brushing. After the primer is dry, apply one of the following:

- Coating of asphalt emulsion by brushing at a rate of three gallons per square.
- An asphalt-base roof coating by brushing at a rate of three gallons per square.
- A fatty acid pitch-based roof coating at a rate of three gallons per square.

When end laps are separated, lift the upper lap high enough to force a liberal amount of plastic cement between plies and press the top firmly in place. When roofing or shingles have become separated from the flashing, gently lift separated shingle or area, force the plastic cement beneath it and press the shingle or roofing firmly into the cement.

Drainage Systems

The drainage system includes all gutters, drains, scuppers, and crickets. The primary function of this system is to remove water from the roof as quickly as possible and to prevent its accumulation. Every roof must have some provision for drainage, including the so-called dead-level decks. It is important that drainage areas be kept free from debris that will interfere with proper drainage. Many roof failures can be traced to inadequately designed or improperly installed drainage systems. Ponding water may indicate structural defects.

Maintenance and Repair of Drains

Drains must be kept clean in order to function properly. When a drain is broken, remove it and install a new one in accordance with specifications for new roof construction. If a separation occurs between the flashing flange and the roof, make repairs. If water stands around the drain and the drain is clean, it is set too high and should be relocated or reinstalled at a low point of the area.

Maintenance and Repair of Gutters

Gutters must be kept clean in order to function properly. When gutters are rusted or corroded, but the metal is not deteriorated, remove all rust, moisture, loose scale, grease, and dirt and apply a new coat of paint.

Gravel Stops and Metal Edge Strips

The primary function of gravel stops and metal edge strips is to finish off all exposed edges and eaves to prevent wind from getting under the edges and causing blow-offs. Another important function of the gravel stop is to prevent the loss of gravel or slag from areas near the edge of the roof. The flashing flange of the gravel stop or edge strip should be nailed securely to the roof deck and double felt strip, and then the finishing coat of bitumen and surfacing or cap sheet should be applied. The lip of the gravel stop should protrude a minimum of 3/4 inch above the roof deck where the lip of the edge strip should be a maximum of 1/2 inch above the deck.

Omission

If metal gravel stops or metal edge strips are omitted during construction, metal gravel stops or edge strips should be installed (in accordance with standard specifications for new roof construction) at the first indication of failure in this area. This practice will prevent costly blow-offs.

Deteriorated Gravel Stops and Edge Strips

When gravel stops and edge strips are slowly deteriorated, remove all rust, moisture, loose scale, grease, and dirt, and apply a fresh coating of paint to ferrous metals.

When the gravel stops or edge strips are severely deteriorated, however, remove and discard the section containing the area and replace in accordance with standard specifications for new roof construction after damaged overhang and facia boards are repaired or replaced with new ones.

Separation of Flashing Flange from Roof Membrane

When the flashing flange becomes separated from the roof membrane scrape off the gravel or slag at least 12 inches from the gravel stop and lift the flange high enough to force plastic cement beneath it. Renail it in place and apply two layers of felt stripping, the top one overlapping the lower one two to three inches. Pour hot bitumen over the bare area and, while hot, embed clean, dry slag or gravel as used for the roof.

Bitumen Flowing

If bitumen flows beneath gravel stops and down facia boards (a failure more common to the coal-tar pitch roofs and usually the result of the original design or construction), little can be done to correct it after installation of the roof. During construction, however, it can be prevented by:

1. Extending the first two plies of felt at least six inches beyond the edge of the roof.
2. Cutting the next two or three plies back even with the edge.
3. Turning the extension of the first two plies back and over the other plies, thus forming an envelope. Caution: Bent edge should not extend past the edge of the roof, because it may be damaged in installing gravel guards.
4. Installing gravel stops in the usual manner over this.

Other Projections from the Roof

The projections included in this section are those of pipes, stacks, ladder struts, flagpoles, and bracings for signs that penetrate the roofing. These projections should not be flashed with felt, as any expansion, contraction, or other movement in these members will crack the felt and a leak will develop. All such projections should be flashed with a metal sleeve flashing with flange or, on built-up roofs, a pitch pocket with flange.

To repair defective flashings at projections, cut away old felt flashings and install metal sleeve flashing with flange or a pitch pocket with flange. The flashing range should be placed on the last ply of felt and securely nailed and cemented to it. The flange then should be stripped with two layers of felt, the top one overlapping the bottom one by two to three inches and cemented with hot bitumen. When a pitch pocket is used, the cup is filled with hot bitumen. On wood decks, a layer of concrete one inch in thickness is placed in the pitch pocket and allowed to sit prior to pouring the hot bitumen.

Skylights

Skylights made of multiple supporting members and panes are often a source of leakage. As a rule, maintenance consists of replacing ceiling strips, cushion strips, and/or skylight compound at joints between glass sheets. When glass, metal, flashings, and joint cappings must be replaced, materials and methods should be similar to those used in the original installation. Wood or steel structural frame members which have deteriorated should be replaced.

Figures 9.3 through 9.18 are historical records and checklists suggested for use during roof inspections.

Means Forms
HISTORICAL RECORD:
BUILT-UP ROOFS

BUILDING _____ USED FOR _____ DATE _____

Permanent _____ Temporary _____ Year roof was applied _____

Kind of Roof Deck:

Wood _____ Concrete slab _____ Concrete block _____

Gypsum slab _____ Gypsum plank _____ Steel _____

Slope of Roof:

Flat _____ in. per foot _____

Area of Roof:

Squares (one square equals 100 sq. ft.) _____

Type of Built-Up Roof:

Asphalt

Surfaced _____ Unsurfaced _____

Cold process _____ Wide selvage _____

Coal-tar pitch _____

King of Surfacing:

Slag _____ Gravel _____ Crushed stone _____

Promenade tile _____ Slate slabs _____ Mineral-surfaced cap sheet _____

Smooth-surfaced cap sheet _____

Other _____

Number of Plies of Felt:

2 _____ 3 _____ 4 _____ 5 _____

Kind of Felt:

Organic (rag) _____ Coated _____ Uncoated _____

Asbestos _____ Coated _____ Uncoated _____

Insulation:

Yes _____ No _____ Thickness _____

Type of insulation _____

Where placed _____

Vapor seal: Yes _____ No _____

Type of vapor seal _____

Flashings:

Base flashing: Metal _____ Kind of metal _____

Composition _____ Kind _____

Other (describe) _____

Counter or cap flashings: Yes _____ No _____

Through wall: Yes _____ No _____ Metal _____

Kind of Metal _____ Composition _____ Kind _____

Other (describe) _____

Flashing block: Yes _____ No _____

Previous Maintenance: (Describe briefly with dates.)

Roof membrane: _____
_____
_____

Flashings: _____
_____
_____

Previous Repairs: (Describe briefly with dates.)

Roof membrane: _____
_____
_____

Flashings: _____
_____
_____

Figure 9.3

253

⚓ Means Forms

HISTORICAL RECORD:
ASPHALT-SHINGLE ROOFS

BUILDING _____ USED FOR _____ DATE _____

Permanent _____ Temporary _____ Year roof was applied _____

Kind of Roof Deck:

Sheathing boards _____ Thickness (in.) _____ S.S. _____

T&G _____ Plywood _____ Thickness (in.) _____

Underlayer:

None _____ Saturated felt _____ Paper _____

Asphalt Shingles _____ Wood shingles _____ Other _____

Slope of Roof (in. per ft.):

_____

Area of roof, squares (one square equals 100 sq. ft.) _____

Type of Shingles:

Strip shingles _____ Exposure (in.) _____

Individual shingles _____ Exposure (in.) _____

Strip shingles: Square butt _____ Hexagonal _____

Square butt shingles: Thick butt _____ Standard weight _____

Heavy weight _____ Class A _____

Hexagonal shingles: Standard weight _____ Heavy weight _____

Individual shingles: Standard weight _____ Heavy weight _____

Method of laying: American _____ Dutch lap _____ Hexagonal _____

Lock down _____ Exposure (in.) _____

Color of Roofing Granules:

_____

Flashings:

Valley flashings: Roll roofing _____ Asphalt shingles _____

Metal _____ Kind of metal _____

Drip edge: Roll roofing _____ Asphalt shingles _____ Metal _____

Kind of metal _____

Vent flashings: Roll roofing _____ Metal _____

Kind of metal _____

Chimney flashings: Roll roofing _____ Metal _____

Kind of metal _____

Previous Maintenance: (Describe briefly with dates.)

Asphalt shingles: _____

_____

Flashings: _____

_____

Previous Repairs: (Describe briefly with dates.)

Asphalt shingles: _____

_____

Flashings: _____

_____

Figure 9.4

Means Forms
HISTORICAL RECORD:
ASPHALT ROLL-ROOFING ROOFS

BUILDING _____ USED FOR _____ Date _____

Permanent _____ Temporary _____ Year roof was applied _____

Kind of Roof Deck:

Sheathing boards _____ T&G _____

S.S. _____ Thickness (in.) _____

Plywood _____ Thickness (in.) _____

Underlayer:

None _____ Saturated felt _____ Paper _____

Asphalt Shingles _____ Wood Shingles _____ Other _____

Slope of Roof (in. per ft.):

_____

Area of roof, squares (one square equals 100 sq. ft.) _____

Type of Roll Roofing:

Mineral surfaced _____

Smooth surfaced _____ Wide selvage _____

Method of Laying:

2-in. lap _____ 4-in. lap _____ 19-in. lap _____

Concealed nails _____ Exposed nails _____

Type of Lap Cement:

Hot applied _____ Cold applied _____

Color of Roofing Granules:

_____

Flashings:

Valley Flashings: Roll roofing _____ Asphalt shingles _____

Metal _____ Kind of metal _____

Drip edge: Roll roofing _____ Asphalt shingles _____ Metal _____

Kind of metal _____

Vent flashings: Roll roofing _____ Metal _____

Kind of metal _____

Chimney flashings: Roll roofing _____ Metal _____

Kind of metal _____

Previous Maintenance: (Describe briefly with dates.)

Asphalt roll roofing: _____

_____

_____

Flashings: _____

_____

_____

Previous Repairs: (Describe briefly with dates.)

Asphalt roll roofing: _____

_____

_____

Flashings: _____

_____

_____

_____

Figure 9.5

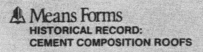

HISTORICAL RECORD:
CEMENT COMPOSITION ROOFS

BUILDING _____ USED FOR _____ DATE _____

Permanent _____ Temporary _____ Year roof was applied _____

Type of Roof:

Corrugated sheets _____ Shingles _____

Type of Shingle:

Dutch lap _____ Hexagonal _____

American _____ Multiple unit _____

Type of Deck:

Shingle: Wood _____ Other _____

Corrugated: Wood _____ Other _____

Purlin spacing (in.) _____

Slope of Roof (in. per ft.):

_____

Area of roof squares (one square equals 100 sq. ft.) _____

Type of Underlayer:

Asphalt-saturated asbestos felt _____ Asphalt-saturated organic felt _____

Other _____

_____

Weight of felt (lb.) _____

Kind of Fasteners:

Shingles: Nails _____ Other _____ Size _____

Corrugated sheets: Clips _____ Type _____

Screws _____ Size _____ Bolts _____ Size _____

Nails _____ Size _____ Other _____ Size _____

Flashings:

Valley flashings: Metal _____ Kind of metal _____

Other _____

_____

Vent flashings: Metal _____ Kind of metal _____

Other _____

_____

Drip edge: Metal _____ Kind of metal _____

Chimney flashings: Metal _____ Kind of metal _____

Other _____

_____

Previous Maintenance: (Describe briefly with dates.)

Cement composition roofing: _____

_____

Flashings: _____

_____

Previous Repairs: (Describe briefly with dates.)

Cement composition roofing: _____

_____

Flashings: _____

_____

_____

_____

Figure 9.6

 **Means Forms**

HISTORICAL RECORD:
METAL ROOFS

BUILDING _____ USED FOR _____ DATE _____

Permanent _____ Temporary _____ Year roof was applied _____

Kind of Roof Deck:

Wood _____ Metal _____ Other _____

Solid _____ Open _____ Purlins _____

Purlin spacing (in.) _____

Slope of Roof (in. per ft.):

_____

_____

Area of Roof:

Squares (one square equals 100 sq. ft.) _____

Kind of Metal:

Terne (tin) _____ Wt. coating _____

Copper (wt. ft.²) _____ Galvanized steel (wt.) _____

Coating _____ Aluminum _____ Gauge _____

Protected metal _____ Kind of protected metal _____

Type of Metal Roof:

Flat sheets _____ Corrugated sheets _____

Special shapes _____ Shingles or tiles _____

Type of Seams: Batten _____ Standing _____

Other _____

Flat _____ Soldered: Yes _____ No _____

Type of Fasteners:

Nails _____ Clips _____ Screws _____

Cleats _____ Other _____

Flashings:

Valley flashings (describe) _____

_____

_____

Vent flashings (describe) _____

_____

_____

Chimney flashings (describe) _____

_____

_____

Previous Maintenance: (Describe briefly with dates.)

Metal roofing: _____

_____

Flashings: _____

_____

Previous Repairs: (Describe briefly with dates.)

Metal roofing: _____

_____

Flashings: _____

_____

Comments: _____

_____

Figure 9.7

Means Forms
HISTORICAL RECORD:
SLATE ROOFS

BUILDING _____ USED FOR _____ DATE _____

Permanent _____ Temporary _____ Year roof was applied _____

Type of Slate:

Approx. thickness (in.) _____

Color of Slate:

Approx. wt./square (lb.) _____

Slope of Roof (in. per ft.): _____

Area of Roof:

Squares (one square equals 100 sq. ft.) _____

Roof Deck:

Wood _____ Thickness (in.) _____

Width (in.) _____ Other (explain) _____

Underlayer:

Asphalt-saturated asbestos felt _____ Weight _____

Asphalt-saturated organic felt _____ Weight _____

Other (explain) _____

_____

Type of Fasteners:

Nails _____ Size _____

Other _____

_____

Flashings:

Valley flashings: Metal _____ Kind of metal _____

Other _____

_____

Vent flashings: Metal _____ Kind of metal _____

Other _____

_____

Drip edge: Metal _____ Kind of metal _____

Chimney flashings: Metal _____ Kind of metal _____

Other _____

Previous Maintenance: (Describe briefly with dates.)

Slate roofing: _____

_____

Flashings: _____

Previous Repairs: (Describe briefly with dates.)

Slate roofing: _____

_____

Flashings: _____

_____

Comments: _____

_____

_____

_____

Figure 9.8

Means Forms
HISTORICAL RECORD:
TILE ROOFS

BUILDING _____ USED FOR _____ DATE _____

Note: For promenade tile on flat decks, use Historical Record for Built-Up Roofs.

Permanent _____ Temporary _____ Year roof was applied _____

Type of Tile: Shingle _____ Spanish _____ Interlocking _____

   Other _____ Thickness (in.) _____ Wt. per square _____

   Color _____

Slope of Roof (in. per ft.): _____

Area of Roof:

Squares (one square equals 100 sq. ft.) _____

Roof Deck: Wood _____ Thickness (in.) _____

   Width of sheathing boards (in.) _____ Other _____

_____

Underlayer: Asphalt-saturated asbestos felt _____ Weight _____

   Asphalt-saturated organic felts _____ Weight _____

   Other (explain) _____

Type of Fasteners: Nails _____ Size _____

   Other _____

_____

Flashings:

   Valley flashings: Metal _____ Kind of metal _____

   Other _____

_____

   Vent flashings: Metal _____ Kind of metal _____ Other _____

   Drip edge: Metal _____ Kind of metal _____

   Chimney flashings: Metal _____ Kind of metal _____

   Other _____

Previous Maintenance: (Describe briefly with dates.)

   Tile Roofing: _____

_____

_____

   Flashings: _____

_____

_____

Previous Repairs: (Describe briefly with dates.)

   Tile Roofing: _____

_____

_____

   Flashings: _____

_____

_____

Comments: _____

_____

_____

_____

_____

Figure 9.9

 **Means Forms**
HISTORICAL RECORD:
WOOD-SHINGLE ROOFS

BUILDING _____ USED FOR _____ DATE _____

Permanent _____ Temporary _____ Year roof was applied _____

Kind of Roof Deck:

Sheathing boards _____ T.G. _____ S.S. _____

Thickness (in.) _____ Shingle lath _____

Width of spacing (in.) _____ Thickness (in.) _____

Plywood _____ Thickness (in.) _____

Underlayer:

None _____ Saturated felt _____

Paper _____ Wood shingles _____

Asphalt roll roofing _____ Asphalt shingles _____

Other _____

Slope of Roof (in. per ft.): _____

Area of roof (one square equals 100 sq. ft.) _____

Kind of Shingles:

Cedar _____ Cypress _____

Pine _____ Other _____

Thickness of Shingles:

4 butts/2 in. _____ 5 butts/2 in. _____

Other (define) _____

Exposure (in.): _____

Shingles prestained: Yes _____ No _____

Flashings:

Valley flashings: Metal _____ Kind of metal _____

Wood shingles _____ Other _____

Vent flashings: Metal _____ Kind of metal _____

Other _____

Drip edge: Metal _____ Kind of metal _____

Wood shingles _____ Roll roofing _____

Chimney flashings: Metal _____ Kind of metal _____

Other _____

Previous Maintenance: (Describe briefly with dates.)

Wood shingles: _____

_____

Flashings: _____

_____

_____

Previous Repair: (Describe briefly with dates.)

Wood shingles: _____

_____

Flashings: _____

_____

Comments: _____

_____

_____

_____

Figure 9.10

260

INSPECTION CHECKLIST:
BUILT-UP ROOFS

BUILDING _____ SHEET _____ OF _____

DATE _____ TIME OF DAY _____

WEATHER _____ INSPECTOR _____

ROOFING MEMBRANE

General Appearance:

Good _____ Fair _____ Poor _____

Watertightness:

No leaks _____ Leaks with long-continued rain _____

Leaks every rain _____

Reported Causes of Leaks:

Weathering of roofing material _____

Faulty material _____ Faulty design _____ Faulty construction _____

Wind damage _____ Hail damage _____ Traffic on roof _____

Other mechanical damage (describe) _____

_____

_____

Low spots (water ponding) _____ Failure of flashings _____ Failure of gravel stops _____

Other causes (describe) _____

_____

_____

Adhension of Mineral Surfacing to Bitumen:*

Good _____ Fair _____ Poor _____

Bare Areas: (Give approximate percentage of total roof area below.)

Bituminous coating exposed* _____ Condition of coating: Smooth _____

Alligatored _____ Cracked _____

Felts exposed _____ Felts disintegrated _____ Edges of felts curled _____

Blisters _____ (Give size, range, and approximate number per square if numerous.) _____

Cracked to allow water to enter: Yes _____ No _____

Buckles _____ Cracked to allow water to enter: Yes _____ No _____

Cracks in membrane _____ Through to roof deck: Yes _____ No _____

Fishmouths _____

General Condition of Roof Membrane:

_____

_____

_____

Treatment Recommended:

_____

_____

_____

Comments:

_____

_____

_____

_____

*Surfaced roofs only.   Page 1 of 2

Figure 9.11

BUILDING INSPECTION
CHECKLIST (Interior and Exterior)

BUILDING _____ SHEET _____ OF _____

FLASHINGS

Base Flashings:

Metal:

    Deteriorated _____ Vertical joints open _____

    Flanges of base metal flashing loose: Yes _____ No _____

    Caused by: Inadequate nailing _____ Not properly sealed with felt strips _____

Plastic:

    Sagged or separated from parapet wall _____

    Buckled _____ Cracked _____

    Failure of base flashing: Weathering _____ Mechanical _____

    Surface coating disintegrated: Yes _____ No _____

    Vertical laps not cemented properly: Yes _____ No _____

Cap Flashings:

Metal:

    Firmly embedded into vertical wall: Yes _____ No _____

    Deteriorated _____ Vertical joints open _____

    Not covering base flashing adequately: Yes _____ No _____

Plastic:

    Surface coating disintegrated _____

    Flashing felt disintegrated _____

Flashing Block:

    Groove pointed sufficiently: Yes _____ No _____

Treatment Recommended: _____

_____

_____

_____

Parapet Walls:

    Mortar joints deteriorated _____ Settlement cracks in walls _____

    Joints in tile coping open _____ Concrete coping cracked _____

    Other defects (describe) _____

Treatment Recommended: _____

_____

_____

_____

Comments: _____

_____

_____

_____

_____

Page 2 of 2

Figure 9.11 (continued)

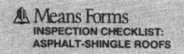

Means Forms
INSPECTION CHECKLIST:
ASPHALT-SHINGLE ROOFS

BUILDING _____   SHEET _____ OF _____

DATE _____   TIME OF DAY _____

WEATHER _____   INSPECTOR _____

Note: Asphalt-shingle roofs should never be walked upon directly. When it is necessary to get on a roof, ladders or boards with cleats nailed to them should be used to distribute the weight.

General Appearance:

Good _____ Fair _____ Poor _____

Watertightness:

No leaks _____ Leaks with long-continued rain _____ Leaks every rain _____

Reported Cause of Leaks:

Weathering of shingles _____ Faulty material _____

Faulty design _____ Wind damage _____

Faulty Application:

Nailed too high _____ Too few nails _____ Exposure too great _____

Hail damage _____ Traffic on roof _____

Other mechanical damage (describe) _____

_____

Failure of flashings _____ Other causes (describe) _____

_____

Condition of Shingles:

Apparently unchanged _____ Buckled _____ Blistered _____

Loss of granules: Slight _____ Medium _____ Severe (bare areas) _____

Curled _____ Tabs missing _____

Asphalt coating damaged (hail or other causes) _____

Coating alligatored or cracked _____

Other defects (describe) _____

General Condition of Asphalt-Shingle Roof:

_____

Treatment Recommended:

_____

Flashings: (Describe condition if defective.)

Chimney flashings: Satisfactory _____ Defective _____

Wall flashings: Satisfactory _____ Defective _____

Ridge flashings: Satisfactory _____ Defective _____

Vent flashings: Satisfactory _____ Defective _____

Valley flashings: Satisfactory _____ Defective _____

Edge flashings: Satisfactory _____ Defective _____

Drainage System: (Describe condition if defective.)

Gutters: Satisfactory _____ Defective _____

Downspouts: Satisfactory _____ Defective _____

Treatment Recommended:

_____

_____

_____

Figure 9.12

 **Means Forms**
INSPECTION CHECKLIST:
ASPHALT ROLL-ROOFING ROOFS

BUILDING _____ SHEET _____ OF _____

DATE _____ TIME OF DAY _____

WEATHER _____ INSPECTOR _____

Note: Asphalt roll-roofing should never be walked upon directly. When it is necessary to get on a roof, ladders or boards with cleats nailed to them should be used to distribute the weight.

General Appearance:

Good _____ Fair _____ Poor _____

Watertightness:

No leaks _____ Leaks with long-continued rain _____ Leaks every rain _____

Reported Cause of Leaks:

Weathering of roofing _____ Faulty material _____

Faulty application: Insufficient lap _____ Lap not cemented _____

   Exposed nails loose _____ Inadequate nailing _____ Wind Damage _____

Hail damage _____ Traffic on roof _____

Other mechanical damage (describe) _____

_____

Failure of flashings _____

Other causes (describe) _____

Condition of Roofing:

Apparently unchanged _____ Buckled _____ Blistered _____

Loss of granules: Slight _____ Medium _____ Severe (bare areas) _____

Asphalt coating damaged (hail or other causes) _____

Coating alligatored or cracked _____

Other defects (describe) _____

_____

General Condition of Roll-Roofing:

_____

Treatment Recommended:

_____

Flashings: (Describe condition if defective.)

Chimney flashings: Satisfactory _____ Defective _____

Wall flashings: Satisfactory _____ Defective _____

Ridge flashings: Satisfactory _____ Defective _____

Wall flashings: Satisfactory _____ Defective _____

Vent flashings: Satisfactory _____ Defective _____

Valley flashings: Satisfactory _____ Defective _____

Edge flashings: Satisfactory _____ Defective _____

Drainage System: (Describe condition if defective.)

Gutters: Satisfactory _____ Defective _____

Downspouts: Satisfactory _____ Defective _____

Treatment Recommended:

_____

_____

Figure 9.13

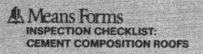

INSPECTION CHECKLIST:
CEMENT COMPOSITION ROOFS

BUILDING _____ SHEET _____ OF _____

DATE _____ TIME OF DAY _____

WEATHER _____ INSPECTOR _____

Note: Cement composition roofs should never be walked upon directly. When it is necessary to get on a roof, ladders or boards with cleats nailed to them should be used to distribute the weight.

General Appearance:

Good _____ Fair _____ Poor _____

Watertightness:

No leaks _____ Leaks with long-continued rain _____ Leaks every rain _____

Reported Cause of Leaks:

Weathering of roofing materials _____ Faulty material _____

Faulty design _____ Faulty construction _____

Wind damage _____ Hail damage _____

Other mechanical damage (describe) _____

_____

Faulty underlayer _____ Failure of flashings _____

Insufficient side lap _____ Insufficient end lap _____

Other causes (describe) _____

_____

Condition of Roofing:

Apparently unchanged _____ Loose shingles or sheets (%) _____

Broken or cracked shingles (%) _____ Broken or cracked corrugated sheets (%) _____

Apparent cause of breakage: Nailed too tightly _____ Mechanical damage _____

Other (describe) _____

_____

Failure of fasteners: Yes (%) _____ No _____

General Condition of Cement Composition Roof:

_____

Treatment Recommended:

_____

Flashings: (Describe condition if defective.) _____

Chimney flashings: Satisfactory _____ Defective _____

Wall flashings: Satisfactory _____ Defective _____

Ridge flashings: Satisfactory _____ Defective _____

Vent flashings: Satisfactory _____ Defective _____

Valley flashings: Satisfactory _____ Defective _____

Edge flashings: Satisfactory _____ Defective _____

Drainage System: (Describe condition if defective.)

Gutters: Satisfactory _____ Defective _____

Downspouts: Satisfactory _____ Defective _____

Treatment Recommended:

_____

_____

Figure 9.14

INSPECTION CHECKLIST:
METAL ROOFS

BUILDING _____   SHEET _____ OF _____

DATE _____   TIME OF DAY _____

WEATHER _____   INSPECTOR _____

Kind of Metal: _____

General Appearance:

Good _____   Fair _____   Poor _____

Watertightness:

No leaks _____   Leaks with long-continued rain _____

Leaks every rain _____

Reported Cause of Leaks:

Corrosion _____   Faulty design _____

Faulty construction _____   Insufficient lap _____

Defective fasteners _____   Broken seams _____

Faulty seams _____   Failure of flashings _____

Other causes (describe) _____

General Condition:

Rust or corrosion: None _____   Slight _____   Severe _____

Percent of total area evidencing corrosion: _____

Condition of protective coating (if any): Good _____   Fair _____   Poor _____

Seams broken: Yes _____   No _____

If yes, location _____   Number _____

Breaks (not in seams): Yes _____ No _____   Location _____

Number _____   Size _____

Holes: Yes _____ No _____   Location _____

Number _____   Size _____

Expansion joints: Too few _____   Sufficient number _____

If too few, indicate where needed _____

Treatment Recommended:

_____

_____

Flashings: (Describe condition if defective.) _____

Chimney flashings: Satisfactory _____   Defective _____

Wall flashings: Satisfactory _____   Defective _____

Ridge flashings: Satisfactory _____   Defective _____

Vent flashings: Satisfactory _____   Defective _____

Valley flashings: Satisfactory _____   Defective _____

Edge flashings: Satisfactory _____   Defective _____

Drainage System: (Describe condition if defective.)

Gutters: Satisfactory _____   Defective _____

Downspouts: Satisfactory _____   Defective _____

Treatment Recommended:

_____

_____

Figure 9.15

INSPECTION CHECKLIST:
SLATE ROOFS

BUILDING _____   SHEET ____ OF ____

DATE _____   TIME OF DAY _____

WEATHER _____   INSPECTOR _____

Note: Slate roofs should not be walked upon directly. When it is necessary to get on a roof, ladders or boards wih cleats nailed to them should be used to distribute the weight.

General Appearance:

Good _____ Fair _____ Poor _____

Watertightness:

No leaks _____ Leaks with long-continued rain _____

Leaks every rain _____

Reported Cause of Leaks:

Weathering of slate _____ Faulty material _____

Faulty design _____ Faulty construction _____

Wind damage _____ Hail damage _____

Failure of flashings _____

Other mechanical damage (describe) _____

_____

Faulty underlayer _____ Other causes (describe) _____

_____

Condition of Slate:

Apparently unchanged _____

Slate disintegrated: Slight _____ Severe _____

Broken or cracked slate (%) _____

Apparent cause of breakage: Nailed too tightly _____ Mechanical damage _____

Other (describe) _____

_____

Failure of fasteners: Yes (%) _____ No _____

Other failures (describe) _____

General Condition of Slate Roof: _____

_____

Treatment Recommended: _____

_____

Flashings: (Describe condition if defective.) _____

Chimney flashings: Satisfactory _____ Defective _____

Wall flashings: Satisfactory _____ Defective _____

Ridge flashings: Satisfactory _____ Defective _____

Vent flashings: Satisfactory _____ Defective _____

Valley flashings: Satisfactory _____ Defective _____

Edge flashings: Satisfactory _____ Defective _____

Drainage System: (Describe condition if defective.)

Gutters: Satisfactory _____ Defective _____

Downspouts: Satisfactory _____ Defective _____

Treatment Recommended:

_____

_____

Figure 9.16

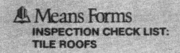 **Means Forms**

**INSPECTION CHECK LIST:
TILE ROOFS**

BUILDING		SHEET OF
DATE		TIME OF DAY
WEATHER		INSPECTOR

Note: For promenade tile on flat decks, use Inspection Form for Built-Up Roofs.

General Appearance:

Good _____ Fair _____ Poor _____

Watertightness:

No leaks _____   Leaks with long-continued rain _____

Leaks every rain _____

Reported Cause of Leaks:

Weathering of tile _____   Faulty material _____

Faulty design _____   Faulty construction _____

Wind damage _____   Hail damage _____

Failure of flashings _____

Other mechanical damage (describe) _____

_____

Faulty underlayer _____   Other causes (describe) _____

Condition of Tile:

Apparently unchanged _____   Broken or cracked tiles(%) _____

Apparent cause of breakage: Nailed too tightly _____   Mechanical damage _____

Other (describe) _____

_____

Failure of Fasteners:

Yes (%) _____   No _____

Other failures (describe) _____

_____

General Condition of Tile Roof:

_____

Treatment Recommended:

_____

Flashings: (Describe condition if defective.) _____

Chimney flashings: Satisfactory _____   Defective _____

Wall flashings: Satisfactory _____   Defective _____

Ridge flashings: Satisfactory _____   Defective _____

Vent flashings: Satisfactory _____   Defective _____

Valley flashings: Satisfactory _____   Defective _____

Edge flashings: Satisfactory _____   Defective _____

Drainage System: (Describe condition if defective.)

Gutters: Satisfactory _____   Defective _____

Downspouts: Satisfactory _____   Defective _____

Treatment Recommended:

_____

_____

Figure 9.17

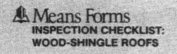

Means Forms
INSPECTION CHECKLIST:
WOOD-SHINGLE ROOFS

BUILDING _____ SHEET _____ OF _____

DATE _____ TIME OF DAY _____

WEATHER _____ INSPECTOR _____

General Appearance:

Good _____ Fair _____ Poor _____

Watertightness:

No leaks _____ Leaks with long-continued rain _____

Leaks every rain _____

Reported Cause of Leaks:

Weathering of shingles _____ Cracked shingles _____

Curled shingles _____ Failure of shingle nails _____

Failure of flashings _____

Other causes (describe) _____

_____

_____

Condition of Shingles:

Apparently unchanged _____ Cracked (%) _____

Curled (%) _____ Loose (%) _____

General Condition of Wood-Shingle Roof (describe):

_____

_____

_____

Treatment Recommended:

_____

_____

_____

Flashings: (Describe condition if defective.) _____

Chimney flashings: Satisfactory _____ Defective _____

Wall flashings: Satisfactory _____ Defective _____

Ridge flashings: Satisfactory _____ Defective _____

Vent flashings: Satisfactory _____ Defective _____

Valley flashings: Satisfactory _____ Defective _____

Edge flashings: Satisfactory _____ Defective _____

Drainage System: (Describe condition if defective.)

Gutters: Satisfactory _____ Defective _____

Downspouts: Satisfactory _____ Defective _____

Treatment Recommended:

_____

_____

_____

Comments:

_____

_____

_____

Figure 9.18

CHAPTER

10 WINDOWS AND DOORS

Windows and doors must be given the same amount of attention as walls, ceilings, and floors. This chapter contains descriptions of suggested maintenance and repair procedures for windows and doors. The major problems with windows and doors are listed below:

- Surface (frame) disintegration
- Swelling and warping
- Inferior hardware fasteners
- General lack of cleanliness
- Weatherstripping (see Chapter 6)
- Workmanship

Windows Windows provide the required amount of natural light and shield the occupants from the elements. Glass, which is utilized in most windows, is a transparent, non-absorbent material that is resistant to most acids and alkalines.

The following types of windows are available: fixed, double-hung, casement, horizontal sliding, awning, and hopper. Items common to all types of windows include:

- Frames
- Sashes (glass frames)
- Hardware
- Weatherstripping (see Chapter 6)

Inspection of Wood and Metal Framed Windows

Windows should be inspected regularly for the following items:

- Loose-fitting frames
- Damaged frames and moldings
- Ill-fitting or broken sash
- Deteriorated putty
- Broken or worn sash cords
- Damaged weatherstripping (see Chapter 6)
- Hardware and other operating parts

When inspecting window frames, peeling paint should be noted and immediately repainted. Be sure to scrape, sandpaper, prime, and finish paint the peeling frame.

Window Washing

The frequency of window washing depends on the environment. The following procedure is recommended:

Equipment	Materials
Sponge-Cellulose Squeegee Chamois or Cloth Safety Equipment for Outside Washing (high places) Ladder Putty Knife	Window Washing Solution (commercially available or inplant mixed) Do not use ammonia since it is destructive to putty. Also do not use harsh abrasives.

1. Prepare the window washing solution or use a commercially available agent.
2. Dip the sponge in the solution and apply on the glass surface covering 100 percent of the area, which includes corners.
3. Remove the water solution with a dry squeegee and chamois. Start at the top corner of the pane and draw the blade along the upper edge of the glass.
4. Wipe the blade of the squeegee with the chamois, sponge, or cloth.
5. Next, draw the squeegee down from the dry areas at the top nearly to the bottom of the pane. Repeat this along the entire width. Overlap the strokes.
6. Wipe the blade after every stroke.
7. Finish the strip along the bottom of the pane with the squeegee blade vertical.
8. Wipe clean all frames and surrounding areas, being careful not to touch the glass.

Labor Productivity	Frequency
300–450 S.F./Hour per side Depending on location and window type and if ladder and other equipment are needed.	Twice a year-minimum

Wood Windows

Wood window failures may be caused by the following:

- Weathering.
- Binding or parting of stop beads.
- Windows forced out of shape by settling, shrinking or twisting of the building frame.
- Swollen or improperly fitted sash.
- Broken or unequal length sash cords.
- Sash struck by paint in the pulley stile.
- Inoperable hardware and other accessories.

Some of the suggested ways of alleviating problems due to the above causes of failure are described in the following section.

Weathering: Protect surfaces of frames, and moldings with paint or similar coatings. Many windows can be purchased with factory baked-on coatings.

Binding: Check for foreign matter embedded in the tracks (guides) of the window. Remove excessive matter with a scraper. Clean with a paint thinner. If the window is stuck, use a floor or glazier's chisel (a wide-body chisel) and hammer. Insert the chisel between the window sash (that part which moves up and down) and the window stop (that part which holds the sash in place). Tap gently. Repeat the tapping in other locations along the window where needed. Do not use a screwdriver.

The window frame may need expanding. If so, lay a small block of wood in the groove below or above the sash. Tap sharply with a hammer on the block. Slide block along the groove, tapping as you go. Repeat on the opposite side of the window.

When it is determined that the sash binds because of pressure against the parting bead, thorough waxing of parts in contact prevents much unnecessary labor. Binding of horizontal sliding sash is also relieved by this procedure. When the inside stop beads of a double hung window press too tightly against the sash, reduce the thickness of beads by planing, sanding, or scraping along the edge adjacent to the sash or move the beads farther from the sash. Carefully plane the top or bottom rail to relieve binding in the horizontal sliding sash. After planing, coat the sash with linseed oil and wash it when the oil has dried.

Settlement: Careful re-nailing frequently restores the shape of frames that are forced out of shape by settling, shrinking, or twisting of the building frame. Cutting the sash is not recommended.

Swelling: Do not cut any window that will resume its original size when properly dry. Cut or plane a sash or frame that is swollen by moisture when it is determined that the member is too large even when dry. Remove and plane a sash whose vertical edge binds against the pulley stile or running face of the window frame.

To locate high spots on the sash, rub the sash stile with chalk and then slide the sash from closed to open position two or three times. High spots on the sash are indicated at points where the chalk has rubbed off.

Bowing: Occasionally a parting bead becomes too long, causing a bow in the middle. In this case, remove the bead from the frame and cut 1/8 to 1/4 inch from one end.

Cords: Broken or missing cords call for careful removal of the stop bead. Remove the lower sash from the frame. Knot the cords to prevent them from running through the pulley after the window is detached. Remove the parting bead to remove the upper sash. Install a new cord, then replace the upper sash, parting bead, lower sash, and stop bead in that order. An economical and efficient substitute for broken cord or pulley is the sash control spring. Proper sash operation requires that the sum of the two edge clearances be 1/8 to 3/16 inch. When installing a control spring on a new sash or old sash removed from the frame, nail a spring to each edge of the sash so that the top of the spring is six inches below the top rail of the upper sash, or six inches below the check rail or lower sash. Use 2 sixteen-gauge, 3/4-inch long brads for each spring. When installing the spring on the in-place sash, raise the lower sash at the high point and hold the spring with prongs toward the frame. Push the spring up between the sash and frame, using enough pressure to flatten out the top spring curve to prevent it from sliding upwards. Install the spring on the opposite side of the sash in the same manner. Lower the upper sash to the low point and follow the same installation instructions as above, except the spring is pushed downward. No nails or screws are required.

Balancing Adjustment: When a sash with spiral or pullman type balances is out of balance, follow the manufacturer's directions for balance adjustment. Attempts to repair such a sash without complying to prescribed directions may result in further damage.

Replacing a Broken Window Pane: The following procedure is recommended for replacing broken window frames for wooden windows.

Equipment	Materials
Clean Cloth	Glass
Brush	Glazing Compound
Gloves	Glazier Points
	Cutter Linseed Oil

1. Remove broken bits of glass with glove-protected hands.
2. Remove window putty by scraping it off.
3. Paint grooves of frame with linseed oil.
4. Lay in bead of putty (glazier's compound) about 1/8-inch thick all around.

5. Measure opening for glass and either cut (see steps below) or obtain pre-cut from a local supplier. The cut size should be 1/8 inch smaller (horizontally and vertically) than the actual measurement of the opening to allow for clearance and easier installation. For small, multiple-paned sash, single strength glass is adequate. For metal sash and for larger panes, double strength glass is recommended.
6. To cut glass, begin by covering the working surface with paper or cloth material.
7. Dust off the glass.
8. Mark the lines to be cut on the reverse side of the glass where the cut is to be made. Use a china marking crayon.
9. Holding the glass cutter firmly, and at more of a right angle to the glass than one would hold a pencil, cut the glass. Use an even pressure and cut toward the body.
10. After the line is scored, gently tap the underside of the cut near the edge of the glass with the reverse end of the cutter.
11. Place glass over a guide, such as a pencil or ruler, and snap with a gentle downward pressure.
12. After running a small ribbon of putty inside opening all around, push the glass into the putty. If the sash is wood, continue with the following steps. If the sash is metal, skip to number 17.
13. Insert two glazier points along each side to hold the glass in position and tap them halfway into the wood.
14. Complete inserting glazier points about four inches along the entire perimeter of the glass.
15. Roll the putty (or glazier's compound) into a 1/2-inch thick rope and lay it along all four sides. Apply putty with the fingers.
16. Use a putty knife to strip off excess putty and obtain a smooth surface. For metal sash, follow the steps below:
17. After pressing glass firmly in place against the putty, place (press-in) special clips into the slots in the frame.
18. Spread final layer of putty around the glass edge (follow steps 15 and 16 above).

Labor Productivity

Dependent on size, type and location of broken glass.

Frequency

As needed.

Metal Windows

Maintenance and repair of metal windows is usually considerably less involved than that for wood. General repair procedures are similar to that for metal doors (discussed later in this chapter). The most common failures of metal windows are:

- Rusting
- Warping
- Sticking of operating devices

It is important to lubricate the mechanisms regularly and to keep fastening devices secure. Problems of alignment caused by building settlement must be adjusted in conjunction with overall corrective measures, which may involve stabilizing the foundation and frame. Caulking must be maintained in good order to prevent leakage of moisture and air.

If a window is stuck, check to see that the sliding track is clear of debris and hardened paint. The sliding window may have to be removed for this examination. Remove all debris and paint with scraper. Wipe clean with paint thinner. Check to see that the tracks are not disturbed. If so, carefully straighten the bent portion. If the window labors in movement, rub candle wax or some similar material on the slider track to allow window to move more easily. If the frame of the window is bent out of shape, attempt to straighten out.

Doors

There are two main door types. The first is the interior door which controls the passage of sound and view between interior living spaces. The second type is the exterior door which provides entrance into the building with control of weather, view, and privacy. The major categories of doors are:

- Flush (hollow-core or solid-core).
- Stile and Rail.
- Accordion and Folding.
- By-pass.

Exterior doors are subject to more abuse and weathering than interior doors. However, for both types, defects and corrective measures are similar.

Doors should be dusted periodically, the frequency depending upon the environmental exposure conditions. At least twice a year the door and frames should be washed using a mild detergent compatible with the material of the door. An expected average labor productivity for cleaning both sides of the door and frames is about seven doors per hour.

Inspection

Doors should be inspected quarterly for the following:

- Installation failures.
- Loose or broken batten and deteriorated or damaged frames.
- Paint deterioration.
- Material damage, such as cracked or broken glass, split or cracked wood panels, warped or dented metal, and warped or broken screening.
- Broken or inoperative hardware (Maintenance of door and window hardware is discussed in Chapter 11).

Butt Hinged Wood Doors

Mechanical injury to mullions, headers, jambs or hardware usually causes trouble with large wood batten framed and braced doors. Decay, resulting from exposure to weather or shrinkage of door members, also causes distortion or failure. Frequently, the free edge of the door sags and causes the door to bind at the bottom and open at the top.

The following checklist should be followed when considering the need for repair:

1. Examine the jamb opening to see that hinge and lock sides are plumb and parallel.
2. Check the door header to see that it is level.
3. Check anchorage of the jamb.
4. Check anchorage of the hinges.
5. Check the lock face plates for projection beyond the face of the door.
6. Check all members for swelling, shrinking, or warping.

The following repair procedures may be followed when the door has shrunk, warped, swollen, or sagged.

Shrinkage: When a door shrinks, remove the hinges and install a filler (cardboard or metal shim) at the outer edge of the jamb and hinge mortise. This forces the door closer to the jamb at the lock edge and, provided hinge pins do not bind, the door should then operate satisfactorily. Each hinge should be shimmed equally to prevent the door from becoming hinge bound. When the door has swelled, place the shims in the inner edge of the hinge mortise.

Warping: Restore a warped door to its normal shape by removing it and laying it flat. Weighing it down may also be necessary. If it is still warped after a reasonable length of time, battens screwed to the door help restore it to a true plane. Screw eyes, rods, and turn buckles help straighten a door by gradually pulling it into place.

Sagging: For a sagging door, install a diagonal batten brace from the top of the lock side to the bottom of the hinge side to repair a sagging door permanently. The diagonal brace must cover the joints between the rail and stile and be securely fastened to both members, at top and bottom, and other intermediate members. Temporary repair is made by installation of a wire stay brace equipped with turn buckles and placed diagonally in the reverse direction from a batten brace.

Other Repairs: Doors or door members may require rebuilding because of neglect or abuse. Remove the door to a flat surface and replace the damaged member. Carpenters' clamps assist in holding door members square while nails or screws are secured in place.

When the preceding methods fail to correct the trouble, trim the door. However, do not cut doors immediately following rain or damp weather. When dry, the door may fit too loosely.

Panel Wood Doors

Failures in panel doors are similar to those in large wood doors as covered above. In addition, doors are subjected to:

- Binding at the hinge edge.
- Friction between the dead bolt and strike plate.
- Friction between the latch bolt and strike plate.

If the trouble is with the hardware, refer to Chapter 11 for a further discussion on the maintenance and repair of hardware.

Rattling: An excessive space between the door and stop head causes the door to rattle. Removal and refitting of the head stop while the door is closed remedies this difficulty. The door may also rattle because of too much play between the latch bolt and strike plate. Correct this condition by moving the plate and replacing it back toward the stop.

Loose Hinges: Hinges become loosened if a door is too tight on the hinged edge and binds against the hinge jamb. If the door has excess clearance on the lock side, and the entire pin seems to move slightly when the door is closed, loosen both hinges at the frame and insert cardboard under the jamb leaves along the outer edges. To make a uniform space between the jamb and the door, insert a strip of cardboard under the inner edge of the top hinge in the leaf which is fastened to the jamb. This usually corrects the trouble by pulling the upper part of the door closer to the jamb. However, considerable space above the door and along the outside lock edge may result. In that case, loosen the screws in the leaf of the bottom hinge which is attached to the jamb and insert the cardboard under the outer edge. If the hinge has been pulled loose, and the wood screws have damaged the wood fibers on either the door or jamb, the holes may be plugged with wood plugs or filled with plastic wood.

Warping: A warped door that has sprung inward or outward at the hinge edge is difficult to close without considerable pressure against the bulging part. This trouble is generally overcome by placing an additional hinge midway between the other two to hold the door straight. If another hinge cannot be obtained, a temporary repair is made by shifting the hinges outward or inward on the jamb. Adjust stops according to the position of the door when closed or latched.

Settling and Shrinkage: Settling of the foundation or shrinkage and deflection of frame members causes problems at door openings. When the greatest settling is on the hinge side of the door, the door will tend to become floor bound on the lock side. When settling is the greatest on the lock side, the door will bind at the head jamb. As a result, the bolt in the lock will not be in alignment with the strike plate, making it impossible to lock the door securely. Vertical settling and horizontal deflection will cause the jamb opening to become out of square. On most wood doors, the simple correction is to plane the surface at either the top or bottom rail for proper clearance.

Latching: When the latch does not operate because of improper alignment, check the face surface of the strike plate. The surface markings will show the points of interference. Remove the plate and file the opening where necessary or the strike plate can be adjusted to the proper setting.

Uneven Side Margins: It is seldom necessary to plane the surface of a door if the trouble is caused by loose or improperly mortised hinges. However, if planing is necessary because the side margins are uneven, and the door strikes at the top or bottom because of settling of the frame or similar causes, locate the points of friction. It is easier and less noticeable to plane the hinge edge. Plane the lock edge of the door approximately 1/8-inch deep. If the door still strikes along the lock edge, plane the hinge edge and cut in the hinge leaves flush with the surface, if necessary.

Sticking: Damp weather causes a door to stick. Absorption of moisture results in swelling of the frame work and door. If the door has an even margin along the top and bottom edges, and if the hinges are firm, it is necessary to plane the hinge edge. It is best to plane the hinge edge surface because hinges are more easily removed and remortised than the lock. Take care not to over-plane the surface. When it is necessary to trim a door for ideal fitting, carefully mark the amount to be planed, allowing clearance on all sides of approximately the thickness of a dime. To determine the correct amount of wood to be removed, set the door hard against the hinged edge. Wedge it plumb, and inscribe the desired thickness of material to be planed off. When planing or cutting is completed, repaint before re-hanging the door.

Beveling: The proper bevel for wood doors is two feet, eight inches to three feet in width, in about 1/16 inch per inch of door thickness. On doors of smaller width that are of equal thickness, the bevel should be increased proportionately to the decreased width of the door. On wider doors, the bevel can be decreased proportionately to the increase in width of the door.

Metal Doors

Because the majority of metal doors and fittings are shop designed and fabricated, it can be assumed that they will maintain their shape and mechanical operating ability provided hinges, locks, and other fittings remain secure in their fastenings. This is accomplished by checking screens, louvers, nuts, and bolts and special fasteners and operating devices regularly and keeping them tight and in good order. Building settlement, mechanical failure, and collision may require investigation and corrective measures for a basic cause of misalignment in the structural framing itself. Frames must be plumb and corners squared so the door fits its opening with proper clearances. Weather-proofing and caulking must be maintained in a workman-like manner. Mechanically operated doors must be removed and straightened, repaired or replaced. Repair material and finishing should match the existing material. Shop repair of metal doors should meet acceptable standards for welding, riveting, and bolting. Replacement of surface metal on fireproof metal-clad wood doors must be weather tight and of material of the same gauge as originally provided. Service doors in areas where persons enter and exit should be fitted with kick plates and bumper protection to prevent damage to adjacent walls.

Doors used in hangars and warehouses are commonly of the rolling type and motorized; they impose extreme loads on the narrow bearing surface of supporting rails. They require specialized maintenance in accordance with manufacturers' recommendations. If major warping or displacement of sliding panels occurs, engineering personnel should be consulted on repair procedures. Settling or failure of roof trussing can cause displacement of overhead rails and guides, imposing stress on the panels. Maintenance of large metal areas of doors is similar to that for metal sidings. Complicated devices for operating large doors should be maintained according to manufacturers' recommendations.

Minor, routine maintenance should be performed as follows:
1. Keep rails, guides, springs and rollers secure and free from dust, dirt, corrosion and obstruction.
2. Keep upper rails and rollers lubricated according to manufacturers' recommendations.
3. Check and maintain alignment of rollers and rails.
4. Check guides for security and alignment. Straighten bent guides.

5. Repair damaged glazed sections promptly.
6. Inspect and lubricate motors according to manufacturers' recommendations.
7. Keep hinges and springs free from dirt, debris, and corrosion, and lubricate them regularly. Replace sprung or broken hinges and springs and other door fittings.
8. Keep locks and latches lubricated and in good repair.

Storm Doors

Storm doors are commonly made of wood or aluminum with appropriate glazing. Wood storm doors should be maintained according to previous instructions on wood doors. Aluminum storm doors require little maintenance. Guide tracks and sash should be cleaned periodically. To keep them operating smoothly a spring loaded safety chain should be provided.

For both wood and aluminum screen doors, a door closer should be employed to keep the door closed when not in use, plus a latch to keep the door in a closed position.

Screen Doors

Screen doors made with wood or aluminum frames should be fitted with wire guards and wood or metal push bars on the interior side to avoid pressure on the screening in opening. Doors should be sized or trimmed to allow proper clearance at the head, jamb and bottom; too much clearance will defeat the purpose of excluding insects. However, doors should not stick or bind. A sagging screen door can be straightened by the use of a metal rod with a turnbuckle. One end of the rod is fastened to the face of the frame, at the center of the intersection of the bottom rail and the outer vertical rail. The other end is fastened as high on the face of the hinged rail as it will reach. The turnbuckle is then turned to shorten the rod and thus lift the bottom rail. Wood door frames are maintained as recommended in the discussion on wood doors.

Moisture Problems in Glass

Glass doors and windows are both subject to moisture problems when incorrectly installed or maintained. The incorrect window or combination of window panes (storm windows) will develop condensation at humidity levels considered best for indoor comfort. The condensation that forms on single glass and non-insulating glass runs down onto window sills and walls. The insulating qualities found in using insulating glass will help prevent costly damage.

A relative humidity of 30 to 40 percent is accepted as the minimum for indoor comfort during the winter months. Since lower and more economical indoor temperatures can be tolerated comfortably at high humidity levels, humidifiers are recommended during the heating season. This, however, in turn often sets up the potential for condensation problems.

Condensation

If storm windows are not being used with single pane, non-insulating glass, and moisture condenses on the inside surface of the window during cold weather, this indicates that, though the building is well sealed, the interior humidity is excessive. Little harm will be done to the window, although some paint finishes may suffer.

To correct this type of moisture condensation problem, follow the procedures below:

1. Reduce the interior humidity of the building by: a) turning down the furnace humidifier; b) installing a dehumidifier.
2. Install storm sash or insulating-glass.

3. Around aluminum windows, install quarter-round adjacent to the casing and the window frame. The quarter-round (wood) acts as a sacrificial building material in that it, instead of the window sill or casing, soaks up any condensation. The quarter-round also helps prevent ice from forming on the window frame. The quarter-round will have to be replaced every year, which is a lot cheaper than replacing the sills and/or casings.
4. Protect window sills by covering them with a plastic type material.

Storm Windows

If storm windows are used and there is condensation on them, the interior window (or sash) is poorly fitted and is allowing warm air to escape into the space between the two windows. If the inside of the interior window sweats, the problem lies in cold air coming through the storm window openings.

To stop the condensing of moisture on the interior side of a storm sash and interior side of the inside window, follow this method:

1. Install felt (or other flexible) weather stripping on all four sides of the interior window if the inside of the storm window sweats.
2. Install weather stripping along storm windows if the inside of the interior window sweats.

Venetian Blinds

The major maintenance problem with Venetian blinds is the excessive wear or inferior material of the cords of tapes. Due to an accident, the Venetian slats may become damaged.

Cleaning of Venetian blinds amounts to a simple dusting or vacuuming of the individual blinds and accessories. An average production rate is about 15 sets of blinds per hour.

Periodically wash the blinds with an all-purpose synthetic detergent. When the film of dirt is light, open the Venetian blinds so the slats are in a parallel position and wipe top surfaces clean with a soft unsoiled cloth. After completion, close the blinds and repeat the procedure on the underside facing. During this operation be sure excessive amounts of water are not allowed to come into contact with the tapes and cords.

When facilities are available and slats must be cleaned independently, it may be accomplished on a production-type setup. Place the slats in a tub or similar type of container which contains lukewarm water and a mild detergent. Wipe each slat with a cloth and wipe dry with a clean cloth. A unit productivity of 15 minutes per set of blinds can be expected.

HARDWARE AND MISCELLANEOUS MATERIALS

A listing of the various types, styles, and manufacturers of hardware available for building and plant maintenance would fill a volume comparable in size to this text. For this reason, only basic principles of hardware maintenance are discussed in this chapter.

Comprehensive Hardware Maintenance Program

The facilities manager should develop a comprehensive hardware maintenance program. This program should prevent, or at least monitor, any potential hardware deterioration through frequent inspection, cleaning, lubrication, and immediate replacement when necessary. The only logical means of bringing the task of hardware maintenance and repair under control is to adhere to the following general guidelines:

1. Select the applicable hardware for the right use and location. This is important not only when the building is in the design stage, but also when replacing deteriorated hardware. Keep in mind that hardware which is not functional cannot, in many cases, be replaced with the exact same type due to manufacturer's improvements, etc.
2. Include a visual and operational check of all hardware. This inspection program should be performed at least once each year, preferably once every six months.
3. Clean and lubricate all visual parts weekly. Clean and oil covered parts every six to twelve months.
4. Replace or repair any deteriorated or inoperative hardware immediately.

Hardware Inventory

Once the hardware is selected, an inventory of each item of hardware is mandatory. This should include:

- Manufacturer.
- Type and style.
- Date purchased.
- Date installed, if obtainable.
- Contractor who installed it.
- Any available manufacturer's recommendations and/or directions for maintaining the hardware.
- Chronological record of repair and replacements.

If the hardware is replaced, the same items listed above should be documented for the new items.

Deterioration of Hardware

The two major problems encountered with hardware are *operational abuse* and *corrosion*. Other problem areas are *inferior manufactured hardware* and *improper installation*. These latter two problems can be solved by following job specifications and completing periodic on-the-job inspections. Of course, operational problems can be the result of inferior material or workmanship that is not up to quality standards. The extent of deterioration depends upon the environment in which the hardware exists. Attention should be given to protecting the surface of an item with a coating to prevent rusting and corrosion. Or use items made of a rust-proof material, such as stainless steel.

If, based on an economical analysis, repair is chosen over replacement, consult one or more of the following sources for recommendations on repair procedures:

- Manufacturer of the hardware.
- Manufacturers' trade association.
- Experienced technical personnel of the company.

Door and Window Hardware

Interior and exterior hinged, sliding, and rolling overhead doors of every type, including storm and screen doors, should be checked for faulty installation and signs of wear. The operation and condition of hinges, lock assemblies, down, tracks, hangers, and other hardware items should be checked. If parts are rusted, they should be cleaned and a light coat of oil applied. The same procedure applies to the mechanical operator for monitor, sash, and skylights. Available keys should be fitted for proper lock operation. Brass- and bronze-surfaced plates and knobs should be cleaned as described later in this chapter.

Door Closing Devices

The first item of inspection is to see that the closing device is properly installed. The manufacturer's instruction sheet should be checked to see if the installation procedure was followed. Often, closers are applied too near to, or too far from, the hinge jamb, which affects operation as well as shortening the life of the closer. It is imperative that the closer and its arm be in their right places and firmly fastened. Familiarization with the adjustment features of the device to be maintained is necessary so that a proper understanding of their functions can be had.

Closing devices that have hydraulic controlling action, such as liquid door closers and checking floor hinges, must be refilled with the correct type of fluid at regular intervals. Keeping closers full of liquid is one way to prolong their life as well as to insure their smooth operation. Finally, all closing devices should be overhauled periodically, which involves complete disassembly and cleaning. Floor type closers for exterior doors in locations subject to ice and snow are usually difficult to maintain and should always be installed on the inside of out-swinging doors. To simplify maintenance, concealed-type door closers should be limited to locations where they are necessary for operational reasons.

Lock Sets

The basic function of a lock set is to lock and/or latch a door in its closed position; therefore, the first order of maintenance should be to determine whether the lock is in alignment with its strike; the latch bolt should seat easily into the strike without any binding or scraping. The dead bolt, when projected by action of key or turn, should also seat without forcing or cutting away wood or other materials behind the strike. In the operation of knobs, set screws must be tight. Before tightening, however, the knob action should be tested to see if it is free. If not, the set screws should be loosened in the shank and thread knobs counter clockwise, just enough to permit easy action without binding; then the set screws should be tightened, being sure that they are against a flat side of the spindle, rather than on the corner.

If the knobs have screwless or dutch fastenings, a spanner wrench will be needed to make adjustments. The wrench should be used to loosen the inside knob shank, and this knob or its shank component threaded up or down the spindle, depending on the adjustment needed. The lubricating agent for locks and cylinders is dry graphite. Do not use grease or oil. To free gummed or corroded parts, kerosene or penetrating oil may be used, but this should be allowed to evaporate before re-assembling the lock. Board or unit types of lock sets represent a different problem in that they have knobs and are assembled in an integral part of the lock mechanism. Adjustment features differ greatly among the various manufacturers; therefore, it is recommended that the manufacturer's instructions be consulted before any attempt is made to correct poor installation. The reader is referred to Figure 11.1 for a summary of common lock problems and remedies.

Thresholds

Loose fastenings are often a source of trouble with thresholds. At each inspection, screws should be tightened and loose anchors, expansion shields, etc., replaced. Check for dirt and gravel, particularly in interlocking types, and the door strip seat in the interlock. Replacement thresholds should always be placed in a new caulking base.

Common Lock Problems and Remedies			
Problem	Cause	Cure	How to Prevent
Lock frozen — can't get the key into the cylinder.	Moisture in the lock expands when frozen.	Warm the key; insert it in the cylinder gradually; alcohol on the key will speed the process.	Keep the inside dry; it is best to spray with graphite; or use fine typewriter oil, but not ordinary lubricating oil.
Bolt stuck — key turns but the bolt won't move open or closed.	Door is out of line.	Check the door alignment.	Check door alignment regularly, especially when it shows signs of jamming.
	Bolt is blocked by paint over the end.	Scrape clean with a knife; use paint remover to assure full removal of paint.	Use extra care when painting the edge of the door to avoid going over the bolt or plate.
Key binds — goes into the cylinder, but won't turn.	Improper mounting.	Remove the lock; check all parts to see that they are aligned.	Make certain to follow instructions when installing a new lock.
	Cylinder in upside-down; tumbler springs won't work.	Remove cylinder; replace it in proper position.	Make certain to follow instructions when installing a new lock.
	Duplicate key fails.	Use original key to check if door works; have duplicate key checked at locksmith.	Don't buy inferior blanks for duplicate keys. See that all burrs are filed.
	Outside elements affected metal.	Remove cylinder. If inoperable, replace.	Make a protective cover; use thin coated metal; cut to any decorative shape.
Key breaks — part of the key remains inside the cylinder.	Key wasn't inserted all the way in before it was turned.	Remove the cylinder; use long, fine pin to push the key out from the shaft end. If no spare key is available, the tip of a screwdriver can be used to open or close the lock if the cylinder is turned properly.	Lubricate cylinder with graphite; insert key fully before turning.
	Incorrect key.		Mark keys for easy identification.

Figure 11.1

Sliding Door Equipment

The various types of hangers have both vertical and lateral adjustments. Find how these work and consult the manufacturer's specifications to obtain the utmost efficiency in keeping the doors rolling easily. Bottom tracks and guide channels should be kept free of debris.

Fire Exit Devices

Particular care must be taken to insure proper operation of exit devices. Under no circumstances should chains, padlocks, or auxiliary locks be used on exit doors. Familiarization with these devices will insure that personnel will keep them functioning in such a way as to lock against ingress while still permitting unimpeded egress. This involves keeping strikes aligned and free of debris so that latches will hold. Cross bars that are loose or flabby in action indicate that service is needed. Check for worn springs, pivots, or spindles. When astragals are present, they should be adjusted to keep center clearance at a minimum.

There are many other types of builders' hardware items not specifically mentioned in this section. All of them must be examined periodically if they are to do their job. Tight fastenings and proper lubrication will keep most of them operational. The applicable hardware manufacturer's instructions for assembly and parts should be available for proper maintenance.

Bolts and Screws

All bolts and screw heads should be carefully inspected. If rust discoloration or rust scale is evident, it should be thoroughly wire brushed. If any bolts or screws have corroded to the extent that they lack good fastening strength, they should be removed and replaced. If the existing bolt or screw holes are too large for new fasteners, new holes should be drilled and new fasteners, of self-tapping type, be provided in an adjoining solid portion of the metal.

Metals

There are numerous types of metals used inside a building, such as trim work, handrails, kickplates, switch cover plates, and corner protectors. The major types of metals used are listed below:

- Aluminum
- Brass
- Bronze
- Copper
- Special alloys
- Stainless steel
- Wrought iron
- Special manufacturer-applied treatments (such as baked-on paint surfaces to aluminum)

General Metal Maintenance

Metal should be dusted (wiped clean) on a daily basis using a soft (or lightly damp) cloth. Do not use any detergents, unless recommended by the manufacturer. The metals should have a polished finish. The frequency depends on the environmental exposure conditions, but should be done at least once a week. Use a commercially available polishing agent that will not harm the metallic surface. Do not use abrasive material on the metal.

General guidelines that are part of any cleaning program for metals are listed below:

- Match the cleaner to the appropriate finish.
- Make a spot test first.
- Follow the manufacturer's directions.
- Do not allow the cleaner to drop on adjacent surfaces.

Aluminum

For routine cleaning of aluminum, the mildest method will usually work easily and well; however, conditions vary and if the surfaces have been neglected for a long time, it may be necessary to experiment with small areas before selecting a method. Three methods for routine cleaning are:

1. Wash with clean water and dry thoroughly. Wash with a synthetic detergent cleaner, rinse and dry. Use a non-etching chemical cleaner. Follow manufacturer's directions.
2. If the aluminum has accumulated a thick coating of foreign matter, it is usually easier to remove the heavy dirt first with a solvent cleaner. Then try one of the following operations.
   a. Use a wax-based polish cleaner with a clean soft rag or pad following the manufacturer's directions.
   b. Use a non-wax-based polish cleaner with a clean and soft rag or pad, following manufacturer's directions.
   c. Use a mild abrasive cleaner scouring powder with a damp, clean cloth, rinse well and dry.
3. If the result of methods 1 and 2 listed above are unsatisfactory, use a stainless steel wool pad of fine texture with liquid wax or one of the cleaners mentioned above under method 2. Mild steel wool can also be used; however, any remaining particles will rust-stain the aluminum, so they should be removed.

In the process of waxing aluminum, the aluminum should be cleaned thoroughly with a solvent cleaner. The wax should be applied with a soft, clean cloth; polishing should be done with another soft cloth. Before lacquering an aluminum surface, old lacquer worn off in spots should be stripped completely with a lacquer remover. The surface should be cleaned with a solvent-type cleaner. For good lacquer adhesion, it is preferred that a good bond to the surface be attained by using an etching-type cleaner. After three to five minutes have elapsed, the surface should be rinsed with clean water, then a thoroughly wet coat of lacquer applied by paint spray equipment, if available. Otherwise, a clean paintbrush should be used, and the lacquer thinned in accordance with manufacturer's recommendations. The first coat should be allowed to dry, then a second coat applied in the same manner. To protect aluminum from stains and other damage, the use of a strong cleaner should be avoided and the aluminum padded against scraping or denting by heavy loads. The amount of time required to clean and repair aluminum will depend upon the quantity and location of the metal.

Brass and Bronze Finishes

Correctly serviced, natural brass and bronze finishes age beautifully and contribute to the beauty of a building. No attention to this hardware is required, other than wiping with a damp cloth, until the original lacquer begins to wear off. When abrasion of the lacquer becomes apparent, the best appearance will be achieved by using a lacquer thinner or some non-acid agent to remove all the remaining lacquer coating. Then the item may be polished with a brass polish. In time, the latter will impart an attractive, soft luster characteristic of old brasses and bronzes.

Chromium- and Nickel-plated Ware

These metals should be cleaned weekly, using a soft, clean cloth which has been dampened with clean water.

On a less frequent basis, clean thoroughly, using a dampened cloth and a suitable polish. Allow the polish to dry and rub off with a clean, dry cloth. Do not use polish on chromium-plated ware. A neutral synthetic detergent can be used to clean chromium-plated ware.

Other Non-ferrous Metals

Solid non-ferrous metals may be polished almost indefinitely without fear of wearing through the metal. Most metals become dark or discolored when exposed to air. This tarnish is a very thin coat of metal oxide. It may be removed with fine abrasive. Coarse abrasives should not be used because they damage metal surfaces.

Stainless Steel

Stainless steel is easy to maintain. Usually, washing stainless steel with clear water and wiping with a damp cloth will keep the surface sparkling. Where chemical deposits pose a problem, periodic rinsing with fresh water is all that is needed. Salt from ocean spray and other chemical deposits should be rinsed off as soon as possible. When acid solutions are used to remove stubborn stains, be careful to rinse the solution off immediately and to follow safety precautions. Discoloration and possible corrosion occur if solder and welding flux are not removed at the time of installation. Flux can be washed off with clean hot water at the time of installation. Yellowish discoloration caused by welding can be removed with a mild abrasive or a phosphoric acid cleaner. Only steel wool should be used to maintain stainless steel. Fingerprints and scratching are caused by heavy pedestrian traffic. Detergent and water will usually remove fingerprints. If prints are stubborn, use a clear white mineral oil or water cleaner. Surface scars can be removed by rubbing with an abrasive or a polishing cloth in the direction of the scratch marks in the finish. The following tips may prove helpful in stubborn cases:

- When using abrasives or detergents, rub only in the direction of the polishing fines.
- Always use the mildest process that will do the job.
- Remove corrosive chemicals immediately with clear water.
- To remove chewing gum, use only wood or plastic, stainless steel, or chrome-plated tools. Knives or scrapers made from other metal may scratch or leave metal particles embedded on the surface.
- Always rinse after using detergents or abrasives, and wipe the surface dry to prevent streaking and spotting.

Lacquered Metal

A clear lacquer usually lasts for several years without yellowing, cracking, chalking, or collecting foreign matter. Merely by wiping with a clean cloth moistened with water, a mild soap, or detergent will keep lacquered metal clean. Lacquer should be applied in the following manner:

1. Strip old lacquer with a lacquer remover.
2. Clean the surface with accepted materials and equipment. If not available, use a solvent cleaner.
3. Apply a uniform, wet coat of lacquer, preferably with spray equipment, according to directions. If spray equipment is not available, a clean paintbrush can be used, first thinning the lacquer with a flow-evaporating solvent.
4. Let the first coat dry thoroughly before applying the second coat.

Metals which require low maintenance are aluminum, bronze, brass, and stainless steel. Be especially careful in the inspection and cleaning process around openings.

In order to establish the most effective maintenance program relative to interior metals, consult the manufacturer's representative for his recommendations on the care and possible repair of the metal.

Repair of Metals

For small items, as for hardware, it would be more economical to replace the item than to repair it. However, for extensive areas of metal, such as hundreds of linear feet of aluminum handrail, the possibility of repair, instead of replacement, should be considered. For a complete discussion on corrosion of metals, refer to Chapter 7.

Plastics

Most plastics are either acrylic, polycarbonate, or butyrate and require special care in the selection of a cleaning process. Many of the strong solvents, alkaline solutions, and various types of alcohol may damage plastics. Plastics are susceptible to damage (scratching) if abrasives are used. The above mentioned plastics may be harmed when washed with trisodium phosphate, naphtha, or mild soap solutions. Acrylic and butyrate plastics soften if exposed to boiling water. The temperature of cleaning solution should be limited to 200 degrees Fahrenheit when cleaning acrylic plastics and 150 degrees when cleaning butyrate. Polycarbonate may withstand temperatures up to 280 degreees.

CHAPTER 12

LANDSCAPING AND DRAINAGE

This chapter presents the necessary information for the maintenance department to adequately maintain landscaping and drainage. It does not cover in detail all types of grasses, shrubs, and trees available for planting. For specific information as to the selection of new plantings for a particular geographical area, consult with a local professional nursery.

Lawns

There are two types of grasses in common use today: these are the warm season grasses and the cool season grasses. The warm season grasses, used as permanent planting through the South, include Bermuda, Carpet Grass, Buffalo, Centipede, and Saint Augustine. The cool season grasses include Kentucky Blue Grass, the Fescues, Redtop, Timothy, Rye Grass, Crested Wheat Grass, and Bent Grass. Cool season grasses are grown successfully in the South at elevations where the temperatures are not too high. Refer to Figures 12.1, 12.2, and 12.3 for characteristics of grasses and plants.

Seeding

A good seedbed must be prepared before any grass seeding is attempted. Under certain physical soil conditions, grass stands can be obtained by merely scattering the seed on top of the ground, but this method is not always dependable. Better stands are obtained when the grass is sowed on a properly prepared seedbed. Poor seedbed preparation and inadequate surface protection for the tender seedlings account for more grass seedling failures than adverse weather conditions.

Late August is the best time to start a new lawn or repair an old one. However, before reseeding old failing lawns, the cause of the trouble should be determined and corrected. Some of these causes are: too much shade, grass cut too short, lack of aeration, or sprinkling too often instead of soaking occasionally. Also, wetting in the evening and leaving the grass wet during the night can be harmful.

Seedbeds: A grass seedbed should be firm, friable, and free of weeds, stones, or other debris. Treat soil with herbicides before seeding where weeds are a serious problem. After the surface has been graded to the specifications, till the soil to a depth of three to five inches unless it has been loosened sufficiently during construction or grading operations. The seedbed must be firm before seeding, but be careful not to produce a finely pulverized surface except where the use of very small seeds makes it advisable. A finely pulverized surface is susceptible to damage by washing, puddling, crusting, and blowing and should be protected by mulch.

Characteristics of Warm Season Grasses

Plant	Shade or Sun	Growing Habit	Seeding Rate[1]		Min[2] Live Pure Seed %	Principal Use
			Improved Grounds	Unimproved Grounds		
Bahia Grass[3]	Sun	Rhizome	2.0	20	60	Unimproved grounds, erosion control.
Bermuda Grass	Sun	Stolon Rhizome	1.0	15	70	Improved grounds, lawns, ball fields.
Bermuda Grass, med fine	Sun	Stolon Rhizome	—	—	—	Improved grounds, lawns, ball fields.
Bermuda Grass, fine leaved	Sun	Stolon Rhizome	—	—	—	Golf greens.
Bermuda Grass, coarse leaved	Sun	Stolon Rhizome	—	—	—	Semi-improved grounds only.
Big Bluestem	Sun	Bunch	—	—	—	Unimproved grounds, erosion control.
Broomsedge	Sun	Bunch	—	—	—	Embankments as mature hay.
Buffalo Grass[3]	Sun	Stolon	3.0	15	50	Improved and unimproved grounds.
Carpet Grass	Sun	Stolon	2.5	20	60	Unimproved grass.
Centipede Grass	Shade	Stolon	—	—	—	Partially shaded lawns.
Dallis Grass	Sun	Bunch	—	20	60	Unimproved grounds, pastures.
Grama Grass	Sun	Bunch	3.0	25	60	Improved and unimproved.
Kikuyu Grass	Both	Rhizom Stolon	—	—	—	Semi-improved grounds.
Little Bluestem	Sun	Bunch	—	—	—	Unimproved grounds, erosion control.
Manila Grass	Both	Stolon	—	—	—	Partially shaded lawns.
St. Augustine Grass	Shade	Stolon	—	—	—	Shaded lawns.
Sand Drop-Seed	Sun	Bunch	—	10	80	Unimproved grounds, erosion control.
Weeping Lovegrass	Sun	Bunch	—	5	80	Unimproved grounds, erosion control.
Zoysia	Sun	Stolon	—	—	—	Lawns

[1]Improved grounds: rate per 1,000 square feet. Unimproved grounds: rate per acre. These are pasture rates and should be doubled for embankments.
[2]Basis for computing plating rates.
[3]For quick germination, use treated seed.

Figure 12.1

Characteristics of Cool Season Grasses

Plant	Shade or Sun	Growing Habit	Seeding Rate[1]		Min[2] Live Pure Seed %	Principal Use
			Improved Grounds	Unimproved Grounds		
Bent Grass, Colonial	Sun	Bunch	2.0	—	70	Golf Courses.
Bent Grass, Creeping	Sun	Stolon	2.0	—	70	Golf Courses.
Blue Grass, Canada	Sun	Rhizome	—	25	70	Barren Ridges, unimproved grounds.
Blue Grass, Kentucky	Both	Rhizome	2.5	25	70	Lawns, physical training areas, general use.
Blue Grass, Roughstalked	Shade	—	2.5	—	70	Shady moist lawns in northeast.
Brome Grass, Smooth	Sun	Rhizome	—	25	70	Unimproved grounds, erosion control.
Fescue, Chewings	Shade	Bunch	3.0	60	70	Lawns, cemeteries, shady mixtures.
Fescue, Red	Shade	Rhizome	3.0	40	70	Lawns, cemeteries, shady mixtures.
Fescue, Tall	Sun	Bunch	5.0	50	80	Unimproved grounds, erosion control.
Orchard Grass	Both	Bunch	—	25	70	Unimproved grounds, erosion control.
Reed Canary Grass	Sun	—	—	10	70	Unimproved grounds, lowlands.
Redtop	Sun	Bunch	—	15	70	Unimproved grounds, erosion control.
Rye Grass, Perennial	Both	Bunch	4.0	40	80	Lawns, cemeteries, shady locations.
Rye Grass, Italian	Both	Bunch	4.0	40	80	Shady lawns, temporary cover.
Timothy Grass	Sun	Bunch	—	15	80	Unimproved grounds.
Wheat Grass, Crested	Sun	Bunch	4.0	25	80	Unimproved grounds, erosion control.
Wheat Grass, Intermediate	Sun	Bunch	—	25	80	Unimproved grounds, erosion control.
Wheat Grass, Western	Sun	Rhizome	—	25	80	Unimproved grounds, erosion control.

[1]Improved grounds: rate per 1,000 square feet. Unimproved grounds: rate per acre. These are pasture rates and should be doubled for embankments.
[2]Basis for computing planting rates.

Figure 12.2

Bare Areas: Bare areas increase the hazards of wind and water erosion. When preparing for grass seedling operations, limit the size of the area to be exposed at one time and schedule the preparation, seeding, and mulching operations to reduce the erosion hazard. Apply an anchor mulch protection immediately after seedbed tillage is completed. Then do the necessary seeding through the mulch cover. Again, investigate the reason for bare areas before repairing the lawn.

Seeding Methods: Grasses are usually seeded with special grass attachments. A uniform distribution of seed is essential for an even cover turf. Half the seed should be planted in one direction, and the remainder at right angles to the first seeding with hand operated seeders. Broadcast-type seeding should be followed with a light harrowing or raking to cover the seed before compacting the soil. On slopes too steep to operate large seeding equipment, seed grasses with ordinary hand seeders or with equipment specially designed for this purpose.

Most grass seed should be planted from one-quarter to one-half inch deep. Seed, either left on the surface or buried too deeply, will not germinate rapidly nor grow as well as seed lightly covered with soil. Proper incorporation of seed into the soil is important. Immediately after seeding and proper incorporation of the seed, the soil should be compacted and mulched.

			Seeding Rate[1]		Min[2]	
Plant	Shade or Sun	Growing Habit	Improved Grounds	Unimproved Grounds	Live Pure Seed %	Principal Use
Alfalfa	Sun	Clumps	Not recommended.	15	80	Areas outleased for hay, firebreaks.
Annual Lespedeza	Sun	Clumps	Not recommended	20	80	Temporary cover, erosion control.
Bur Clover	Sun	Winter Annual	Not recommended.	Not seeded.	—	Improved grounds, shady slopes.
Chinese Lespedeza	Sun	Clumps	Not recommended.	15	80	Erosion control on slopes.
English Ivy	Both	Creeping	Not seeded.	Not seeded.	—	Improved grounds, shady slopes.
Ice Plant	Sun	Creeping	Not seeded.	Not seeded.	—	Ground cover, erosion control.
Japanese Honeysuckle	Both	Creeping	Not seeded.	Not seeded.	—	Erosion control, unimproved grounds.
Kudzu	Sun	Creeping	Not seeded.	Not seeded.	—	Erosion control, unimproved grounds.
Lippia	Sun	Creeping	Not seeded.	Not seeded.	—	Ground cover.
White Clover	Sun	Stolon	Only in mixtures.	Only in mixtures.	—	Mix with grasses for better cover.

Characteristic of Plants Other than Grasses

[1]Improved grounds: rate per 1,000 square feet. Unimproved grounds: rate per acre. These are pasture rates and should be doubled for embankments.
[2]Basis for computing planting rates.

Figure 12.3

Sprigging and Sodding

Vegetative grass planting include solid, strip, mulch, or spot sodding, plugging and sprigging. Each method involves the transplanting of topsoil and roots from one area to another. The method used depends upon the type of grass involved and the time available to establish a protective cover. Most of these methods of grass establishment are costly and can be justified only where seeding will not succeed or produce results required in the time available. Vegetative planting extends the period in which grasses can be successfully established. Vegetative planting materials should be selected and handled carefully to assure a weed-free vigorous growth. Grass over two inches high should be mowed before lifting for transplanting. Protect the materials to be planted from the hot sun and drying winds, and plant immediately after harvest in a moist, well-prepared, and fertilized soil. Keep new plantings moist and protect them from run-off and erosion damage until their roots become reestablished.

Solid Sodding: Complete sodding is the quickest way to establish a vegetative ground cover. Although costly, it is often the most practical method for repairing vegetative failures on critical areas, such as steep slopes and waterways exposed to heavy runoff. The sod should be cut about one and one-half inches thick in convenient sizes for handling. With a mechanical cutter, the sod should be cut into long strips and rolled into bundles to facilitate transportation and transplanting. Start laying strips at the bottom of the slope at right angles to runoff flow. Work upgrade. Stagger joints of sod strips and fill any open joints with loose soil. Water and roll or tamp all sod after laying. Where there is danger of newly layed sod washing out, it should be pegged down with wooden pegs, wire staples, or chicken wire.

Sprigging: Grasses such as Bermuda, Carpet, Centipede, Saint Augustine, Buffalo, and others that spread by surface or underground root stock can be propagated by transplanting the root stock only. Sod of the grass is lifted, chopped, or shredded to secure a supply of root stocks with little or no soil attached. These are planted in a well-prepared, moist seedbed before they dry out. For row sprigging, they are planted in furrows 18 to 24 inches apart and 2 to 6 inches deep, depending upon the type of soil. When using machines that open the furrows, drop the sprigs through metal tubes and cover them. Sprigging with machines costs much less than hand sprigging. As the grasses and the rows develop, they spread over the intervening spaces. In broadcast sprigging, spread the sprigs uniformly over a prepared surface, either by hand or by a manure spreader. Cover with a disc or similar implement to a depth of two to three inches in clay soils and four to five inches in sandy soils. The method will depend upon how quickly a complete cover must be established. A combination of sprigging and mulching is necessary on slopes susceptible to erosion.

Other Methods of Establishing Cover

A number of modified forms of sodding can be used to meet local needs or to reduce costs. If a comparatively long period is available for establishing a complete cover.

- **Mulch Sodding:** This is spreading a mixture of topsoil and sprigs over the area to be vegetated. This mixture can be prepared by discing the sodded area. Prepare the mixture immediately before use so that it will not dry out.
- **Strip Sodding:** Plant strips of sod at regular intervals over the area to be treated. If the area is seeded before the strips are placed, a complete covering of the area by grass will be speeded up.
- **Plugging:** This is similar to strip sodding, except that small plugs of sod are spaced at regular intervals in rows running at fight angles to each other instead of strips. Grass spreads out from the strips or plugs to fill the intervening spaces.

Lawn Maintenance

Mowing

Mowing is usually the most costly phase of grounds maintenance; thus, it should be done as part of a routine maintenance program by trained personnel. Keep the mower blades sharp and the engines properly adjusted and in good operating condition. Avoid mower damage by keeping areas free of discarded wire, sticks, bottles, rocks, stumps, or other harmful objects.

Mow grass according to height rather than on a fixed timetable. Mow often enough so that clippings can remain on the ground without smothering the existing stand of grass. Mow the grass to a height consistent with the use of the area and with specified company maintenance standards. Regular mowing checks weed growth and encourages the development of low-growing, sod-forming grasses.

Mow improved areas, such as lawns, to a height of not less than one and one-half to two inches for warm season grasses and two to two and one-half inches for cool season grasses. Close mowing starves the grass roots because of inadequate top growth, permits soil moisture to evaporate more rapidly, and aids the growth of crab grass, which spreads aggressively once it gets a good start. Shallow, weak-rooted turf also provides poor cover and permits invasion of weeds.

Cut semi-improved areas, such as archery ranges, to a height of three to four inches. Where weed control is a major problem, mow the area before weed seeds mature. Allow newly planted grass to become well established before mowing. In other cases, safety or fire control requirements may regulate the maximum height of grass permitted.

Fertilizing

Fertilizers are plant foods that are applied to soil to establish and maintain healthy and vigorous growth. Commercial fertilizers generally contain nitrogen, phosphate, and potash. The amount of each element required depends upon the type of soil, the vegetation desired, and the climate. An analysis of each type of soil is usually necessary. Types of fertilizers, times of application, and the amount to be applied each time should be included in the exterior maintenance plan or landscaping maintenance plan.

Fertilizers containing one or more of the three main elements are available commercially and the percentage of each element is indicated on the tag or label of the bag. The first number in the series indicates the percentage of available nitrogen, the second number gives the percentage of available phosphate and the third shows the percent of available potash. For example, a 10-6-4 fertilizer contains 10 percent nitrogen, 6 percent phosphate, and 4 percent potash. The remaining 80 percent consists of inert materials.

It is important that the principal plant food elements be properly balanced when growing grasses and that this balance be maintained after the vegetation is established. The rate of application, therefore, varies and should be governed by the results of soil tests and the amount of rainfall.

The best time to apply commercial fertilizer is when the seed is sowed. For established lawns, however, fertilizer should be applied frequently. Nitrogen is the primary need for grass. When more grasses are to be maintained in a stand, the basic plant food should be nitrogen. Cool season grasses can be top dressed with fertilizer in the spring, just as growth is beginning or in the early fall, when the plants are still growing. Warm season grasses are fertilized in the early summer, and one application per year is usually sufficient. Improved grounds should receive an application of complete fertilization at least every three years, generally in the fall. Improved areas require two to three pounds of nitrogen per one thousand square feet. Divide the annual amount into two or three applications per year when fertilizing cool season grasses on improved areas. With two applications, apply one in early spring and the other in early fall. If three applications are made, apply in early spring, late spring, and early fall.

Liming

Acid soils generally require liming to reduce their acidity and to increase the activity of important soil organisms. Lime is applied in the form of ground limestone, hydrated or burned lime, marl, oyster shells, and similar materials. Soil tests will indicate how much lime, if any, is required.

Lime is applied by use of a conventional lime broadcaster, which resembles a seeder, a fan-type spreader trailed behind a truck; or a fertilizer spreader. Lime can be applied whenever convenient; however, fall is the best time of the year.

In humid regions, periodic liming of soil is necessary to maintain the desired type of soil reaction. Determine the amount of lime required to correct excessive acidity by testing representative soil samples.

Mulching

The term mulching refers to the application of straw, hay, gravel, cinders, specially prepared from cutback asphalt, asphaltic emulsion and similar materials. Mulch forms a protective covering on the soil, increasing water intake, reducing evaporation, and checking erosion. Use mulch that permits the water to percolate through, rather than one that mats and sheds water.

Weeds

Weeds and brush must be controlled to permit vigorous growth of desired plantings, to prevent erosion of drainage ways and reservoirs, eliminate objectionable plants, and reduce fire hazard. Where it is necessary to control weeds for protection of desirable covers, complete eradication is seldom justified. About half the weeds will be annuals and the others will be perennials living two or more years. Noxious weeds are usually creeping perennials, spreading from roots or seed, and are difficult to kill. Various techniques may be employed to effectively control weeds. The techniques employed will vary according to climate, season, soil, and types of plants. Some of the techniques used to control weeds are listed below:

- Use of chemicals (herbicides).
- Mowing and chopping.
- Burning.
- Tilling the soil.
- Planting the area with other types of vegetation that have superior competitive characteristics.
- Changing the soil's fertility or acidity so that problem weeds will be less able to compete with the desired vegetation.
- Manipulation of water level at critical times to destroy some water weeds.

Although it is usually a relatively simple matter to kill most weeds, it is often difficult to prevent their return from seed or unkilled portions of plants. There are so many highly specialized matters involved in weed control that a complete program should be developed in cooperation with specialists before weed control projects are undertaken.

Herbicides: Herbicides are chemicals which kill or damage plants. Herbicides generally are used for weed control along fences, railroads, and other places which are impossible or difficult to mow. They should not be used where cheaper methods of weed control are effective.

Mowing and Chopping: Mowing is usually the most economical method of weed control in areas where mowers can be used. Weeds should be mowed before they go to seed. More frequent mowing may be necessary for fire protection.

Burning: Burning should be used for weed control only where it has definite advantages over other methods and when it is safe to do so. It usually damages the vegetative cover and results in dust and erosion problems. Plants such as poison ivy should not be burned, because the sap vaporizes, floats in the air, and affects the skin of susceptible persons in the area.

Tilling: Till the soil primarily before seeding. Do not till if it adds to a dust problem, or if mowing is more economical.

Planting: It is sometimes possible to smother weeds by planting grasses or shading them with vines or shrubs. Plant only when it is economical to do so.

Chemicals: The fertility or acidity of the soil can be changed by the addition of chemicals, such as lime and plant food elements. By these means, the growth of certain types of plants is restrained, while the growth of other types is favored. For example, the addition of organic matter will heighten a heavy soil and promote growth of plants.

Changing Water Levels

If the water level can be regulated, cattails and other weeds extending above water level can be killed, first by cutting off their tops to weaken them and then by raising the water level to reduce the oxygen supply.

Weed Control for Lawns

Weeds frequently invade turf as a direct result of inadequate maintenance. Control is generally a minor problem on well-managed turf. Timely mowing, proper height of cut, adequate fertilizing, and adequate use of water generally keep weeds in check without any special treatment. Weed eradication is simplified if a minimum cutting height of one and one-half to two inches is maintained on improved grounds during periods of high temperatures, and cool season grasses are not mowed during summer months. Under certain conditions when weed growth is so vigorous that control by ordinary maintenance practices is not adequate, the use of chemicals such as herbicides may be needed. Care should be taken in their use not to harm adjacent shrubbery or landscaping. Consult with your local nursery professional for the type of chemical needed.

Brush Control

There are various methods that can be used to control brush. Mechanical methods utilize mechanical equipment to control brush. Chemical methods are also commonly used to control brush.

Weeds in Reservoirs and Drainage Systems

Weeds in reservoirs and drainage systems should be controlled or removed, although large ditches may have a mat of perennial grass or of vines for erosion protection. Where weeds are difficult to control, rearrangement of structures and drainage channels often is advisable. Properly cleaned culverts of correct size and drainage channels with side slopes graded 3:1 or flatter, and established in perennial grasses, help to prevent silting in the bottom of channels, thus eliminating the encouragement of weed growth. Using a dragline is one of the most effective mechanical methods of controlling aquatic weed growth on channel bottoms, as it removes silt as well as vegetation. Equipment with buckets on a continuous chain also is satisfactory. Both weeds and silt can be removed from dry ditches with graders, scrapers, bulldozers, or similar equipment. To eliminate undesirable aquatic weeds, use herbicides or soil sterilants when water is low or the ditch is dry. These can be applied either as foliage sprays, or directly on the soil. Use underwater sprays for submerged aquatic weeds.

Poisonous Weeds

Poison ivy and poison oak, which are harmful to personnel, require special control methods. The most effective method of controlling these weeds is by the use of spray chemicals that will kill the weeds on contact.

Replanting Trees and Shrubs

It is important to know when to plant the specific types of trees and shrubs in order that they are not damaged. Plant deciduous trees and shrubs while vegetative growth is inactive. This generally excludes April, May, June, and July for northern latitudes and March, April. May, June, July, and August for southern latitudes. Plant evergreens in northern zones in late summer after new growth has hardened and in spring prior to the time new growth begins. Plant evergreens in southern zones in early fall to permit root development before cold weather, and in the spring before new growth begins.

Planting Procedures

The following methods are suggested for storage, digging and planting trees and shrubs.

Temporary Storage: For temporarily storing plants follow the procedure suggested below.

1. Dig the tree pits before the tree arrives so trees are out of the ground no longer than necessary. Large balled and burlapped trees should be placed in the holes directly from the truck.
2. If the plants must remain out of the pound for a few days, cover the balls of earth with soil to prevent the roots from drying near the surface.
3. Cover the roots of bare root trees and shrubs with wet burlap or similar material as soon as they are unloaded, if planting is to take place within a few hours.
4. Heel in other bare root trees and shrubs at once in a trench deep enough and wide enough for the roots of the plant.

Digging Pits: It is suggested that the following procedure be observed when digging pits.

1. Separate the soil into three piles: soil, top soil, and subsoil. If the soil consists of sand or gravel or contains excessive building refuse, discard the material, remove from holes, and use good quality clay or silt loam.
2. Dig pits wide enough to accommodate all the roots without crowding or twisting. Prepare all pits with straight sides. Dig tree pits at least two feet wider than the spread of roots or ball of earth, with at least six inches of compacted top soil in the bottom of the pit.

Planting Balled and Burlapped Plants: Set the plants so that the top of the ball will not be more than one inch below the surface. Leave the burlap in place, but cut it loose at the top and bend it downward into the pit. Back fill in three or four layers, and tamp each layer firmly against the ball and sides of the pit. Water each layer as it is placed.

Plants and Containers: The planting methods for plants shipped in containers is similar to that for balled plants. Remove the plants from their containers, making sure that the balled soil enclosing the roots is not broken. Remove the plants from their pots by placing one hand against the soil, inverting the pot and tapping the ridge of the pot sharply on a fairly solid or rigid object, such as the handle of a spade inserted in the ground. Remove plants from one-gallon and five-gallon cans by cutting the sides of the can and forcing the halves apart. Never pull plants from the container or carry or move them by taking hold of the trunk or stem. Such handling will dislodge the ball of soil and harm or kill the plant.

Bare Root Plants: Bare root plants that are not to be replanted immediately should be cared for as indicated above. Before planting, remove all damaged or broken roots and cut back the branches to compensate for loss of roots. Plant them immediately after taking the plants out of the packing in which they are shipped or after removing them from the healing-in trench or from another hole. Spread the roots out to their fullest extent. Backfill and water simultaneously and, as the pit fills, grasp the plant close enough to the ground and move it up and down to encourage soil to flow under and between the roots. During the last few inches of filling, omit the watering and press the soil gently down to a depth of about two inches below the surrounding ground surface. This will leave a basin into which subsequent watering can be made.

Seedlings and Rooted Cuttings: These are used in landscaping, erosion control and where large areas are to be planted. Dig holes deep enough to accommodate the roots in a vertical position with the plant set slightly below the surface elevation. The vertical side of this hole should be on the downhill side of the slopes. Tamp the backfill in on the incline side of the hole. Where seedlings are in flats or plant bands, a small wooden tamper should be used to compact the fill after it is brought up to the proper level for setting the plant.

Maintenance of Trees and Shrubs

The normal care of established shrubs and vines includes fertilizing, watering and, in some cases, spraying. Valuable trees often require surgery, cabling, and bracing.

Protection From Equipment

Young trees, as well as mature ones, are easily damaged by maintenance equipment, especially power mowers. To minimize this problem, it is best to remove any lawn or soil away from the trees and install a division strip in order that the equipment is kept away from the tree.

Watering

Water trees, shrubs, and vines only as necessary to keep them healthy and to promote vigorous growth. Where rainfall is insufficient, the following procedures should be observed.

Trees: Trees should be encouraged to root deeply because the water supply will be more consistent and the roots will be less apt to feed on lawns and gardens or to spread surface roots which damage curbs and walks. Therefore, deep-water trees, especially trees located in areas vulnerable to damage by surface roots. Once young or newly planted trees have been established, or until after the first winter has passed, water should be withheld. This will encourage the roots to follow the receding water table. When necessary, water near the drip line. One way to deep-water a tree is with a pointed pipe about four feet long and perforated along the lower one foot. Attach the pipe to a hose and jet it into the soil for watering at the four-foot depth. Watering should be done at intervals of two feet along the drip line of the tree. Watering or deep-watering also may be done in a trench which roughly follows the drip line. The object is to assure penetration. When a trench is used, probe with the rod to ascertain the depth to which the water has penetrated. If, in the future, the grade of the land is raised, a dry well must be built around the tree. In the feeder root zone of a mature tree or in the case of a young tree, install three- to five-foot long vertical pipes in the ground. These sections of pipe, whether made of cast iron, transite, or soil tile, should be about four inches in diameter, with tops at ground level and filled with clean rock. Trees can then be watered by filling the pipes with water.

Shrubs: Because shrubs root at or near the surface, surface watering is needed. Here again, amounts and frequency depend upon the rainfall. When shrubs need water, the same procedure applies as in the case of trees. It is better to water less often and to insure penetration to three feet than to give a daily spraying that only wets the surface. Where summer watering of shrubs is a routine undertaking and shrubs are not immediately adjacent to lawns, a small soil basin should be constructed at the base of each shrub. To aerate a shrubbery planting, cultivate the entire area in spring and follow this by working the loose soil away from each shrub with a hoe. The diameter of the basin created is more important than the depth. Basins should be large enough to hold five to 20 gallons of water, depending upon a shrub's size. If properly constructed and filled with water at ten-day intervals, such basins will serve to keep shrubs in good condition. They will be able to take nourishment from the soil and resist pests and diseases which often attack a plant. It should be noted that basins should never be created in this manner around plants like azaleas, rhododendrons, and other similar acid-loving plants. These types of plants should be mulched with pine needles or leaf mold and never permitted to go dry for even short periods.

Vines: Vines of the climbing or clinging type are aggressive plants and generally require less water than shrubs. Their roots will range far, feeding under lawns and shrubs. Vines that do not climb, but which spread on the soil to make permanent ground covers, sometimes form roots at each joint, but these roots are almost invariably shallow. Their foliage shades the ground, keeps it cool and thus curtails evaporation. Nevertheless, they will require some watering in dry regions. When they show distress by wilting of the leaves and by discoloration, a good overhead sprinkling is in order, but not while bright sun is on the foliage, because this can cause burning. Because roots of these plants are shallow, depth of watering should be less than for shrubs, and watering is generally more frequent than for lawns.

Fertilizing

The amount of fertilizing necessary is governed by the nature and condition of the soil in which plants are grown. After planting, the use of commercial fertilizer should be avoided. Instead, old manure or a little bone meal should be mixed with the soil in the bottom of the plant pits or worked into the area intended for permanent ground covers. When the foliage of trees, shrubs, and ground covers takes on a yellow tinge and the leaves become sparse and small, a need for fertilizing is usually indicated.

When to Fertilize: Apply a fertilizer, if needed, in the spring when growth is just beginning. The plant then may make full use of the nutrients.

Applying Fertilizers: The following guidelines should be observed when fertilizing trees, shrubs and vines:

- **Trees:** Apply recommended fertilizer (depending upon the type of tree, condition of soil, etc.) per inch of trunk diameter at the drip line of the tree. Punch holes with a crowbar about 18 inches deep at a three foot spacing along the drip line, so the fertilizer can get to the roots. For large trees, punch holes every three feet along a circular path three feet inside the drip line, in addition to those along the drip line. Stagger holes in the two circles.
- **Shrubs:** Apply recommended amounts of fertilizer to the shrubs, the amount depending upon their size. Scatter the fertilizer at the base of the shrub, work it into the soil and then water to carry it into the soil (except in seasons of heavy rainfall).
- **Vines:** While the foliage is dry, scatter fertilizer by hand or with a broadcast seeder in quantities and types depending upon the types of vine.

Cultivation and Mulching

Cultivation at the base of trees and shrubs is required where there is little summer rainfall. This will control weed growth and conserve moisture. Mulching is superior to cultivation in humid sections, particularly for shrubs and shrub groups. Mulching is a practical means to reduce evaporation, prevent excessive soil temperatures and also provide an acid condition of surface soil for acid-loving plants. Peat, wood mold, decomposed sawdust, and similar materials are suitable for mulching. These are applied in surface layers to two to three inch depths around the base of the plant, extending as far from the central stems as the leaf canopy. The mulch must be replenished each spring since it decays and disappears.

Pruning

Existing trees (except conifers) should be pruned as needed to thin out an overly dense crown, to shorten long or weak limbs, and to remove dead, dying, diseased, or injured wood. Young trees are pruned at planting time to compensate for loss suffered in transplanting. Shrubs and vines need pruning to remove undesirable growth and dead branches and, in some cases, to develop a desirable shape.

Pruning Young Trees (except conifers): Prune trees at the time of planting to improve structure and to reduce top growth to compensate for roots lost in moving. Do not prune trees that have been pruned at the nursery. Start at the top of the tree and work down. Remove closely parallel branches, bruised and broken limbs, and superfluous growth at the base of the main branches.

1. When removing a branch, make the cut flush with the main branch. Do not leave a short stub, because the healing callus cannot close over the stub, which will decay and may prematurely injure the tree. When cutting back a branch, cut to a bud so as not to leave a stub. Paint all pruning wounds over one inch in diameter with a tree wound compound to retard checking and decay of exposed wood.
2. Do not cut back the central trunk.
3. Hold back the spread of the limbs by pruning as the young transplant grows until the trunk develops and becomes able to support a large crown.

Pruning Existing Trees (except conifers): Most shade and ornamental trees may be pruned at any time of the year except as indicated herein. When severe pruning is necessary to correct extensive damage or neglect, prune during dormant periods or in the early spring to permit recovery while growth is rapid. To avoid the need for frequent pruning, anticipate tree growth for two years and prune accordingly.

The following method is suggested for pruning trees.

1. Use pole shears and pole saws for cutting branches which cannot be readily reached with short hand tools.
2. Prune young shade trees to produce a sturdy framework.
3. Remove lower branches of shade and deciduous trees gradually, as the tree develops, to encourage a well-developed crown.
4. Remove branches broken by winter ice.
5. Do not remove the lower branches of broad-leaved evergreens.
6. Prune shade trees to resist damage by wind and ice.
7. Remove branches that extend over building and endanger roofs, eaves and windows or hang within eight feet of sidewalks and private drives. Prune trees along streets to provide clearance for buses, moving vans, and similar equipment.
8. Cut back branches that overhang or grow into power lines. Anticipate the effects of ice and wind on branches which might fall on power fines. Shape the entire tree, rather than notch the top.
9. Remove dead or broken branches and those that turn back toward the center of the tree. Thin out branches that interfere with each other.
10. Plan cuts to leave wide crotches rather than narrow ones. Wide crotches are more resistant to damage from wind and ice.
11. Paint wounds more than one inch in diameter. Use tree wound paint, if considerable repair of large trees is anticipated; otherwise, an outside lead paint is satisfactory.
12. Prevent multiple leaders from developing in a central stem tree, such as holly.

Pruning Conifers: Do not prune conifers except for the removal of broken, dead and diseased limbs. Conifers used as accent plants can be sheared to preserve compact, symmetrical form, but in all other locations they are at their best when left to be developed as nature intended them to be.

Pruning Shrubs and Vines: Prune shrubs and vines as follows:

• **Evergreen shrubs and vines:** Remove undesirable growth periodically. Cut dead branches back to live wood. Remove exceptionally long branches to well inside the main body of the shrub. This encourages the growth of new shoots appearing at the base of the shrub so they can renew the shrub as the weak branches lose vigor and are pruned out. Shrubs can be made to remain constantly fresh, neat, and limited to relatively new wood by selection and thinning, rather than by overall shortening. Shearing of shrubs is a common error which leads to excessive density, sparseness of flowers, and unnatural outline. Some shrubs should be naturally dense and of sharply defined contour. Most shrubs should be encouraged to grow in sprays, either somewhat stiffly, or arching outward (depending on the type of shrub), covered with flowers or berries and thus obtaining the greatest natural beauty. Plants of prostrate habit, planted as ground coverage, should be forced to remain prostrate by cultivating their lateral growth and by pruning away all erect branches.

- **Deciduous shrubs:** Prune these immediately after the flowering stage. Old wood, easily recognizable by rougher, darker colored bark, should be pruned back or taken out entirely. These shrubs bloom on the new growth and this growth should be encouraged to reach its greatest strength and fullness before a new spring arrives.
- **Vines:** Vines, in the sense of changing and climbing vines, require little pruning except the removal of loose ends and brandies that encroach upon windows and doors and threaten to work their way under roofing tile or into trees. Climbing roses are not vines. They must be lead and fastened on trellises or fences and require attention. Pruning consists chiefly of limiting the number of canes rising from the roots and the cutting of lateral branches to two or three buds.

Repair of Tree Injuries
To repair extensive damage caused by wind and ice storms:
1. Sever broken limbs smooth to provide satisfactory conditions for the wood to heal.
2. Remove loose or torn bark cleanly.
3. Shape the edges carefully to permit drainage of rain water from the wound. This helps to prevent rot and permits rapid healing of the injury. Paint the raw edges of the bark with shellac, immediately after making a cut, to prevent drying. Paint the wound with tree wound paint.

Grading
Most trees are sensitive to any important change in the depth of soil over their roots, particularly in high rainfall areas and in clay soils. Two or three inches or less of fill over the root areas may kill the tree. If fill is required under a tree, first spread a continuous layer of crushed rock or gravel about six inches deep from the trunk out to the drip line. To provide aeration, install vertical breathing tubes from the surface of the fill down into the layer of aggregate. Breathing tubes should be a four-inch clay tile. Provide a minimum of four under small and medium trees and six to eight for large trees. To provide breathing space for the tree trunk, construct a masonry well around it from the aggregate layers to the top of the finished grade. Allow ample room in the well for trunk growth.

Control of Plant Diseases, Insects, and Rodents

Plants are considered diseased when their structures or functions deviate from what is normal under local environmental conditions. Organisms or agents which are direct causes of plant diseases are fungi, bacteria, and viruses. Diseased plants exhibit one of several characteristic symptoms, depending on the species of plant, the causal agent, and the environment of the plant. Disease symptoms can be classified into three general groups: dying tissue, dwarfing or underdevelopment, and overgrowth.

The control of plant diseases is based primarily on protection and eradication. For expert guidance and proper cultural practices, and the diagnosis and treatment of the disease, consult with your local conservation department or local professional nurserymen.

The most satisfactory method of combating disease is the utilization of good cultural practices. Plants maintained in vigorous growing conditions are less susceptible to attack by disease. Proper fertilization, protective spray programs, and adequate water are basic requirements.

Eliminate disease agents primarily by:
- Removal and destruction of fungus or bacterial parasites, either at their sources or at the points where infections occur.
- Application of dormant or delayed dormant sprays.
- Removal of decayed wood in trunks and branches.

Grass Diseases

The best way to control grass diseases is to follow sound maintenance procedures. Most of the damaging grass diseases can be controlled by:
- Using types of grasses that are adapted to the climate.
- Adequate and systematic fertilization programs.
- Adjustment of soil reaction to pH 6.0 or higher.
- Provision of adequate surface and subsurface water drainage.
- Correction of surface compaction with suitable aerating equipment.
- Adjustment of watering practices to provide long intervals between application. When soil is watered to the proper depth, roots are strengthened and watering is required at less frequent intervals. Frequent light watering encourages shallow root growth.
- Mowing the grass to proper height.

The major troublesome grass diseases are listed below:
- Brown Patch
- Snow Mold
- Leaf Spot
- Copper Spot
- Mildew
- Fungi, such as mushrooms and toadstools

Specific guidance in the establishment of, or for the correction of, cultural procedures should be obtained from a soil expert, from a local soil conservation office, or a local nurseryman. The diagnosis of grass diseases, and the method of application, type, and dosage of chemical treatment in all instances should be determined or confirmed by these specialists.

Insect Damage: Insects, in addition to causing direct damage to plants, are also disease-carrying agents. Insects may be divided into two groups, according to their method of feeding:
- Sucking insects, which obtain their food by destroying the plant tissue and sucking the sap.
- Chewing insects, which bite off and eat portions of the plant.

The methods used to combat and control insects by artificial means are largely determined by their feeding habits and, to a lesser extent, on their life cycles. For details of control measures at any individual locality, consult with a specialist in the field.

Drainage and Erosion Control

The important factors involved in drainage are grades or slopes, soil permeability, and vegetative covering. Good drainage is essential for the proper distribution of moisture in the soil and for the control of excessive runoff water. Improper drainage will result in wet and dry spots and soil erosion, and possibly building damage. Good drainage results in a balanced growth of vegetation, operational safety and efficiency, and protection of capital investment in land and structures.

Types and Functions of Drainage

Drainage can be classified into two broad categories: surface or subsurface drainage. Classification depends on whether the water is on or below the surface of the ground when it is first collected.

Surface Drainage: Surface drainage provides for the collection, control, and disposition of water from the land surface as well as large quantities of runoffs from roads, runways, and other paved areas.

Subsurface Drainage: Subsurface drainage is the collection and removal of water from locations below the ground surface. Properly designed and maintained, surface drainage should reduce and, in many cases will eliminate, the need for special underground facilities.

Types of Erosion

Erosion can be caused by natural agents such as wind and water or can result from human activities. Chief sources of erosion from human activities are uncontrolled pedestrians and vehicular traffic.

Wind Erosion: Wind erosion is most often prevalent in hot, dry areas where the ground surface is usually dry in summer or winter and protective vegetation is limited, or totally lacking. The minimum wind that is required to start the movement of soil particles depends upon the size and the weight of the particles. Generally, a velocity of only eight or nine miles per hour is required.

Water Erosion: Water erosion may result from the impact of individual raindrops, or from the uncontrolled flow of water down a slope. Splash and scour erosion may result separately or simultaneously. At the upper end of a watershed, splash is the main cause of erosion, while in a valley channel, most erosion results from scour.

Erosion Control: Erosion can be decreased by reducing velocity of surface water runoff. Vegetative cover provides one of the best ways of doing this. This is particularly important in natural or artificial waterways. It is also important to reduce the volume of flowing water from higher ground. Use diversion ditches to intercept the flow of water from above and carry it into the drainage system where needed. Terraces or contour ridging are necessary in exceptional cases. In cases where serious erosion problems result from slopes or grades that are too steep, it is often more economical to reduce the slopes or grades than to maintain them as originally constructed. This is an engineering problem, which can be solved best by a civil engineer or landscape architect.

The immediate effect of erosion is the loss of weathered topsoil containing valuable organic matter and plant food elements, and subsequent channelization and gullying which damage surface structures. Consequently, erosion control practices should be designed to prevent damage, as well as correct it. In many areas, the source of dust is neighboring fields. In these cases, some relief can be obtained with wind breaks or snow fences, but a more effective, long-range solution will involve cooperation with all persons involved. The most effective way of checking dust erosion is to anchor the soil, usually with a vegetative cover. Other general methods are by surface treatments, such as calcium chloride, oil, mulches, asphalt, rip rap, or mortar.

Vegetative Treatments

Vegetative covers are the cheapest and most effective dust control method and also assist in the prevention of water erosion. They cushion the impact of the raindrops and prevent blasting of the soil particles into the air. The stems slow up the movement of water and the roots bind the soil particles together. This promotes the seepage of rainfall into soil. Local or native grasses can usually be found which will grow even in arid regions and sandy soil.

The vegetative covers available are as follows:

- **Perennial grasses:** The essential requirement of the vegetative treatment with grasses is the selection and use of grasses which are best adapted to the soil and climatic condition of the region in which the installation is located. Summer or winter annuals are used for temporary control of erosion on barren areas until the planting season for perennial vegetation.

- **Trees, shrubs, and vines:** These fit into the erosion control plan because they provide protection for steep, rough areas, embankments, and non-use areas. They also provide windbreaks and barriers for sound abatements. There are many types of trees, shrubs, and vines available for this use. Consult with a local professional nurseryman.

Drainage and Erosion Control Structures

Frequent inspection and maintenance of drainage systems are required to prevent costly breakdowns and repairs. The frequency of these items will, however, vary with the amount and tendency of rainfall and with the adequacy and condition of the drainage facilities. Inspection should take place at least once each year.

In general, drainage facilities should be kept free of silt bars, snags, vegetation, logs, tree branches and other debris that encourage silt disposition. Where silt must be removed, trace the source of the silt and take corrective measures to prevent reoccurrence. Control of erosion of the watershed will frequently prevent excessive siltation in the ditch. Logs, boards, or rocks haphazardly thrown into ditches for foot crossing will usually lead to trouble. Also, they cannot be used during periods of high water. Piers and bridge abutments should be maintained as needed to avoid harmful erosion and trash or silt accumulation. Maintenance and corrections normally required include:

- Cleaning inlets, catch basins, culverts, storm drains, ditches and under-drains.
- Restoring and re-stabilizing.
- Modification of existing waterways and structures, as may be needed.

Open Earth Channels
Open earth channels serve to intercept or divert and carry surplus water or concentrated surface runoff away from specific ground sites. When stabilized by vegetation, channels with side slopes of 3:1 or flatter can be maintained by mechanized equipment. Recurrent maintenance (seasonal and emergency) of open earth channels is required to remove plant growth, rock, soil deposits and to make other repairs to maintain flow. Maintenance projects involving open earth channels, lined channels, or other structures require design data and flow analysis for technical review.

Vegetated Channels
Vegetative waterways are required as a part of the water disposal system on practically all sloping areas. They are used to reduce gully erosion and to convey runoff. In virgin lands, there are natural swales and draws leading down hillsides. Many of these are plowed out and, soon after the vegetation is destroyed, they become large gullies. If damaged areas or vegetative waterways are to be re-seeded or re-sodded, insure that all fill materials are properly compacted to provide a good seedbed and to reduce erosion during the period of vegetation establishment. When these waterways are to be vegetated by seeding, securely anchored mulch or other protection is usually required until vegetation becomes established.

When establishing new turf by seeding, divert the water from the newly seeded area until it is well established. Where water diversion is not feasible, fit and tamp sod blocks into place. Anchor sod pieces with stakes or wire. Annual fertilization of the planted sections may be needed, at least until the planting has developed a mature root system. Wherever feasible, seeding is preferred to sodding.

Natural swales should be properly vegetated. Grasses are generally recognized as the most efficient type of plant for vegetative lining of channels. Each region has one or more sod-forming species that are satisfactory for use alone or in mixture according to local conditions. Growths that obstruct stream flow should be removed by treatments with selected chemicals. Care should be taken, however, to insure that the chemical treatment will not adversely effect the water or life in the water.

Lined Channels

The function of channel linings is to facilitate and control transportation of water to the point of disposal as rapidly as possible without damage to grounds by loss of soil or water.

Linings commonly employed are concrete, bituminous materials, brick, stone, and split pipes. The type will vary with local conditions, topography, and expected peak loads. Depending on local conditions, such as soil type and permeability, weep holes may be installed at the bottom of the lining to prevent uplift from hydrostatic pressure. They may be from one to five inches in diameter and placed at intervals of up to ten feet. All lined ditches must be kept clean and free from debris, silt, and weed growth. Lining is expensive and is usually not used where unlined channels perform satisfactorily.

When installing rip rap or pipe lining, be sure that it is high and wide enough to prevent washing and undermining of the edges.

Bank Protection

Special measures are frequently required for the protection of drainage channels against undercutting and caving. These types of failures result from scour erosion and removal of soil materials from the base of the slope. Vegetative cover should be established for the protection of bank slopes. Under some conditions, the banks can be stabilized by planting willows or other vegetation. Grass covers usually are most desirable, but vines or shrubs are more satisfactory for larger channels. Once established, the vegetation should be maintained as prescribed for semi-improved areas. Large ditches, which carry continuous flows, should be cleared of any vegetative growth in the bottom. However, do not remove vines, mats, or perennial grass covering the sides, which offer protection against erosion and bank cutting. Allow vegetation to grow in the back of revetments, jetties, and areas where silt deposition occurs, as well as on slopes protected by brush mats. Types of bank protection are divided into two categories: those which divert flow and those which act as a buffer or shield. The type of protection best adapted to a specific site is largely determined by the nature of the problem and the materials and equipment available.

Dikes and Levies

Dikes and levies are earthen embankments used principally to protect low-lying lands against flooding. Levy maintenance is the same as for most fill and/or slopes, but give special attention to inspection to forestall failures. Also important are rodent control and consideration of automatic plug gates or pumps to remove excess water from protected areas. During emergency periods, daily inspections may be needed, but in any case, inspection should be made at least twice a year.

Tile Drains

Tile drains remove excess free water from the soil, but do not ordinarily remove water needed by vegetation. Tile drains also establish a better moisture-soil-air relationship for growing vegetation. If used, maintain a chart showing locations of lines and outlets of installed drain tile. Use sod augers and probes to locate lines for which records are not available. Repair sections of these lines and outlets which have failed to prevent damage to other parts of the system. Indications of failure are poor drainage after rains in some sections of the tiled area and cave-ins over the lines. Install flap covers at outlets to keep rodents out. Pay special attention to all outlets to insure that they are open and that adequate head walls and catch basins are constructed. Open and clean plugged lines or, where necessary, replace with larger tile and correct the tile grade. Clean out tile lines filled with tree roots. If the trees and shrubs causing difficulty cannot be killed or removed, replace the section with sewer line installed with carefully mortared joints. Failures in tile lines may be caused by siltation in the tile, broken tiles, roots or other obstructions to flow, and improper grading. To correct these failures, uncover the tile line and remove the obstruction, and relay to proper grade. Savings can be effected by digging up and reusing old sections of tile that are still in good condition. This applies especially to the larger and more expensive sizes of tile having diameters of eight inches or more.

Check Dams

A check dam, as the term is used here, is a small, low dam constructed in a drainage channel or gully to decrease the rate of flow, check scouring erosion, and stabilize the drainageway. For repair of damage to major structures, determine the causes of failure and develop engineering plans for the repair work. Plant turf in channel bottoms after the grades have been corrected. Temporary check dams consisting of logs, branches, or loose rock can be used on gentle slopes while vegetation is being established. Steeper grading and large runoff flows usually require permanent masonry structures.

Causes of Failure: Most check dam failures are caused by heavy or prolonged rainstorms, which may result in the following types of damage:

- Overflow of entire dam.
- Flow around ends of dam.
- Excessive seepage under the dam.
- Erosion of the downstream side of the dam, thus underscoring of foundation.
- Floating logs and other debris lodge in the notch openings, thus overflow and failure.

Checklist: Inspect check dams after major storms which tax them to capacity. Some of the important items to check include:

- Upstream side of fill for evidence of wave action which can be corrected by rip rap or other methods.
- Entire fill for deep cracks or holes made by burrowing animals.
- Growths of trees, vines, or other vegetation which may penetrate the fill with large root systems.
- Fill for proper crest elevation and for evidence of sloughing or erosion.
- Emergency spillway for debris and silt deposits, for rank growing vegetation which may impair capacity and for erosion.
- Vegetative cover. The fill should be well vegetated for erosion control.

Maintenance: Following inspection, make these immediate repairs:

- Remove rank growing vegetation.
- Repair erosional damages.
- Place fill to raise low points in the crest of the dam.
- Fill holes and cracks in the dam.
- Relieve seepage pressures by laying a blanket of impervious material on the upside of the dam. Where seepage or recurring erosion problems are encountered, an engineering study is advisable to develop plans for the repair.

Manholes, Inlets, and Catch Basins

Manholes, inlets, and catch basins together are used for collection and disposal of runoff into subsurface pipes. These inlets consist of open top concrete or masonry box structures fitted with removable covers. This type of structure is commonly used in conjunction with approved ground, roadside, and similar storm drain systems. Their principal function is acceleration of water runoff in low or flat areas where pools of water will collect.

Maintenance: The following procedure is recommended for maintenance of manholes, inlets, and catch basins.

1. Check the inlet and barrel of the culvert. Keep them free of logs and debris.
2. Check the channel at the outlet end of the culvert for excessive erosion. This can cause a deep gully and weaken the foundation of the structure.
3. Check for erosion around edges of inlet. The grounds should slope gently downward so that water flows toward the basin. Fill the holes in the inlet and plant vegetation.
4. Check rank growing vegetation in the vicinity of the inlet for interference from large roots.

5. Carefully check for cracks, chips, and settlement. If small cracks are located, they can be repaired with a cement-paint mixture which can be made of waterproof cement and water. Larger cracks will require a more permanent repair and may at times require a rather extensive overhaul. Repair or replace deteriorated mortar and loose, broken, or displaced brick. Remove all sand, mud, grit, or other debris.
6. Metal ladders and steps should be kept free of corrosion and periodically painted with a rust-resistant paint. Loose, broken, or otherwise damaged ladders and steps should be repaired or replaced immediately.
7. Damaged, missing or corroded frames and covers must be repaired or replaced, manhole covers should be properly seated.
8. Where storm sewer piping joins the manhole wall, check for defective bonding and make necessary repairs.
9. If settlement is found, the pipes should be checked for misalignment and corrected, if defective.

Culverts and Outlet Structures

Culverts should be kept free of debris, and excess silt should be cleaned out as often as necessary, to maintain maximum flow during peak periods of storm water runoff. When metal culverts are cleaned, some silt may be left. A small amount of silt left in the bottom serves as a protective measure against excessive flow line wear. In some cases, the installation of head walls and cutoff walls will be required to prevent the undermining and clogging of openings. Usually, this can be handled effectively with the proper installation and upkeep of rip rap or sod. Culverts should be carefully examined for possible blockages, either during or immediately after heavy rains or excessive snow meltage, and cleaned as needed.

Although some expedient machine methods have been utilized, cleaning of culverts is usually accomplished by hand. In this way, better protection can be given to rip rap or sod. Dislodgement of rip rap or sod permits side scouring and undermining at the culvert entrance.

Proper maintenance of outlet structures is similar to that utilized for other drainage facilities. The type and degree of maintenance will vary with the type of outlet.

Where outlets enter a drainage channel at, or slightly above, the water level, there is normally little need for maintenance other than to prevent blockage at the outlet. An exception occurs when the outlet carries rapidly flowing water into a relatively narrow drainage channel. Under these circumstances, some provisions must be made to prevent scouring and erosion at the far bank of the channel. In most cases, any one of the various types of rip rap will suffice.

Where water is dropped from one level to another, control structures are needed when there is a pronounced difference between the levels. By use of drop inlet spillways and chutes or pipes, water is controlled by dropping it into spilling basins. If water is dropped over a retaining wall without means of controlling the force of the fall, a depression will form in the bottom of the drainage channel and undermining of the footing of the headwall or bank erosion will result. In either case, the condition should be corrected by lengthening the outlet pipe and projecting it out from the bank, by installing rip rap or by constructing an apron and wing wall outlet. If inspection of a concrete spillway with wing walls and an apron discloses cracks, chipping, or settling, immediate correction should be made. This can normally be accomplished by filling the cracks and chips and by reinforcing the footing or foundation areas. Failure to correct these faults, especially in northern areas, may lead to the accumulation of moisture and accompanying freezing, expansion, and breakage during cold spells. Steps should also be taken to correct any erosion that may be located behind or to the side of the structures.

Tide Gate

Although it is not a common occurrence, some storm drainage lines discharge directly into larger salt water bodies. When this occurs, considerable damage may result if proper maintenance precautions are not taken. In most cases, salt water can be prevented from leaking into natural water courses by the installation of a tide and spill gate. These gates may be installed at the outlet end of open ditches, or at the outlet end of culverts that empty into similar locations. Care should be taken to keep the channel clear of all debris so that the gate will not become jammed open and fail to operate properly with the incoming tide. The seaward side of the gate must also be kept clear so that the gate will open and release the retained water. When it is necessary to extend the outlet end of the ditch into the water, it should be provided with a V-type spearhead to prevent clogging of the outlet by wave action. Care must be taken to prevent accumulation of debris both in the channel and in the seaward side of the spearhead.

Drop Spillways

Properly constructed drop spillways are usually less susceptible to clogging and easier to maintain than most other types of erosion control structures. When either the standard drop spillway with the straight rectangular weir inlet or box inlet is used, little difficulty is encountered with sediment and debris, other than the removal of large obstacles that may become lodged and prevent proper flow. Regardless of the inlet type used, care should be taken to keep the apron and the spilling basin free of all foreign elements that may prove detrimental to maximum efficiency. In other than the special cases just mentioned, general maintenance for drop spillways will parallel that for inlet and outlet structures. Items to be considered include:

- Repair chips and cracks in the concrete (see Chapter 5).
- Settlement of part or all of the structure.
- Erosion of inlets and aprons or spilling basins.
- Erosion of the soil around inlets, aprons, head walls, and wing walls.

Storm Sewers

Storm sewers or drains are used to collect and dispose of surface water through the use of underground structures. A storm sewer usually consists of an underground layer of pipe, box, or arch-shaped structures supplemented by the necessary manholes, inlets, catch basins, and outlet structures. In general, the routine maintenance of storm sewers includes cleaning, flushing, and all the necessary repairs to defective lines and structures. Plans should be made for cleaning (that requires such equipment as steel buckets and cutters) at least one-third of the entire system on an annual basis.

Cleaning: Storm sewers may be partially or completely blocked by broken or collapsed pipe, infiltration of tree roots, accumulation of settlement, or other obstructions. If the stoppage results in either broken or collapsed pipes, the damaged areas must be excavated and the pipe replaced. Almost all other types of blockages can be removed with one of the more frequently utilized standard or expedient cleaning techniques. Among these methods are the following:

Storm sewers may be flushed by blocking all pipe opening in the manhole, filling it with water, and then quickly removing the block at one of the outlets by means of a rope attached to it. The sudden discharge of the released waters will normally cleanse the pipes of all the sediment contained therein. If this does not fully accomplish its intended purpose, the process may be repeated with a rubber ball, slightly smaller than the interior of the pipe, which is wrapped in burlap and inserted so that it will be flushed through the pipe. This increases the scouring action of the water working around the ball as it passes through the pipe. Another method, although not as effective as commonly believed, is to use a fire hose at each manhole to wash silt and loose debris down to the next manhole where it can be cleaned out. When using this method, always work from the upper end of the storm sewer to the lower or outlet end.

Mechanical devices for cleaning storm sewer lines of silt, sand, or tree roots include various cutters, brushes, scoops, scrapers, and screws that are drawn through by hand or power-operated devices. Sectional sewer rods with various types of cutting devices and scraping and flushing heads may be used alone or in combination. These items can be used to remove blockages, or to make holes in obstructions for the utilization of other types of equipment. Flexible steel rods or tubes, which may be coupled with a continuous line, are also effective in removing much of the debris normally found in storm sewers. Screws, augers, and sand cups are utilized as end tools. The tool and rod are fed into the drain line until an obstruction is reached, at which time they are rotated to remove or break up the blockage. Rotation may be accomplished by means of hand ratchet, portable powered unit or other mechanical means. In some extreme cases, it may become necessary to employ steel cutters or buckets attached to a steel cable or wire rope. Before doing this, either a sectional rod or rotating shaft must be used to break an opening through the obstruction. The cable is then threaded through the opening, buckets or cutters are attached, and the cable is pulled through the pipe with a winch. Care must be taken not to damage the pipe or loosen the joints when using this method. To remove difficult obstructions and grease and to clean pipes larger than 15 inches in diameter, a turbine-type rotating cleaner, equipped with cutting knives and wire brush attachments, is recommended. A final problem confronting maintenance personnel is drain pipe blockage that results from misaligned or collapsed pipes. Minor variances in alignment are seldom serious, but where the displacement has been great enough to cause significant or complete blockage, the pipes must be removed and relayed. If the obstruction results from collapsed pipes, the damaged pipe must be removed and new pipes installed.

Hillside Diversions

These diversions include terraces, furrows, and diversion ditches. They provide an outlet into natural hillside draws that are well vegetated, or into special erosion-free outlet channels that are designed and constructed to receive the runoff and convey it down steep slopes. The outlet channels may be either vegetated waterways or flumes lined with an erosion-resistant material, such as concrete. They should be checked after major storms that tax their capacity. Where vegetated waterways are used, the velocities down steep grades are retarded by controlling the conditions of vegetative cover. To provide uniform resistance, it is important that the vegetation be maintained so that the erosion and silt deposits are controlled. Special erosion control grasses are used in these waterways. They require heavy fertilization and frequent mowing if a uniform erosion resistant cover is to be maintained throughout the year. Where concrete or other erosion-resistant linings are employed in these channels, it is important to prevent erosion along the sides and underneath the flume lining. By maintaining small berms which divert surface water away from each side of the flumes, side erosion is reduced. When checking hillside diversion, important items to watch are:

- Low points in the berm.
- Holes burrowed by animals or temperature cracks in the sod which can release water through the berm and cause it to fail.
- Rank growth of weeds, seedling trees or vines which can impair the capacity of diversion.
- Erosion in the waterway.
- Deposits which curtail channel capacity. Check at least twice each year, and always after a major storm that produces runoff approximately equal to the design capacity.

Repairs: In making repairs, some of the important considerations include the following:

1. When filling in low points on the berm, cover fill materials with topsoil and either seed or sprig, and cover seed and springs with mulch or cover the new fill with sod. Maintain the berm about eight inches higher than the flow crest at maximum runoff.

2. Fill holes and temperature cracks. When filling these underground holes, do not fill the outlet ends. At the inlet end, fill to approximately one foot of the surface with small stones and sand, and fill above the point with tamped soil.
3. Destroy rank growth and woody plants.
4. Do not dump silt deposits on the berm.

Diversion Ditches

Vegetation should be maintained in areas above the diversion ditches. Thus, runoff will not carry large amounts of silt into the channel. Regular maintenance will prevent bushy or woody growth from obstructing flow and causing silt to accumulate.

Construction Slopes

Construction slopes are maintained in the same manner as natural ground slopes. However, because they normally result from cut and fill operations in construction grading, they usually are not as stable as natural slopes. Typical construction slopes are the cut and fill sections in building areas, highways, railways, and airfield runways. In most of these construction slope cases, many of the maintenance techniques employed in normal drainage maintenance and erosion control will be incorporated. Dikes and flumes to protect cut and fill slopes from washing are typical techniques. Slopes can be damaged by erosion, slipping or sliding, or by a combination of these. The problem of slipping and sliding differs widely on various slopes. Some of the variables to consider are the soil characteristics, the geological structures, and the nature and extent to which underground waters may be concentrated, diverted, or dispersed. Before attempting to establish a maintenance program for slopes subjected to sloughing and sliding, consult with a soil engineer.

In many cases, maintenance can be decreased by modifying the slope. However, before modifying the slope, compare the cost of this work with estimated savings in periodic maintenance. Precise recommendations cannot be made as methods vary with the type of soil, condition and size of the slope, and prevailing climatic conditions. One of the methods most frequently used is the growing of vegetation as a protective cover. In establishing such new covers, the seedbed should be protected with some type of organic mulch until the vegetation is firmly rooted. Methods consist of changes in the slope ratio, construction of terraces, or installation of concrete chutes or flumes with interceptor ditches.

Maintenance: Erosion damage to construction slopes can result from either surface or subsurface water. Therefore, it is important that provisions be made for proper drainage facilities so that maintenance costs can be kept to a minimum..

Surface Water: Since surface water erodes the soil by scouring action, it should be diverted from the tops of construction slopes. The use of diversion ditches, dikes, gutters and flumes will, in most cases, control surface runoff.

Subsurface Water: Subsurface water can cause slippage and sliding on construction slopes. A slip is caused most commonly by movement of an earth mass down an inclined plane which has been lubricated by seepage of storm runoff or by ground water. The plane is usually the surface of an impervious soil or rock layer where ground water has corrected and is trapped. A slide is a sloughing or collapse of a slope which is steeper than the natural angle of repose (the natural slope on which the soil sits). Slides result in construction of steep cut or fill slopes that have not been properly stepped. Slides also are caused by washing or saturation of the slope by surface or ground water, frost action, weathering, vibration from blasting, excavation at the toe of the slope, overloading at the top of the slope or other mechanical disturbances. With proper maintenance, soils can be stabilized by removing the water before it causes damage to the slope. Some removal may be accomplished with vegetation but, in many cases, underground drains must be employed. No matter what method is used, it is important that the vegetation and/or drains be properly maintained in order that they do not lose their functionality.

Miscellaneous Exterior Structures

Antennas and Utility Towers

Antenna towers are used to support one or more antennas, while utility towers are normally used as bases for fuel and water tanks, large banks or floodlines, and similar structures. Since most personnel dislike working on such structures, inspection and maintenance are difficult. Therefore, except where confident and experienced personnel are available, it is advisable to have the inspection and maintenance work done by contractors experienced in this area. Inspection of antenna towers, unlike that of most utilities, is required only once every two years, unless damage results from particularly heavy storms.

Foundations

Due to the heavy weights involved, some minor settlement may be expected immediately after the erection of an antenna or utility tower. This settlement is rarely serious and seldom requires any action. However, irregular settlements tending to tilt the structure, or erosion at footings that partially undermines them call for immediate corrective action. Settlements that continue over a long period of time should be investigated to determine the cause. The remedies depend upon the cause of the settlement; no fixed rules can be given. Erosion should be controlled, and eroded areas at footings should be backfilled with suitable material that has been well tamped into place. Tighten anchor bolts and see that all other connections of tower to foundation are firmly secure.

Guidelines, Anchors, and Bases

Anchors must provide adequate and rigid support to the ends of the guidelines. Insure that firm, well-tamped material surrounds the anchor and that it is adequately protected from erosion or other conditions that would tend to loosen it. If anchors show evidence of being too small or too light for the tension of the attached guidelines, an immediate structural analysis needs to be made. In such cases, it may be necessary to use a deadman or other support to provide temporary reinforcement until the anchor can be repaired or rebuilt. Anchor rods must be protected by coatings or wrappings to prevent the encroachment of rust and corrosion. For temporary structures, or for light work, other devices, such as a deadman or commercial metal anchors, may also be needed.

Guys must be kept tight and in good condition. Ends are secured with clamps, wedge sockets, or eye splices. Use thimbles where necessary to prevent shaping or sharp ends in the cable. Check nuts on cable clamps periodically and tighten any that might have worked loose. If the cables have become slack, they must be tightened up by means of block and tackle or some other tension device. If the guys are located where they are subjected to traffic damage or where pedestrians might be injured by running into them, provide wood guards marked with black or yellow stripes.

Where either free-standing or guide towers are used as antennas, they are installed on insulated bases. In order to be effective, such bases must be clean and free from any cracks or other defects that would impair proper functioning. Control the vegetation around the bases, and do not permit any accumulation of trash or debris.

Towers normally will be painted only to prevent rust and corrosion. Painting for the sake of appearance will not be accomplished unless conditions are such as to reflect unfavorably on the installation or activity. Because of the special hazards involved in painting towers, and the special experience required, painting should be done by experienced contractors. However, where qualified personnel and suitable equipment are available, the work may be accomplished by in-house maintenance personnel.

Winches and hoisting devices must be maintained in good condition at all times. Ladders must be checked at regular intervals and any loose parts and fastenings must be securely tightened. All ladders must be provided with safety cages. Rest platforms should be installed on ladders for high towers. Check ground wires for continuity and space gaps on lightening protection equipment for proper width of gap.

Vegetation in the area under, and immediately around, towers must be controlled. Where mowers can be used, mow the vegetation when it gets to be about eight inches high, and cut it to a height of two to three inches. In places where mowers cannot be used, control vegetation with herbicides and soil sterilants.

PART IV

ELECTRICAL AND MECHANICAL MAINTENANCE AND REPAIR

The subject of maintaining electrical and mechanical equipment is unique in that the equipment is comprised of many different materials. In addition, each company which manufactures the equipment has its own standards. This necessitates a separate set of operational, maintenance, and repair procedures. An example is the maintenance and repair of a 15-ton air conditioning unit. Even though the rated output of the unit is essentially the same no matter which manufacturer produces it, and even though the basic operation of the units is the same, the physical components of the different units may be such that the same maintenance and repair procedures cannot be used.

The following chapters in this part of the book focus on those components and systems common to the various mechanical and electrical apparatus on the market today. Since there is such a wide variety of equipment available, specific, step-by-step repair and maintenance procedures are not included in this book. The reader is referred to the manufacturer for this kind of information, along with information concerning any special types of equipment.

Chapter 13 provides answers in managing the maintenance of electrical systems and components. Chapter 14 presents maintenance and repair procedures for plumbing systems. Finally, Chapter 15 focuses on heating and air conditioning and other mechanically related systems.

Many components of both electrical and mechanical systems are common. For example, bearings can be found on electrical motors and turbines. So as not to duplicate information, typical components are presented only once.

Eliminating Unnecessary Maintenance Time

In order to repair any electrical or mechanical apparatus, the maintenance personnel must be able to read and understand various types of drawings, specifications, manufacturer's information, and, of course, have experience in the operation, control, and maintenance of the equipment. Effective training programs and published literature, such as those discussed in Chapter 16, are available to insure the development of an efficient maintenance staff.

Once technically trained personnel are available, the next step is to compile the documents necessary for each piece of equipment and system in the plant. These include, but are not necessarily limited to, the following:

- Drawings depicting the locations of all the equipment.
- Schematic drawings of the equipment.
- Specifications of the individual pieces of equipment.
- Manufacturer's manuals for each piece of equipment.

Attacking the Problem

No matter how technically well-trained a person is, and how much written material is available to maintain or repair the equipment, much time may be wasted if a systematic problem-solving technique is not applied. Of the many procedures available, most effective in repairing and maintaining mechanical and electrical equipment is the *process of elimination.*

Prior to initiating the actual process of elimination, the maintenance worker must know the following:

- What the function of the system is and how it operates under normal conditions.
- What each component of the system contributes and how it functions.
- Where each part is located and how it receives or exerts power or pressure onto an adjacent component.
- Where the most effective places for access to test a system are.
- A list of probable reasons why the system or its components may fail to function.

The process of elimination begins by considering or listing all the possible reasons why the equipment is not functioning as designed. Once this has been done the technician should begin to eliminate those items, beginning with the most obvious, one at a time until the cause is found.

CHAPTER
13

ELECTRICAL SYSTEMS

The focus of this chapter is the maintenance and repair of the electrical apparatus listed below.
- Distribution systems, including lighting
- Switch gear, including circuit breakers, fuses, control switches, and disconnecting switches
- AC and DC motors
- Transformers
- Batteries

Preliminaries to Maintenance and Repair

The electrical maintenance group needs to know the location of substations, main feeders, subfeeders, panel boards, and branch circuits, and all electrical equipment, such as motors. This information should be developed in the form of specifications and drawings at the time of the construction of the facility. If the whereabouts are not presently documented, time and resources should be allocated for obtaining such documentation; or much time will be wasted locating items when maintenance is required.

Any information to be compiled should be done in the form of a drawing, such as a line diagram. The diagram should include the following information.
- Location of all panel boards, feeders, and circuits
- Sizes, types, and numbers of wires for each feeder and circuit
- Location and rating of all transformers
- Location, type, and rating of all motors
- Location, size, type, and setting of all protective devices, such as relays

To prevent electrical breakdowns and thus minimize the amount of repair necessary, follow the general maintenance guidelines below.
- Perform routine inspections based on formal procedures at regular intervals.
- Keep the environment adjacent to electrical equipment dry and at controlled humidities according to the manufacturer's instructions.
- Regularly clean all electrical apparatus.
- Maintain connecting electrical parts in tight condition.
- Adhere to all safety precautions relating to the maintenance and repair of the equipment.

Work Safety

It is extremely important when working with electrical equipment to adhere to the safety rules listed below.

- Do not touch bare wires until a check is made to be certain that all lines have been de-energized.
- Shut off all power before working on any piece of equipment.
- Only work on energized equipment when making a test that requires power to be delivered to the equipment.

When working on equipment follow the National Electrical Safety Code.

Tools

Figure 13.1 contains a list of tools and equipment for electrical maintenance and troubleshooting. Maintain tools and equipment in proper condition, and repair when necessary.

If established maintenance schedules are not followed, under-maintenance results in one or more of the following defects.

- Deterioration of cable insolvation from moisture, oil, heat, or other causes.
- Excessive or low voltages, caused by deficiencies in the system.
- Loss of illumination, inadequate power and voltage.
- Corrosion of metals.
- Hot cables, caused by unbalanced loading.
- Power failure.

Distribution System

The distribution system consists of the substations, main feeders, subfeeders, panel boards, and branch circuits, along with any needed switch gear, and fusing devices.

Wiring

The main feeders, subfeeders, and branch circuits are nothing more than two or more wires enclosed in some sort of protective covering such as plastic, metal, or both in the care of a conduit. Particular attention must be given to electric wiring in the building. Loose wires, poor connections, bare conductors, defective convenience outlets and switches, defective attachment cords and other unsafe conditions should be corrected. Any type of wiring, provided it is in accord with the National Electric Code and satisfies the functional requirement of the building, will be satisfactory for repair and replacement. Make all repairs to wiring in a permanent manner using approved materials. Ascertain that wiring has been done in a workmanlike manner, circuit arrangement is correct, circuits are not overloaded, and work conforms with the National Electric Code. Wiring systems should be checked frequently to reduce fire hazards, correct defective wiring, and eliminate oversized lamps and overloaded circuits.

Check the following items when inspecting wires and cables.

- Dirty, poor ventilation, detrimental ambient conditions, presence of moisture, grease, oil, chemical fumes.
- Improper or unauthorized connections and dangerous temporary connections.
- Damaged wiring devices, defective insulators, cleats, and cable supports, broken or missing parts, or exposed live parts.
- Excessive cable sag and vibration, crowded cable spacing, excessive number of conductors in conduit and raceways.
- Evidence of overheating, grounds, and short circuits; overheated splices, damaged or defective insulation.
- Need for painting of noncurrent-carrying parts subject to corrosion.
- Unsafe, unreliable cable and wire to fighting and power panels.
- Remove fuses, switches, and other sources of discontinuity in the neutral wire of ground AC systems.

Equipment and Tools for Electrical Maintenance

Equipment	Use
Air Gap Feeler Gage	To check motor or generator air gap between stator and rotor
Cable Fault Locator	For location of faults in underground cable
Circuit Breaker, Over Load Relay Test Equipment	Calibrates and tests operation circuit breakers and relays
Electrician's Tools (wire cutters, screw driver, fuse pullers, wrenches, bushings and lock nuts, etc.)	For assembly, disassembly or repair of electrical equipment
Hand Stones (all grades); Grinding Rig; Canvas Strip	Grind, smooth and finish commutators or slip rings
Infrared Temperature Detector Gun	Detects hot and over-heated areas in conductive equipment
Insulation Resistance	Test and monitor insulation resistance;
Tester, Thermometer, Psychrometer	Use thermometer and psychrometer for temperature-humidity correction
Light Meter	Measure light intensity
Low Resistance Tester	Checks low resistance paths of electrical contacts, switches, etc.
Meters, such as multi-meters, volt-meters, ohmmeters, clamp-on ammeters, watt meters, clamp-on PF meters	Measure circuit voltage, resistance, current and power
Motor Rotation Tester	Checks direction of motor rotation before connection
Oscilloscope, Vacuum Tube Volt Meter, High Impedance Meters	Test electronic circuits
Portable Capacitance and Resistance Bridge	Provides accurate readings of capacity and resistance
Portable Transistor Radio	Indicates areas of excessive arcing or RF interference
Portable Oil Dialetric Tester; Portable Oil Filter	Test OCB, transformer oil or other insulating oils. Recondition used oil.
Potential and Current Transformers, Meter Shunts	Increase range of test measurements or meters to permit reading of high voltage and high current circuits
Recording Meters, Instruments	Provide permanent record of voltage, current, power, temperature, etc., on charts for study
Safety Equipment (rubber gloves, boots, mats, etc.)	Provides safe and efficient electrical maintenance
Spring Tension Scale	Checks brush pressure on DC motor commutators or on AC motor slip rings; tests electrical contact pressure and relays, starters or contactors.
Tachometers	Checks rotating machinery speeds
Variable Auto Transformer	Allows testing of circuits at reduced voltage
Vibration Analyzers	Detect and help sources of rotating machine vibration

Figure 13.1

Ducts

Ductwork that is exposed should be inspected on a frequent basis. Any ductwork that is found to be damaged due to such things as corrosion or operational abuse should be replaced.

All systems of the underfloor duct type require checks for evidence of oil and water. Entrances and fittings should be checked and corrected as necessary to prevent entrance of liquids, insects, and rodents. External heat and heat caused by overloaded circuits will cause cracking of insulation and drying of tape splices which, in turn, allow moisture to enter with resultant grounding. Under-floor conduits and duct systems should be kept sufficiently clear of electrical and hot water floor-heating systems to prevent undue heating of the enclosures. This is also true for exposed ductwork.

Bus ducts, trolley ducts and open busways require annual cleaning and removal of oil substances and dirt. Examination should reveal whether adequate areas for ventilation are provided, clearances are sufficient to prevent contamination from oil and water, and equipment is placed so as to prevent damage from moving equipment. Ventilated type busways should have the buses blown off annually with clean, dry, compressed air.

Fuses, Circuit Breakers, and Switch Gear

Fuses protect the wiring system and electrical equipment from short circuits or overloads. It is important that the fuse be of an amp rating that is consistent with the rating of the circuit which it is protecting. The evidence of a fuse that has failed is that the strip of metal in the fuse has been dissected. When a fuse is no longer effective, it cannot be repaired, but it must be replaced.

Fuse Inspection and Maintenance: A fuse consists of many parts, some current-carrying and some noncurrent-carrying, all subject to atmospheric conditions. The frequency of inspection will necessarily be a function of the conditions at a given fuse location. The following are things that an operator can watch for when he is replacing a fuse or while inspecting an installation:

- Inspect insulators for breaks, cracks, or burns.
- Inspect contact surfaces for pitting, burning, alignment, and pressure.
- Examine fuse holder before installing a renewable element for excessive erosion of the inside of the fuse tube, tracking, and excessive dirt on the outside of the fuse tube.
- See that bolts, nuts, washers, pins, and terminal connectors are in place and in good condition.

Circuit Breakers: A circuit breaker is like a fuse, but it can be repaired. In most instances if a circuit breaker has been opened due to an overload, all that is necessary is to reset it. However, at times the overloaded circuit breaker has to be replaced, which is a more expensive proposition than for a comparable fuse.

If a breaker fails to trip, serious damage may result. Circuit-breaker failures may be minimized by frequent visual inspections, periodic external inspections, and regularly scheduled complete maintenance inspections. During regular patrol trips through a switchyard or substation, the operator should observe any unusual condition that may be an indication that maintenance is needed. When inspecting circuit breakers in service, any of the following unusual conditions would serve as a warning that maintenance is needed.

- Low bushing oil gage reading
- Low breaker tank oil level
- Oil or air leaks
- Broken porcelain
- Loose hardware
- Broken or missing cotter pins
- Air supply compressor running excessively
- De-energized cabinet heaters

- Excessive corrosion or rust
- Loose gasketed covers
- Unusual noise, smoke, or temperature

When planning a maintenance schedule for circuit breakers, the following factors must be taken into account.

- Accumulated increments of kva interrupted
- Time since last inspection
- Number of operations switching and testing
- Number, location, and magnitude of faults
- Condition of oil
- Cleanliness of local atmosphere
- Vintage, age, and type of breaker

Circuit breakers of earlier designs are less efficient than modern breakers. The interrupters used in older breakers require longer interrupting time, resulting in more mechanical wear, and insulation deterioration. Inspect breakers in this class after each fault interruption, taking into account the following additional factors.

- Experience with specific breakers and duty imposed.
- Preliminary inspection records.
- Any unusual condition that may demand immediate attention, such as noise, temperature, smoke, oil or air leakage, or friction.

Refer to Figure 13.2 for circuit breaker maintenance and repair.

Circuit Breaker Maintenance and/or Repair Procedures

Problem	Cause	Maintenance/Repair Procedure
Failure to trip	Blown fuse in control circuit.	Replace blown fuse.
	Loose or broken wire in trip circuit.	Repair faulty wiring. See that all binding screws are tight.
	Dirty contacts on tripping device. (Control switch, protective relays or auxiliary switch.)	Clean dirty contacts.
	Failure to control power.	Investigate.
Failure to close	Blown fuse in control circuit.	Replace blown fuse.
	Loose or broken wire in trip circuit.	Repair faulty wiring. See that all connections are tight.
	Dirty contacts on tripping device. (Control switch or auxiliary switch)	Clean dirty contacts.
	Failure to control power. Insufficient air pressure.	Investigate.
Unnecessary tripping (that is tripping should not occur)	Setting of relays or calibration setting of direct tripping device too low.	Set relay or device for proper value according to ampere load of circuit.

Figure 13.2

Switches: Check switches for heating, oxidation, and adequate capacity. Knife switches may require realignment of blades and jaws. If the jaws of a switch become annealed as a result of overheating, poor contact will result; they should be replaced at once. Annealed jaws may be protected by tapping with an insulated object when the blade is open. If the jaw rings on being tapped, it is not annealed; if a dull, dead sound results, it is annealed. The frequent operation of the switch will prevent oxidation of the blades.

Inspection Checkpoints for Disconnecting Switches

OPERATING GEAR
_____ Group-Operated Switches: rust, corrosion, loose brackets and holding bolts, nonrigid bearings and supports.
_____ Grounding Cables, Clamps, and Straps: weak supports, broken or frayed portions of conductors, loose connections.
_____ Insulating Section of Operating Rod: indications of cracks or signs of flashovers.
_____ Movable Connections: inadequate lubrication, rust, corrosion, other conditions resulting in malfunctioning.
_____ Switch: gears stiff or adjustment needed.
_____ Locking and Interlocking Devices and Mechanisms: functional inadequacy to prevent unauthorized operation.

MOUNTINGS AND BASES
_____ Rust, corrosion; twisted, bent, or warped; loose or missing ground wire.

INSULATORS
_____ Cracks, breaks, chips, or checking of porcelain glaze: more than thin or transparent film of dirt, dust, grease, or other deposits on porcelain.
_____ Damage indicated by streaks of carbon deposits from flashovers.
_____ Loose, broken, or deteriorated cement holding insulator to other parts.

BLADES AND CONTACTS
_____ Excessive discoloration from overheating; roughness and pitting from arcing.
_____ Misalignment of blades with contacts.
_____ Arcing Horn Contacts: burns, pits, failure to contact each other throughout their length when switch is opened and closed.
_____ Inadequate tension of bolts and springs.
_____ Inadequate blade stop.
_____ Lack of hinge lubrication; insufficient nonoxide grease for blades and contacts.

CONNECTIONS
_____ Cable or Other Electrical Connections: loose bolts, discolorations indicating excessive heating at connection points.
_____ Corrosion, particularly that resulting from atmospheric conditions.
_____ Electrical Clearances of Cable or Other Conductor: inadequate to other phases or to ground for applicable circuit voltage.
_____ Flexible Connections: frayed, broken, or brittle.
_____ Cable from Grounding Switch to Grounding System: frayed, broken strands, loose connections.

Figure 13.3

Heating of switches can be attributed to oxidation, overload, unbalanced loading, and exterior forces such as steam lines, furnaces and poor ventilation. Loose or oxidized screws, nuts, and fittings in poor contact pressure should generally be investigated immediately at any indication of undue heating. Excessive deposits of dust and dirt in the operating parts of the switch invariably cause binding of shafts, triggers, rollers and pins as well as the operating levers. All insulating surfaces must be kept clean in order to prevent electrical breakdown.

Refer to Figure 13.3 for a list of items to check for disconnecting switches.

Cabinets: Check distribution panels or cabinets for rust, dirt, oil, debris, moisture, loose connections, and unauthorized or nonstandard attachments. All disconnecting devices should be labeled to indicate circuit or feeder designation and the unit should be labeled to show all current and voltage characteristics. Cabinets should be kept locked to vent tampering by unauthorized personnel. In large cabinets, the presence of moisture or humid conditions may require installation of strip heaters to maintain a temperature that will inhibit condensation when the equipment is not operating. When panel enclosures are exposed to corrosive fumes, they should be painted with an acid-resisting paint. The areas around all enclosures should be kept free for emergency servicing. Circuits should be numbered in a suitable directory provided on the inside of the lower door of the cabinet.

Lighting Systems

Each lighting system is designed to produce a specific level of illumination adequate for those working in the area. The amount of illumination provided starts to decrease almost as soon as it is put into operation. The rate of decrease is infinitesimal at first, but increases as time passes. This increase is due to dirt on the lamps, with lamp output decreasing along with dirt on the room walls or ceiling which decreases the reflectability. Illumination should be maintained at recommended levels (as prescribed by the Illuminating Engineering Society) to conserve eyesight, improve morale, increase safety, improve housekeeping, decrease fatigue, reduce headaches, and increase production, all of which are directly reflected in lower operating costs.

Incandescent Lamps

Fixtures are designed for particular lamp sizes and finishes. The current trend to reduce lamp size has made possible higher-wattage lamps. Excessive heat generated by the new high-wattage lamps can cause damage, and care should be exercised not to exceed the wattage of the particular fixture. Incandescent lamps will yield maximum lighting economy when line voltage is at or near the designated voltage printed on the bulb.

Fluorescent Lamps

There are three types of 40-watt fluorescent lamps: instant-start; preheat, rapid-start; and preheat lamp. They all have particularly the same physical dimensions but different internal construction. The preheat, rapid-start lamp will operate satisfactorily in either the preheat or rapid-start circuits. It has a very short lamp life in an instant-start circuit. The instant-start will operate satisfactorily with an instant-start ballast, will burn off the ballast in a rapid-start circuit, and the lamp will not light in a preheat circuit. A preheat lamp has a very short lamp life in an instant-start circuit and has poor starting in a rapid-start circuit.

Mercury Lamps

Mercury lamps have the best-maintained light output. Long average life is a primary characteristic of most mercury lamps. Mercury lamps are also extremely economical to use where high footcandles of illumination are desired.

Lamp Size

As light sources are designed to operate most efficiently and economically at the rated voltages, special emphasis should be given to use the lamps to suit the voltage of the circuit. The effect on lamps operated over or under the rated voltage range is as follows:

- **Fluorescent lamps:** line voltage higher than the maximum range will reduce illumination and may cause uncertain starting of fluorescent types.
- **Incandescent lamps:** Line voltage higher than the maximum lamp range will increase the light output, but will shorten lamp life. Line voltage below the minimum range will extend lamp life, but reduce light output approximately 3% percent for each 1% percent in voltage drop.
- **Mercury lamps:** Line voltage higher than the maximum lamp range will shorten lamp life. Line voltage below the minimum range will reduce illumination and may cause uncertain starting.

Lamp Maintenance and Repair

Careful observation by all maintenance personnel is needed to discover and report any evidence of defects in lighting systems. Repair any deficiencies promptly to prevent progressive deterioration of the system. Maintain the required illumination intensity by keeping lamps, fixtures and reflective areas clean and in good repair; replacing defected lamps; and keeping the voltage steady.

Refer to Figure 13.4 for maintenance and repair procedures for lamps.

The progressive decrease of light caused by accumulation of dirt necessitates periodic cleaning of lighting equipment. The frequency of cleaning depends on local conditions.

The cleaning schedule for a particular installation should be determined by a light meter reading after the initial cleaning. When subsequent footcandle readings have dropped 20 to 25 percent, the fixture should be cleaned again.

Lighting equipment should be washed, not just wiped off with a dry cloth. Washing reclaims 5 to 10 percent more light than dry wiping and reduces the possibility of marring or scratching the reflecting surfaces of the fixture.

Removable glassware, counter reflectors and diffusing louvers should be cleaned by immersing them in a solution of synthetic detergent cleaner and scrubbing with a soft brush or sponge. If steel wool needs to be used, it should be of a number 0 or finer grit. The fixture should then be rinsed in warm clear water and dried with a clean cloth. If the equipment is fixed it should be wiped with a moist cloth or sponge, using a solution of synthetic detergent cleaner. If required, use number 0 steel wool to remove dirt film. Wipe off the excess moisture with a clean cloth. Enamel, chrome, aluminum or silver-plated reflecting surfaces that cannot be adequately cleaned and polished should be replaced.

Lamp Replacement

Neglected lamp outages reduce illumination. If burned-out lamps are not promptly replaced, illumination may drop to unsafe footcandle levels in a short time. In some cases it may be satisfactory and more economical to clean lamp surfaces and fixture interiors only at the time of lamp replacement. Lamp replacement is done either by an individual or group method.

Maintenance and/or Repair of Lamps		
Problem	Cause	Maintenance and/or Repair Procedure
Fluorescent Lamp Equipment		
Lamp fails to start or flashes on and off	Lamp pins not contacting. Lamp at end of life. Starter defective. Low line voltage. Fault in circuit of luminair. Low temperature of surrounding air.	Seat lamp firmly and correctly. Replace with tested lamp. Replace with tested starter. Match lamp rating to line voltage or increase line voltage. Check wirings and lamp holders. Check ballast. Shield lamp from drafts. Enclose lamp to conserve heat. Maintain voltages within the rated voltage range of the lamp. Use thermal-type starters.
Lamp flickers, swirls or flutters	Cold or too rapid starting.	Allow a new lamp to operate a few hours for seasoning. Turn off a few moments, then turn on. Change lamps and, if flicker remains, replace starter.
Ends of lamp glow	Poor ballast. Faulty starter. Improper wiring or ground.	Check ballast. Replace starter. Check wiring and ballast for ground.
Lamp darkens early in life	Improper starting. Low line voltage. Poor lamp holder contact.	Replace starter. Increase voltage. Seal lamp firmly in lampholder. Check ballast and wiring.
Short lamp life	Low or high line voltage. Lamps turned on and off.	Maintain branch circuit voltage within the range specified on ballasts. Frequency of starting affects lamp life. Long periods of burning give long life. Short periods of burning reduce lamp life.
Radio interference	Not installed properly. Line feedback. Radiation direct from lamp.	Auxiliary equipment should be enclosed in a steel channel. Wiring should be made up with tight connection: clamps and starters should be firmly installed in sockets and fixture grounded. Install filter at radio. Locate radio antenna system at least 10 ft. from fixtures.
Noise from ballast	Fluorescent equipment is not noiseless type.	Take special precaution in locating ballasts. If unit is particularly noisy, replace ballast.
Incandescent Lamp Equipment		
Lamp not burning	Lamp loose. Lamp burned out. Loose or broken connections.	Tighten in socket. Replace with new lamp. Secure terminals. Repair wiring.
Lamp burns dim	Low voltage.	Match lamp rating to line voltage or increase line voltage.
Short lamp life	High voltage. Lamp failure due to mechanical shock. Incorrect lamp. Excessive vibration.	Match lamp rating to line voltage. Improve voltage regulation and avoid surges. Replace lamp. Be sure water does not drip on bulb. Use rough service lamps if required. Replace with lamp of size for which luminaire is rated. Use vibration or rough service lamps.
Lamp breakage	Water contacts bulb. Bulb touches luminaire glassware.	Use enclosed, vapor-tight luminaire if exposed to moisture. Use correct lamp size.

Figure 13.4

Individual Method: Burned-out lamps are replaced by electricians on request. To prevent reduced illumination from lamp outages follow the maintenance procedures below.

1. Instruct employees to report burnouts as soon as they occur.
2. Replace blackened or discolored lamps even though they are still burning.
3. Replace fluorescent lamps as soon as they begin to flicker.
4. Replace any lamp with the same type wattage and voltage as that of the lamp removed. If frequent burnouts occur, voltage rating of the lamps may be too low. Lamps of higher wattage than called for in the lighting design plans should not be used.

Group Method: Group replacement of lamps before they burn out is considered the most economical method for replacement in large areas. Whenever possible, group replacement should be accomplished simultaneously with fixture cleaning.

Replacement of lamps by this method is accomplished by installing new lamps in all fixtures in the prescribed area after they have been in use 70 to 75 percent of their rated life. Lamps thus removed should not be reused since their light output and expected life have been reduced greatly.

Rated life of lamps can be obtained from manufacturers. Frequency of turning on and off has little effect on incandescent lamps. Extended service incandescent lamps should be used for lamps in continuous use since turning them on and off will shorten their life. Mercury vapor lamps and fluorescent lamps also have a longer life when operated continuously than when used for intermittent service.

Inspection
Refer to Figure 13.5 for items to check when making an inspection of a lighting system.

Maintenance and/or Repair of Lamps (continued)		
Problem	Cause	Maintenance and/or Repair Procedure
Mercury Lamp Equipment		
Lamp fails to start	Lamp loose.	Tighten in socket.
	Lamp burned out.	Replace.
	Low voltage.	Connect line to transformer tap closest to line voltage. Increase line voltage by changing transformer tap.
	Wiring fault.	Check wiring. Tighten connections.
	Low temperature.	Lamps may not start when temperature is below 32°F.
	Fluctuating voltage.	Check line voltage. Momentary dips of 10%, or more, often cause lights to go out.
Lamp frequently goes out	Wiring fault.	Tighten connections. Check wiring. Separate lighting circuits from heavy power circuits.
Annoying stroboscopic effect	Cyclic flicker.	Where there is a 3-phase supply, connect luminaires on alternate phases. On single phase, add incandescent luminaires to the system.

Figure 13.4 (continued)

Inspection of Lighting System Checklist

LIGHTING FIXTURES

____ Inadequately supported, insecure, and improperly located, evidence of unauthorized removal and relocation.

____ Incorrect types installed in hazardous locations; change in facility class requires replacement.

____ Improperly located in clothes closets. (Should be above door or in ceiling and not serviced with cord pendants.)

____ Cracked or broken luminaires and fixture parts, missing pullcords, metal pullchains not provided with insulating links.

____ Indications of objects being supported from, hung on, or stored in fixtures.

____ Evidence of overheating, undersized, or other damage to sockets, exposed or damaged connecting wiring.

____ Repair or replace exposed live wires, undersized or overheated sockets, broken or missing pullcords, cracked glass luminaires, and missing fixture parts.

LAMPS

____ Oversized, blistering, loose base, thermal cracks from contact with fixture, bare lamps in hazardous locations, poor socket lamp connections, improper types for special applications.

____ Operation of fluorescent fixtures shows poor burning and starting characteristics, and loud humming ballasts.

SWITCHES

____ Defective operation, broken or missing parts, arcing noises.

CONVENIENCE OUTLETS

____ Dirty, inadequate, defective contacts, difficult plugging, overheating, evidence of overloading on multiple sockets servicing lamps or appliances, lack of grounding terminal. Revamp local wiring to eliminate dangerous and improper conditions.

CORDS, CORD EXTENSIONS, PORTABLE AND APPLIANCE CORDS

____ Inadequate, unsafe, unreliable, incorrect types being used.

____ Lengths too long, poor insulation, twisted, spliced, exposed to damage underfoot, lying on floor or across heated surfaces or lamps; lamp types used for portable extensions that are subject to moisture, oil, and grease.

____ Plugs: cracks, breaks, loose connections, wires improperly attached and in danger of pulling away from plug when removing from outlet, missing protective cover on male ends, no grounding terminal or ground wire with clamp.

LIGHTING VOLTAGE

____ Spot measurement at fixture and lighting outlets indicates measured voltage in excess of 5 percent of nominal lamp voltages.

____ Unauthorized connections of hot plates, coffee pots, heating devices, and other electrical equipment on lighting circuits.

____ Interference with branch circuits for power and lighting from motor starting or stopping such as excessive light flicker, or excessive voltage dips causing fluorescent and mercury lamps to drop out.

ILLUMINATION LEVELS

____ Ambient conditions such as dirty walls and ceilings.

____ Spot measurement of light levels using accurate foot candle meter indicates depreciation of 20 to 25 percent of level obtainable from clean fixtures and new lamps seasoned for 100 hours.

____ Failure to keep continuous record of illumination levels at established checkpoints.

Figure 13.5

Motors

Good maintenance of motors consists basically of cleanliness and lubrication. A dirty motor will run, but at higher than normal temperatures which tend to shorten the life of both the insulation and bearings. Motor failures are caused by loading, age, vibration, contamination, or commutation problems.

An overloaded motor will fail earlier in its expected lifetime than one which is not overloaded. Therefore, the easiest way to solve the problem of overloaded motors would be either to eliminate overloads or replace the motors with larger units. To prevent a sudden overload, check all protective devices, such as overload relays, fuses, circuit breakers, field loss relays, and voltage relays periodically.

Failures due to age can be delayed by thorough periodic cleaning; dipping and baking the windings; replacing fast wearing bearings and brushes; also by tightening, turning and overcutting commutators on DC motors.

Failures, due to vibrations, can be in the form of sprung or broken shafts, bearing or seal failures, insulation breakdown, broken electrical connections, damage to the motor's mechanical structure, broken brushes, and a variety of commutator or slip ring problems. The vibration itself can be caused by misalignment, bearing problems, or inbalance.

Problems with bearings stem from such items as contamination, lack of lubrication, improper fits either internal or external; also overloading, stray electric current, slip ring damage, and vibration itself.

The most common cause of motor failure is contamination. Dirt coats motor windings and cuts down on heat dissipation, or blocks ventilation passages and increases insulation temperatures so that failure ultimately occurs. Dirt also causes wear in moving parts, especially bearings. Moisture in combination with dirt also causes insulation failure by shortening the windings or shortening to ground, and causes rust, a deteriorating factor.

Commutation problems for DC motors (that is, problems with commutators and brushes) are evident in such conditions as rapid brush wear, chipped or broken brushes, burned brushes, copper feathering or drag, commutator burning, slot or pitch patterns that are overfilming.

Inspection should be programmed to include the items noted on Figure 13.6.

Maintenance and Repair of AC and DC Motors

Refer to Figure 13.7 for causes and solutions to motor trouble in the field. Also refer to Figure 13.8 for actions to take to correct motor problems in the shop.

Motor Failures

It would be helpful if some sort of formal program could be devised to determine or predict which motors are likely to fail first. This sort of program would have to be developed by each individual company. The company would have to consider the following items in developing such a program:

- **Type of motor:** The less complicated in construction the motor is, the less
  likely it is to fail.
- **Age:** The older the motor is the more likely it is to fail.
- **Operation:** The more time a motor is fully loaded (or overloaded) the shorter lifetime it will have.
- **Environment:** The more corrosive the environment is to the motor and its housing, the earlier failure will occur.
- **Testing and Inspection:** Inspection by experienced personnel can uncover many items such as bad installation, loose windings and other such conditions. Also, performing standard tests on the insulation, temperature, and vibration will help evaluate and predict the lifetime and failure of the motor.

Motor Inspection Checklist

GENERAL
____ When practicable, start, run, and cycle motor and generator equipment through load range. Take care in starting motors and generators. On standby or infrequently operated equipment, check rotor freedom and lubrication. At humid locations check records for evidence of regular exercise; if not found arrange for drying out windings; megger windings before starting motor.

RUNNING INSPECTION (While Equipment Operates)
____ Log or Operator Records: evidence of motor or generator overload, low power factor of load, excessive variations in bearing temperature, operating difficulties.
____ Exposure: unsafe accessibility for maintenance of instrumentation; exposed to physical or other damage from normal plant functions, processes, traffic, and radiant heat; inadequate personnel guards, fences; insufficient, missing, or illegible signs, identification, or operating instructions.
____ Housekeeping: dust, dirt, airborne grit, sand; dripping oil, water, other fluids, vapors; rust, corrosion; peeling, scratches, abrasions or other damage to painted surfaces. Remove oil and solvent cans, oil or solvent soaked rags and waste, other combustibles, particularly those near commutating machinery; remove obstructions that may interfere with rotation or ventilation.
____ Machine Operation: noisy, unbalanced, rubbing, excessive vibration, rattling parts.
____ Structural Supports: inadequate, cracks, settlement; defective or inadequate vibration pads, shockmounts, dampers; loose, dirty, corroded bolts and fittings.
____ Ventilation: dirty, inadequate amount of air passing through machine; dirty, clogged, stator-iron air slots causing excessive temperature. (Too hot to touch. Measured temperature should not exceed 80 degrees C for open frames, or 90 degrees C for enclosed frames. Compare with manufacturer's data.)
____ Motor and Generator Leads: exposed bare conductors; frayed, cracked, peeled insulation; poor taping; moisture, paint, oil, grease; vibration, abrasions, breaks in insulation at entrance to conduit or machines; arcs, burns, overheated, inadequate terminal connections; lack of resiliency, lack of life, dried-out insulation; exposure to physical damage, traffic, water, heat, for semipermanent, temporary, or emergency connections.
____ Bearings: improper lubrication (check lubrication schedules for lubricant used and frequency), improper oil level in oil gages, incorrectly reading gages, noisy bearings, overheated bearing caps or housings. (If bearings are too hot to touch, determine causes. A slow but continuous rise in bearing temperature after greasing indicates possible over-lubrication or under-lubrication, improper lubricant, or deteriorated bearings. Under normal conditions, the temperature of ball or roller bearings will vary from 10 to 60 degrees F above the ambient temperature).

Figure 13.6

Motor Inspection Checklist (continued)

_____ Collector Rings, Commutators, Brushes: excessive sparking, surface dirt, grease (check cleanliness with clean canvas paddle); sparking or excessive brush movement caused by eccentricity, sprung shaft, worn bearings, high bars of mica, surface scratches, roughness; end-play resulting from magnetic-center hunting of rotor; inadequate brush freedom; nonuniform brush wear; poor commutation, improper brushes, incorrect brush pressures. Adjust brush spring pressure to between 1-3/4 to 2-1/2 psi of brush-commutator contact area for light metallized carbon or graphite brushes; for pressure for other type brushes, check manufacturer's data. (Measure with spring scale.)

_____ Starters, Motor Controllers, Rheostats, Associated Switches: damaged or defective insulation, loose laminations, defective heater or resistance elements, worn contacts, shorts between contacts, arcing, grounds, loose connections, burned or corroded contacts. Replace worn contacts and defective heater or resistance elements.

_____ Protective Equipment: dirty, signs of arcing, symptoms of faulty operation, improper condition of contacts, burned-out pilot lamps, burned-out fuses.

SHUTDOWN INSPECTION (while equipment is not in operation and is electrically disconnected. A shutdown inspection includes a running inspection.)

_____ Stators: dirt, debris, grease; coils not firmly set in slots; burns, tears, aging, embrittlement, moisture in insulation; clogged air slots; rubbing, corrosion, loose laminations of stator-iron; charred or broken slot wedges; abrasion of insulation or chafing in slots; signs of arcing or grounds.

_____ Rotors: difficult turning, rubbing, excessive bearing friction, end play, overheating, looseness of windings, charred wedges, broken, cracked, loosely welded or soldered rotor bars or joints; cracked end rings in squirrel cage motors; loose field spools and deteriorated leads and connections in synchronous motors; deteriorated insulation in wound rotors.

_____ Rotor-Strator Gaps: Check gaps on 5 hp or larger induction motors, particularly of the sleeve bearing type. Where practicable, measure and record gaps on the load, pulley, or gear end of the motor. Measure at 2 rotor positions, 180 degrees apart, 4 points for each rotor position. If there is more than 10% variation in gaps, arrange for realignment.

_____ Mechanical Parts: corrosion, improper lubrication, misalignment, end play, interference, inadequate chain or belt tension.

_____ Insulation Resistance: Test insulation resistance of motor and generator windings. Compare results with Maintenance Information. Insulation resistance values are arbitrary and should be correlated with operating conditions, exposure to moisture, metallic dust, age, length of time in service, severity of service, and maintenance levels.

Figure 13.6 (continued)

Maintenance and/or Repair Procedures for AC–DC Motors

Problem	Cause	Maintenance and/or Repair Procedure
Hot bearings; general	Excessive belt pull.	Decrease belt tension.
	Pulley too far away.	Move pulley closer to bearing.
	Pulley diameter too small.	Use larger pulleys.
Hot bearings; sleeve type	Oil too heavy.	Use a recommended lighter oil.
	Oil too light.	Use a recommended heavier oil.
	Insufficient oil.	Fill a reservoir to proper level in overflow plug with motor at rest.
Hot bearings; ball type	Insufficient grease.	Maintain proper quantity of grease in bearing.
	Excess lubricant.	Reduce quantity of grease; bearing should not be more than 1/2 filled.
	Heat from hot motor or external source.	Protect bearing by reducing motor temperature.
	Overloaded bearing.	Check alignment, side thrust and end thrust.
Oil leakage from overflow plugs	Stem of overflow plug not tight.	Remove, recement threads, replace, and tighten.
	Cracked or broken overflow plug.	Replace the plug.
	Plug cover not right.	Requires cork gasket or if screw type, may be tightened.
DC Motor		
Fails to start	Circuit not complete.	Switch open, leads broken.
	Brushes not down on commutator. Brushes worn out.	Held up by brush springs, need replacement. Brushes worn out.
	Brushes stuck in holders.	Remove and sand, clean up brush boxes.
	Power may be off.	Check line connections to starter with light. Check contacts in starter.
Motor starts, then stops and reverses direction of rotation	Reverse polarity of generator that supplies power.	Check generating unit for cause of changing polarity.
	Shunt and series fields are bucking each other.	Reconnect either the shunt or series fields in order to correct the polarity. The connect armature leads for desired direction of rotation. The fields can be tried separately to determine the direction of rotation individually, and connected so that both give same rotation.
Motor does not come up to rated speed	Overload.	Check bearing to see if in first class condition with correct lubrication. Check driven load for excessive load or friction.
	Voltage low.	Measure voltage with meter and check with motor nameplate.
	Motor off neutral.	Check for factory setting of brush rigging or test motor for true neutral setting.
	Motor cold.	Increase load on motor to increase its temperature, or add field rheostat to set speed.
Motor runs too fast	Neutral setting shifted off neutral.	Reset neutral by checking factory setting mark or testing for neutral.
	Part of shunt field rheostat or unnecessary resistance in field circuit.	Measure voltage across field and check with nameplate rating.
Motor continually runs too slow	Voltage below rated. Overload.	Measure voltage and try to correct to value on motor nameplate. Check bearings of motors and the drive to see if in first class condition. Check for excessive friction in drive.
	Neutral setting shifted.	Check for factory setting of brush rigging or test for true neutral setting.

Figure 13.7

Maintenance and/or Repair Procedures for AC–DC Motors

Problem	Cause	Maintenance and/or Repair Procedure
Motor overheats or runs hot	Overloaded and draws 25 to 50 percent more current than rated.	Reduce load by reducing speed or gearing in the drive, or loading in the drive.
	Voltage above rated.	Motor runs drive above rated speed requiring excessive horsepower; reduce voltage to nameplate rating.
	Inadequately ventilated.	Location of motor should be changed, or restricted surroundings removed; if covers are too restricting they should be modified or removed. Open motors cannot be totally enclosed for continuous operation.
	Armature rubs pole faces due to off-center rotor, causing friction and excessive current.	Check brackets or pedestals to center rotor and determine condition or bearing wear for bearing replacement.
Hot armature	Brush tension too high.	Limit pressure to 2 to 2½ psi. Check brush density recommended by the brush manufacturer.
Hot commutator	Brushes off neutral.	Reset neutral.
	Brush grade, too abrasive.	Get recommendation from manufacturer.
	Inadequate ventilation.	Check as for hot motor.
Hot fields	Voltage too high.	Check with meter and thermometer and correct voltage to nameplate value.
	Inadequate ventilation.	Check as for hot motor.
Motor vibrates and indicates unbalance	Loose or eccentric pulley.	Tighten pulley on shaft or correct eccentric pulley.
	Belt or chain whip.	Adjust belt tension.
	Foundation inadequate.	Stiffen mounting place members.
	Motor loosely mounted.	Tighten holddown bolts.
	Motor feet uneven.	Add shims under foot pads to mount each foot tightly.
Motor sparks at brushes or does not commutate	Neutral setting not true neutral.	Check and set on factory setting or test for true neutral.
	Brush grade wrong type. Brush pressure too light, brushes stuck in holders. Brush shunts loose.	See brushes.
Brush wear excessive	Brushes too soft.	Blow dust from motor and replace brushes with a grade as recommended by manufacturer.
	Off neutral setting.	Recheck factory neutral or test for true neutral.
	Brush tension excessive.	Adjust spring pressure not to exceed 2 to 2½ psi.
Motor noisy	Motor loosely mounted.	Tighten foundation bolts.
	Foundation hollow and acts as sounding board.	Coat underside with soundproofing material.
	Strained frame.	Shim motor feet for equal mounting.
	Belt slap or pounding.	Check condition of belt and change belt tension.
AC Motor		
Motor stalls	Wrong application.	Change type or size. Consult manufacturer.
	Overloaded motor.	Reduce load.
	Low motor voltage.	See that name plate voltage is maintained.
	Open circuit.	Fuses blown, check overload relay, starter, and push-button.

Figure 13.7 (continued)

Maintenance and/or Repair Procedures for AC–DC Motors		
Problem	**Cause**	**Maintenance and/or Repair Procedure**
Motor runs and then dies down	Power failure.	Check for loose connections to line, to fuses and to control.
Motor does not come up to speed	Not applied properly. Starting load too high.	Consult supplier for proper type. Check load motor is supposed to carry at start.
Motor takes too long to accelerate	Excess loading. Poor circuit. Defective squirrel-cage rotor.	Reduce load. Check for high resistance. Replace with new rotor.
Motor overheats while running under load	Check for overload. Wrong blowers or air shields, may be clogged with dirt and prevent proper ventilation. Unbalanced terminal voltage. High or low voltage.	Reduce load. Good ventilation exists when a continuous stream of air leaves the motor. If not, check with manufacturer. Check for faulty leads, connections, and transformers. Check terminals of motor with voltmeter.
Motor vibrates after corrections have been made	Driven equipment unbalanced.	Rebalance driven equipment.
Unbalanced line current on polyphase motors during normal operation	Unequal terminal volts. Brushes not in proper position in wound rotor.	Check leads and connections. See that brushes are properly seated and shunts in good condition.
Contactor or relay does not close.	No supply voltage. Low voltage.	Check fuses and disconnect switch. Check power supply. Wire size may be too small.

Figure 13.7 (continued)

Shop Procedure for Maintenance and/or Repair of AC–DC Motors

Problem	Cause	Maintenance and/or Repair Procedure
Hot bearings; general	Bent or sprung shaft.	Straighten or replace shaft.
	Misalignment.	Correct by realignment of drive.
Hot bearings; sleeve	Oil grooving in bearing obstructed by dirt.	Remove bracket or pedestal with bearing and clean oil grooves and bearing housing; renew oil.
	Bent or damaged oil rings.	Repair or replace oil rings.
	Too much end thrust.	Reduce thrust induced by driven machine or supply external means to carry thrust.
	Badly worn bearing.	Replace bearing.
Hot bearings; ball	Deterioration of grease or lubricant contaminated.	Remove old grease, wash bearings thoroughly in kerosene and replace with new grease.
	Broken ball or rough.	Replace bearing; first clean housing thoroughly.
Motor dirty	Ventilation blocked, end windings filled with fine dust or lint.	Clean motor will run 10° to 30°C. cooler; dust may be cement, sawdust, rock dust, grain dust, coal dust, and the like. Dismantle entire motor and clean all windings and parts.
	Rotor winding clogged.	Clean, grind, and undercut commutator. Clean and treat windings with good insulating varnish.
	Bearing and brackets coated inside.	Dust and wash with cleaning solvent.
Motor wet	Subject to dripping.	Wipe motor and dry by circulating heated air through motor. Install drip or canopy-type covers over motor for protection.
	Drenched condition.	Motor should be covered to retain heat and the rotor position shifted frequently.
	Submerged in flood waters.	Dismantle and clean parts. Bake windings in oven at 105°C. for 24 hours or until resistance to ground is sufficient. First make sure commutator bushing is drained of water.

DC Motor

Problem	Cause	Maintenance and/or Repair Procedure
Fails to start	Armature locked by frozen bearings in motor or main drive.	Remove brackets and replace bearings, or recondition old bearings.
Motor does not come up to rated speed	Starting resistance not all out.	Check starter to see if mechanically and electrically in correct condition.
	Short circuit in armature windings or between bars.	Check commutator for blackened bars and burned adjacent bars. Check windings for burned coils or wedges.
	Starting heavy load with very weak field.	Check full field relay and possibilities of full field setting of the field rheostat.
Motor runs too fast	Voltage above rated.	Correct voltage or get recommended change in air gap from manufacturer.
	Load too light.	Install fixed resistance in armature circuit.
	Shunt field coil shorted.	Install new coil.
	Shunt field coil reversed.	Reconnect coil leads in reverse.
	Series coil reversed.	Reconnect coil leads in reverse.
	Series field coil shorted.	Install new or repaired coil.
	Motor ventilation restricted causing hot shunt field.	Hot field is high in resistance; check causes for hot field, in order to restore normal shunt field current.
Motor gaining speed steadily and increasing load does not slow it down	Unstable speed load regulation.	Inspect motor to see if oc neutral. Check series field to determine shorted turns. If series field has a shunt around the series circuit, it may be removed.
	Reversed field coil shunt or series.	Test with compass and reconnect oil.
	Too strong a commutating pole or commutating pole air gap too small.	Check with factory for recommended change in coils or air gap.

Figure 13.8

332

Shop Procedure for Maintenance and/or Repair of AC–DC Motors (continued)

Problem	Cause	Maintenance and/or Repair Procedure
Motor continually runs too slow	Motor operates cold.	Motor may run 20 percent slow due to light load. Install smaller motor, increase load, or install partial covers to increase heating.
	Armature has shorted coils or commutator bars.	Remove armature to repair shop and make corrections.
Motor overheats or runs hot	Draws excessive current due to shorted coil. Grounds in armature, such as two grounds which constitute a short.	Repair armature coils or install new coil. Locate grounds and repair or rewind with new set of coils.
Hot armature	Core hot in one spot indicating shorted punchings and high iron loss.	Sometimes full slot metal wedges have been used for balancing. These should be removed and other means of balancing investigated.
	Punchings uninsulated. Punchings have been turned or band grooves machined in the core. Machined slots.	No-load running of motor will indicate hot core and drawing high no-load armature current. Replace core and rewind armature. If necessary to add ban grooves, grind into core. Check temperature on core with thermometer; it should not exceed 90°C.
Hot commutator	Shorted bars.	Investigate commutator mica and undercutting, and repair.
	Hot core and coils that transmit heat to commutator.	Check temperature of commutator with thermometer to see that total temperature does not exceed ambient plus 55°F. rise; total not to exceed 105°C.
Hot fields	Shorted turns or grounded turns.	Repair or replace with new coil.
	Resistance of each coil not the same.	Check each individual coil for equal resistance within 10 percent and if one coil is too low, replace coil.
	Coils not large enough to radiate its loss wattage.	New coils should replace all coils if room is available in motor.
Motor vibrates and indicates unbalance	Armature out of balance.	Remove and statically balance or balance in dynamic balancing machine.
	Misalignment.	Realign.
	Mismanting of gear and pinion.	Recut, realign, or replace parts.
	Unbalance in coupling.	Rebalance coupling.
	Bent shaft.	Replace or straighten shaft.
Motor sparks at brushes or does not commutate	Commutator rough.	Grind and roll edge of each bar.
	Commutator eccentric.	Turn and grind commutator.
	Mica high; not undercut.	Undercut mica.
	Commutating pole strength too great, causing over-compensation or strength too weak indicating under-compensation.	Check with manufacturer for correct change in air gap or new coils for the commutating coils.
	Shorted commutating pole turns.	Repair coils or install new coils.
	Shorted armature coils on commutator bars.	Repair armature by putting into first-class condition.
	Open circuited coils.	Same as above.
	Poor soldered connection to commutator at high speeds.	Resolder with proper alloy of tin solder.
	High bar or loose bar in commutator at high speeds.	Check commutator nut or bolts, and retighten and grind commutator face.
	Brushes chatter due to dirty film on commutator.	Resurface commutator face and check for change in brushes.
	Vibration.	Eliminate cause of vibration by checking mounting and balance of rotor.

Figure 13.8 (continued)

Shop Procedure for Maintenance and/or Repair of AC–DC Motors (continued)

Problem	Cause	Maintenance and/or Repair Procedure
Brush wear excessively	Commutator rough.	Grind commutator face.
	Abrasive dust in ventilating air.	Reface brushes and correct condition by protecting motor.
	Bad commutation.	See corrections for commutation.
	High, low, or loose bar.	Retighten commutator motor bolts and resurface commutator.
	Electrical wear due to loss of film on commutator face.	Resurface brush faces and commutator face.
	Threading and grooving.	Same as above.
	Oil or grease from atmosphere or bearings.	Correct oil condition and surface brush faces and commutator.
	Weak acid and moisture laden atmosphere or bearings.	Protect motor by changing ventilation air, or change to enclosed motor.
Motor noisy	Brush singing.	Check brush angle and commutator coating, resurface commutator.
	Brush chatter.	Resurface commutator and brush face.
	Armature punching loose.	Replace core on armature.
	Armature rubs pole faces.	Recenter by replacing bearings, or relocating brackets or pedestals.
	Magnetic hum.	Refer to manufacturer.
	Excessive current load.	May not cause overheating, but check chart for correction of shorted or grounded coils.
	Mechanical vibration.	Check chart for causes of vibration.
	Noisy bearings.	Check alignment, loading of bearings, lubrication and get recommendation of manufacturer.
Motor stalls	Incorrect control resistance of wound rotor.	Check control sequence. Replace broken resistors. Repair open circuits.
Motor connected but does not start	One phase open. Motor may be overloaded.	See that no phase is open. Reduce load.
	Rotor defective.	Look for broken bars or rings.
	Poor stator coil connection.	Remove end bells, locate with test lamp.
Motor does not come up to speed	Voltage too low at motor terminals because of line drop.	Use higher voltage on transformer terminals or reduce load.
	If wound rotor, improper control operation of secondary resistance.	Correct secondary control.
	Low pull-in torque of synchronous motor.	Change rotor starting resistance or change rotor design.
	Check to see whether all brushes are riding on rings.	Check secondary connections. Leave no leads poorly connected.
	Broken rotor bars.	Look for cracks near the rings. A new rotor may be required as repairs are usually temporary.
	Open primary circuit.	Locate fault with testing device and repair.
Motor takes too long to accelerate	Applied voltage too low.	Get power company to increase voltage tap.
Wrong rotation	Wrong sequence of phases.	Reverse connections of motor or at switchboard.
Motor overheats while running under load	Motor may have one phase open.	Check to make sure that all leads are well connected.
	Grounded coil.	Locate and repair.
	Shorted stator coil.	Repair and then check watt-meter reading.
	Faulty connection.	Indicate by high resistance.
	Rotor rubs stator bore.	If not poor machining, replace worn bearings.

Figure 13.8 (continued)

Shop Procedure for Maintenance and/or Repair of AC–DC Motors (continued)

Problem	Cause	Maintenance and/or Repair Procedure
Motor vibrates after corrections have been made	Motor misaligned	Realign.
	Weak foundations.	Straighten base.
	Coupling out of balance.	Balance coupling.
	Defective ball bearing.	Replace bearing.
	Bearings not in line.	Line up properly
	Balancing weights shifted.	Rebalance rotor.
	Wound rotor coils replaced.	Rebalance rotor.
	Polyphase motor running single phase.	Check for open circuit.
	Excessive end play.	Adjust bearing, or add washer.
Unbalanced line current polyphase motors during normal operations	Single phase operation.	Check for open contacts.
	Poor rotor contacts in control wound rotor resistance.	Check control devices.
Magnetic noise	Air gap on uniform	Check and correct bracket fits or bearing.
	Loose bearings.	Correct or renew.
	Rotor unbalance.	Rebalance.
Motor Starter		
Contactor or replay doesn't close	Open-circuited coil.	Replace
	Pushbutton, interlock, or relay not making contact.	Adjust for correct movement, ease of operation, and proper contact pressure.
	Loose connections or broken wire.	Check circuit with flashlight (turn power off first).
	Pushbutton not connected correctly.	Check connections with wiring diagram.
	Overload relay contact open.	Reset relay.
	Damaged, worn, or poorly adjusted mechanical parts.	Clean and adjust mechanically. Align bearings and free the movement. Repair or replace worn or damaged parts.
Contactor or relay does not open	Pushbutton not connected correctly	Check connections with wiring diagram.
	Shim in magnetic circuit (DC only) worn, allowing residual magnetism to hold armature closed.	Replace shim.
	Pushbutton, interlock or relay contact not opening coil circuit.	Adjust for correct movement, ease of operation, and proper opening.
	Contacts weld shut.	See "Contacts weld shut" (below).
	Damaged, worn, or poorly adjusted mechanical parts.	Clean and adjust mechanically. Align bearings and free the movement. Repair or replace worn or damaged parts.
Excessive corrosion of contacts; contacts weld shut; contacts overheat	Insufficient contact spring pressure causing contacts to overheat or draw an arc on closing.	Adjust, increasing contact pressure. Replace spring or worn contacts if necessary.
	Rough contact surface causing current to be carried by too small an area.	Dress up contacts with sand paper or fine file. Replace if badly worn.
	Sluggish operation.	Clean and adjust mechanically. Align bearings and free movement.

Figure 13.8 (*continued*)

Shop Procedure for Maintenance and/or Repair of AC–DC Motors

Problem	Cause	Maintenance and/or Repair Procedure
	Chattering of contacts as a result of vibrations outside of controller cabinet.	Check control switch contact pressure and replace spring, if it does not give rated pressure. Tighten all connections. If this does not help, mount or move control, so that vibrations are decreased.
	Abnormal operating conditions.	Check rating against load. If conditions are too severe for open-type contactors, replace with oil-immersed or dust-tight equipment. Instruct operator in proper manipulation of manually operated device.
Arc lingers across contact	If no blowout is used, note travel of contacts.	Increasing travel of contacts will increase rupturing capacity.
	If blowout is serious, it may be shorted. If blowout is shunted it may be open circuited.	Low at wiring diagram and see kind of blowout and then check to see if circuit through blowout is correct.
	Ineffective blowout coil.	Check rating and, if improperly applied, replace with correct coil. Check polarity and reverse coil if necessary.
	If used, arc box may have been left off or not in correct position.	See that arc box is fully in place.
	Overload.	Check rating against load.
	Creepage or voltage breakdown over or through arc box wall.	Clean; dry out in oven, or replace.
Noisy AC magnet	Improper seating of the armature.	Adjust mechanical parts and clean pole faces.
	Broken shading coil.	Replace.
	Low voltage.	Check power supply. Wire size may be too small.
Abnormally short coil life.	High voltage.	Check supply voltage against rating of controller.
	Gap in magnetic circuit. AC only.	Check travel of armature and adjust so that magnetic circuit is completed; clean pole faces.
	Too high an ambient temperature.	Check rating of controller against ambient temperature. Get coil of higher ambient rating from manufacturer if necessary.

Figure 13.8 (continued)

Replacement versus Repair of Electrical Motors

Upon the failure of an electrical motor, a decision must be made whether to replace or to repair it. A motor with unusual electrical or mechanical features will most probably be repaired. This decision is harder to to make for standard motors. Repair or replacement should not be based only on the cost difference between a rewind and a new motor. Operating cost penalties and the cost of modification accessories should also be considered.

Items to in the decision to repair or replace a motor are listed below.

- Cost
- Characteristics of the motor
- Mounting and connecting conditions
- Mechanical interchangeability
- Available alternatives
- Cost of down time
- Rerating possibilities
- Nameplate data
- Replacement criteria

Repair prices for smaller sizes of both protected and totally enclosed AC motors can exceed cost of a new motor. However, in larger horsepower ratings, the repair cost of a protected motor can be less than 65% of a new unit, and for a totally enclosed motor the repair cost can be less than half that of a new drive.

Refer to Chapter 14 for the maintenance and repair procedures for bearings.

Batteries

As for any other type of electrical equipment, batteries should be placed under a preventive maintenance program. The batteries should be inspected on a periodic basis in order to insure optimum efficiency. Measures that should be included in a battery maintenance program are listed below:

- The battery should be kept properly charged. Overcharging or undercharging should not be done. Ideally, the battery should be charged on a daily basis.
- Water should be kept at proper levels according to manufacturer's recommendations; low-level point is at the top of the separators and the high point usually one-sixteenth of an inch below the bottom of the filling tube. The level of water should be between these points by adding approved battery water. The water level should be checked on a weekly basis and added as needed.
- The battery should be kept clean and dry and be washed down on a weekly basis.
- Battery temperatures should be kept at the manufacturer's recommended levels.
- Keep metal objects and tools off batteries to prevent shorts.
- Keep open flames away from top of batteries to prevent explosion.
- On a monthly basis, clean the vent plugs and neutralize.
- If using acid, add only on the direction of a battery manufacturer.

Other measures that are pertinent to battery maintenance include:

- Do not take specific gravity readings from the same cell each time.
- Add only water to storage batteries. Never add acid, electrolytes or any special powders, solutions or jellies. They can be harmful and can reduce the voltage and capacity of the cells.
- All cells in a battery should take the same amount of water. If one cell takes more than the others, it should be examined for leakage.
- Batteries should be charged as soon as possible after discharge. Do not allow a discharged battery to remain uncharged for more than 24 hours.
- When the maximum difference in cell voltages reaches .05 volts, the battery should be replaced.

Transformers

Transformers require very little, and receive less, attention than most other electrical apparatus. The maintenance required will be governed by the size, importance and location on the system, ambient temperatures and the surrounding atmosphere. Therefore, dampness, dust, and corrosive atmospheres can be causes of transformer problems.

Listed in Figures 13.9 and 13.10 are some of the critical points that should be examined during inspection relative to power and distribution-type transformers.

Transformer Failure and Repair

When a transformer winding fails, it should be disconnected from the power source immediately. Indications for such a failure are smoke or liquid coming from the tank, which may be accompanied by noise. Do not re-energize the transformer if it has been established that a winding has failed. Fire hazard is always present in transformers other than the askarel types.

After neutralizing the transformer from its power source, make careful inspection and tests according to the manufacturer's instructions. Figure 13.10 is an inspection checklist for distribution transformers. This inspection and testing may have to be done by an outside consultant. However, before spending money for a consultant, check the following items:

- Check for external damage to brushings, leads, etc.
- Check the level of liquid in all compartments.
- Check the liquid temperature.
- Check for evidence of leakage.
- Check for evidence on the primary and secondary sides of the transformer for the cause of failure.

Grounds

Periodic inspection of the electrical grounding system can be done utilizing Figures 13.11 and 13.12.

Lightning Arrestors

Figure 13.13 is an inspection checklist for lightning arrestors.

Inspection Checklist—Power Transformers

De-energized State

BUSHINGS AND INSULATORS:
(Remove all grease, dirt, and other foreign materials by washing and then drying)

_____ Insulators and Porcelain Parts: indications of cracks, checks, chips, breaks; where flash-over streaks are visible, re-examine for injury to glaze or for presence of cracks. Chipped glaze exceeding 1/2 inch in depth or an area exceeding one square inch on any insulator or insulator unit, report for investigation by a qualified electrical engineer.

_____ Severe cracks, chipped cement, or indications of leakage around bases of joints of metal to porcelain parts at terminal and transformer ends.

_____ Terminal Ends: mechanical deficiencies, looseness, corrosion, damage to cable clamps.

_____ Improper oil level in oil-filled bushings. (Fill bushing if oil is below proper level.)

_____ Heating evidenced by discolorations, and corrosion indicated by blue, green, white, or brown corrosion products, on metallic portions of all main and ground terminals, including terminal board and grounding connections inside transformer case.

_____ Pipe, Bar Copper, and Connections: indications of overheating or flashover fusing.

_____ Cable Connections: broken, burned, corroded, missing strands. (Fused portions of connectors, cables, pipe, or bus copper should be filed smooth and all projections removed; clean metalwork, disconnecting if required and cover with thin coating of nonoxide grease; if connections are disassembled, rough spots on contact surfaces should be filed smooth, and all projections removed; see that all bolted and crimped connections are tight by setting up nuts or recrimping when looseness is evident or suspected; clean and tighten all corroded or loose connections; repair or replace cables with frayed and broken strands; repair frayed or broken cable insulation.)

ENCLOSURE AND CASES

_____ If case is opened for any reason, examine immediately for signs of moisture inside cover and, where present, look for plugged breathers, inactive desiccant, enclosure leakage. (Protect tranformer liquid from dust, dirt, and windblown debris by covering open tank with temporary cover made of wood, kraft paper, plastic sheeting, or other suitable dust-tight material; clear plugged openings; if desiccant is inactive replace with fresh material or reactivate for proper functioning; if rust or corrosion is evident on inside cover, clean and paint with preservative.)

COILS AND CORES

_____ When cover is open, examine interior for deficiencies, dirt, and sludge. If feasible, probe down sides with glass rod, and if dirt and sludge exceed approximately 1/2 inch, arrange to change or filter insulating oil, and have coils and cores cleaned.

BUSHING-TYPE INSTRUMENT TRANSFORMERS

_____ Indications of deteriorated insulation: overheating evidenced by excessive discoloration of terminals and other visible copper; physical strains indicated by bent or distorted members.

_____ Terminals, including secondaries: corrosion, loose connections.

Figure 13.9

Inspection Checklist—Power Transformers (continued)

____ Secondary Leads: visible broken, cracked, or frayed insulation.

____ Conduit and Associated Fittings Carrying Secondary Leads: rust, corrosion, other deterioration, loose joints in conduit fittings and around terminal boxes. (Clean, tighten, or repair terminals; tighten all loose or defective conduit-supporting clamps; clean and paint conduit and associated fitting areas showing rust and corrosion; if fuse box for bushing-type potential transformers is installed, check fuses for proper size, as specified by manufacturer or station engineers; assure that fuses have not been shorted out or bridged; replace blown or improperly sized fuses).

AUTOMATIC TAP-CHANGERS (load ratio control apparatus)

____ Make inspection in accordance with manufacturer's instructions. (Clean and lubricate all moving parts and contacts in accordance with manufacturer's recommendations.)

FORCED-AIR FANS AND FAN CONTROLS

____ Fans and Motors: defective bearings, inadequate lubrication, presence of dirt, bent or broken fan blades or guards, lack of rigidity of mountings, indications of corrosion or rust. (Make minor repairs as necessary to assure dependable and continuous service until next inspection.)

____ Starting and Stopping Devices: improper functioning as determined from operating once or twice.

____ Fan Speed: not in accordance with nameplate requirements.

WATER COOLING SYSTEMS

____ Water not being delivered in required quantity.

GAGES AND ALARMS

____ Liquid Level Gage and Alarm System: dirty, not readable, improper frequency of calibration.

____ Pressure Gages and Valves on Inert Gas Systems: improper frequency of gage calibration; leaks in piping both before opening and after closing tanks; apply soap bubble test to all joints and connections when pressure is unsteady. ENGINEERING TESTS should be performed under the supervision of a qualified electrical engineer before, during, or after inspection is applicable. Assistance of inspectors and craft personnel is required, and arrangements should be made with the proper authority to assure coordinated effort by everyone taking part.

____ Test grounding system.

____ Measure load current with recording meter over period of time when load is likely to be at its peak; measure peak-load voltage; make regulation tests and tests of operating temperature during peak-load current tests; test and calibrate thermometers or other temperature alarm systems.

____ Test dielectric strength of insulating liquid.

____ Test insulation resistance.

Energized State

CONCRETE FOUNDATIONS AND SUPPORTING PADS

____ Settling and movement, surface cracks exceeding 1/16 inch in width, breaking or crumbling within 2 inches of anchor bolts.

Figure 13.9 (*continued*)

Inspection Checklist—Power Transformers (continued)

____ Anchor Bolts: loose or missing parts, corrosion, particularly at points closest to metal base plates and concrete foundations resulting from moisture or foreign matter, and exceeding 1/8 inch in depth.

MOUNTING PLATFORM, WOODEN

____ Cracks, breaks, signs of weakening around supporting members; rot, particularly at bolts and other fastenings, holes through which bolts pass, wood contacting metal.

____ Burning and charring at contact points, indicating grounding deficiency.

____ Inadequate wood preservation treatment.

MOUNTING PLATFORMS, METALLIC

____ Deep pits from rust, corrosion, other signs of deterioration likely to weaken structure.

HANGERS, BRACKETS, BRACES, AND CONNECTIONS

____ Rust, corrosion, bent, distorted, loose, missing, broken, split, other damage; burning or charring at wood contact points caused by grounding deficiency.

ENCLOSURES, CASES, AND ATTACHED APPURTENANCES

____ Collections of dirt or other debris close to enclosure that may interfere with radiation of heat from transformer or flashover.

____ Dirt, particularly around insulators, bushings, or cable entrance boxes.

____ Leaks of liquid-filled transformers.

____ Deteriorated paint, scaling, rust; corrosion, particularly at all attached appurtenances, such as lifting lugs, bracket connections, and metallic parts in contact with each other.

NAMEPLATES AND WARNING SIGNS

____ Dirty, chipped, worn, corroded, illegible, improperly placed.

GASKETS

____ Leakage, cracks, breaks, brittleness.

INERT GAS SYSTEMS

____ Incorrect pressure in system. (Maximum: 3 to 5 pound, Minimum 1/4 to 1 pound.)

____ Pipe and Valve Connections: leaking gas (indicated by liquid oozing out of joints or by soapsuds test).

____ Loose gas tank fastenings, loose valves.

____ If previous arrangements were made with operating forces, bleed a little gas from system by means of pressure-regulating device; note evidence of leaks.

BUSHINGS AND INSULATORS

____ Cracked, chipped, or broken porcelain, indication of carbon deposits, streaks from flashovers, dirt, dust, grease, soot, or other foreign material on porcelain parts, signs of oil or moisture at point of insulator entrance.

GROUNDING AND PHASE TERMINALS

____ Overheating evidenced by excessive discolorations of copper, loose connection bolts, defective cable insulation, no mechanical tension during temperature changes, leads appear improperly trained, and create danger of flashovers from unsafe phase-to-phase or phase-to-ground clearances caused by deterioration of leads or expansions during temperature changes.

Figure 13.9 (*continued*)

Inspection Checklist—Power Transformers (continued)

INSTRUMENT TRANSFORMER JUNCTION BOXES AND CONDUITS

___ Loose or severely corroded components, including secondary lead connections.

BREATHERS

___ Holes plugged with debris; desiccant-type breathers need servicing or replacement.

TEMPERATURE-INDICATING AND ALARM SYSTEMS, INCLUDING CONDUIT AND FITTINGS

___ Loose fastenings, rust, severe corrosion, deteriorated paint, other mechanical defects, loose electrical connections.

MANUAL AND AUTOMATIC TAP CHANGERS

___ Loose connections, rust, severe corrosion, other mechanical defects, lack of lubrication, signs of burning around conducting and nonconducting parts of terminal boards.

LIQUID LEVEL INDICATORS

___ Rust, corrosion, lack of protective paint, cracked or dirty gage glasses so that liquid level not discernible, plugged gage-glass piping, liquid level below permissible level indicated by mark for gaging, signs of leakage around piping, gage cocks, gage glasses, or other indicating devices.

FANS AND FAN CONTROLS FOR AIR-COOLED TRANSFORMERS

___ Lack of rigidity in mounting fastenings.
___ Motors (external): dirty, moisture, grease, oil, overheating, detrimental ambient conditions.
___ Apparent deterioration of open wiring and conduit that may cause malfunctioning of either fans or controls.
___ Improper functioning when manual (not automatic) fan controls operated.

WATER COOLING SYSTEMS

___ Leaks in piping, fittings, or valve; visible drainage system plugged; open ditches for drainage water fouled with vegetation.
___ Bearings: evidence of wear, indications of corrosion, external deterioration, leaks.
___ Incipient deterioration, corrosion, rust, loose fastenings, other mechanical deficiencies, loose electrical connections for all components of alarm system.
___ Temperature Devices: signs of deterioration that might cause malfunction or difficulty in taking readings.
___ When pressure gage readings on each side of strainer varies more than a pound or two, look for cause, such as plugged strainer.

GROUNDING

___ Visual Connections: loose, missing, broken connections; signs of burning or overheating, corrosion, rust, frayed cable strands, more than one strand broken in 7-strand cable, more than 3 strands broken in 19-strand cable.

Figure 13.9 (*continued*)

Inspection Checklist—Distribution Transformers

De-energized State

BUSHINGS AND INSULATORS:
(Remove all grease, dirt, and other foreign materials by washing and then drying.)

_____ Insulators and Porcelain Parts: indications of cracks, checks, chips, breaks; where flashover streaks are visible, re-examine for injury to glaze or for presence of cracks.

_____ Chipped glaze exceeding 1/2 inch in depth or an area exceeding one square inch on any insulator or insulator unit, report for investigation by a qualified electrical engineer.

_____ Severe cracks, chipped cement, or indications of leakage around bases of joints of metal to porcelain parts at terminal and transformer ends.

_____ Terminal Ends: mechanical deficiencies, looseness, corrosion, damage to cable clamps.

_____ Improper oil level in oil-filled bushings (fill bushing if oil is below proper level).

_____ Heating evidenced by discolorations, and corrosion indicated by blue, green, white, or brown corrosion products on metallic portions of all main and ground terminals, including terminal board and grounding connections inside transformer case. (Clean metalwork, disconnecting if required, and cover with thin coating of nonoxide grease. If connections are disassembled, rough spots on contact surfaces should be filed smooth, and all projections removed. See that all bolted and crimped connections are tight by setting up nuts or recrimping when looseness is evident or suspected. Clean and tighten all corroded or loose connections. Repair or replace cables with frayed and broken strands. Repair frayed or broken cable insulation.)

ENCLOSURE AND CASES

_____ If case is opened for any reason, examine immediately for signs of moisture inside cover, and where present, for plugged breathers, inactive desiccant, enclosure leakage. (Protect transformer liquid from dust, dirt, and windblown debris by covering open tank with temporary cover made of wood, kraft paper, plastic sheeting, or other suitable dust-tight material. Clear plugged openings; if desiccant is inactive, replace with fresh material or reactivate for proper functioning. If rust or corrosion is evident on inside cover, clean and paint with preservative.)

COILS AND CORES

_____ When cover is open, examine interior for deficiencies, dirt, and sludge. If feasible, probe down sides with glass rode, and if dirt and sludge exceed approximately 1/2 inch, change or filter insulating oil, and have coils and cores cleaned. (Use low-pressure air, if available, to blow out dust from air-cooled transformers, or pull out dust with vacuum equipment.)

CAGES AND ALARMS

_____ Liquid Level Gage and Alarm System: dirty, not readable, improper frequency of calibration. ENGINEERING TESTS should be performed under supervision of qualified electrical engineer before, during, or after inspection, as applicable. Assistance of inspectors and craft personnel is required, and arrangements should be made with proper authority to assure coordinated effort by everyone taking part.

Figure 13.10

Inspection Checklist—Distribution Transformers (continued)

_____ Test grounding system.
_____ Measure load current with recording meter over period of time when load is likely to be at its peak; measure peak-load voltage; make regulation tests and tests of operating temperature during peak-load-current tests; test and calibrate thermometers or other temperature alarm systems.
_____ Test dielectric strength of insulating liquid.
_____ Test insulation resistance.

Energized State

CONCRETE FOUNDATIONS AND SUPPORTING PADS
_____ Settling and movement, surface cracks exceeding 1/16 inch in width, breaking or crumbling within 2 inches of anchor bolts.
_____ Anchor Bolts: loose or missing parts, corrosion, particularly at points closest to metal base plates and concrete foundations resulting from moisture or foreign matter, and exceeding 1/8 inch in depth.

MOUNTING PLATFORMS, WOODEN
_____ Cracks, breaks, signs of weakening around supporting members; rot, particularly at bolts and other fastenings, holes through which bolts pass, wood contacting metal.
_____ Burning and charring at contact points, indicating grounding deficiency.
_____ Inadequate wood preservation treatment.

MOUNTING PLATFORMS, METALLIC
_____ Deep pits from rust, corrosion, other signs of deterioration likely to weaken structure.

HANGERS, BRACKETS, BRACES, AND CONNECTIONS
_____ Rust, corrosion, bent, distorted, loose, missing, broken, split, other damage; burning or charring at wood contact points caused by grounding deficiency.

ENCLOSURES, CASES, AND ATTACHED APPURTENANCES
_____ Collections of dirt or other debris close to enclosure that may interfere with radiation of heat from transformer or flashover.
_____ Dirt, particularly around insulators, bushings, or cable entrance boxes.
_____ Leaks of liquid-filled transformers.
_____ Deteriorated paint, scaling, rust; corrosion, particularly at all attached appurtenances, such as lifting lugs, bracket connections, and metallic parts in contact with each other.

NAMEPLATES AND WARNING SIGNS
_____ Dirty, chipped, worn, corroded, illegible, improperly placed.

GROUNDING
_____ Visual Connections: loose, missing, broken connections; signs of burning or overheating, corrosion, rust, frayed cable strands, more than 1 strand broken in 7-strand cable, more than 3 strands in 19-strand cable.

BUSHINGS AND INSULATORS
_____ Cracked, chipped, or broken porcelain, indication of carbon deposits, streaks from flashovers, dirt, dust, grease, soot, or other foreign material on porcelain parts, signs of oil or moisture at point of insulator entrance.

Figure 13.10 (continued)

Inspection Checklist—Distribution Transformers (continued)

GROUNDING AND PHASE TERMINALS
——— Overheating evidenced by excessive discolorations of copper, loose connection bolts, defective cable insulation, no mechanical tension during temperature changes, leads appear improperly trained, and create danger of flashovers from unsafe phase-to-phase or phase-to-ground clearances caused by deterioration of leads or expansions during temperature changes.

BREATHERS
——— Holes plugged with debris; desiccant-type breathers need servicing or replacement.

GRILLS AND LOUVERS FOR VENTILATION OF AIR-COOLED TRANSFORMERS
——— Plugged with debris or foreign matter interfering with free passage of air. (Openings located near floor or ground line can be inspected with small nonmetallic framed mirror having long insulated handle, used in conjunction with light from hand flashlamp having insulated casing. Throw light beam onto mirror and reflect upward into openings.)

Figure 13.10

Information Checklist for Electrical Ground

——— Visual Connections: loose, missing, broken connections; signs of burning or overheating, corrosion, rust, frayed cable strands, more than one strand broken in 7-strand cable, more than 3 strands broken in 19-strand cable.

——— Underground Connections: unsatisfactory condition or defects uncovered when 4 or 5 connections are exposed to view by digging.

——— Test Measurement: Permissive Resistance (see table below)

From	To	Permissive Resistance
Point of connection on structure, equipment enclosure, or neutral conductor.	Top of ground rod	See Figure 13.12
Ground rod, mat, or network	Ground (earth)	See Figure 13.12
Gates	Gateposts	1/2 ohm
Operating rods and handles of group-operated switches	Supporting Structure	1/2 ohm
Metallic-cable sheathing	Ground rod, cable, or metal structure	1/2 ohm
Equipment served by rigid conduit	Nearest cable attachment on conduit runs of less than 25 ft.	10 ohms

——— When total resistance in check point 4 or 5 exceeds allowable, measure resistances of individual portions of the circuits to determine the points of excessive resistance and report.

——— Substandard resistance values resulting from poor contact betwen metallic portions of grounding system and earth.

——— Structural steel, piping, or conduit run exceeding 25 ft. used as a current-carrying part of grounding circuit for protection of equipment.

——— Absence of ground-cable connections.

Figure 13.11

Permissive Resistance for Grounding System

Maximum permissive resistance for grounds and grounding systems between equipment or structure being grounded and solid ground (earth):

Maximum Permissive Grounding System	Resistances (ohms)
a. For generating stations:	1
b. For main substations, distribution substations, and switching stations on primary distribution systems:	3
c. For secondary distribution system (neutral) grounding, Noncurrent-carrying parts of the distribution system itself, and enclosures of electrical equipment not normally within reach of other than authorized and qualified electrical operating and maintenance personnel:	10
d. For individual transformer and lightning-arrester grounds on distribution system:	10

Figure 13.12

Inspection Checklist—Lightning Arrestors

FOUNDATIONS AND SUPPORTS
___ Signs of weakness, cracked or broken concrete, burns, loose holddown bolts, rust, corrosion, mechanical damage.

GROUNDING CABLES FOR POLE-MOUNTED LIGHTNING ARRESTERS (where accessible to public)
___ Cracks, breakage, splintering, defective paint, evidence of tampering, other weakness in protective moldings.

TREATED CERAMIC-GAP TYPE (such as Thyrite and Autovalve)
___ Porcelain Insulators: signs of flashovers and serious flashover marks; scarring, chipping, or cracking of porcelain; dirt, grease, or other film on porcelain.
___ Metal Bases, Caps, and Intermediate Section Couplings: indications of loose bolts, rust, corrosion, or loose cement.
___ Connections to Line, Equipment, or Ground Lead: looseness, corrosion, breakage, or misalignment that may put undue mechanical strain on porcelain.
___ Ground Cable Connection to Ground Mat: loose or corroded connectors where visible.

OXIDE-FILM TYPE
___ Accumulation of dirt, particularly on edges of cells, deterioration of paint, rust, corrosion.
___ Loose connection and mounting bolts, badly corroded connection posts. Tighten all loose bolts; remove cable connection, clean post, and reconnect cable; inspect internal parts, and clean if accumulation of dirt is noticeable.

PELLET-TYPE (Obsolete, not acceptable for replacement)
___ Indications or burns and scars on porcelain bodies from flashovers, cracked or broken bodies and caps.
___ Mounting Clamps: rust, corrosion, loose bolts at arrester and supporting point of bracket.
___ Poor physical condition of ground cable from arrester to point of connection to ground rod or grounding system, where visible; loose or corroded connectors.

Figure 13.13

Inspection Checklist—Lightning Arrestors (continued)

CAPACITOR-TYPE
____ Signs of flashover on porcelain insulators and metal enclosures resulting in cracking, breaking, or burning.
____ Connection Points: looseness, corrosion, frayed ground cables, evidence of mechanical strain.
____ Enclosures: excessive rust and corrosion.
____ Porcelain: dirt accumulations in appreciable amounts.
____ Tighten all loose connection and mounting bolts; clean dirt from insulator and enclosure.

EXPULSION-GAP TYPE
____ Looseness of mounting, flashovers, damage to tubing, corrosion, loose ground connections, signs of burning and apparent damage from visual check of gap opening between arcing horn and line being protected.
____ Poor physical condition of ground cable from arrester to point of connection to ground rod or grounding system; loose or corroded connectors.
____ Signs of burning on external air gaps.
____ Check opening of air gap; examine tube carefully for damage from flashovers and burnouts; corrosion of metal mounting parts.
____ Adjust air gap, where deficient, to conform to following gap distances, in accordance with voltage of line being protected. Interpolate to obtain proper air gaps for voltages not listed.

Volts	Minimum External Gap in Inches (Up to 3,300 feet altitude*)
13,800	3/4
23,000	1-1/2
34,500	2
46,000	3-1/4
69,000	5-1/2
92,000	8-1/2
115,000	11
138,000	14

*Above 3,000 feet, spacing should be increased.

Figure 13.13 (continued)

CHAPTER 14

PLUMBING SYSTEMS

Proper maintenance saves money by sustaining the useful life of fixtures and by conserving water and utilities. Given proper treatment, the period of efficient use of metal faucets, traps, piping, and other fittings may be doubled. For example, good maintenance of faucets eliminates dripping and saves water. One cold water faucet, leaking one drop every second, wastes about 2,300 gallons in one year. This chapter covers the items listed below.

- Fresh water systems
- Waste water systems
- Gas distribution systems
- Fixtures
- Septic tanks
- Valves
- Pumps
- Turbines
- Bearings
- Special plumbing equipment

Preliminaries to Maintenance and Repair

The plumbing maintenance group should know the location of all types of pipes, fixtures, fittings, valves, and other related items to be maintained. The locations should be noted during the time of construction. If the information was not obtained during the construction phase, steps should be taken to obtain such information and place it on a line diagram.

The information in this chapter is presented in general terms, since it is impossible to cover in detail the various types of plumbing materials and all of the component accessories produced by the numerous manufacturers. These are omitted because of the readily available instructions, diagrams, drawings, and photographs that most manufacturers provide with their plumbing products.

A fundamental maintenance program for water distribution systems includes the items listed below.

- Information on equipment and systems
- Corrosion prevention (including protective coating, painting, cathodic protection)
- Proper tools and test equipment
- Proper safety measures

Information pertaining to lubrication types and frequencies should also be on hand. In order to avoid failures due to faulty lubrication, observe the following precautions.

- Do not over lubricate.
- Do not lubricate totally enclosed or insufficiently guarded items while in operation.
- To prevent contamination, keep lubricant containers tightly closed except when in use.

Refer to Chapter 6 for more information on the prevention of corrosion and surface deterioration and the correct type of surface protections to be used.

Tools and Equipment

Effective maintenance requires the proper tools in order to service the facility. Listed below are tools that should be available to the maintenance group. Specific tools and test equipment may be required for specialized equipment.

- Pipe wrenches
- Pipe cutters
- Reamers
- Socks and dyes
- Vises
- Drain and trap augers
- Force cup-plunger
- Snakes
- Closet augers
- Caulking tools
- Tube cutters
- Flaring tools
- Tube benders
- Torches
- Small tools, including tape, chisels, level hammers, brace and bit, electric drill, saws, and wrenches.

Fresh and Waste Water Service Systems

To ensure an effective piping system, follow the procedures below when inspecting equipment and performing follow-up maintenance for exposed piping, underground piping, location of leaks, pipe repairs, and leaking joints.

Complete replacement is recommended for damaged tubing.

Exposed Piping

Damage to exposed piping includes leakage, corrosion, loose connections, defective caulked joints on bell and spigot pipe, loose bolts on flanged pipe and clamp type connections, and damage to protective coating. Recaulk defective joints; replace defective gland nuts and bolts at expansion joints and clamp-type couplings; repair defective anchorage of expansion joints; stop leakage by takeup within limits of packing gland adjustment; replace defective packing; repair or replace defective hangers and supports.

Underground Piping

Evidence of leakage, ponding, erosion, or settlement of areas adjacent to piping; excessive supply pressure, water hammer or vibratory noises in line are typical problems with underground piping. (When exposed for alteration or repair, examine for deficiencies similar to those listed above for "Exposed Piping".) Refer to Figure 14.1 for maintenance of water distribution systems.

Location of Leaks

Locate and eliminate leaks in water piping systems as quickly as possible to prevent serious damage to footings, walls, floors, and other parts of the structure and finished materials adjacent to the pipe. The early location of the leak will conserve water. The location of leaks can be done by systematically inspecting exposed piping and valves, and by examining walls, floors, and ceiling around concealed piping. In addition, check gages, meters, and other water flow recording devices for evidence of abnormal flow, which might indicate loss through leakage.

Pipe Repairs

Flow must always be stopped before repair work is done. This is done by closing the shut-off valve in back of the point to be repaired. In circulating piping systems, the section to be repaired should be isolated by valving off the section. If it is not possible to close the line for a short time, the pipe can be temporarily frozen as described below.

If the pipe is covered, expose about 18 inches of its length. In any case, build a box around the pipe. Pack the box with 40 to 60 pounds of ice mixed with coarse salt, two parts ice to one part salt.

When thawing frozen pipes wrap the frozen section in rags and apply steam or hot water over the rags, or wrap an electric thawing apparatus around the frozen portion of the pipe. When thawing a frozen water pipe, open the faucet or outlet and work from the faucet or outlet toward the supply. Do not thaw the middle of the frozen line first. If the pipe is inaccessible, apply heat at the nearest point below the frozen section.

The type of pipe connections used for repairs depends on the working condition and the kind and size of pipe. The most commonly used connectors are screwed ends, flanged ends, brazed ends, soldered ends, flared ends, and hub ends.

Distribution Systems: Maintenance		
Item	**Suggested Maintenance**	**Frequency**
Flow tests	Make tests as conditions indicate need.	7
Pressure tests	Make tests as conditions indicate need.	7
Loss-of-head tests	Determine, using pitometer.	6
Cleaning	Employ outside contractor.	7
Cleaning and lining	Employ outside contractor.	7
Operating procedure	Change procedures as conditions dictate.	7
Leak location	Use standard techniques.	7
Leak remedies	Plan and prepare in advance.	7
Leak repair	Apply standard techniques.	7
Thawing	Apply standard techniques.	7
Backflow preventers		
Double check value	Test tightness of unit.	3
Reduced pressure	Test tightness of unit.	7

Key to Frequency: 1. Daily   5. Semi-annually
                               2. Weekly   6. Annually
                               3. Monthly   7. Variable — see manufacturer's specifications or current operation standards.
                               4. Quarterly

Figure 14.1

In most cases, damaged or deteriorated pipe can be repaired most economically by removing and replacing the affected section of piping and fittings. If the plumbing system has been functioning properly, the same size piping and fixtures should be reinstalled.

If it is found to be more economical to repair a leak or if the pipe cannot be replaced easily, follow this repair procedures described below.

Small Leaks: Use wood plugs to stop small holes temporarily. They can be replaced later with metal plugs, or repairs may be made by other means. Temporary wood plugs may be used to plug the ends of a pipe up to eight inches in diameter, but such pipes must be braced to withstand existing pressure.

Large leaks: Follow the procedure below to repair large leaks.
1. Split sleeve.
2. Cut out a section of the cracked pipe and replace it with a piece of pipe and mechanical couplings. Consult the manufacturer's instructions for the proper installation. If the break is too long for a short insertion piece, a whole length of pipe can be inserted. All systems that carry potable water should be properly flushed and disinfected before being placed in service once the repair has been made.

Leaking Joints
Leaking joints are usually caused either by faulty connections or improper lubricants. Most often the fault is with the connection; for instance, steel pipe which was cut with dull or improperly adjusted threading tools, causing shaved, chewed, rough, or unnecessarily deep threads. Threads in this condition are likely to cause a leaky connection. The connection should be replaced with properly threaded piping.

Sanitary Piping Systems

Stoppages in sanitary piping systems are usually caused by improper materials lodging in the drain, trap, or waste line. Obstructions may be removed by manually operated devices, by chemicals, or when necessary, combinations of both. The method depends upon the seriousness and nature of the stoppage. The obstruction should be entirely removed, and not merely moved from one place to another in the pipe line. After the stoppage has been relieved, pour boiling water into the fixture to ensure complete clearance. Some of the methods to remove stoppages are discussed below.

Plumber's Force Cup
Properly fill the fixture with boiling water and place the cup over fixture opening. Work the handle up and down several times to start alternate compression suction actions.

Auger
A coiled spring steel auger is used for opening clogged water closet traps, drains, and long sections of water pipe lines. Another device frequently used is a coiled wire plumber's snake. Manually operated devices such as these are most frequently used in clearing stoppages in lavatories, service sinks, and bathtubs. To clear obstructions, first use a plumber's force cup. If the obstruction is in the trap and has not been cleared by the action of the plunger, clear the trap by inserting a wire or snake through the clean-out plug at the bottom of the trap. If the trap is not fitted with such a plug, remove the trap. Protect the finish of the packing nut with adhesive tape or wrap cloth around the jaws of the wrench. Do not use a heavy steel spring coil snake to clear traps under lavatory sinks or bathtubs. Use a flexible type of wire spring snake which will easily follow the bends in the trap.

Chemicals
If manually operated devices fail to clear stoppages, several types of chemicals may be used to dissolve or burn them out. Some of these are described below.

Potassium Hydroxide (Caustic Potash): Stoppages can be burned out by pouring a strong solution of this chemical in hot water into the lines of the fixture opening. Pour the mixture slowly into the pipe through a funnel. Since this solution causes serious burns, personnel should wear goggles and rubber gloves. It will also damage glazed earthenwear, porcelain and porcelain enamel surfaces.

Sodium Hydroxide (Caustic Soda): When the stoppage is caused by grease congealing and acting as a binder for solid particles, effective cleaners such as caustic soda will help break it up. Adding water to the chemical releases ammonia gas, which assists conversion of grease to soap. It forms a gas which causes a boiling and heating action and further assists in dissolving the grease. In clearing a partially blocked drain, a small quantity of cleaner is dropped into the open drain, followed by a quantity of scalding hot water. Such cleaning agents cannot be satisfactorily used when the drain is completely plugged since some flow is required to loosen the chemicals. A completely blocked drain must first be partially cleared with a plumber's snake before using the chemical cleaner.

Interceptors and Traps

Traps provide water seals which prevent the entrance of sewer gas into the building. Water must run into all drains at frequent intervals to ensure the adequacy of the water seal. Traps sometimes lose their water seal from siphoning which is caused by inadequate venting of the line and a new vent must be installed as near the trap as possible. Frequent checks must be made to ascertain that the seal is in working order.

Back Flow

Back flow of sewage or waste results in mixing impure water with the potable water supply system. Back flow is caused by siphon action through a branch outlet when submerged below the flood level of the fixture, producing negative pressure and consequent back flow. Some of the ways to prevent back flow are listed below.

- Install vacuum breaker or back flow preventers.
- Make frequent checks of supply lines located below fixtures.
- Bathtub hand sprays, kitchen sink sprays, and similar hose connections and extensions used elsewhere should be inspected frequently for back flow and not submerged in the liquid held in the fixture.
- Ensure that there are no direct interconnections between potable water and nonpotable water piping systems.
- Check any tanks which are directly connected with potable water supply and sewer connections and correct the possibility of back flow by insulation of indirect waste and vacuum breakers.

Sump Pumps and Septic Tanks

Check sump pumps frequently to see that automatic operations are satisfactory, that screens are not clogged, and that bearings and motor are lubricated. Lubricate pump bearings through a grease or oil cup. Repair or service pumps and impeller shaft in accordance with the manufacturer's instructions.

Sump Pumps

Use the checklist shown in Figure 14.2 to inspect and maintain sump pumps.

Septic Tanks

Use the checklist in Figure 14.3 to inspect and maintain septic tanks.

Fixtures

Because they are the most exposed part of the plumbing system, fixtures must withstand daily wear and tear. They are especially vulnerable to the effects of poor maintenance and operating practices. A fixture should be cleaned at regular intervals (daily or weekly). Cleaning agents used should not be harmful to the surface.

Water Closets

Moisture on the floor at the base of a water closet usually indicates that the seal or gasket between the closet and its outlet has failed. It may, however, also result from condensation on the tank or piping, or from leakage of the tank, flush valve, or piping. When it has been determined that the leakage is from the seal, remove the water closet and install a new seal to prevent damage of the building and the possible entry of sewer gas into the room.

Inspection and Maintenance Checklist for Sump Pumps

_____ **Sumps:** floating objects, accumulated deposits in sump bottom.

_____ **Wiring and Controls:** loose connections, breaks and other damage to wiring and insulation, short circuits, loose or weak contact springs, worn or pitted contacts, defective float switch. Tighten connections and parts; replace or adjust contact springs; clean contacts; make minor repairs.

_____ **Floats:** bent rod, binding, other damage. Straighten bent rod, relieve binding, apply light oil to moving parts, check operation of alarm system.

_____ **Pump Operation:** failure to start when switch makes contact; rough operation; fads to empty basin at normal rate; inadequate suction, discharge, and shutoff heads against normal operating standards. Rotate by hand to determine drag or misalignment; service all pumps.

_____ **Packing Gland:** evidence of leakage. Remove packing; replace grooved or scored shaft sleeve; replace packing a strip at a time, tamp each strip thoroughly and stagger joints, and place lantern ring in proper position; where grease sealing is used, fill lantern ring completely with grease, place in proper position, then place remaining rings of packing.

_____ **Bearings:** inadequate lubrication, unavailability of lubricating instructions, oil rings do not turn freely with shaft. Drain lubricant; wash oil wells and bearings with kerosene, unless specifically directed otherwise; repair or replace defective oil rings; refill oil wells with proper lubricant.

_____ **Motor:** unusual noise, vibration, end play of shaft, overheated bearings, lubrication leaks, improper lubrication practices, presence of dirt, moisture, other accumulations.

_____ **Supports:** unsound, ineffective, misalignment of shafting indicated by check made with straightedge and thickness gage or wedge. Make minor adjustment.

_____ **Piping System:** evidence of strain on pump casing.

_____ **Cleanup:** remove dirt, moisture, and other accumulations from pedestal, switch support, piping, and valves, remove rust spots and spot paint all bare spots.

Figure 14.2

Water Closet Tanks

Water continually running into the closet bowl after the toilet has been flushed usually indicates leakage from the supply valve, or that the rubber tank ball is not seating properly. In order to correct the situation, first determine whether the tank ball is operating properly. The lift wire may be bent or the ball may be worn out or misshaped, and thus, fails to drop tightly into the hollow seat. Replace the lift wire and/or the ball. If the hollow seat of the discharge opening is corroded or grit-covered, sand the surface of the seat smooth with fine sandpaper. It is more economical to replace a leaky tank float than to repair it.

Flush Valves

Flush valves (for water closets, urinals and other fixtures) usually fail in operation through clogging of the by-pass orifice, wear of the valve seats, and/or failure of the diaphragm or piston packing. The symptoms are the same: that is, continual leakage through the valves. The remedy is to open the valve, blow out the orifice, examine each part, replace defective parts, and reassemble the valve. If the valve still operates improperly, repeat the disassembly and inspection until the defective part is found. In some cases, it may be more economical to replace the component assemblies rather than the individual parts.

Inspection and Maintenance Checklist for Septic Tanks

____ **Manhole Frames and Covers:** rust, corrosion, poor fit, missing, physical damage.

____ **Concrete and Masonry Surfaces:** cracks, breaks, spalling, deteriorated mortar joint.

____ **Inlet and Outlets:** clogging, high concentration of suspended solids, which may clog subsurface disposal fields.

____ **Flooding:** wall surfaces above normal liquid level show signs of frequent or occasional flooding. Determine source of infiltration on influent side.

____ **Sediment:** check depth. When sediment is two feet or less from effluent invert, septic tank should be pumped out; flush off concrete and masonry surfaces; deposits of grease or oil scum indicate improper operation of interceptors.

DOSING TANKS

____ **Tank:** if liquid level is less than three inches below level of liquid in septic tank, defective operation of siphon or clogged drainage field is indicated.

____ **Siphon:** overflow clogged or blocked, unsatisfactory operation. Remove or break up and flush clean any material interfering with operation.

DISTRIBUTION BOXES

____ **Stopboards and Gates:** improper functioning, undue leakage. Remove accumulations that might cause odors or interfere with proper seating.

____ **Water Levels:** Water level above invert of outlet or slow drainage indicates poor functioning of drain fields.

TILE DISPOSAL FIELDS

____ **Ground Surface:** ponding; indications of heavy trucking loads or other traffic, which may break drains or force them out of alignment.

Figure 14.3

Urinals

Maintenance of urinal fixtures follows the same procedures as those for water closets. If the urinal fixture has a pipe trap, service the trap and drain as for lavatories and sinks.

Shower Heads

Shower heads which supply an uneven or distorted stream can usually be repaired by removing the perforating face plating and cleaning the mineral deposits from the back of the plate with fine sandpaper and steel wool. Free clogged holes with a small diameter tool or by blasting with compressed air.

Lavatories, Wash Fountains, Service Sinks, Bathtubs

Drain stoppages usually result from obstructions in the trap or in the line beyond the trap. First use a plumber's force cup; if the obstruction is stiff in the trap and is not cleared by the plunger, clean the trap by inserting a wire or snake through the clean-out plug at the bottom of the trap. If the trap is not fitted with such a plug, remove it. Protect the finish of the packing nut with adhesive tape or wrap a cloth around the jaws of a wrench. Do not use a heavy steel coil spring snake to clear under lavatories, wash fountains, sinks, or bathtubs. Use a flexible type of wire or spring snake which will easily follow the bends in the trap.

Valves

Valves control the flow of fluid in pipes. There are certain types of valve problems that tend to be more frequent. They are as follows:

- Bad control elements
- Broken or cracked parts
- Broken or distorted springs
- Excessive leakage
- Sticking or jamming

The principal types of valves are gate valves, globe and angle valves, check valves, diaphragm valves, and plug or ball valves. Wear occurs more frequently in globe and check valves; however, certain features are built into these types to provide for maintenance and renewal of parts. Gate valves are used where then is minimal operations; for this reason, they do not wear out quickly, and are not, as a rule, provided with maintenance features of globe and check valves. The diaphragm valve may be considered low in maintenance; normally the only part subject to wear can be replaced without removing the valve body from the line. Valves that are seldom used often become hard to operate, especially in hot water lines, hard water lines or other lines in which deposits of scale or solids tend to collect. They should be operated periodically to prevent scale and passage of solids that may have collected on the disk, plug, or ball closure seat. Valves that will not properly open, close or throttle not only waste fluids, but are hazardous. They should be corrected by checking the bypass or by providing relief in the plumbing systems.

Refer to Figure 14.4 for inspection and routine maintenance procedures for valves.

Valve Failures

If a valve leaks slightly when shut off, open it a little to flush particles that may keep it from seating. Do not try to force it shut with a bar, because this can completely ruin the seat. Small globe valves in bypass assemblies should not be shut off abruptly. Inlet and outlet pressures should be equalized before closing or opening large valves since uneven pressure may become great enough to warp the gate or disk and jam the valve. Solids cause clogging and internal wear in valves; solids and fluids make proper valve closure almost impossible. The particles become embedded in nonmetallic disks and linings, causing an imperfect seal. If solids are present in the fluids, the piping system must be equipped with fittings that can handle the solids or with an adequate strainer. If a strainer is used, it should be located ahead of the valves and cleaned on a periodic basis. Corrosion, rust, and a build-up of deposits will freeze the moving parts and then the valves may not work.

Valves and Accessories

Item	Suggested Maintenance	Frequency
Manually Operated Valves:		
Gate Valves:		
Distribution system valves	Locate, check operation, lubricate stem packing; if packing leaks, dig up valve and tighten packing gland or replace packing; check stem alignment; check for broken stem or stripped stem or chewed nut.	5
Valve bypass	Check for position, inspect and lubricate.	5
Gears	Check for position, correct any deficiencies.	5
Vault	Check condition, clean, check masonry; make repairs, as necessary.	5
Treatment Plant Valves:	Operate inactive valves.	4
	Lubricate as required (including gears).	6
	Replace or resurface leaking valve seats.	7
	Lubricate chain wheels.	4
Butterfly Valves:	Check valve stem for watertightness, and adjust, if necessary.	5
	Check operation and inspect for tight closure.	6
Rotary Valves:		
Cone valves (and ball valves)	Operate; lubricate metal-to-metal contacts in pilot mechanism; lubricate packing glands; lubricate all parts of seating and rotating mechanisms.	3
Plug valves	Lubricate with lubricant stick.	3 or 4
	Operate all valves; check for corrosion and foreign matter between plug and seat; lubricate gearing.	
	Inspect; dismantle if necessary; clean, wire brush, remachine plug and body or replace if condition is beyond remachining.	6
Corporation cocks	Remove and replace whenever necessary.	7
Globe Valves	Operate valve to prevent sticking; check for leakage, adjust packing nut and replace packing if necessary.	4
	Check valve closure for tight shutoff; if valve does not hold, remove valve stem and disk and regrind seat and disk.	5
Diaphragm Valves	Operate valve; check valve stem and lubricate as necessary; check for tight closing.	4
	Check diaphragm for cracks; renew as necessary.	6
Sluice Gates	Operate inactive gates; lubricate stem screws and gears.	3
	Clean valve with wire brush and paint with corrosion-protective paint.	6
	Check seating wedges on valves seating against pressure.	6
Power-Operated Valves:		
Hydraulic-cylinder-operators:		
Hydraulic cylinder	Check through one valve operation cycle.	3
Piston rod and tell-tale rod	Oil packing; tighten packing gland if leakage exists; replace packing if necessary.	3
Wasteline discharge	Check for water flow when valve is wide open and shut; if leakage occurs, disassemble valve and piston, check leathers for wear and replace as necessary.	3
Cylinder and piston	Disassemble; inspect for scoring and corrosion; check cup leathers; polish any scored areas; remove corrosion products from piston surfaces and cylinder heads.	6
Pneumatic valve operators	Check packing and air hose; lubricate as necessary.	3
	Check piston, cylinder and leathers; clean and maintain similar to hydraulic valve operators.	6
Valve operator pilot controls	Check control through one full cycle of operation.	3
	Lubricate pins, linkage, packing glands and adjustment rod threads as necessary; remove corrosion products; check for leakage and repair.	3
	Disassemble; inspect unit and clean strainers; examine diaphragm for failure; regrind or replace worn valve seats.	6

Figure 14.4

357

Valves and Accessories

Item	Suggested Maintenance	Frequency
Automatic Valves:		
Air-release valves, valve unit	Remove valve from service; inspect float for leaks, and pins and linkage for corrosion; remove corrosion products; clean orifices.	6
Vault	Inspect for condition of masonry, steps and manhole covers; repair as necessary.	6
Altitude Valves:		
Pilot controls	Inspect and lubricate.	3
Valve unit and operator	Disassemble; inspect hydraulic cylinder and repair; inspect valve, repair and paint, as necessary.	6
Check Valves	Inspect the closure control mechanism (if any); clean and adjust as necessary; check pin wear if balanced disk type; check seating on ball type. Disassemble; clean, reseat and repair, as necessary.	7
Float Valves	Inspect float; repair as necessary. Inspect valve and valve operating mechanism.	3 6
Pressure-Regulating Valves	Inspect, clean, adjust, disassemble and repair as necessary (see manufacturer's instructions).	6
Valve Accessories:		
Gear boxes	Lubricate gears (see manufacturer's instructions). Check gear operation through full operating cycle; listen for undue noise, etc. Check housing for corrosion; paint as necessary.	3 or 4 5 6
Valve boxes	Clean debris out of box; inspect for corrosion; check alignment and adjust as necessary.	5
Floor stands	Lubricate stem and indicator collars. Inspect condition; clean and paint.	4 6
Valve position indicators:		
Post indicators	Lubricate.	4
Electric position indicators	Check contact points, wiring, etc.	6

Key to Frequency: 1. Daily 5. Semi-annually
2. Weekly 6. Annually
3. Monthly 7. Variable — see manufacturer's specifications
4. Quarterly

Figure 14.4 (*continued*)

Repair Procedures for Valves

When repairing a valve, follow the steps listed below.

1. Disassemble the valve according to the manufacturer's recommendations.
2. Clean the valve according to the manufacturer's recommendations.
3. Inspect and make a decision as to whether to repair or to replace the entire valve or any one or all of its components or parts.
4. Reassemble the valve according to manufacturer's recommendations.
5. Test according to manufacturer's recommendations.

If the cost of the parts requiring replacement in the valve is 40 percent or more of the cost of a new unit, replace the entire unit with a new one.

Valve Packing

Listed below are some of the common reasons for valve packing failure.

- Dirt
- Poor stem finish
- Wrong packing size
- Improper packing installation
- Loss of lubricant from the packing

To pack a valve, follow the procedure listed below.

1. Carefully disassemble the valve, remove handle, follower nut, and gland.
2. Measure the outside diameter of the stem and of the gland. Subtract these two dimensions and divide by 2. If this dimension falls between the standard sizes, select a packing to the next largest 1/16 of an inch.
3. Remove the old packing with a packing hook. Face the hook away from the stem to avoid shaft damage.
4. Clean the stem, stuffing box, and gland. If oil, grease, and graphite have been used as lubricants, it may be necessary to use a solvent to remove all traces of gum deposits.
5. Inspect the stem and stuffing box for damage and replace or repair damaged parts.
6. Wrap the packing stocks snugly around the shaft, cut the packing ring with a sharp cutting tool on the stem or on a metal rod of the same diameter. A 45 degree cut will make an excellent packing joint if carefully done, but straight cuts can also be made. Always inspect this first ring to ensure the joint meets. Undersized packing rings provide a leakage path.
7. Cut the remainder of the rings on a miter board using the master ring as a pattern. Twist the rings about one and one-half turns in the direction of coiling on the spool.
8. Install the first ring; stagger the joint of each succeeding ring 90 or 120 degrees relative to the first joint so that no joint is directly behind the joint of the preceeding ring.
9. Carefully tamp each ring in place and compress each layer of the same amount. Tamping tools for various sized valves will reduce repacking time.
10. Reassemble the valve.

During packing, the valve should be in a neutral position to prevent damage that may result if the valve is completely open or closed. Examination of the used packing will give clues as to the condition of the valve, and may be a means of solving future packing problems. Wear on the outside diameter of the packing reveals the rings have been rotating with the shaft or were too closely packed. Bulges in the side of the ring or rings reveal adjacent rings were cut too short. Rings next to the gland which are badly deformed reveal improper installation of packing and excessive pressure use.

Faucets

Continual turning on and off of the faucet wears the washer on the stem. This eventually causes the faucet to drip. Repair dripping faucets by replacing the washer. First, shut off the water supply to the faucet at the nearest valve and then remove the faucet handle, bonnet, and stem. Put adhesive tape around the bonnet or wrap a piece of cloth around the jaws of the wrench to prevent them from marrying the finish to the bonnet. Inspect the washer on the end of the stem. If the circular impression of the washer is centered around the bib screw, replace the washer with one of the same size and type. If the impression is off center, trim and recenter the seat with a seat-reforming cutter of the proper size and type. When installing a new washer, be sure to use a bib screw that will secure the washer firmly without bottoming in the hole and with ample threading engagement. If the bib screw is too long, its head will not hold the washer securely and may result in noise when the faucet or valve is operated. A bib screw that is too short may loosen and result in vibration and noise when the faucet or valve is operating, or if tightened too much, may pinch and distort the washer. Stop leakage at the bonnet, replace the gasket stem and the bib gasket.

Pumps

Pump maintenance procedures vary according to the type of pump. Generally, bearings and sealing devices operate in similar ways, regardless of the pump type. Maintenance for bearings usually means periodic lubrication only. Packing glands require only minor adjustments. Mechanical seals require only a visual inspection for leakage. The only time seals require special attention is when they are used in chemical or special pumps. See Figure 14.5 for typical pump problems and causes. Refer to Figures 14.6 through 14.10 for methods of pump maintenance for specific problems.

Centrifugal pump maintenance problems are generally confined to the casings, wearing rings, and impellers.

Pump casings cause problems when corrosion or incompatible materials are handled in the pump. It is important that pump casings are designed of materials that are corrosion-resistant. The casings that are corroded completely through can be patched without any decrease in pump capacity.

The problems encountered with pump impellers are corrosion, abrasions, and physical damage. Selection depends on the shape of the material of the impeller for the fluid that is being pumped.

Pump wearing rings require the most maintenance in a centrifugal pump. The most important way to minimize ring maintenance is to select the proper wearing ring material and ensure that it is in complete adjustment.

Vertical Turbine Pumps

Vertical turbine pump maintenance problems are similar to those of centrifugal pumps.

Rotary Pumps

The impellers of rotary pumps are subjected to very little corrosion by abrasive materials. Impeller wear is caused by the fluids being pumped, if for any other reason. The major maintenance problems with rotary pumps are in the bearings and seals. To extend bearing life, maintain proper alignment between the incaps and pump casing.

Reciprocating-type rotary pumps major maintenance problems are focused on the packing glands or seals of the connecting rod and the piston ring. Worn piston rings or packings are quickly indicated by the lack of fuel pressure while the pump is running. First, check the pump valve before replacing the piston rings. A leaky packing gland at the end of the connection rod might also cause the pump to act as if it had worn piston rings.

Refer to the information shown in Figure 14.11 for methods of correcting typical pump problems.

Pumps: Problems and Causes

Problem	Causes
Pump does not deliver water	Pump not primed.
	Pump or suction pipe not completely filled with water.
	Suction lift too high.
	Air pocket in suction line.
	Inlet of suction pipe insufficiently submerged.
	Suction valve not open or partially open.
	Discharge valve not open.
	Speed too low.
	Wrong direction of rotation.
	Total head of system higher than design head of pump.
	Parallel operation of pumps unsuitable for such operation.
	Foreign matter in impeller.
Insufficient capacity delivered	Pump or suction pipe not completely filled with water.
	Suction lift too high.
	Excessive amount of air or gas in water.
	Air pocket in suction line.
	Air leaks into suction line.
	Air leaks into pump through stuffing boxes.
	Foot valve too small.
	Foot valve partially clogged.
	Inlet of suction pipe insufficiently submerged.
	Suction valve only partially open.
	Discharge valve only partially open.
	Speed too low.
	Total head of system higher than design head of pump.
	Parallel operation of pumps unsuitable for such operation.
	Foreign matter in impeller.
	Wearing rings worn.
	Impeller damaged.
	Casing gasket defective permitting internal leakage.
Insufficient pressure developed	Excessive amount of air or gas in water
	Speed too low.
	Wrong direction of rotation.
	Total head of system higher than design head of pump.
	Parallel operation of pumps unsuitable for such operation.
	Wearing rings worn.
	Impeller damaged.
	Casing gasket defective permitting internal leakage.

Figure 14.5

Pumps: Problems and Causes (continued)

Problem	Causes
Pump loses prime after starting	Pump or suction pipe not completely filled with water. Suction lift too high. Excessive amount of air or gas in water. Air pocket in suction line. Air leaks into pump through stuffing boxes. Inlet of suction pipe insufficiently submerged. Water-seal pipe plugged. Seal cage improperly located in stuffing box, preventing sealing fluid from entering space to form the seal.
Pump required excessive power	Speed too high. Wrong direction of rotation. Total head of system higher than design head of pump. Total head of system lower than pump design head. Foreign matter in impeller. Misalignment. Shaft bent. Rotating part rubbing on stationary part. Wearing rings worn. Packing improperly installed. Incorrect type of packing for operating conditions. Gland too tight resulting in no flow of liquid to lubricate packing.
Stuffing box leaks excessively	Seal cage improperly located in stuffing box, preventing sealing fluid entering space to form the seal. Misalignment. Shaft bent. Shaft or shaft sleeves worn or scored at the packing. Packing improperly installed. Incorrect type of packing for operating conditions. Shaft running off center because of worn bearings or misalignment. Rotor out of balance resulting in vibration. Gland too tight resulting in no flow of liquid to lubricate packing. Excessive clearance at bottom of stuffing box between shaft and casing, causing packing to be forced into pump interior. Dirt or grit in sealing liquid, leading to scoring of shaft or shaft sleeve.
Packing has short life	Water-seal pipe plugged. Seal cage improperly located in stuffing box, preventing sealing fluid from entering space to form the seal.

Figure 14.5 *(continued)*

Pumps: Problems and Causes (continued)

Problem	Causes
	Misalignment.
	Shaft bent.
	Bearings worn.
	Shaft or shaft sleeves worn or scored at the packing.
	Packing improperly installed.
	Incorrect type of packing for operating conditions.
	Shaft running off center because of worn bearings or misalignment.
	Rotor out of balance resulting in vibration.
	Gland too tight resulting in no flow of liquid to lubricate packing.
	Failure to provide cooling liquid to watercooled stuffing boxes.
	Excessive clearance at bottom of stuffing box between shaft and casing, causing packing to be forced into pump interior.
	Dirt or grit in sealing liquid, leading to scoring of shaft of shaft sleeve.
Pump vibrates or is noisy	Pump or suction pipe not completely filled with water.
	Suction lift too high.
	Foot valve too small.
	Foot valve partially clogged.
	Inlet of suction pipe insufficiently submerged.
	Operation at very low capacity.
	Foreign matter in impeller.
	Misalignment.
	Foundations not rigid.
	Shaft bent.
	Rotating part rubbing on stationary part.
	Bearings worn.
	Impeller damaged.
	Shaft running off center because of worn bearings or misalignment.
	Rotor out of balance resulting in vibration.
	Dirt or grit in sealing liquid, leading to scoring of shaft or shaft sleeve.
	Excessive grease or oil in antifriction-bearing housing, or lack of cooling, causing excessive bearing temperature.
	Lack of lubrication.
	Improper installation of antifriction bearings (damage during assembly, incorrect assembly of stacked bearings, use of unmatched bearings as a pair, etc.)
	Dirt getting into bearings.

Figure 14.5 (continued)

Pumps: Problems and Causes (continued)

Problem	Causes
	Rusting of bearings because of water getting into housing.
	Excessive cooling of water-cooled bearing resulting in condensation in the bearing housing of moisture from the atmosphere.
Bearings have short life	Misalignment.
	Shaft bent.
	Rotating part rubbing on stationary part.
	Bearings worn.
	Shaft running off center because of worn bearings or misalignment.
	Rotor out of balance resulting in vibration.
	Excessive thrust caused by a mechanical failure inside the pump or by the failure of the hydraulic balancing device, if any.
	Excessive grease or oil in antifriction-bearing housing or lack of cooling, causing excessive bearing temperature.
	Lack of lubrication.
	Improper installation of antifriction bearings (damage during assembly, incorrect assembly of stacked bearings, use of unmatched bearings as a pair, etc.).
	Dirt getting into bearings.
	Rusting of bearings because of water getting into housing.
	Excessive cooling of water-cooled bearing resulting in condensation in the bearing housing of moisture from the temperature.
Pump overheats and seizes	Pump not primed.
	Operation at very low capacity.
	Parallel operation of pumps unsuitable for such operation.
	Misalignment.
	Rotating part rubbing on stationary part.
	Bearings worn.
	Shaft running off center because of worn bearings or misalignment.
	Rotor out of balance resulting in vibration.
	Excessive thrust caused by a mechanical failure inside the pump or by the failure of the hydraulic balancing device, if any.
	Lack of lubrication.

Figure 14.5 (continued)

Pumps and Priming Systems: Maintenance

Item	Suggested Maintenance	Frequency
Rotary Pumps:	General pump maintenance.	7
Overhaul	Check clearances of moving parts, packing, bearings and alignment; paint exterior and interior surfaces.	6
Reciprocating Pumps:		
Drives	See Figure 14.24.	
Piston water pumps	Determine slippage, i.e., check for percent delivery below 90%.	6
	Dismantle and inspect; check valves, valve seats and springs; reface or replace worn parts; remove packing and repack; check alignment; check plunger or rod for scoring; clean and paint interior and exterior.	6
Diaphragm pumps	Dismantle; clean diaphragm and check for cracks or leaks; check drive mechanism.	6
Sludge Pumps:		
Packing	Lubricate.	1
	Renew packing; clean cylinder walls.	7
	Check packing gland adjustment.	2
Bearings & gear	Lubricate.	1
Transmission	Check and fill gear transmission with oil; drain moisture.	3
	Change oil.	4
Shear pins	Check adjustment; set eccentric.	2
	If pin fails, check and remove cause; replace pin.	7
Ball valves	Check balls for wear, replace, if necessary.	4
Gear motor and electric motor	Check and service.	5
Eccentric	Remove brass shims and take up babbitt bearing; run it for one hour.	6
Screw Pump:		
Seals	Check lubrication.	1
Bearings and filters	Lubricate.	2
	Replace wick filters; drain and flush bearings, refill with oil.	6
Packing glands	Check for leakage and adjust or repack as necessary.	2
Float mechanism	Clean and adjust.	3
Sump and Bilge Pump:		
Sumps	Clean out floating objects and accumulated deposits.	2
Pump operation	Check switch contacts, sump emptying rate, pumping heads, misalignment; service pump as necessary.	2
Packing gland	Check for leakage; adjust or repack as necessary.	2
Wiring and controls	Check connections for breaks or other damage to wiring and insulation; check for short circuits, worn contacts, defective float switch, tighten connections and parts; clean, repair or replace parts as necessary.	3
Floats	Check for bent rod, binding or other damage; correct undesirable conditions; apply light oil to moving parts; check alarm system.	3
Bearings	See section on bearings	3
Supports	Check and adjust as necessary; check alignment.	3
Motor	See Chapter 13.	
Priming Systems:		
Pressure	Check valves.	6
Ejector	Check ejector, electrodes, solenoids, relays, valves; repair or replace as necessary.	3
Vacuum pump	Check air compressors, pressure controls, relief valves, float valves, diaphragm cutouts, floats, etc.; repair or replace worn parts as necessary.	5
Automatic	Check general performance and adjust; repair or replace parts as necessary.	2

Key to Frequency:
1. Daily
2. Weekly
3. Monthly
4. Quarterly
5. Semi-annually
6. Annually
7. Variable — See manufacturer's specifications

Figure 14.6

Well Pumps: Maintenance

Item	Suggested Maintenance	Frequency
Turbine Well Pumps:		
Oil-lubricated pump bearings	Make certain tubing and lubricators are filled; check solenoid oilers for operation, and see that they are filled; check oil lubricator for underwater bearings; oil feed should be 3 to 4 drops per minute.	1
Water-lubricated pump and bearings	Check for level of water in prelubricating tank.	1
Impeller	Check impeller for efficiency; adjust if necessary.	4
	Check for pitting; check clearances.	7
Bowls and Waterways	Check for pitting, wear and corrosion.	7
Overhaul	Check clearances, binding, vibration and loss of capacity; paint.	7
Air Lift Pumps	Lubricate compressor; blow down air receiver; check air-receiving tank valve; check cooling water supply; balance air supply to wells.	1
	Check compressor for oil pumping; check air filters; replace as necessary; check for air contamination; check for oil in well.	2
Ejector Pumps:		
Centrifugal pump	See Figure 14.7.	6
Ejector assembly	As directed, remove ejector, foot valve and screen; check for wear and corrosion; check foot valve seating; check ejector throat; repair as necessary, or consult manufacturer.	6
Body	Paint exterior and interior, as necessary.	6
Reciprocating Well	Check pump delivery; remove pump and repair, if delivery has decreased; check valves and cup leathers.	5
	Check pump jack.	6

Key to Frequency:
1. Daily
2. Weekly
3. Monthly
4. Quarterly
5. Semi-annually
6. Annually
7. Variable — See manufacturer's specifications

Figure 14.7

Turbine Well Pumps: Maintenance

Problem	Possible Cause	Solution
Pump won't start	Impeller locked.	Check for sand; try raising or lowering impellers; backwash; if unsuccessful, pull pump.
	Trash in casing.	Check for such material; pull pump and clean.
	Corrosion or growths in pumps out of service for long periods.	Back flush with acid, chlorine and/or Calgon.
	Bearing friction	Change oil; check tube tension nut for tightness; check for bent shaft; replace if necessary; check anchoring of pump to make certain it hasn't caused bending or distortion of pump.
	Faulty motor or wiring.	Check circuit breaker or fuses for open line; if starter, reset; disconnect motor from pump to see if it starts; check motor wiring against wiring diagram.
	Burned out fuses.	Test voltage on phases of motor terminals.
	Tripped overload relay.	Reset.
	Low voltage, defective motor, defective starting equipment.	Repair as necessary.
	Impellers not adjusted properly.	Set high enough to allow room for shaft stretch due to hydraulic thrust; adjustment should allow shaft to turn freely.
	Well cave-in	Call outside contractor.
Pump doesn't deliver water	Wrong rotation.	Switch power loads.
	Speed too slow.	Check voltage and frequency of power supply; check for excessive bearing friction, corrosion or obstruction of impeller; if belt drive, check for slippage or wrong pulley size.
	Pump not primed.	Vent well to atmosphere to eliminate vacuum at pump suction; make certain there is a net positive suction heat on the pump.
	Pump parts failure.	Look for broken shaft, broken down assembly and loose column-pipe joints; tighten all impellers.
	Pumping head too great.	Make sure discharge valves are open and that check valves don't stick; also, check water table level and other factors.
	Clogged suction.	Clear clogged suction pipe, strainer or impeller by backwashing; if well screen is plugged.
Pump uses too much power	Wrong pump.	Check model number, capacity, rating, etc.
	Overspeeding.	Check for high frequency, voltage, wrong gear ratios.
	Wrong lubricant.	Check oil, motor bearings; check water-lubricated bearings.
	Tight packing.	Adjust to allow enough leakage to lubricate shaft.
	Impeller run.	Adjust impeller height.
	Wrong rotation.	Adjust two power leads.
	Misalignment, tight bearings, vibration in pump or casing.	Check and adjust.

Figure 14.8

Centrifugal, Mixed-Flow, Axial-Flow and Turbine Pumps: Maintenance

Problem	Possible Cause(s)	Solution
Pump Performance	Run field tests of pump operation, capacity, etc.	7
Centrifugal-Type Pumps (horizontal and vertical)		
Lubrication		
Hand-oiler	Check daily; adjust according to manufacturer's specifications.	1
Solenoid-oiler	Check leads, oil flow, rate of oil flow and adjust as necessary.	1
Ball-bearings	Check oil level in housing oil well; fill when necessary.	1
Packing glands (grease sealed)	Check grease cup; maintain proper pressure	1
Enclosed-shaft type bearings	Refill oil cup when necessary.	2
Ball-thrust bearings	Add fresh grease.	3
	Change grease where pump operates more than 50 times daily.	4
Guide bearings	Add grease when necessary.	3
Ball bearings on open shaft Type O.	Add grease when necessary.	3
Shaft bearings	Drain lubricant, wash out wells with kerosene and add fresh lubricant.	4
Bearing housing	Open housing, flush clean with kerosene and add fresh, clean grease.	4
Universal joint couplings	Lubricate when necessary	5
Bearings		
Antifriction bearings	Check bearing temperature with a thermometer; change, or adjust lubrication, if temperature so indicates.	3
	Drain lubricant; wash wells and bearings with kerosene.	4
	When pump is dismantled, check condition of bearings and bearing rate; replace if necessary.	6
Sleeve bearings	Check bearing temperature with thermometer; take necessary corrective step if bearing is too hot.	3
	Drain lubricant; wash out wells and bearings with kerosene; measure bearings with feeler gauge; take corrective action if necessary; refill with oil.	4
	Check oil rings; replace if necessary.	4
	When pump is dismantled, check condition of bearings; replace as necessary	6
Packing		
Stuffing box	Check for free movement of gland; oil the gland and bolts; check for excessive leakage in order that gland adjustment will not reduce; repack, if necessary.	5
Gland assembly	Check; if leakage is excessive, tighten gland; if tightening does not alleviate condition, remove packing and inspect sleeve.	2
Sealing Water System		
Packing glands	Check for leakage; adjust to product slow, constant drip.	1
	Disassemble sealing-water lines and valves to make certain that water passages are clear.	6
Rotary Seals	Consult manufacturer's instructions.	6
Shafts	When pump is dismantled, check for alinement and distortion; examine for pitting, erosion and scoring; repair or replace when necesary.	6
Shaft sleeves	Inspect for wear; replace as necessary.	6
Wearing or Sealing Rings	Check for wear and clearance; replace if clearance is greater than 0.05 inch on a diameter of the ring.	6

Figure 14.9

Centrifugal, Mixed-Flow, Axial-Flow and Turbine Pumps: Maintenance (continued)		
Problem	**Possible Cause(s)**	**Solution**
Impeller	Remove and inspect thoroughly; check for deposits or scaling and remove; check for erosion and cavitation effect; remedy cause of cavitation if possible; weld cavitation pits if severe; machine and balance.	6
Casing	Clean and paint with enamel-like finish coating; set up routine program	6 or 7
Pump Shutdown	For normal shutdowns, or for overhaul, shut all valves, drain pump; disconnect motor switch and remove fuses; before startup, drain oil and grease and refill; for dismantling, drain oil and remove grease.	7
Overhaul	Drain pump, etc; dismantle with care, check all parts, repair or replace worn parts; check valves (foot and check); reassemble parts with care.	6 or 7
Mixed-Flow Pumps	See manufacturer's instructions.	7
Axial-Flow Pumps	See manufacturer's instructions.	7
Turbine-Type Pumps	See manufacturer's instructions.	7

Key to Frequency:
1. Daily
2. Weekly
3. Monthly
4. Quarterly
5. Semi-Annually
6. Annually
7. Variable — See manufacturer's specifications

Figure 14.9 (*continued*)

Pumping Station: Maintenance

Item	Suggested Maintenance	Frequency
Buildings	See Chapter 1–12.	
Valves	Check operation.	4
	Check packing glands, tighten or replace packing as necessary.	4
	Paint.	4
	Dismantle and repair as necessary.	6
Piping	Paint, color code or otherwise identify	6
Pump controls		
Manual		
Valves	See above.	
Switches	Check for binding and arcing; clean, repair or replace.	5
Instruments	See Figure 14.20.	
Automatic	Observe operation through one complete cycle and adjust as necessary.	2
Water level	Check float, cable (or rod); adjust and oil, where necessary.	4
	Check switches, solenoids, contractors, relays and alarms; clean, adjust and repair or replace as necessary.	5
	Overhaul and test all parts for proper functioning; disassemble and clean components; check packing of valves and pivots; repack as necessary; reassemble, test and return to service.	6
	Apply hydrostatic test to pressure tank if one is used; check water level.	5
	Check bubble pipe, compressor system, and diaphragm switch; clean, repair or replace worn parts.	4
Pressure	Check all components for proper operation.	4
	Overhaul and test all parts.	6
	Check instruments	
Standby Equipment	Check on same schedule as operating units; overhaul.	7
Housekeeping	Clean up after all maintenance operations on all equipment.	7

Key to Frequency:

1. Daily	5. Semi-annually
2. Weekly	6. Annually
3. Monthly	7. Variable — see manufacturer's specifications
4. Quarterly	

Figure 14.10

Methods of Correcting Typical Pump Problems

Cause	Correction
1. Failure of Bearings	
A. Chips or other contaminants in bearings.	Clear.
B. Coupling misalignment	Realign pump and motor shafts.
C. Electric motor shaft end play. Driving or hammering coupling on or off pump shaft.	Eliminate all end play on electric motors. Couplings should be a slip fit onto the pump shaft.
D. Excessive or shock loads.	Reduce operating pressure.
E. Inadequate lubrication.	Consult with manufacturer's specifications.
F. Overhung load.	Pumps are not designed to handle high overhung loads or side thrust on the drive shaft. Provide outboard bearings.
G. Pump running too fast.	Reduce speed.
2. Lack of Volume	
A. Dirt of chips blocking pump.	Clean.
3. Motor Overloading	
A. Inadequate motor-overload protection.	Install larger capacity unit and bigger heaters.
B. Low voltage.	Larger wire leads.
C. Motor not correctly sized for pressure and volume requirements.	Check and correct.
D. Motor wired for wrong voltage.	Check motor leads for proper voltage connections.
E. Starting pump at full pressure.	Reduce pump pressure before starting motor. Readjust pressure after motor is up to required speed.
F. Starting pump at full pressure and volume.	Use higher starting torque motor or start pump with valve closed so no oil will flow.
4. Oil Seal Leakage	
A. Abrasives on pump shaft.	Protect shaft material.
B. Damaged packing.	Replace oil seal assembly.
C. Improper fluid.	See manufacturer's specifications.
D. A misalignment of coupling.	Realign pump and motor shafts.
E. Temperature and oil exceed recommendation.	Provide cooling.

Figure 14.11

Methods of Correcting Typical Pump Problems (continued)

Cause	Correction
5. Pump Noise Excessive	
A. Air bubbles in intake line.	Provide reservoir with baffles. All pressure return lines to reservoir must be below oil surface, and on opposite side of the baffle from intake lines.
B. Air leak in suction side of pump.	Tighten as required. Replace seal if needed.
C. Coupling misalignment.	Realign pump and motor.
D. Filter or strainer restricted.	Clean or replace filter or strainer.
E. Low oil level.	Fill reservoir to specified point
F. Pressure ring worn.	Replace.
G. Pump airbound.	Stop and bleed pump.
H. Pump rotating in wrong direction	Correct—check nameplate.
I. Pump running too fast.	Reduce speed.
J. Reservoirs not vented.	Vent reservoir through breather air filter.
K. Restricted flow through suction and piping.	Check suction piping fittings for blockage.
L. Type of oil incorrect.	Use recommended quality of oil.
6. Pump Not Delivering Oil	
A. Air leak in suction line.	Tighten joints and apply good pipe compound, nonsoluble in oil.
B. Bleed-off in other section of circuit.	Check for open center valves or other controls connected with a tank port.
C. Broken pump shaft or rotor.	Replace broken parts.
D. Key sheared at rotor or coupling.	Replace key.
E. Loose pressure on adjusting screw.	Tighten.
F. Not enough oil in pump.	Check volume of oil that will free-flow through pump inlet.
G. Oil too heavy for proper priming.	Thinner oil should be used, per recommendations for given temperatures and service.
H. Oil level low.	Maintain at recommended level.
I. Pump running in wrong direction.	Correct; check nameplate.
J. Pump airbound.	Back out cylinder oil flows freely, and pump free of air.
K. Pump running too slowly.	Increase speed.
L. Suction line or suction filter blocked.	Clean lines and blocked filter. Periodic checks should be made as a preventive maintenance precaution.

Figure 14.11 (continued)

Bearings

Bearing failures are one of the principal causes of the failure of the motor, pumps, and other mechanical equipment. Causes of bearing trouble are varied, although the best indication is overheating or abnormal noise. Faulty bearings are a cause or a result of more extensive trouble.

Bearing Maintenance

Bearing maintenance is not limited to the replacement of worn out bearings, but requires periodic inspections, lubrication, and proper bearing protection.

It is important that all bearings be lubricated properly. Proper lubrication means not only routine lubrication, but also using the specified lubricant in the correct amount in accordance with the manufacturer's instructions.

Figure 14.12 outlines the causes of and recommended corrections for overheated and noisy bearings.

Other Plumbing Systems

Following are checklists for use during maintenance and inspection of other plumbing systems.

Gas Distribution Systems

Refer to the checklist in Figure 14.13, as well as other sections in this chapter, for the inspection, maintenance, and repair of gas distribution systems.

Earthquake Valves and Pits

Do not tap or jar valve and keep all wrenches and metal objects from making contact with valve. Refer to the checklist shown in Figure 14.14 for inspection and maintenance of earthquake valves and pits.

Methods of Correcting Typical Pump Problems (continued)

Cause	Correction
7. Pump Pressure Not in Accordance with Specifications	
A. Defective pressure gauge.	Repair or replace.
B. Oil bypassing to reservoir.	Test circuit pressure progressively look for open-center valves or other valves open to reservoir.
C. Pressure adjusting screw set too low.	Reset.
D. Pump not delivering oil.	See specifications.

Figure 14.11 (continued)

Methods of Correcting Typical Bearing Problems

Cause	Correction
1. Bearings—Overheated	
A. Bearings selected with inadequate internal clearance for conditions.	Replacement bearing should have identical marking as original bearing for proper internal clearance.
B. Housing bow out of round. Housing warped. Excessive distortion of housing. Undersized housing bore.	Relieve pinching of bearing. Be sure pedestal surface is flat, and shims cover entire area of pillow block base.
C. Incorrect type of grease or oil, causing lubricant breakdown.	Comply with manufacturer's specifications.
D. Incorrect linear or angular alignment of two or more coupled shafts with two or bearings.	Correct alignment by pillow blocks. Be sure shafts are more coupled in straight line—especially when three or more bearings operate on one shaft.
E. Insufficient grease in housing.	Lower half of pillow block should be filled.
F. Low oil level. Loss of lubricant through seal.	Oil level should be at center of lowest ball or roller in bearing.
G. Unbalanced load. Housing bore too large.	Rebalance machine. Replace housing with one having proper bore.
2. Noisy Bearings	
A. Bearings selected with inadequate internal clearance for conditions	Replacement bearing should have identical marking as original bearing for proper internal clearance.
B. Distorted shaft and other parts of bearing assembly.	Replace the bearing.
C. Foreign matter entering bearing housing.	Clean out bearing housing Replace worn-out seals or improve seal design to obtain adequate protection of bearing.
D. Housing bore out of round. Housing warped. Excessive distortion of housing. Undersized housing bore.	Relieve pinching of bearing. Be sure pedestal surface is flat, and shims cover entire area of pillow block base.
E. Rotating seals rubbing against stationary parts.	Check running clearance of rotating seal to eliminate rubbing. Correct alignment.
F. Rubbing of shaft shoulder against bearing seals.	Remachine shaft shoulder to clear seal.

Figure 14.12

Turbines

Items to be included in the inspection and routine maintenance of turbines, both small and large, are described in the checklist shown in Figure 14.15.

Filtration, Aeration, Clarification, and Chlorination Equipment

Figures 14.16 through 14.29 include inspection and maintenance items for filtration, aeration, clarification, and chlorination equipment.

Figure 14.13

Inspection, Maintenance, and Repair Checklist for Gas Distribution Systems

_____ **Exposed Piping:** leakage, loose connection, rust, corrosion, other damage (use soap solution for leakage test).

_____ **Underground Piping:** signs of leakage such as brown grass strips across lawns, dead trees, and shrubs when other plants are green in areas of buried piping. Determine condition when buried piping and valves are exposed for alteration and repair.

_____ **Pressure-Regulating and Reducing Valves:** leakage, loose connections, rust, corrosion, defective operation; possible damage from freezing resulting from water infiltration into valve pits; clogged vent to outside of building, defective screen.

_____ **Gas Cutoff Valve on Face of Building:** improper operation, difficult access, not clearly visible.

_____ **Meters:** loose connections, leakage, corrosion, rust, broken glasses, moisture behind glasses, defective gaskets, dirty or difficult to read, settlement.

_____ **Pits:** debris, water, cracks and leakage, rust and corrosion. Clean debris and water from pit; caulk all cracks; clean and paint all exposed parts, except moving parts.

Figure 14.14

Inspection and Maintenance of Earthquake Valves and Pits

_____ **Pit:** debris, water, cracks, leakage, rust, corrosion. Sweep, cover and surrounding area clean before removing cover.

_____ **Valve:** not absolutely level, leakage around all joints, gaskets, and cap plug on top.

_____ Prior to operational check and tripping of valve, examine appliances having pilot lights and record their location; open all windows in rooms where these appliances are located.

_____ Close remote shutoff valve, if so equipped, or tap lightly by hand only (never with metal object) until by listening close to valve, the sound indicates the valve pendulum has dropped and is not in closed position.

_____ Light gas burner at fixture nearest valve and test for leakage through valve. If flame dies out after a few minutes, the valve is tight.

_____ Remove cap plug at top of valve, and nut valve by lifting up on exposed stem.

_____ Replace plug, tighten, and test for leakage around plug.

_____ Make sure that all pilot lights are relit.

Inspection and Maintenance Checklist for Turbines

Small Turbines
Running Inspection
____ **Vibration: excessive.**
____ **Bearings:** excessive temperature.
____ **Lubrication:** dirty strainer, dirty or emulsified oil, improper level of oil or grease.
____ **On Rings:** improper operation.
____ **Carbon Ring Seals:** excessive steam leakage.
____ **Constant Speed Machines:** improper speed (check with tachometer).
____ **Excess Pressure-Governed Turbines:** improper pump discharge pressure.
____ **Trip Valve:** improper operation (overspeed unit and check speed with tachometer).

Shutdown Inspection
____ **Dismantle:** remove turbine casing, bearing covers, governor housing, throttle valve, trip valve, and housings or covers that enclose moving parts.
____ **Clearances and Moving Parts:** dirty, no freedom of movement.
____ **Rotor:** remove deposits on blading.
____ **Buckets and Blades:** misalignment, corrosion, pitting, erosion. Clean, repair, or replace as required.
____ **Carbon Rings or Other Types of Seals:** excessive wear, breakage; journal in contact with carbon rings for wear and corrosion (particularly in units that have been idle for extended periods and those in which steam leakage has been noted around shaft).
____ **Reassemble** and perform running inspection as noted above.

Large Turbines
Running Inspection
____ **Vibration:** excessive.
____ **Lubrication:** dirty or emulsified oil, improper level in sump, excessive temperature at bearing inlet and outlet, inadequate operation of emergency oil pump, inadequate operation of governor, throttle trip valve, and bleeder nonreturn tripping mechanism when lube oil pressure is low.
____ **Steam Leakage:** excessive steam leakage through carbon rings or labyrinth seals.
____ **Pressure:** improper stage pressures, improper gland seal pressure under varying loads.
____ **Clearances:** measure axial and radial clearances when built-in micrometer is installed.
____ **Rankine (nonbleeding) Steam Rate:** check at rated load and half load, compare with manufacturer's standards.
____ **Exhaust Casing Relief Valve:** inadequate operation.
____ **Varying Loads and Extraction Demands:** inadequate operation of governor.
____ **Reduced Load:** inadequate operation of bleeder nonreturn valves.
____ **No Load:** inadequate operation of governor and trip valve when speed is raised to rpm of overspeed tripping mechanism. (Usually 10 percent above rated speed.)

Figure 14.15

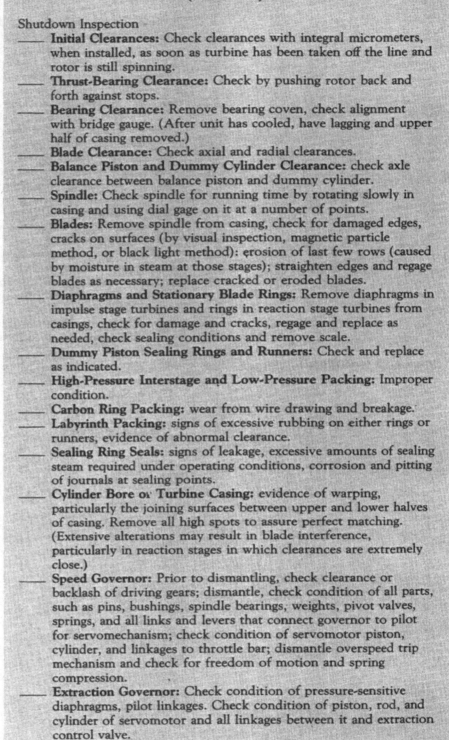

Inspection and Maintenance Checklist for Turbines
(continued)

Shutdown Inspection

_____ **Initial Clearances:** Check clearances with integral micrometers, when installed, as soon as turbine has been taken off the line and rotor is still spinning.

_____ **Thrust-Bearing Clearance:** Check by pushing rotor back and forth against stops.

_____ **Bearing Clearance:** Remove bearing coven, check alignment with bridge gauge. (After unit has cooled, have lagging and upper half of casing removed.)

_____ **Blade Clearance:** Check axial and radial clearances.

_____ **Balance Piston and Dummy Cylinder Clearance:** check axle clearance between balance piston and dummy cylinder.

_____ **Spindle:** Check spindle for running time by rotating slowly in casing and using dial gage on it at a number of points.

_____ **Blades:** Remove spindle from casing, check for damaged edges, cracks on surfaces (by visual inspection, magnetic particle method, or black light method); erosion of last few rows (caused by moisture in steam at those stages); straighten edges and regage blades as necessary; replace cracked or eroded blades.

_____ **Diaphragms and Stationary Blade Rings:** Remove diaphragms in impulse stage turbines and rings in reaction stage turbines from casings, check for damage and cracks, regage and replace as needed, check sealing conditions and remove scale.

_____ **Dummy Piston Sealing Rings and Runners:** Check and replace as indicated.

_____ **High-Pressure Interstage and Low-Pressure Packing:** Improper condition.

_____ **Carbon Ring Packing:** wear from wire drawing and breakage.

_____ **Labyrinth Packing:** signs of excessive rubbing on either rings or runners, evidence of abnormal clearance.

_____ **Sealing Ring Seals:** signs of leakage, excessive amounts of sealing steam required under operating conditions, corrosion and pitting of journals at sealing points.

_____ **Cylinder Bore or Turbine Casing:** evidence of warping, particularly the joining surfaces between upper and lower halves of casing. Remove all high spots to assure perfect matching. (Extensive alterations may result in blade interference, particularly in reaction stages in which clearances are extremely close.)

_____ **Speed Governor:** Prior to dismantling, check clearance or backlash of driving gears; dismantle, check condition of all parts, such as pins, bushings, spindle bearings, weights, pivot valves, springs, and all links and levers that connect governor to pilot for servomechanism; check condition of servomotor piston, cylinder, and linkages to throttle bar; dismantle overspeed trip mechanism and check for freedom of motion and spring compression.

_____ **Extraction Governor:** Check condition of pressure-sensitive diaphragms, pilot linkages. Check condition of piston, rod, and cylinder of servomotor and all linkages between it and extraction control valve.

Figure 14.15 (continued)

Inspection and Maintenance Checklist for Turbines
(continued)

____ **Throttle Trip Valve:** Prior to dismantling, determine time required for valve to become fully seated from instant of tripping hand lever; dismantle valve, check condition, take measurements of valve seat, guide, and bushings to determine amount of wear, and trueness of valve stem; valves showing signs of leaking valve seat should be ground in with valve in place.

____ **Throttle and Expansion Control Valves:** Dismantle chests and remove throttle bar from housing; check condition of valve disks and seats; check distances between seating surfaces and points of engagement on lift bar; grind in seats when necessary; check clearances in guides and bushings because excessive clearances at these points causes valve vibration when valve is in throttling position; on reassembling glands, adjust carefully to avoid friction.

____ **Bearings:** Measure clearances between journal and bearing bore, using strips of lead wire; note any irregularities in bearing metal, or wiping or scoring of journal; repour and machine bearing metal as necessary; halves should be machined down if clearance is too great; if insufficient clearance, shims should be inserted in joint; check adjustment at bearings with bridge.

____ **Couplings:** defective wearing parts, clogged oil channels, and holes in flexible couplings.

____ **Operational Check:** After completion of shutdown inspection, reassemble and place in operation. Perform running inspection noted above.

Figure 14.15 (continued)

Filtration Equipment: Maintenance

Item	Suggested Maintenance	Frequency
Gravity Filters		
Filter media	Inspect surface for unevenness, sink holes, cracks, algae, mud balls, or slime.	3
	Dig out sand and gravel at craters of appreciable size; locate and repair underdrain system breaks.	7
	Chlorinate to kill algae growths.	7
	Probe for hard spots and uneven gravel layers; if present, treat filter with acid.	4
	Check wash water rise rate and sand expansion during backwashing.	5
Filter media	Check sand condition for grain size growth; sample sand, determine weight loss on acid digestion, and run sieve test; acid-treat if necessary, or replace sand, if necessary.	5
Gravel	Check elevation of gravel surface.	3
	Examine gravel for crustation, cementation, alum penetration, mud balls, if necessary, remove, clean and relay gravel.	
Underdrain system	Remove sand from area of 10 sq. ft., and inspect 2 sq. ft. area of gravel (or more); if underdrains are deteriorated, remove all sand and gravel, repair underdrains, replace gravel and sand.	6
	If porous underdrain, clogged by alum floc, treat with 2% NaAH solution for 12 to 16 hours.	7
Wash water troughs	Check level and elevation, adjust.	4
	Check for corrosion; if present, dry troughs, wire brush, and paint.	5
Operating Tables	Clean table (console or panel) inside & out.	2
Cables	Adjust tension	7
Hydraulic lines (or pneumatic)	Check for leakage.	7
4-way valves	Adjust, tighten packing glands or add new packing.	3
Transfer valves	Lubricate with grease.	3
Valve position indicator	Adjust, if necessary.	3
4-way transfer valves	Disassemble, clean, lubricate and replace worn parts.	6
Table	Paint inside.	6
Rate Controllers		
Direct-acting	Clean exterior, check diaphragm leakage, tighten packing, check freedom of movement and zero differential.	2
Diaphragm pot	Disassemble, clean and replace.	6 or 7
Controller mechanism	Disassemble, service; clean ventury; paint surfaces needing protection.	7

Key to Frequency:
1. Daily
2. Weekly
3. Monthly
4. Quarterly
5. Semi-annually
6. Annually
7. Variable — see manufacturer's specifications

Figure 14.16

Aeration Equipment: Maintenance

Item	Suggested Maintenance	Frequency
Waterfall Type Aerators (Cascade)	Inspect aerator surfaces; remove algae; clean.	1
Waterfall Type Aerators (Tray)	Clean and repair trays; clean coke or replace.	5
Waterfall Type Aerators (Cascade)	Repair or replace surfaces as necessary.	6
Injection Aerators		
Porous ceramic plate or tube	Check discharge pressure; if clogging is evident, dewater tank, clean diffusers.	7
Porous ceramic plate or tube	Drain aeration tank, check for joint leaks, broken diffusers, clogging.	5
Water side of ceramic diffusers	Clean with acid, in place, or remove and soak in acid.	5
Air side of ceramic diffusers	If plates are clogged by iron oxide, treat with HCl; if clogged by soot, oil, etc., remove diffusers and burn.	5
Saran wound diffusers	Clean by scrubbing with soap or detergent.	5
Nozzles	Clean nozzles inside and out.	5
Blowers and Accessory Equipment		
Compressor or blower	Lubricate, check output pressure for indications of clogging.	1
Air filters	Clean, repair or replace.	2
Compressor or blower	Open, inspect, clean repair and paint exterior surfaces.	6

Key to Frequency:
- 1. Daily
- 2. Weekly
- 3. Monthly
- 4. Quarterly
- 5. Semi-annually
- 6. Annually
- 7. Variable — see manufacturer's specifications

Figure 14.17

Clarification Equipment: Maintenance

Item	Suggested Maintenance	Frequency
Mixing Basins	Drain, wash down walls, flush sediment to waste line; repair spalled spots on walls and bottom; check valves on sluice gates; lubricate and paint valves as necessary.	5
Baffled Mixing Chambers	Clean baffles and repair as necessary.	5
Flocculator Basins	Check paddle rotation to ascertain if any flocculators are inoperative.	3
	Clean and lubricate drive, bearings, gears and other mechanical parts; check underwater bearings for silt penetration; replace scored bearings.	5
Rapid (or Flash) Mixers	Check paddles; clean bearings and drive shaft; lubricate and paint as necessary.	5
Revolving Sludge Collector	Drain tank, check submerged parts.	5
Operating parts	Lubricate	1 or 2
Speed reducers & oil baths	Remove water and grit, replace oil, as necessary.	2
Drive head	Lubricate — do not overlubricate.	1
Worm-gear	Check oil level.	2
	Drain water from housing.	3
Turntable bearings	Lubricate.	3
	Change oil.	3
Chains	Drain off water, add oil as necessary.	3
	Change oil.	5
Annular ball bearings	Lubricate.	
Center bearings, shaft bearings, bushings, etc.	See manufacturer's instructions.	4
Tank Equipment	Tighten bolts and nuts; check for excessive wear, flush and backblow sludge line, check motors, couplings, and shear pins, check rakes, clean and paint equipment.	6
Conveyor-Type Collector Basins	See above, and consult manufacturer's instructions.	7
Upflow Clarifier	See manufacturer's instructions.	7
Cathodic Protection Rectifier-type	Check exterior and interior for condition; see manufacturer's instructions; repair, replace, or paint as necessary.	3
Sacrificial anodes	Check anode condition and all connections, and replace as necessary.	3

Key to Frequency:
1. Daily
2. Weekly
3. Monthly
4. Quarterly
5. Semi-annually
6. Annually
7. Variable — see manufacturer's specifications

Figure 14.18

Feeders, Softening Units, and Drives

Refer to Figures 14.20 through 14.24 for information on the inspection and maintenance of feeders, softening units and drives.

Fire Hydrants

Figure 14.25 covers the inspection and routine maintenance of fire hydrants.

Automatic Controls and Flow Measuring Devices

Information on the inspection and maintenance of automatic controls and instruments and flow-measuring devices can be found in Figures 14.26 and 14.27.

Chlorination Equipment: Maintenance

Inspection	Action	Frequency
Operation Maintenance	Insert new lead gasket in chlorine valves or tubes to cylinder or equipment.	7
Condensation on Chlorine Cylinders	Ventilate.	7
Chlorine Leak Detection	Use unstoppered bottle of acqua-ammonia to detect leaks; repair immediately.	1
Gas System	Disassemble, clean and replace faulty parts in piping, meters, valves and tubing.	1
Chlorine Valves	Open and close valves to assure that none are frozen in a set; check stuffing boxes, and repair or replace faulty valves or packing.	1
Chlorine Solution Tubes	Look for locations of potential leaks, and for iron and manganese deposits; if Fe or Mn are present, treat with a solution of hexametaphosphate in makeup water.	6
Chlorine Feeder Water Supply	Clean water strainers and pressure reducing valves; adjust float valves and ejector capacity.	3
Hard-Rubber Threads, Valves and Parts	Disassemble or operate; use graphite grease to prevent freezing; hand tighten only, do not use tools.	4
Vacuum Relief	Clean out any obstructions.	1
Cabinet and Working Parts	Clean all parts where accumulation may interfere with proper operation.	2
Overhaul	Disassemble and clean all parts thoroughly; paint cabinet inside and out; examine parts and repair or replace as needed; use care in choice of cleaning agents and lubricants.	6
Direct-Feed Chlorinators	Use same procedures as for solution feed machines where they apply.	

Key to Frequency:		
	1. Daily	5. Semi-annually
	2. Weekly	6. Annually
	3. Monthly	7. Variable — see manufacturer's specifications
	4. Quarterly	

Figure 14.19

Dry Chemical Feeders: Maintenance

Inspection	Action	Frequency
Dry Feeders	Remove chemical dust accumulations; check feeder performance; check for loose bolts; clean solution tank of accumulated sediment; lubricate moving parts.	1
Drive Mechanisms and Moving Parts	Service and lubricate.	4
Calibration	Check feed-rate accuracy and adjust, as necessary.	3
Overhaul Feeders	Thoroughly clean feeder and feeding mechanism; paint; service and lubricate drive mechanisms, bearings; clean and paint solution tanks.	6
Feeders Out-of-Service	Clean, remove all chemicals from hopper and feeder mechanism.	7
Disk Feeders	Clean rotating disk and plow.	3
Oscillating Feeders	Check and adjust mechanism and adjustable stroke rod.	3
Rotary Gate Feeders	Clean pockets of star feeder and scraper.	3
Belt-Type Feeders	Check vibratory mechanism, tare-balance, feeding gate, belt drive and belt; calibrate delivery.	3
Loss-in-Weight Feeders	Check feeder scale sensitivity, tare-weight and null balance.	3
Screw Feeders	Clean screw, check ratchet drive or variable speed drive.	3
Lime Slakers	Clean dust-removal and vapor-removal equipment; remove clinkers.	1
	Clean equipment; wipe off feeder; check operation of vapor removal equipment; clean compartments.	2
	Repair agitators, stirrers and heat exchanger baffles.	3
	Overhaul slakers; drain and clean; check sides and bottom for wear and repair; paint lids, exterior, inside top edges; check for leaks and incrustation in heat exchanger; check thermometer accuracy; clean and lubricate bearings; clean and repair controls, floats, piping, screens, valves and vapor removal equipment; paint where necessary.	6
Dust Collectors		
Motors	Lubricate motors.	3
Filter bags	Check condition and attachment. Securely attach bags; replace damaged or torn bags.	3

Key to Frequency:
1. Daily
2. Weekly
3. Monthly
4. Quarterly
5. Semi-annually
6. Annually
7. Variable — see manufacturer's specifications

Figure 14.20

Liquid and Solution Chemical Feeders: Maintenance

Item	Suggested Maintenance	Frequency
Pot Feeders		
Flow through pot	Determine amount of chemical fed to ascertain if flow through pot is effective.	1
Sediment trap	Clean trap and check needle valve.	3
Chemical pot	Clean pot and orifice.	5
Overhaul	Clean and paint pot feeder and appurtenances.	6
Differential Solution Feeders		
Chemical storage tank	Inspect and clean.	5
Oil volume	Check and replenish.	5
Pitot tubes and needle valve	Check and replace as necessary.	6
All equipment	Paint as necessary.	7
Decanter or Swing-Pipe Feeders		
Swing-pipe	Check to make certain it does not bind.	3
Motor ratchet, pawl, reducing parts	Check and lubricate.	5
Overhaul	All parts to be inspected, cleaned, repaired and painted as necessary.	6 or 7
Rotating Dipper Feeders:		
Motor	Follow manufacturer's instructions.	7
Transmission	Change oil after 100 hr. operation.	7
	Drain and flush, clean interior and refill.	5
Shaft bearings	Lubricate.	2
Drive chain	Clean, check alignment; check sprocket teeth; lubricate chain and sprockets.	3
Agitator	If used, clean and lubricate according to manufacturer's instructions.	7
Belt drives	Check alignment, tension and inner cords of belt drives.	3
Dipper and float valve	Check dipper clearance and adjust float valve setting.	5
Proportioning pumps (hypochlorinators)		
Operator inspection	Inspect sight feeders, rate of flow, piping, joints	1
Feeder	Clean feeder.	2
Solution tank	Clean.	3
Linings	If cracks occur, special linings should be repaired.	6
Overhaul	Disassemble, clean and overhaul.	6

Key to Frequency:
1. Daily
2. Weekly
3. Monthly
4. Quarterly
5. Semi-annually
6. Annually
7. Variable — see manufacturer's specifications

Figure 14.21

Ion-Exchange Softening Units: Maintenance

Item	Suggested Maintenance	Frequency
Softener Unit		
Shell	Clean and wire brush; paint.	6
Valves and fittings	Check for obstructions, corrosion and fastness.	4
Multiport valves	Lubricate with grease; follow directions for lubrication procedure.	5
Ion-exchange medium	Check bed surface for dirt, fines and organic growths; remove foreign matter and add resin to desired level.	4
Gravel	Probe through resin to determine gravel surface; level gravel surface with rake during backwash flow; replace gravel when caked, or if resin is being lost to effluent; wash and grade gravel and place in four separate layers; use new lime-free gravel at discretion of inspector.	4
Underdrains	Check pressure drop through underdrains; if necessary, remove manifold or plate underdrains; clean and replace.	6 or 7
Regeneration Equipment		
Salt storage unit	Clean tank as necessary to remove dirt.	7
Brine tank	Clean out dirt and insolubles; allow to dry; paint both exterior and interior surface.	5
Ejector	Clean, disassemble, check erosion and corrosion; clean clogged piping; assemble and replace.	6
Operating Conditions		
Flow rates	Check rate of flow through bed; adjust controls to optimum rate, depending on type of resin.	4
Backwash rates	Check rate and adjust controls to optimum rate.	4
Pressure	Check difference between inlet and outlet pressures; if undesirable changes in pressure drop have occurred, seek cause and remedy.	4
Efficiency	Compare total softening capacity with previous inspection; determine cause of decrease, if any, and remedy situation.	4
Out-of-Service Softeners	Drain; keep synthetic resins damp; do not regenerate before draining.	7

Key to Frequency:
1. Daily
2. Weekly
3. Monthly
4. Quarterly
5. Semi-annually
6. Annually
7. Variable — see manufacturer's specifications or current operation standards.

Figure 14.22

Lime-Soda Softening Equipment: Maintenance		
Item	**Suggested Maintenance**	**Frequency**
High Rate Softening Devices:		
Motors	See Chapter 13.	
Valves and piping	Check for leaks; and check operation.	7
Time clock and controls	Check for proper operation and adjust.	3
Tank unit	Drain, clean and inspect wearing parts.	5
Chemical feed lines	Clean, if necessary, to maintain flow.	5
Recarbonation Units		
Burners, Compressors, Gauges, and Traps	Check and adjust to ensure optimum operation.	1
Drier, scrubber and traps	Check absorbent and replace as necessary; adjust spray; clean out connecting piping; clean gas traps.	3
Entire unit	Check for internal and external corrosion; repair as necessary; paint or apply protective coatings.	5

Key to Frequency:

1. Daily
2. Weekly
3. Monthly
4. Quarterly

5. Semi-annually
6. Annually
7. Variable — see manufacturer's specifications

Figure 14.23

Drives: Maintenance

Item	Suggested Maintenance	Frequency
Flat Belts		
Operator inspection		7
Renewal	Check belt condition; renew as necessary	4
Alignment	Check pulley alignment; adjust as necessary.	7
V-Belts		
Operator inspection		
Tension	Check and adjust as necessary.	2
Alignment	Check and adjust as necessary.	7
Renewal	Check belt condition and renew as necessary.	4
Chain Drives	Check slack and adjust.	4
	Check alignment and adjust.	5
	Lubricate exposed, enclosed, underwater chains, as directed; do not lubricate elevator or conveyor chains handling dusty chemicals.	5
	Check condition of all parts; turn chain or sprocket or renew parts as necessary.	5
	Check spare part storage.	5
Right-Angle Drives	Check lubricant flow when starting drive.	7
	Drain and change lubricant.	4
Variable-Speed Drives	Operator inspection, cleaning and drying as necessary.	1
V-belt	Check belt; be sure it runs level; adjust as necessary; adjust tension for shutdown if necessary.	2
Drive mechanism	Lubricate shifting screw and variable shaft.	2
	Lubricate thrust bearings on constant speed shaft.	2
	Lubricate frame bearings on variable-speed shaft.	4
Reducer	Check oil level; refill or change oil as necessary.	3
Gear Reducers, Gear Motors and Speed Changers:		
Flooded type	Check operation, oil level, etc.	1
	Add oil as necessary.	2
	Change oil.	5
	Check for oil leaks, corrosion and mechanical defects; remedy undesirable conditions.	6
Nonflooded types	Check operation; clean unit.	1
	Check gear fit; align gear; check for wear, wash with kerosene; adjust, repair or renew parts as necessary; relubricate.	6
	Lubricate according to manufacturer's instructions.	7
Drive Bearings:	Check for vibration ; adjust or repair as necessary.	1
	Inspect; disassemble, wash with kerosene; check surface defects; check clearance; adjust, repair or renew as necessary.	6
	Check for looseness in equipment not overhauled annually.	6
Temperature checks	Check by hand.	1 or 2
	Check with thermometer; correct undesirable condition.	4
Lubrication	Lubricate bearings without reservoirs or wicks.	1
	Add oil to oil-lubricated bearings.	3
	Drain, flush and refill.	4
	Lubricate oil wick bearings.	4
	Check and lubricate grease-lubricated bearings.	4
Water-cooled bearings	Drain, clean cooling jacket, refill.	6
Ball and Roller Bearings	Check seals, renew leaking seals.	3
	Clean and repack bearings.	6
Sleeve Bearings	Inspect and lubricate; avoid excessive oiling.	4

Figure 14.24

Drives: Maintenance

Item	Suggested Maintenance	Frequency
Thrust Bearings:		
Ball and Roller Thrust		4
Kingsbury	Lubricate.	3
	Check clearance between shoes and thrust sleeve; adjust or renew.	5
	Drain oil, flush, refill.	6
	Disassemble, clean, repair or renew disk.	7
Underwater Bearings	Check clearance, lubricate, and adjust.	5
Couplings	Align, check, clean and adjust.	5
Shear Pins	Renew and sheared pins after determining and eliminating the cause of failure.	7
	Remove pin; check condition; operate drive to wear off corrosion on shear surfaces.	4
	Check pin storage.	6

Key to Frequency:
1. Daily 5. Semi-Annually
2. Weekly 6. Annually
3. Monthly 7. Variable — See manufacturer's specifications
4. Quarterly

Figure 14.24 (continued)

Fire Hydrants: Maintenance

Item	Suggested Maintenance	Frequency
Dry-Barrel Hydrants	Check drain valve to be sure it opens.	3
	Where ground water level rises into barrel, plug drain valve and dewater barrel by a pump.	3
Wet-Barrel Hydrants	Check packing glands and valve seats; repair as necessary.	4
Pit-Type Hydrants	Check for water accumulation; dewater as necessary.	3
All Hydrants		
On dead ends	Flush; check barrel after flushing.	5
Not on dead ends	Flush; check barrel after flushing.	6
	Check water flow.	6
	Repair as necessary; if main shutdown is required, notify fire department.	7
In winter	Check for freezing; thaw, if necessary.	2
Leakage tests	Inspect all places where leaks might occur; repair as necessary.	6
Valve Parts		
Operating nut	Check for rounded corners; replace as necessary; lubricate.	6
Nozzle threads	Check for damage; replace as necessary.	6
Chains	Check for paint fouling; clean.	6
Flow Tests	Determine hydrant flow.	6

Key to Frequency:
1. Daily 5. Semi-Annually
2. Weekly 6. Annually
3. Monthly 7. Variable — See manufacturer's specifications
4. Quarterly

Figure 14.25

Automatic Controlled Instruments: Maintenance

Item	Suggested Maintenance	Frequency
Sensing Devices:		
Flow measurements	Check and clean.	4
Level measurement:		
Floats	Check and clean.	3
Bubble pipe	Check and clean.	3
Electrical probe	Check contacts, wiring and electrical connections; repair as necessary.	4
	Check probe surfaces and calibration; clean, repair or renew as necessary.	5
Diaphragm	Check and clean.	3
Pressure measurement:		
Diaphragm	Disassemble and check for condition and leaks; also, clean, adjust, repair or renew as necessary; check calibration.	6
Bourdon tube	Check calibration, clean and adjust as necessary.	6
Manometer	Clean tubes and gauge unit as necessary.	5
	Check mercury level and add mercury if necessary; clean mercury if necessary.	6 or 7
Electrical quantities measurement	Clean necessary; check zero setting and adjust as necessary; consult manufacturer's instructions.	3 or 4
Position determination	Set floats; check contact points for arcing; clean as necessary.	3
	Polish contacts; renew if necessary.	5
Weight measurement	Clean equipment.	2
	Check knife edges; also where applicable, check poise, belt adjustment and lubricate.	3
	Disassemble, clean, reassemble, lubricate, adjust and calibrate.	6
Temperature measurement	Clean.	4
Transmission System:		
Mechanical	Direct links — make certain pulleys, drums, cable, etc., work freely and are not corroded; clean, lubricate and adjust.	4
Hydraulic	Pressure links — blow down pressure lines, make certain there are no restrictions; correct adverse conditions.	5
Pneumatic Transmitter	Flush liquid side of air relay units; clean, if necessary check diaphragm; check air-input orifice, clean, blow out moisture traps.	1
	Disassemble, repair or renew as necessary.	7
Link	Check connecting tubing for condition; check nozzle system for leaks.	5
Electrical Transmitter	Service transmitter; check signal interval length over instrument range.	3
	Check mercury switch and magnet; adjust as necessary.	4
	Remove old lubricant, add new.	5
Link	Check wires whenever necessary.	7
Terminal Equipment:		
Indicators	Clean cover and glass of gauges.	5
	Check zero setting and calibration.	6
Mechanical transmission	Inspect and service as for transmitter.	4
Hydraulic transmission	Vent air from mercury wells; check pulley shaft, chain, cam, stuffing box and other parts.	2
	Check mercury wells; add new mercury if necessary; clean mercury if necessary.	6
Pneumatic transmission	Service on same schedule and in same manner as transmitter.	
Electrical transmission	Service generally on same schedule as transmitter.	
	Clean unit, especially dials.	5
	Check operation, adjust and repair as necessary.	6

Figure 14.26

Swimming Pools

Clean each day or at some interval established by the maintenance department. The promenade floor should be flushed down with a hypochlorite solution, and the pool should be cleaned as noted below.

1. First, brush the pool sides down with a stiff brush, disturbing the bottom dirt as soon as possible. Clean the bottom with a vacuum cleaner. Place the vacuum cleaner on the promenade floor and attach it to a float supporter hose with a tow-type suction head on its free end. Lower the head to one end of the pool floor and by means of its attached ropes, slowly and carefully pull from side to side (an operator on each side of the pool). The vacuum cleaner discharges into the promenade floor discharge. Excessive floating material can be overflowed to the gutter and diverted to waste disposal.

2. At intervals of about three to six months, or as necessary, empty the pool and scrub thoroughly on the sides and bottom with strong soap solution. All soap traces must be flushed out of the system before the pool is ready for filling.

To repair a specific swimming pool component or material, follow the procedures outlined in this book for the specific item.

Automatic Controlled Instruments: Maintenance		
Item	**Suggested Maintenance**	**Frequency**
Recorders	Clean pen; check ink flow; check cam cycle and pulley freedom.	2
	Check zero position; adjust and lubricate.	4
	Check contact points, armature, clutch, clutch cups, etc; clean, adjust, repair, or renew parts.	5
	Renew modular unit if necessary.	7
	Renew illumination lamp as necessary.	7
Totalizers	Inspect, clean, adjust or repair on same schedule as recorders.	
Combination T-I-R	Check, clean, adjust or repair on same schedule as individual components.	
Automatic Controls:		
Level	Check and clean.	3
Flow	See level and pump control.	
Pressure	See pump control.	
Treatment plant:		
Chemical feed	Consult manufacturer's instructions.	7
Filter rate	Consult manufacturer's instructions.	7
Filter backwash	For automatic controls, consult manufacturer's instructions.	7
Time cycle	Consult manufacturer's instructions.	3
Supervisory	Consult manufacturer's instructions.	3

Key to Frequency:
1. Daily
2. Weekly
3. Monthly
4. Quarterly
5. Semi-Annually
6. Annually
7. Variable — See manufacturer's specifications

Figure 14.26 (continued)

Flow Measuring Devices: Maintenance

Item	Suggested Maintenance	Frequency
Venturi-Type Devices:		
Annular chamber	Flush and clean annular chamber, throat and inlet; purge trapped air from chamber and connecting piping; flush piezometer pressure taps.	4
Exterior	Clean and paint as necessary.	6
Interior	Check interior for corrosion; dismantle, clean and restore; for flanged joints, check possible intrusion of gasket into interior; replace if necessary.	6
Orifice Plates	Remove plate, dress off roughness; flush sediment traps.	4
Pitot Tube	On permanent installations, check tips and clean.	4
Gentile flow tube	Check instrument taps; flush if necessary.	4
Velocity-Type Meters	Check operation: check for noise.	3
Meter pit	Clean, remove water before freezing season.	5
Exterior	Paint as necessary.	6
Interior	Check for worn parts; repair or replace as necessary.	7
Proportional Meters	Check on same program as velocity meters.	7
Magnetic Meters	Check electrical connections.	6
Compound Meters	Check large-flow component on same schedule as velocity-type meters. Check small-flow component on same schedule as volume meters.	3
Head-Area Meters		
Weirs	Check weir edge to make certain it is clean.	1
	Check and open breather pipe, if any.	3
	Drain weir to check evenness of water break-over; check for tuberculation or corrosion; dress-off rough spots.	6
Parshall flume	Check throat section to be sure it is clean and free of growths.	3
	Clean stilling well and connecting pipe.	4
Open-flow nozzles	Check and clean as necessary; purge connecting taps.	3
Volume Meters	Check operation; check for noise	3
	Check mounting and alignment.	6
Meter Pit	Check, clean, remove water to protect against freezing.	6
Measuring Unit	Check for possible hot water damage.	6
Unit Parts	Check for worn parts, repair or replace as necessary; clean and brighten.	7

Key to Frequency:
1. Daily
2. Weekly
3. Monthly
4. Quarterly
5. Semi-Annually
6. Annually
7. Variable — See manufacturer's specifications

Figure 14.27

HEATING AND COOLING AND OTHER MECHANICAL SYSTEMS

This chapter covers the inspection, maintenance, and repair of the items listed below.

- Ventilating and Exhaust Systems
- Heating Systems
- Space Heaters
- Portable Fans
- Water Heaters
- Pneumatic Systems
- Steam Distribution Systems
- High Temperature Water Distribution Systems

Mechanical Ventilation Systems

Duct Systems

When it is necessary to replace or change sections of ducts, use the same-size new duct unless design provisions specify a change. When determining whether a different size can be used, check the air pressure in the duct. A major source of trouble in duct systems, often resulting in the complete unbalancing of the system, is tampering with dampers and grills by unauthorized persons.

Use the instruments listed below to determine whether a system is in balance or not.

- A swinging vein anemometer (veolometer)
- A rotating vein anemometer
- A pilot tube
- Manometer

Maintenance of Ducts: A preventive maintenance program for ducts includes checking for the items listed below.

- Leakage loss caused by loose clean-out doors, warped surfaces, broken joints, holes worn in elbows or other parts, and poor connections in two fans.
- Accumulation of foreign matter, such as dirt, lint, and condensation of oil or water vapor.
- Duct hangers, anchors, and supports. Correct all loose connections so that the ducts are properly and tightly supported.
- Check corrosion. Protective paints prevent corrosion (any commercial paint with a red lead base is good for this purpose).
- Strippable coatings of plastic-like, fire-retardant chemicals are available for protecting interior surfaces of ducts and for easy cleaning.

- Passage of ducts through combustible partitions should be checked closely for clearance and insulation to isolate ducts from combustible materials and reduce or eliminate a fire hazard.

Periodic checking is necessary for correction of air leakage at joints and for wearing of the flexible material. Control leakage by tightening connections and replacing worn fabric.

Fans and Blowers

Extent of servicing depends on severity of operation and location of the unit. Fans operating under dusty conditions, indoor locations, or corrosive atmospheres require more frequent attention than those handling clean air in a dry location. Check for the items listed below during the inspection process.

- Excessive vibration
- Lubricant level leaks
- Loose screws
- Proper tension and belt drives
- Bearing operation temperatures (refer to information on bearings in previous chapter)
- Present condition of unit relative to the accumulation of dirt
- Signs of corrosion
- Axial clearance of each centrifugal fan

If the problem appears to be the motor or any of its components, refer to the chapter on electrical maintenance.

For fans and blowers having adjustable sheaves, adjustment should be made in accordance with the manufacturer's instructions.

Couplings: Couplings require little attention if they are properly installed and kept in alignment. Most problems arise from poor original installations, which can damage the coupling, fan motor shaft, and bearings. Parallelism must be checked with a feeler gage at four points between the coupling valves at top, bottom, and both sides. A straight edge used at corresponding points on the outer surfaces of the two halves confirms concentric alignment.

Dampers and Louvers

Dampers are used most often in duct systems. Their purpose is to control air direction and volume. Maintenance personnel should learn the location of all dampers in a system so that future changes can be handled quickly. Furthermore, if damper settings are changed by unauthorized persons, maintenance persons should be able to locate the change and correct the condition.

Gravity Dampers: Maintenance personnel should periodically check the operation of gravity dampers. If for some reason the damper does not open when the fan is on, improper ventilation may damage the fan motor. Usually, cleaning and lubricating with grease is all that is required.

Automatic Dampers: Maintenance for automatic dampers is usually performed when air conditioning and heating equipment are checked. The damper should be cleaned and lubricated with grease, and the motor and thermostatic control device should be checked for proper operation.

Louvers: The terms louver and damper are often used synonymously. The above information on automatic dampers also pertains to automatic louvers. Louvers should be periodically checked and screens should be cleaned. Replace damaged parts and paint when necessary.

Fan Vibration: For causes and methods of correcting fan vibration, refer to Figure 15.1 and the manufacturer's instructions.

Fan Vibration: Problems and Solutions

Problems	Solutions
Hopping or bent pulley Spring shaft	Hold a rigidly supported pencil or piece of chalk close enough to slowly rotated pulley or shaft to barely touching. If line is continuous, pulley or shaft is okay.
Misalignment of pulleys or couplings	Check with a straight edge.
Dirt in pulley grooves	Pulley grooves must be smooth and concentric. Remove dirt with solvents.
Dirt or lumps on V-belts	Defective belts will have rough joints and should be replaced.
Dirty fan wheels	If fan is handling dirty air, wheel should be cleaned frequently.
Unbalanced motor or motor pulleys	Remove V-belt or disconnect coupling, and run motor to check for motor vibration.
Insufficient rigidity of fan platform	Platform must be steady enough to resist sway or give.
Loose set screws on fan wheel or pulleys Loose foundation bolts Loose bearing bolts Bearings loose on shaft	Tighten all bolts and set screws. Make certain pulley bore is not egg-shaped or over-size.
Wheel unbalance due to wear	Look for evidence of abrasion and corrosion. Replace parts that have become worn by friction. Scrape off corrosion, and recoat damaged surfaces with matching paint. Provide protective coatings for fan blade surfaces. Set screws, bolts, and shafts.
Damaged wheel	If wheel damage is due to foreign material passing through fan, wheel should be returned to factory for proper repair.

Figure 15.1

V-Belt Drives: The following list provides general instructions covering the maintenance of V-belt drives.

- Check sheaves for perfect alignment. Fans and motor shafts must be parallel, and the centers of the grooves on both sheaves must be in perfect alignment.
- Belt drives depend on friction and must be under some tension to operate properly. If loose, belt drives slip and overheat. If too tight, shafts may be strained and bearings stretched excessively. Make any adjustments according to the manufacturer's instructions.
- Protect V-belt drives from oil and grease. Use guards that allow some circulation of air around the drives.
- Purchase and install replacement drives for multi-type V-belt drives in matched sets and lengths in accordance with the manufacturer's instructions.
- All belts should be checked for equal tightness or the entire load may be carried by one or two belts.
- Selecting drives with too few belts or with sheaves that are too small is uneconomical. Frequent adjustment results in more rapid belt stretching and wear, increasing maintenance costs.
- Keep in mind that variable-pitch sheaves are intended to provide for speed regulation, not belt tension adjustment.

Exhaust Hoods: Maintenance personnel should make the following periodic checks, along with the corresponding follow-up action:

- Remove any accumulation of dust, dirt, or grease by spraying or washing with hot water, steam, or solvent.
- Repair or replace broken or cracked surfaces and poor connections to exhaust ducts. Check air pressure periodically. Reduced air pressure may be attributed to: belt slippage; wear on rotor or casing; accumulation of material on the fan wheel or in the fan housing; obstruction of air flow; incorrect direction of exhaust fan rotations; defects in exhaust piping, such as leakage losses and ducts; exhausted adhesive material due to condensation of oil or vapors on duct walls; and loss of suction due to additional exhaust openings having been added to the system since the original design installation.
- Clean greasy hoods and ducts using a solvent.

Filters: Dirt that accumulates in the filters lessens the amount of air passing through them. Therefore, it is important that the filters be serviced regularly. If using electronic filters, manufacturer's instructions should be followed exactly, both for safe operation and for maintenance procedures.

Natural or Gravity Ventilation Systems

Gravity ventilation operates effectively where high heat and good wind velocities are present. These ventilators are usually made of wood, or they may be made of galvanized iron, aluminum, or other types of metals. Apply the correct maintenance procedure for the material of which the ventilation system is comprised.

Wall, Pedestal, and Portable Fans

Wall, pedestal, and portable-type fans are designed with a four-blade propeller, are driven by a low-horsepower motor, and are powered by electricity. Baffles, dampers, veins, axes, doors, louvers, registers, protective grills, and bird and insect screens should be checked frequently. Replace missing parts and tighten loose connections which may cause fan vibration and noisy operation. Scrape corroded surfaces and repaint them with matching paint. Keep fans free of dust and dirt. Remove accumulations of grease and oil with solvent. Make sure that the protective grill does not interfere with the blades as the fan rotates. Refer to Chapters 13, 14, and other sections of this chapter for procedures for maintaining motor bearings and lubrication.

Refer to the checklist shown in Figure 15.2 when inspecting and maintaining plant ventilating and exhaust air systems.

Checklist for Inspection and Maintenance of Plant Ventilating and Exhaust Air Systems

_____ **Fire Hazards:** dust, dirt, soot, drippings, and grease deposits on hoods, on filters, and in systems; flammable materials on fans, guards, ducts, dampers, and discharge louvers.
Remove grease filters, degrease with solvents, sanitize in water at 180 degrees farenheit, and air dry.

_____ **Lubrication:** excessive bearing temperatures, inadequate lubrication of bearings and moving parts.
Lubricate as required, including grease dispensers or oil cup when less than half full. Do not overlubricate.

_____ **Rust and Corrosion:** damage from acid, chemical fumes, and rust. Remove rust and paint bare spots and corroded areas, where applicable.

_____ **Motors, Drive Assemblies, and Fans:** dust, dirt, and other accumulations; worn, loose, missing, or damaged connections and connectors; bent blades; worn or loose belts; unbalance or misalignment; excessive noise and vibration; end play of shaft; ineffective sound isolators; poor condition of motor windings and brush rigging.
Remove obstructions; tighten loose connections and parts; tighten loose belts, or replace multiple belts in sets when one is worn; replace defective brushes; make other minor repairs and adjustments.

_____ **Wiring and Electrical Controls:** loose connections; charred, frayed, broken, or wet insulation; short circuits; loose or weak contact springs; worn or pitted contacts; defective operation; wrong fuses; other deficiencies.
Tighten loose connections and parts; replace cords having wet insulation or where broken in two or more places, or braid that is frayed more than 6 inches; replace or adjust contact springs; clean contacts; replace defective or improper fuses; make minor repairs.

_____ **Fire Protective Devices:** incorrect temperature fusible links; improper setting of thermal unit or releasing device; excessive high-temperature setting of fan-stop device.

_____ **Steam and Hot Water Coils:** clogging; rust and corrosion; scale; leaking; loose connections and parts; bent fins; misalignment; water hammer; air-binding; nonuniform heat spread; open bypass valves; below-normal temperature readings; defective valves, traps, and strainers.
Remove rust and scale from heat-transfer surfaces; tighten loose connections and parts; replace leaking valve packing; close bypass valves; clean valves, traps, and strainers.

_____ **Electrical Heating Units:** pitted, burned, or dirty electrical contacts; short-circuited sections of elements; low voltage in electrical circuits; dirty reflective heat-transfer surfaces.
Test heating output at each step of multiple-step heating levels; clean contacts; clean heat-transfer surfaces; replace defective elements; tighten loose connections; make minor repairs and adjustments.

_____ **Ducts, Collectors, Smokepipes, and Hoods:** clogging; soot, dirt, grease, and other deposits; loose connections and parts; abrasions, wear, and deformations; lack of weathertightness of seams and joints.
Remove deposits; tighten loose connections and parts; make minor repairs and replacements.

_____ **Ducts in Heads and Showers:** leaking, broken, or poorly soldered seams and joints.

Figure 15.2

Heating Systems

Forced Warm Air Systems

To maintain blower motors, fans, and other moving parts, refer to Chapters 13, 14, and sections in this chapter. The main maintenance problem is balancing the system for continuous air circulation, which should be done at the time the forced warm-air heating system is installed. If the system becomes out of balance, either in-house personnel, using manufacturer's instructions, or an outside heating and air conditioning contractor should be contacted to rebalance the system. The main items to check when balancing the system are listed below.

- The two thermostats (one in the heated room and one in the warm air supply plenum).
- The capacity of the heat source.
- The adjustment of the registers.
- The speed of the blower motor.
- The temperature limit control or fuel limit switch.

Checklist for Inspection and Maintenance of Plant Ventilating and Exhaust Air Systems (continued)

_____ **Thermal Insulation and Protective Coverings:** open seams, breaks, missing sections; missing or loose fastening.
Replace or tighten fastenings; make minor repairs.

_____ **Baffles, Dampers, Vanes, Access Doors, Registers, Louvers, Bird and Insect Screens:** soot, dirt, dust, grease, and other deposits; broken, loose, or missing connections and parts; material defects; defective operation of movable parts; loose or poor fit of flashings in surrounding wall surfaces; hinge parts failure; improper seasonal or operating settings of dampers; inadequate air distribution at branch ducts.
Remove deposits; tighten loose or replace defective connections and parts; caulk around flashings and make weathertight; adjust damper; make other minor repairs and replacements.

_____ **Air Filters:** dust, grease, and other deposits; mining; improper fit.
Replace throw-away filters that are dirty or missing or that have an improper fit; wash permanent types; restore viscous coating in accordance with manufacturer's instructions.

_____ **Exhaust Air Systems:** lack of air and weathertightness; inadequate separation of solids from air stream; incorrect suction at intakes; defective operation of blast and interlocking gates, motor-driven dampers, solenoids, and like parts; vegetation destruction, air pollution or fire hazard from exhaust air; inadequacy of protective guards or warning signs.
Make minor repairs.

_____ **Guards, Casings, Hangers, Supports, Platforms, and Mounting Bolts:** loose, broken, or missing parts and connections; deformations; improper level; ineffective sound isolation.
Tighten loose connections and parts; adjust level; replace defective sound isolators; make minor repairs and replacements.

_____ **Bag Collection:** tears; leakage; need of cleaning.

_____ **Wet Collectors:** plugged spray nozzles or drain pipes; sludge accumulations; improper feedwater levels.

_____ **Air Velocity:** measure, compute velocity in cfm and compare with nameplate requirements.

Figure 15.2 (continued)

Gravity Warm Air Systems

Since there are no moving parts in the gravity warm air system except the dampers, maintenance problems are minimal. Most gravity flow systems now in service are old coal burners converted to gas or oil. The foremost consideration when maintaining these systems is ensuring proper firebox conditions, such as draft, fuel quality, and fire size. Since the burner is never built for the firebox in converted systems, unsafe conditions are more likely to occur. Consequently, check all units of this type periodically for proper draft and for the condition of the firebox. Although oversizing rarely occurs in conversion installations, the insulation of an old furnace frequently deteriorates and an overheat cutoff switch should be installed.

Steam Systems

Refer to the discussion on Steam Space Heaters in this chapter for maintenance procedures of steam systems.

Hot Water Systems

The items that should be checked in maintaining a hot water system are as follows:

- Entrapped Air—Each radiation unit should be checked periodically and accumulated air forced out the air vent. If this has to be done more than once a heating season, further investigation of the system should be made.
- Corrosion and Scale.
- Pumps and Mechanical Equipment—Refer to Chapters 13, 14, and other sections in this chapter.

Oil-Fired Burners

Oil-fired burners are common in both warm-air and hot-water heating systems. The degree of maintenance required depends on many factors, such as the quality of the fuel, type of burner, controls, and care by the occupant.

Pressure-Atomizing Burners

The problems with pressure-atomizing burners are (1) mechanical, (2) electrical, and (3) combustion-related.

Mechanical Problems: Refer to the manufacturer's instructions for routine inspection check points.

Oil pumping failure is a common problem and frequently causes trouble. When the burner is running and no fire results, perform the systematic check below.

- Check all valves on suction and feedlines to be sure they are open.
- Check supply of oil tank.
- Check vacuum on suction line.
- Check lines for obstructions.
- Check the viscosity of the oil. Be sure it is within the manufacturer's recommendations.
- If performing the above items does not locate the cause, break a union or otherwise open the discharge line from the pump. If no air or oil can be observed, prime the pump by filling the strainer with heavy lubricating oil or fuel oil. After the pump has emptied the strainer, shut off the burner and repeat, making sure each time to replace cap tightly.
- If the pump loses prime, check the suction for leaks.
- Check the line for tight unions between pump and tank to prevent possible air entrance.
- Check the suction stub into the tank, especially the connections. If the stub can be removed, cap and pressurize the entire suction line, thus revealing any leaks.
- Finally, remove and test the pump.

Failure of Electrical Apparatus: Refer to the manufacturer's instructions. If the motor does not run, follow the maintenance and repair procedures discussed in Chapter 13.

Combustion Trouble: Combustion-related problems and the maintenance procedures to correct them are noted below.

- If the fire is smoky, a dirty or worn atomizer, inadequate primary air, insufficient primary air pressure, or inadequate secondary air mixture are likely causes.
- Irregular stack temperature is the result of a down draft, which is caused by a leaky stack or an obstruction near the top of the stack. Other causes are a clogged burner nozzle, water in the oil, or pressure regulation valve set so low that oil delivery is not uniform.
- Causes of improper flames for high or low oil pressures are oil too viscous and not being properly preheated; or low primary air pressure.

Vertical Cup Burners
Check vertical cup burners during the off-heating season as outlined below.

- Remove the motor from the burner and clean the fire bowl pilot light.
- Remove and clean oil pot.
- Test trip bucket and check trip bucket valve packings.
- Clean main strainer mesh and strainer of the pump.
- Inspect burner and pump motor bearings. If it is noisy, lubricate and replace according to the manufacturer's instructions.
- Complete motor inspection, maintenance, and repair procedures according to Chapter 13.
- Check and clean the atomizer and vanes. When replacing, be sure the atomizer is in the center of the baffle plate.
- Adjust starting gas pilot light, electric ignition points, oil reserve flows, and any other special equipment that is installed in accordance with the manufacturer's instructions.

Gas-Fired Burners
Adjust burners before lighting and adjust orifice pressure in accordance with the manufacturer's instructions. Also refer to the American Gas Association Publications for the installation, maintenance, and repair of gas-fired burners.

Heating Systems Space Heaters

Use the checklist shown in Figure 15.3 to inspect and maintain heating systems in buildings.

Gas Fired Units
For instructions on motor lubrication, refer to Chapter 13 for an overview of motor maintenance. Refer to the previous section for maintenance considerations for burners.

Heat exchanger tubes may be cleaned through the bottom of the unit by removing the bottom pan and burners. Use a wooden lath or a wooden stick shaped to fit, pushing it up and down along the sides of each tube to free the soot.

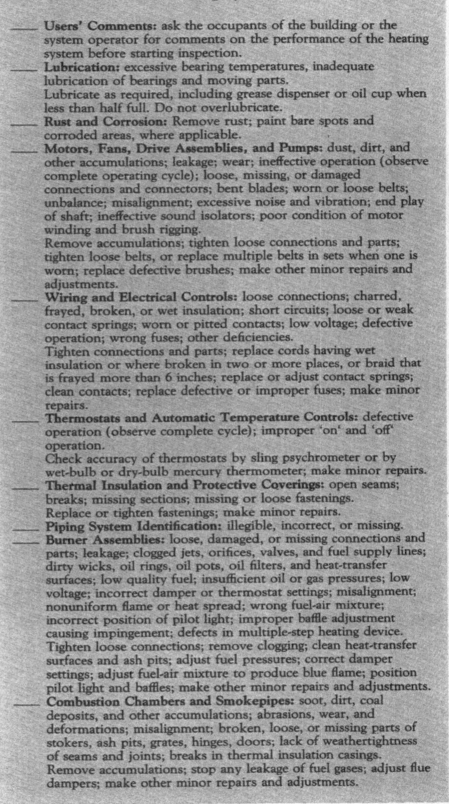

Checklist for Inspection and Maintenance of Building Heating Systems

_____ **Users' Comments:** ask the occupants of the building or the system operator for comments on the performance of the heating system before starting inspection.

_____ **Lubrication:** excessive bearing temperatures, inadequate lubrication of bearings and moving parts.
Lubricate as required, including grease dispenser or oil cup when less than half full. Do not overlubricate.

_____ **Rust and Corrosion:** Remove rust; paint bare spots and corroded areas, where applicable.

_____ **Motors, Fans, Drive Assemblies, and Pumps:** dust, dirt, and other accumulations; leakage; wear; ineffective operation (observe complete operating cycle); loose, missing, or damaged connections and connectors; bent blades; worn or loose belts; unbalance; misalignment; excessive noise and vibration; end play of shaft; ineffective sound isolators; poor condition of motor winding and brush rigging.
Remove accumulations; tighten loose connections and parts; tighten loose belts, or replace multiple belts in sets when one is worn; replace defective brushes; make other minor repairs and adjustments.

_____ **Wiring and Electrical Controls:** loose connections; charred, frayed, broken, or wet insulation; short circuits; loose or weak contact springs; worn or pitted contacts; low voltage; defective operation; wrong fuses; other deficiencies.
Tighten connections and parts; replace cords having wet insulation or where broken in two or more places, or braid that is frayed more than 6 inches; replace or adjust contact springs; clean contacts; replace defective or improper fuses; make minor repairs.

_____ **Thermostats and Automatic Temperature Controls:** defective operation (observe complete cycle); improper 'on' and 'off' operation.
Check accuracy of thermostats by sling psychrometer or by wet-bulb or dry-bulb mercury thermometer; make minor repairs.

_____ **Thermal Insulation and Protective Coverings:** open seams; breaks; missing sections; missing or loose fastenings.
Replace or tighten fastenings; make minor repairs.

_____ **Piping System Identification:** illegible, incorrect, or missing.

_____ **Burner Assemblies:** loose, damaged, or missing connections and parts; leakage; clogged jets, orifices, valves, and fuel supply lines; dirty wicks, oil rings, oil pots, oil filters, and heat-transfer surfaces; low quality fuel; insufficient oil or gas pressures; low voltage; incorrect damper or thermostat settings; misalignment; nonuniform flame or heat spread; wrong fuel-air mixture; incorrect position of pilot light; improper baffle adjustment causing impingement; defects in multiple-step heating device.
Tighten loose connections; remove clogging; clean heat-transfer surfaces and ash pits; adjust fuel pressures; correct damper settings; adjust fuel-air mixture to produce blue flame; position pilot light and baffles; make other minor repairs and adjustments.

_____ **Combustion Chambers and Smokepipes:** soot, dirt, coal deposits, and other accumulations; abrasions, wear, and deformations; misalignment; broken, loose, or missing parts of stokers, ash pits, grates, hinges, doors; lack of weathertightness of seams and joints; breaks in thermal insulation casings.
Remove accumulations; stop any leakage of fuel gases; adjust flue dampers; make other minor repairs and adjustments.

Figure 15.3

Checklist for Inspection and Maintenance of Building Heating Systems (continued)

_____ **Registers, Grills, Dampers, Draft Diverters, Plenum Chambers, Supply and Return Ducts:** soot, dust, and other deposits; clogging; deformations; broken, loose, or missing parts; loose seams and joints; breaks in vapor barriers; hinge parts failure; improper air distribution at branch ducts; improper seasonal damper or register settings.
Remove deposits inside ducts; tighten loose connections; make minor repairs.

_____ **Electrical Heating Units:** burned, pitted, or dirty electrical contacts; short-circuited sections of elements; low voltage in electrical circuits; dirty reflective heat-transfer surfaces.
Clean contacts and heat-transfer surfaces; tighten loose connections; make minor repairs and adjustments.

_____ **Guards, Casings, Hangers, Supports, Platforms, and Mounting Bolts:** loose, broken, or missing parts and connections; deformations; improper level; ineffective sound isolators.
Tighten loose connections and parts; adjust level; replace defective sound isolators; make minor repairs and replacements.

_____ **Steam and Hot Water Heating Equipment:** dust, scale, and other deposits; clogging; leaks; airbinding or water hammer; misalignment; improper slope of unit, resulting in inadequate drainage and heating efficiency.
Remove deposits; clean strainers; stop leaks at packing glands and repair other minor leaks; release entrapped air; make other minor repairs.

_____ **Accessible Steam, Water, and Fuel Oil Piping and Valves (steam and condensate return, hot water heating, humidifier water supply, fuel oil systems):** defective operation; leaks; clogging; casting blowholes; material defects; moisture; vibration.
Repair or replace any defective valves; remove clogging; clean orifices; resurface seats of defective globe valves and replace disks; get permission to shut off hazardous valves or fittings; blow off moisture and entrapped air to relieve air-binding; repair minor leaks.

_____ **Traps:** leakage, defective operation. Disassemble and repair.

_____ **Humidifier Assemblies:** dust; leaking pans; solids in water; clogged piping, inoperative valves; danger of water overflow.
Clean assemblies; remove clogging; adjust floats; add water to empty pans.

_____ **Air Filters:** dust, grease, and other deposits; missing; improper fit.
Replace throw-away filters that are dirty or missing or that have an improper fit; wash permanent filters; restore viscous coating in accordance with the manufacturer's instructions.

Figure 15.3 (continued)

Pilot Burner Orifice: If the pilot burner orifice becomes plugged, remove in order to clean by detaching the gas supply tube at the pilot. The fitting on the pilot burner to which the gas supply line is attached contains the orifice. Clean the orifice with compressed air or pass a wire or straw through it. Be careful not to enlarge or distort the orifice. If the gas has a high content of impurities, it may be necessary to clean the solenoid gas valve occasionally.

Valves: This type of burner contains a type of solenoid valve. Use the following checklist to inspect and service.

- Turn off the gas supply to the valve.
- Turn off the electric power to the valve.
- With a large wrench, unscrew the union nut holding the solenoid assembly to the valve body, being careful not to loosen or damage the metal gasket.
- Remove all foreign matter from the valve disc and plunger assembly, as well as from the plunger tube inside the coil and from the valve seat. Reassemble the valve and check for operation. If any of the original wire leading from the circuit or conduit box to the fan and limit switches must be replaced, use only Type AF wire or its equivalent.

Pilot Burner Failure: When the pilot burner does not light, check the following items:

- Make sure that all valves are open in the gas line from source of gas supply to the pilot burner. An audible hiss can usually be heard when gas is escaping from the pilot burner.
- Check for air in the gas line. Allow the air or the air-gas mixture to escape until an odor of gas is detected, then shut off the manual valve. After five minutes, open the manual valve and light the burner.
- If the flame does not burn properly, check to make sure that the pilot burner and orifice are free of dirt particles restricting the flow of gas. Particularly examine the primary air port, which may plug up from lint or other causes, and produce a flame with a yellow tip. Also check the screen in the gum filter in the pilot gas line for restriction. The pilot burner should burn with a soft blue flame.

Pilot Safety Valve: If a pilot safety valve will not function, follow the instructions below:

- Check the pilot burners. Clean the pilot orifices and air openings to ensure a steady blue name.
- Check the location of the thermocouple with reference to the pilot front flame; be sure the flame touches the thermocouple from 3/8 of an inch to 1/2 of an inch at the end.
- Clean the cone of the couple lead and socket of the connector stud on the hood of the valve to ensure good electrical contact. Reassemble the couple lead to the valve body.
- Replace the couple lead and repeat the test.
- If this test fails, send the complete valve to the manufacturer for overhauling.

Pilot Safety Switch: If the pilot safety switch will not function, follow the instructions below:

- Check the pilot burner. Clean the pilot orifice and air openings to ensure a steady blue flame.
- Check the location of the thermocouple with reference to the pilot flame; be sure the flame touches the thermocouple.
- Check to see that the pilot switch is plugged securely in the valve.
- Check the bulb in the capillary for breaks or cracks.
- If the above noted measures do not result in a correction, send the complete valve to the manufacturer.

Main Burner Failure: When the main burner fails to function, make sure that the pilot burner is lighted and that the flame is beyond the reach of the main burner ports. If the manual gas valve is open and gas is not being supplied by the main burner, check the wiring; be sure that the current is reaching the solenoid valve. To check the action of the solenoid valve, turn off the electric power and close the manual gas valve. The pilot should then be lighted. Hold the valve and apply the line voltage. The opening of the valve can be detected by a slight vibration. Check to see if the limit switch is in the closed position.

Limit Switch: If the overheat control switch cuts off the main burners during normal operation, try the following corrective measures:

- Check that the front louvers are open.
- Check actual BTU/hr. input of the unit against its rated input.
- Check the interchanger on both the flue gas side and air side.
- Be sure that the fan is operating at full speed.
- Replace the overheat control switch.

Motor Does Not Run: Refer to Chapter 14 for methods of correcting motor problems.

Excessive Fan and Motor Noise: Excessive fan and motor noise may be caused either by a damaged fan (bent blade) or by too much play in the motor shaft. Check blade misalignment by manually rotating the fan. Blades can be straightened by hand if the condition is not too serious. Serious damage requires replacement.

Automatic Valve Hums or Flutters: Humming or fluttering of the valve may mean it is installed backward. Check the arrow on the body of the valve. If the arrow is pointing in the correct direction, noise may be a result of a poor electrical connection or a faulty solenoid. Check the arrow direction on the pressure regulator or on liquefied petroleum gas units.

Burners Do Not Ignite: If burners do not ignite when the fan motor is running, check the pilot burner to see whether it is ignited. Check the electrical connections in the junction box. Check the thermocouple connections to the pilot relay or the automatic safety pilot shut-off valve.

Check the operation of the valve by imparting current across the leads of the valve and by determining whether there is a positive clicking noise, which indicates that the valve is operating satisfactorily when current is supplied.

On liquefied petroleum gas units and on units with a 100 percent safety pilot valve, the thermocouple generates a millivoltage current to hold the safety pilot valve open. If the thermocouple connections to the valve terminals are not tight, the flow of electrical current is insufficient to hold the valve open. Check the terminal connections if the main burners fail to operate.

Delay in Main Burner Operation: If the fan operates for two to three minutes before the main burners come on and if there is a positive click at the solenoid valve at the instant the main burners fire, the limit switch is not operating satisfactorily. To correct such a condition, order a replacement switch.

Symptoms of Burner Trouble: If the burners produce a yellow flame or if there is a popping flashback or noise of extinction, either the primary air shutters require adjustment or the burners have incorrect port size. Make adjustment of primary air shutters after the unit has been running for 10 to 15 minutes. Loosen the lock nut, close the shutter down until a yellow tip appears in the flame, then open the shutter until the yellow tip disappears and tighten the lock nut.

Check the size of the ports by removing the burner and checking with a suitable twist drill. Refer to the manufacturer's instructions for the proper port size for the gas being used.

High Flames, Scorched Paint, or Low Flames: A high flame, scorched paint, or very low flame indicates incorrect orifice size or improper gas pressure. Gas pressure can be checked with a suitable manometer. Refer to the manufacturer's instructions for the required pressures for the gas being used.

Oil-Fired Units

Inspect oil-fired units frequently to make sure that the equipment is clean, because accumulations of carbon and soot can cause disastrous fires. After cleaning burner pots, check fuel line connections for leaks. Keep all connections tight. Remove any excess carbon found in the combustion chamber or on the burner equipment. Remove and clean the unit, both inside and out, and remove accumulations of soot in the flue pipes. Undetected leaks can occur in the supply line. To avoid fires, be sure the supply line is always clean and that there is no oil on the outside surface.

Remove and service the burner pilot assemblies. Clean all air holes of the pot using a tough copper wire.

Observe the operation of the heater through a complete cycle. If the heater fails to operate satisfactorily, check the constant-level control valve. Follow the manufacturer's instructions for disassembling the valve. Be sure that the linkages are operating freely; replace any defective parts. Check the thermostatic attachments. If the thermostat and associated devices fail to open and close the constant-level control valve at the proper setting, adjust them in accordance with the manufacturer's instructions. Replace the thermostatic device with a new unit if these adjustments do not result in satisfactory operation.

Check the pot for warped or deformed surfaces and burned-out sections, which can cause loose connections and dangerous leaks. Readjust such surfaces and, if heat and wear have caused irreparable deformation, replace the parts. Touch up worn exterior surfaces with matching paint.

Check the automatic draft meter for defective hinge action and for warped surfaces in all working parts. Correct the deformations and replace the damaged parts.

Blower-Type Oil-Fired Heaters

Except for the oil supply and burner mechanism, blower-type oil-fired heaters operate in a manner similar to blower-type gas-fired heaters. Refer to other sections in this chapter for maintenance instructions.

Electric Heaters

Check all electrical connections for worn or damaged parts and wires for frayed surfaces. Check and calibrate the thermostat according to the manufacturer's instructions. Check the operation of the safety switch on portable units. Check the safety light, which indicates when the unit is operating, and replace any burned-out bulbs. Protect all terminals from moisture, such as condensation or spray.

Check all fuse clips and wire connections for tightness. Check the heating units for pitting and for broken parts, and replace defective parts. Do not allow dust or other foreign matter to collect on the heating units.

Check inner surfaces for corrosion, rust, and wear. It is important to keep parabolic reflectors polished to a mirror-like brightness. Check the heat chamber for collection of dust and dirt. Keep the safety grill free of dust and corrosion; and straighten individual bent rods of the grill. Check frequently for corrosion if the heaters are operating in spaces where gasoline fumes are present.

Check safety guard for proper operating conditions. Lubricate motors and fans according to the manufacturer's instructions. In insulations that use belt drives, make sure that the lower pulley is in alignment at all times and that the motor shaft and blower shaft are parallel in order to have the grooves of each pulley in realignment. All belts stretch after a short period of operation, and thus the belt tension should be checked periodically. The belts are under tension when finger pressure moves the belt approximately 1/2 of an inch from the straight line of the belt to the pulleys.

Steam Units

Steam units are similar to gas-fired heaters except that, instead of a gas burner, there are coils within the unit which use steam supplied from a separate source.

Check unit heaters frequently if operated in a location where the air is dusty or sooty or where corrosive fumes or oil sprays are present. Check the motor for proper operation and cleanliness (refer to Chapter 14).

See that the fan assembly is in its proper position in the shaft and that the hub set screws are tight. Check for loose fan blades, cracks, and excessive vibration. During each inspection, wipe fan blades clean. A dirty fan can get out of balance, causing noise and excessive vibration that damages the motor bearings.

Dirt reduces the capacity of the unit heater. To maintain maximum performance, keep heating elements free of dirt, grease, and other foreign particles in the air. Clean heating elements at least once a year; if operating conditions are poor, clean more often. If corrosive fumes are present, examine and clean the coil frequently.

There are several methods for cleaning heating elements. One method uses the fan action of the heater: dirt is loosened by brushing the entry air side, turning on the fan, and blowing out the dirt, which is collected in a bag located in front of the unit. Another method which is frequently used with heavy accumulation is to blow out the dirt with a high-pressure air hose or with portable blowers. If a more thorough cleaning is required, use a steam gun to spray and rinse heating elements. Remove the motor and spray the heating element with a mild alkaline cleaning solution, then after several minutes, rinse it with hot water.

Periodically check all fastenings on the unit for looseness or wear, which can cause excessive noise and vibration. Remove the strainer basket and clean annually. Replace damaged strainer baskets immediately, since large holes permit scaled dirt and rust into the steam trap.

Proper operation of the steam trap is essential if the unit heater is to produce its rated output. Check the operation of the trap if the heater fails to deliver the proper amount of heat. Also be sure that adequate steam pressure enters the heater. Disassemble the trap according to the manufacturer's instructions. Check the operation of the float linkages; if they do not operate freely, replace defective parts. If the heater continually becomes air-bound, the thermostatic element of the trap may be defective and may need to be replaced.

Check all thermostats for proper operation. Clean the contacts with fine sandpaper. Replace any thermostat that cannot be adjusted properly.

Water Heaters

The service life of water heaters and accessories varies according to the quality of the water, the means for inhibiting corrosion, and the operating conditions. Storage tanks with glass linings or renewable cathodic rod protection should last about twice as long as unprotected units. Useful lives for water heaters are as follows:

- Oil-fired water heaters (unprotected equipment)—8 years
- Gas-fired units—12 years
- Electrical element types—15 years
- Steam coil and instantaneous heater type—20 years

Components may have to be replaced before the end of the useful life of the complete unit. Replacement of an entire unit is more economical when the annual repair cost for labor and material exceeds 40 percent of replacement cost. When replacements are indicated, consider replacing obsolete units with improved equipment instead of buying outdated replacements.

Means of Protection

In rooms where temperatures fall below 40 degrees Fahrenheit, water heaters not in use may freeze. Corrective measures include using thermal insulation on the equipment, draining hot water systems when not in use, and constructing frost-free enclosures around water heaters.

Protective Devices: In order to reduce the possibilities of personal injury, all water heaters must have pressure relief valves installed on them either in the factory or in the field. These valves should be kept in perfect operating condition. Make frequent trial operations and checks, as well as calibrations of relief valves for all units.

Corrosion Protection for Tanks: Hot water storage tanks and water compartments of boilers must be maintained for leaks, scale, rust, contamination, and material defects; make sure all pipe connections, valves, and openings are maintained open and clear. At least once a year, examine the corrosion protection for tanks, including metallic coatings, nonmetallic linings, and replaceable items. Repair or replace defective parts immediately. Annually drain tanks and remove rust. When practicable, use wire brushes and solvents to remove products resulting from corrosion. If hot water is still discolored after thoroughly cleaning tank, make hydrostatic tests of the tanks at 1.5 times the working pressure. Failure to pass the hydrostatic test indicates the need to replace the tank.

Systems Leaks

Keep a lookout for any leaks in faucets, valves, piping, flues, tanks, heaters, and accessories. Repair all defects upon inspection.

If the equipment is old, evaluate the leaky piping and equipment for replacement of the entire system.

Thermal Insulation

Replace broken, water-soaked, or missing insulation with the same or equivalent materials and to the original thicknesses. Insulate cracks and open gaps between segments of insulation with asbestos paste.

Automatic Controls

Check the control device which automatically regulates water temperature. Check for evidence of water leaks at the control devices and correct any defects. Observe the operation of the gages. When the gage needle is frozen or otherwise inoperative, repair the gage and clean the siphon.

Examine water temperature regulators and thermostatic devices for performance and defects. Clean contacts with fine sandpaper. Keep linkages lubricated and wiring adequately insulated. Check accuracy of thermostats and temperature regulators with dry-bulb mercury thermometers. Repair or replace defective thermostats.

Gas Hot-Water Heaters

Make frequent checks of the gas burner and controls. Clean the openings of jets and adjust fuel-air mixture to produce a blue flame. Examine the gas pilot and adjust the pilot flame when too high. The correct height of the pilot flame depends on the gas pressure and on the draft, and is determined by trial-and-error. The minimum height, about 1/2 of an inch, provides a constant pilot flame and unfailing ignition for the burner. Check the operation of the automatic controls for a full operation cycle; examine the water temperature regulator, solenoid or motorized valve, gas pressure regulator, pilot flame failure device, and automatic gas shut-off valve controlled by excess temperature of hot water. Check the draft diverter and clean the openings when clogged.

Typical Problems with Gas Hot-Water Heaters

If there is an insufficient amount of or *no hot water*, check the following items:

- **Pilot flame:** Clean the hot water, pilot orifice, and tube. Check the pilot flame for correct adjustment. Determine the gas input to the heater, compare it to the input specification of the model and rating plate, and adjust accordingly. Be sure a down draft diverter is installed in the hot water heater.
- **Thermostat setting too low:** Adjust thermostat to the desired temperature.
- **Thermostat differential too high:** The thermostat differential should be 10 to 25 degrees. If the off and on periods are not within the differential, replace the thermostat and return the defective unit to the factory for recalibration.
- **Gas input too low:** If the gas input is not within the input specifications on the model and rating plate, increase or decrease gas consumption by adjusting the valve provided for this purpose or by replacing the orifice.
- **Heater too small:** Substitute a larger-size heater to avoid over-firing.

For *high gas consumption*, check the following items:

- **Thermostat setting too high:** If ample hot water is provided, but the result is a high gas bill, reduce the thermostat setting. Set the indicator at not less than 120 degrees and not over 140 degrees for normal consumption and for most economical operation.
- **Flue baffle missing:** If the flue baffle is omitted during the installation of the water heater, the unit operates at lower efficiency. Check the heater flue to ascertain if the flue baffle is in place.

Backfiring or *burning* in the mixer tube is generally caused by (1) an insufficient gas supply to the burner, (2) delayed ignition, or (3) the use of too much primary air. Any of the following defects may cause the problem:

- Gas leakage past the thermostat valve seat
- Pilot light not large enough for proper burner ignition
- Pilot light too far away from the main burner for proper ignition
- Burner out of adjustment

A *thermostat gas leak* is determined by sustaining a flame at the gas orifice spud while the thermostat is in the off position. Check the inside of the thermostat for any foreign material lodged on the valve seat. Remove such material with a soft cloth. In most cases, gums or tar may be removed with a commercial solvent. Also, the valve stems may be too long, causing a leak. Adjust by advancing the valve stem and turning the screw 1/8 of a revolution at a time. If the leak persists, the valve seat is probably imperfect, and it is advisable to return the thermostat to the factory for repair.

A reduction in or outage of the *pilot flame* may occur with a temporary reduction in the gas pressure, or with a collection of foreign material in the pilot filter or pilot burner orifice. Lowering the pilot light results in delayed ignition. Increase the pilot flame to correct any gas pressure variations, or clean the pilot burner orifice to ensure proper gas flow. A 1/2 cubic foot/hr. pilot minimum is necessary for a 1/4 of an inch pilot flame and proper ignition of the main burner.

If the pilot assembly is not fastened properly to its brackets, center the pilot tube in the spot provided in the burner, placing the pilot light not more than 1/2 of an inch away from the nearest port. Loosen the union nut on the pilot control, recenter the tube, and tighten the union nut.

Adjust the burner for medium flame. A yellow flame indicates too much gas and too little air, resulting in poor combustion and causing carbon deposit on the heating surfaces and discharge of odors into the room. A mixture that is too lean (indicated by a hard blue flame) causes backfiring of manufactured gas. To correct this defect, run the air shutter to where the blue part of the flame is barely visible.

Excessive condensation is caused by the chilling of water vapor in the latent flue gases as they leave the combustion chamber heater. Excessive condensation fills the pan at the bottom of the heater and sometimes runs out on the floor. Condensation is formed more often in the vent pipe than in the heater flue. If it is not possible to decrease the length of the vent pipe, install an inverted tee as close to the heater as possible to divert the condensation from the heater flue.

Odors of the flue gases indicate poor combustion, caused by flue stoppage; improper burner adjustment; or poor chimney construction.

Check the following if there is a noticeable *burner flame noise*:

- **Incorrect orifice size:** An oversized orifice may cause the burner to over-fire, resulting in combustion noise. Install the correct gas spud recommended for the type of gas being used, or reduce the main burner input by means of the burner input adjustment, which is provided on most controls used in fiberglass water heaters.
- **Oversupply of air:** More than sufficient primary air causes a hard blue flame with a tendency to roar slightly. Readjust the air shutter on the burner to reduce the primary air to where the flame cone is barely distinguishable.
- **Excessive draft:** Check the installation for excessive draft, which causes the main burner flame to roar.

To correct *improper combustion*, after the unit is cleaned, observe the following precautions to ensure proper adjustment for good combustion:

- Check and clean the burner if needed.
- Check the burner orifice for proper size and true bore.
- When reinstalling the burner, make sure that the burner orifice fitting is not inserted too far into the face of the burner.
- Start the heater and adjust for proper flame. Compare gas input to the specifications on the rating plate.
- Check venting for proper flue action.

Oil-Fired Heaters

Oil burning equipment easily accumulates carbon and soot, which, if not removed, can cause disastrous fires. Inspect often to make certain the equipment is clean and all connections are tight. Remove excess carbon in the combustion chamber or on the burner equipment. After the burners have been cleaned, check the fuel line connections for leaks. Remove accumulations of soot from the flues. To avoid fires, be sure that the supply pipe is always clean and that there is no oil on its outer surface. This applies to all the piping and tubing, starting at the oil tank and reservoir and extending to the burner.

The following sections describe maintenance considerations for specific oil-fired heater components.

Inlet Cleaner: Some heaters contain an oil inlet cleaner located directly behind the burner access door. The device consists of a plunger, which must be pulled out and rotated. After rotating clockwise, push the plunger back into its normal position. Repeat this operation three or four times at each cleaning. After cleaning, the plunger must be pushed all the way in to prevent air leakage into the burner. Failure to perform this operation at least once a week may cause a flame to go out unnecessarily.

Burner Cleaning: When cleaning, leave a small accumulation of carbon at the bottom of the burner to act as a wick and help maintain the pilot fire. Before replacing the burner, make sure that the pilot housing is in place between the two metal guides on the inside base of the burner. In replacing a burner assembly, make sure that both sides of the burner are tightened equally so that the top of the burner and gasket are set firmly against the flue protection. The gasket must be in place at the top of the burner.

Control Valve Strainer: Remove and clean the oil strainer located in the top of the control valve during periodic inspection.

Cleaning: To clean the constant-level control valve, metering stem, and inlet needle, remove the cover plate from the top of the valve and the housing enclosure. Be certain to remove the needle assembly in accordance with the manufacturer's instructions and to follow the same instructions when reassembling the unit. Use only soft materials, such as a toothpick, to clean the metering stem groove and the inlet needle, and wipe them with a lint-free cloth. Clean the outlet seat and metering edge at the same time. Always use care when cleaning the delicately finished surfaces. If foreign matter lodges under the inlet needle and prevents the oil level to rise 1/4 of an inch above normal, the safety trip mechanism operates automatically to shut off the inlet to the valve until oil flow is reestablished. Trip and reset this mechanism occasionally as a precaution against automatic tripping due to accumulation of foreign matter on the inlet seat.

Surfaces: Intense heat is created in the combustion chamber. Watch for warped or deformed surfaces and burned-out sections in the pot, since these can cause loose connections and dangerous leaks. Readjust such surfaces, and, if heat and wear have caused deformation to the extent that repair is not feasible, replace the parts. Touch up worn exterior surfaces with matching paint.

Automatic Draft Meter: The automatic draft meter corrects excessive draft and prevents heat loss in the chimney. Check the meter for defective hinge action and for warped surfaces in all working parts. Correct the deformations and replace the damaged parts.

The following sections contain instructions for general maintenance and repair of problems specific to oil-fired heaters.

Burner Noise: When operating properly at a high fire, the burner makes a slight sound; at a medium fire, the burner is noiseless. Excessive draft causes excessive noise. Adjust the draft meter to maintain a draft of .02 to .03 inches of water column, or as specified by the heater manufacturer. Check the draft with a draft gage as a guide; if a draft gage is not available, adjust the draft meter by observing the color of the flame through the opening punched in the draft tee. This opening is directly above the internal flue on top of the draft tee. When the draft is correctly adjusted, the color of the flame is a brilliant yellow-white. Too much draft results in a low rumbling sound when the burner is in operation; too little draft makes the flame a smoky yellow streaked with red. To change the draft adjustment, rotate the adjusting weight to increase or decrease the draft as indicated by the arrows on the adjusting weight. Another reason for burner noise is that the inlet cleaner has not been replaced in the proper position.

Burner Goes Out: When the burner goes out, first check for air traps. The pipe from the oil tank should drop gradually to the floor and, from this low point, it should be sloped gradually upward to the oil connection at the control valve of the heater.

Another reason for the burner going out is that the pilot is set too low. If this is the case, adjust the pilot fire according to the manufacturer's instructions.

Chimney down draft may also be the cause of extinguishing the burner flame. Down draft is caused when the chimney does not extend above the highest point of the roof or is close to obstructions, such as higher adjoining buildings or by trees. To correct this condition, a proper down draft hood must be put on top of the chimney or the chimney may be extended. Often a stone slab over the top of the chimney with sufficient venting on all four sides is an effective means of stopping chimney down draft.

If the water heater is connected to a flue which also serves a large furnace or space heater, a back-pressure condition from this burner may be created. Furnace burner ignition sometimes causes blow-back into the water heater burner, extinguishing the burner flame. If this occurs, install a back-pressure relief damper. The water heater should vent into the chimney above the furnace flue connection.

Lack of oil supply to the heater may be another reason for the burner to go out. Check to see if the oil tank is empty or nearly empty. If no oil is reaching the burner, check the reset level on the control valve; if it has dropped down, reset it by lifting. Also check for oil flow to the drain tee; and, if no oil flows at this point, clean the strainer valve, metering stem, and inlet needle. Oil flow at the drain tee indicates that the stoppage is in the burner feed pipe and that it is necessary to clean the burner inlet. Water or dirt in the control valve or in a fuel line, which will cause endless trouble until removed, could be the result of washing the floor with a hose and flooding the floor, allowing water into the oil supply pipe. Also check the control valve strainer and clean it if dirty.

Carbon Forms on Burner: About 1/4 of an inch of carbon ordinarily accumulates across the bottom of the burner each year, if the proper oil is used. It is usually possible to clean the burner sufficiently without removing the carbon, by drawing out about half a tank of hot water in order to make the burner operate at high fire. Remove the inlet cleaners assembly and allow the burner to operate no longer than 30 minutes at high fire with excess air.

Electric Heaters

Inspect the electrical heating elements, wiring, and controls. Observe the operation of the automatic controls for a complete operation cycle and adjust them when not satisfactory. Clean the electrical contacts with a fine emery cloth, tighten loose connections, and repair broken or poorly insulated wiring. Replace defective heating elements. Check the voltage when the heating output is not adequate.

If electrical contacts are not enclosed in a dust-type cover, an accumulation of lint and dust may prevent them from making a good contact. To clean the contact, place a piece of hard surface paper between them, close the contact against the paper, and slide the paper back and forth between them until they are free of dust. Never use a file or an abrasive of any kind, as this will damage the contact surfaces. Never use newspaper or other paper which will shred or stick to the contacts. Clean the sensing element with trichloroethylene or a similar cleaner. Dust and dirt on these elements cause faulty thermostat operation.

General Inspection and Maintenance

When inspecting and maintaining all water heaters, use the checklist in Figure 15.4:

Checklist for Inspecting and Maintaining All Water Heaters

_____ **Lubrication:** Lubricate the motor of the oil burner and the moving parts of mechanical devices.

_____ **Rust and Corrosion:** Remove rust; paint bare spots and corroded areas, where applicable.

_____ **Automatic Controls:** improper operation through complete cycle; improper "on" and "off" operation.
Check accuracy of thermostats by sling psychrometer or by wet-bulb or dry-bulb mercury thermometer; make adjustments to attain normal performance.

_____ **Safety and Flame Failure Devices:** unsatisfactory when tested through complete "on" and "off" cycle of operation.

_____ **Thermal Insulation and Protective Coverings:** open seams, breaks, and missing sections; missing or loose fastenings. Replace or tighten fastenings; make minor repairs.

_____ **Burner Assemblies:** loose, damaged, or missing connections and parts; leakage; improper fuel-air mixtures; incorrect height and position of pilot light; improper baffle adjustment causing impingement; dirty heat-transfer surfaces; nonuniform flame spread; misalignment; clogged jets, orifices, and valves; dirty oil filters; defective oil wicks, oil rings, and pots.
Test heating output at each step of multiple-step heating levels; tighten loose connections; repair or replace damaged or missing parts; adjust fuel-air mixture to produce blue flame; adjust pilot light; replace black or burnt-out pilot lamps; reset baffles; clean heat-transfer surfaces and ignition devices; correct misalignment; clean openings in jets, orifices, and valves; replace dirty oil filters, defective oil burner, other parts of assemblies.

_____ **Combustion Chambers:** deformations, breaks, cracks, wear; water and flue gas leakage; burnt-out grates; defective coal feed mechanisms; broken latches and hinges; door misalignment and poor fit; soot deposits, clinkers, ashes.
Tighten loose connections; clean ash pits and grates; make minor repairs.

_____ **Electrical Heating Elements and Controls:** loose connections; chaffed, frayed, or broken insulation; short circuits; loose or weak contact springs; worn or pitted contacts; defective operation; wrong fuses; other deficiencies.
Tighten loose connections; repair contact springs; clean contacts; replace defective fuses; make other minor repairs.

_____ **Gages:** Remove and calibrate gages, instruments, and protective devices. Use standard thermometers and gages if proper test facilities are not available.

_____ **Water Compartments and Tanks:** leaks; loose connections; chipped enamel finish; cracked cement linings; exposed bare metal or interior surfaces; defective manhole gaskets; dripping; corrosion; damage from freezing; defective operation of drain valves; deteriorated anodic rods or other devices provided for limiting corrosion.
Tighten connections where feasible; replace deteriorated anodic rods or segments.

_____ **Water Relief and Steam Safety Valves:** rust, corrosion, scale; mechanical defects; defective operation.

_____ **Supports:** unstable; material defects; loose, missing, or broken puts.

_____ **Steam Coils and Instantaneous Water Heaters:** improper steam pressure and water temperature; steam trap blowing; mechanical defects; clogged strainers; scaled heat-transfer surfaces; open bypass valves; leaking pipe connections.

Figure 15.4

Pneumatic Systems

Deviations from normal pressures and in temperatures of air and water of the compressor, intercoolers and aftercoolers, and compressor lubricating oil indicate that corrective action must be taken. On two-stage compressors, low intercooler pressure may indicate malfunctioning of the low-pressure cylinder valves; high intercooler pressure may be due to improper operation of high-pressure cylinder valves. Locate the defective valve by feeling the valve cover plate and determining which valve is the hottest. Leaking high-pressure suction valves cause the intercooler pressure to fluctuate above normal values. Leaking high-pressure discharge valves cause the intercooler pressure to build up steadily until this safety valve releases it. Low-pressure discharge valves that leak cause intercooler pressure to fluctuate below normal intercooler pressure. Keep the compressor clean at all times. The operator should wipe the machine daily with a cloth. Dirt on the machine works its way into the lubricating system. On air-cooled compressors, dirt accumulation forms an insulating blanket, causing increased temperatures within the machine and access to moving parts. Clean the intake air filter regularly to prevent atmospheric dust from entering the compressor cylinders. Keep piston rod packing tight enough to prevent air leakage, but do not over tighten.

Positive Displacement Air Compressors

For reciprocating compressors, check for the following at the noted intervals:

Daily

- Unusual noise or vibrations
- Abnormal pressures or temperatures of compressed air, cooling water, and lubricating oil
- Proper operation
- Hot stuffing box
- Hot bearings
- Correct lubricating oil levels

Checklist for Inspecting and Maintaining All Water Heaters (continued)

____ **Hydrostatic Pressure Test:** Gag or remove relief valves and cover the openings with blind flanges or pipe caps. Remove drain valves and other devices not designed for the test pressure. Do not perform hydrostatic tests if the surrounding atmosphere is at a freezing temperature. Test tank at 1.5 times maximum allowable operating pressure, holding the test pressure for not less than 2 hours. Drain water from tank after the test.

Figure 15.4 (continued)

Quarterly

- Inspect compressor valves for wear, dirt, and proper seating.
- Check operation of all safety valves.
- Check packing for wear and piston rods for scoring.
- Examine crank case for sludge accumulation.
- Check cylinder heads for tightness.
- Inspect connecting rods and cross heads for wear.
- Inspect bearings for wear and dirt.
- Check operation of lubricators and oil cups.

Annually

- Check cylinders for wear, scoring, corrosion, and dirt.
- Inspect pistons for leakage, wear, and scoring. Check the security of the piston rod. Check head clearances.
- Examine piston rings for damage, wear, tightness, and dirt.
- Check piston rods for wear at packing glands and security to cross head in piston.
- Inspect crank case and crank shaft bearings for wear and proper operation.
- Check crow heads, crow head guides, wedges, and pins for wear and proper operation.
- Check flywheel for security to shaft. Inspect flywheel bearing for wear and dirt.
- Check alignment of compressor with drive.

Maintenance

For intervals and quality of *lubrication* of the air compressor, follow the manufacturer's instructions.

Pack according to the manufacturer's instructions.

Periodically *clean* cylinder jackets of a water-cooled compressor. Dirt accumulations interfere with the water circulation. Clean by using a small hose nozzle to spray water into the jackets. To clean a compressor fitted with mechanical lubricators, feed a strong soap solution into the system through the lubricator. Run the compressor at slow speed for about an hour with the drains at the separator and receiver open in order to remove the water and oil. After the cleaning operation, feed oil for about 30 minutes to prevent rusting.

Replace all defective *valve* parts as required. For other valve maintenance, refer to the chapter on plumbing.

When replacing *worn piston rings*, the new rings must be tried in the cylinder for fit. If the cylinder wall is badly scored or out of round, rebore the cylinder; replace the cylinder linings if they are pitted. Before installing the new rings, clean the ring grooves and remove all carbon deposits . To install the new rings, place several metal strips not more than .02 inches in thickness between the rings and the pistons. Slide the new rings over these strips until they are centered over the grooves, and then pull out the strips. Make sure the ring is free to rotate in its groove. Stagger the splits of succeeding rings so that they are not in a line. Use a ring clamping device when reinstalling the piston. If replacing carbon rings, consult the manufacturer's instructions.

Refer to Chapter 14 for information on *bearing maintenance*.

Always check *piston* and *clearance* after replacing pistons or after adjustment or replacement of main, crank pin, wrist pin, or cross head bearing. Consult the manufacturer's instructions for proper clearances and method of clearance adjustment.

Repair of Reciprocating Air Compressors

For causes of and solutions to reciprocating air compressor problems, refer to the chart in Figure 15.5.

Reciprocating Air Compressor Problems—
Causes and Solutions

Cause	Solution
1. Low air pressure	
A. Air filters plugged.	Remove and clean the air filter.
B. Compressor too small for air demand.	Install a second or larger compressor.
C. Piston rings worn.	Recondition cylinders and replace the piston rings.
D. Speed of unit not up to specifications.	Check speed of motor and drive belt ratio.
E. Suction unloading valves partly	Adjust unloading valves. open.
F. System air leaks.	Tighten all leaking connections.
2. Blower—Receiver Relief Valve	
A. Air to control shut-off.	Open the valve to the pressure control valve.
B. Control stuck; check for dirt or varnish.	Clean or replace the pressure control valve.
C. Broken unloading valve plunger spring.	Replace broken plunger spring.
D. Diaphragm ruptured in HP unloading valves.	Replace HP diaphragm.
E. Filter plugged in control line.	Clean or replace the filter.
F. Unloader line leaking.	Tighten or replace the line.
3. Pressure in intercooler low during full load operation.	
A. Badly worn LP piston rings.	Recondition the LP cylinder and install new piston rings.
B. Intercooler relief valve leaky.	Recondition or install a new relief valve.
C. Leaking diaphragm in HP suction valve allowing high pressure air to backflow through unloading lines and partly unloading LP valve.	Clean or replace HP suction valve.
D. LP unloader plunger spring broken, allowing plunger to rest on valve disc, retarding valve action.	Replace unloader plunger spring.
E. LP suction and/or discharge valves leaking.	Clean or replace leaking valves.
F. Intercooler drain valve open.	Close drain valve.
G. Plugged air filter.	Clean or replace the air filter.
4. Pressure on intercooler high during full load operations.	
A. HP suction and/or discharge valve. (Hottest valve is probable cause.)	Clean or replace leaking valves.

Figure 15.5

Rotary Sliding Vein Compressors

Any deviation from normal operating levels of air and water temperatures and pressures and of lubricating oil additives may mean trouble. On two-stage machines, low interstage pressure indicates a malfunction in the first stage or stoppage in the intake filter. High interstage pressures may indicate that the second stage is not operating properly or that the air from the first stage is not being cooled sufficiently by the intercooler. The operator must keep the machine clean at all times by wiping the machine down daily with a cloth. Dirt accumulations work their way into the machine and cause undue wear. Do not operate a compressor beyond its rated capacity. Overload operation results in overheating and damage to running surfaces. Do not over tighten packing glands, as this may result in rapid packing in shaft scoring.

Maintenance Inspections: Check for the following at the noted intervals.

Daily

- Proper lubricating oil levels
- Abnormal pressures or temperatures
- Unusual noise or vibration
- Hot stuffing box
- Hot bearings
- Motor overheat

Reciprocating Air Compressor Problems— Causes and Solutions (continued)

Cause	Solution
B. HP unloader plunger spring broken, allowing plunger to rest on valve disc, restricting valve action.	Replace unloader plunger spring.
C. Plugged intercooler core.	Remove and clean intercooler.
5. Noisy valves	
A. Assembly loose.	Tighten the valve assembly.
B. Broken or weak valve spring.	Replace the valve springs.
C. Maladjustment of unloader.	Adjust unloader plunger. plunger travel.
D. Broken unloader spring.	Replace the unloader spring.
6. Shortened valve life.	
A. Corrosive vapors entering unit.	Move or modify intake line.
B. Dirt entering unit.	Use more effective filters.
C. Maladjustment of suction unloading plunger.	Adjust unloader plunger.
D. Worn valve seats uneven.	Replace valve seat.
E. Valve spring weak or collapsed.	Replace valve spring.
F. Lift of valve too great.	Adjust valve travel to proper limits.

Figure 15.5 (continued)

Semiannually

- Check the alignment of the compressor to the drive. On two-stage units also check the alignment of the outboard compressor to the inboard one.
- Check condition of packing.

Annually

- Check bearings for wear and dirt.
- Check shaft for wear of seals.
- Examine the mechanical seals for damage.
- Remove rotor blades and inspect for wear.
- Examine cylinder board for wear and scoring.
- Examine all gaskets for damage.

Replace *rotor blades* if the blade thickness at any point is found to be less than 85 percent of the width of the rotor slot or if the blade width is found to be less than 90 percent of the rotor slot depth.

Annually, thoroughly clean the *cylinder* and *terminals*. Blow out all oil holes and make sure that they are open and free of sludge. Flush out cylinder jackets with a water hose to remove dirt accumulation. Stone rough spots and cylinder walls, cylinder heads, and rotor. Replace any defective gaskets.

Replace worn or defective *bearings* as required in accordance with the manufacturer's instructions. Also refer to Chapter 14 for additional information on bearings.

Each time the compressor is inspected internally or disassembled for repair, inspect and maintain the *clearances* according to the manufacturer's instructions.

For intervals and quality of *lubrication*, follow the manufacturer's instructions.

Rotary Twin Lobe Compressors

Do not operate the compressor at a higher than rated capacity, as this results in overheating of the compressor and drive. On units fitted with oil coolers, do not allow the temperature of the oil in the gears and bearings to exceed the temperature recommended by the manufacturer. Maintain oil levels within the limits indicated on the oil level gage. Insufficient oil results in improper lubrication. Too much oil causes overheating of the bearings and gears.

Inspection: Check for the following at the noted intervals.

Daily

- Abnormal suction
- Discharge pressures and oil temperature pressure
- Unusual noise or vibration
- Hot bearings
- Motor overheating
- Oil leaks

Annually

- Corrosion or erosion of parts
- Proper clearances
- Correct alignment
- Worn or broken timing gears
- Timing gear setting
- Operating setting of relief valves
- Shaft or wear at seals

Maintenance: *Timing gears* must be securely locked to their shafts in proper position. Clearances must be in accordance with the manufacturer's instructions.

Keep *seals* free of dust and dirt to ensure a long service life. Carefully follow the manufacturer's instructions when replacing mechanical seals.

Rotary Liquid Piston Compressors

Never run a liquid piston compressor dry. Operating without sealing water results in serious damage to the compressor. Maintain sealing water flow rate carefully. Insufficient sealing water results in a loss of capacity; excess sealing water overloads a compressor drive. For proper sealing water quantities, follow the manufacturer's instructions. On units with stuffing boxes, do not over tighten the packing, as this results in rapid packing wear and scoring of the shaft.

Inspection: Check for the following at the noted intervals.

Daily
- Abnormal discharge pressures
- Unusual noise or vibration
- Motor overheating
- Hot bearings
- Correct sealing water flow and pressure

Semiannually
- Check alignment.
- Examine packing for wear and shaft for scoring at packing or seals.
- Check setting and operation of relief valve.

Annually
- Bearings for wear and dirt
- Corrosion or erosion of parts
- Correct clearances
- Compressor internals for scale deposits
- Gaskets for damage
- Mechanical seals for damage

Maintenance: For frequency and quality of *lubrication*, follow the manufacturer's instructions.

When replacing *packing*, thoroughly clean the stuffing box of old packing and grease. Cover each new piece of packing with the recommended lubricant. Install according to the manufacturer's instructions.

Replace worn or defective *bearings* according to the manufacturer's instructions. Also refer to Chapter 14 for additional information on bearings.

Repair: For causes of and solutions to rotary compressor problems, refer to the chart in Figure 15.6.

Dynamic Compressors

Keep centrifugal and axial compressors clean at all times by wiping down daily with clean, lint-free rags. Never operate the compressor at the critical speed range, as this may damage the machine.

Inspection: Check for the following at the noted intervals:

Daily
- Abnormal pressures or temperatures of lubricating oil, compressed air, and cooling water
- Correct lubricating oil levels
- Unusual noise or vibration
- Bearing temperatures
- Proper operation of auxiliary oil pump

Semiannually
- Check condition of lubricant.
- Check for deterioration of the oil and for presence of water in the oil.
- Test all safety controls.

Repair Methods for Rotary Compressors

Cause	Solution
1. Excessive air temperature on first-stage discharge.	
A. High inlet air temperature.	Relocate intake filter to clean, cool air source.
B. Intake filter clogged.	Clean intake filter.
C. Incorrect lubrication.	Use oil and feed rates in accordance with manufacturer's instructions.
D. Insufficient or high-temperature water.	Increase water gpm or inlet provide cool water supply.
E. Problem with intercooler, causing high first-stage discharge pressure.	Clean intercooler tubes and shell.
F. Scale or residue building water jacket.	Clean water jacket, and filter or treat water supply as required.
G. Swelled or warped rotor blades	Dry out or replace rotor blades.
H. Unloading valve not fully opened, or clogged.	Clean valve and replace any worn part.
2. Excessive air temperature on second-stage discharge.	
A. First-stage malfunction	Follow first-stage sequence.
B. Incorrect lubrication.	Use oil and feed rates in accordance with the manufacturer's instructions.
C. Insufficient or high-temperature inlet water.	Increase water gpm or provide cool water supply.
D. Problem with intercooler, causing high inlet air temperature.	Clean intercooler tubes and shell.
E. Problem with aftercooler, causing high second-stage discharge pressure.	Clean aftercooler tubes and shell.
F. Scale or residue buildup in water jacket.	Clean water jacket; filter or treat water supply as required.
G. Swelled or warped rotor blades.	Dry out or replace rotor blades.
3. Ineffective unloading of compressor.	
A. Clogged control air line or ports in 3-way valve.	Clean air line and 3-way valve.
B. Faulty pressure switch or solenoid valve.	Repair or replace switch or valve.
C. Unloading valve dirty or internal parts worn.	Clean or replace unloading valve parts.

Figure 15.6

Annually

- Check journal and thrust bearings for wear.
- Check alignment and coupling condition.
- Examine compressor cases for corrosion and peeling paint.
- Tighten compressor bolts.

Maintenance: If the compressor is operating inadequately, or if noise or overload conditions indicate internal trouble, dismantle the compressor and inspect as follows:

1. Examine the casing or stator for corrosion, erosion, and dirt. Check for evidence of leaks if diaphragm cooling is provided.
2. Inspect rotor for corrosion or erosion to impellers or blades. Check rotor for balance.
3. Check clearance or rotor bearings and seals.

Clean *oil filter* and *strainers* every three months or as needed. Replace filter cartridges when required. Use the quality of oil recommended and replace oil at the intervals specified by the manufacturer. Also check oil pumps for corrosion, erosion, and wear annually. Repair or replace parts as required.

Check the *alignment* of the unit with a dial indicator. To correct axial misalignment, add or remove shims under the driver to bring the compressor drive center lines into alignment. To correct angular misalignment, parallel the coupling faces by shifting the drive on its base and adding or removing shims under the driver base.

Refer to Chapter 14 for information on the maintenance of *bearings*.

Auxiliary Equipment

Intake Filters

Inspect the air filter once each month for sludge accumulations, clogged filter elements, and proper oil level.

Repair Methods for Rotary Compressors (continued)

Cause	Solution
4. Vibration in excess of normal, knocking noise.	
A. Bearing worn.	Replace bearing.
B. Blade wear excessive.	Replace blades and check lubrication.
C. Cylinder wear erratic.	Rebore, redowel, and check lubrication.
D. Insufficient lubrication.	Increase lube feed rate.
E. Misalignment.	Realign units.
F. Rotor contacting cylinder	Check temperature and or heads. pressure conditions; check internal clearances.
G. Swelled or warped rotor blades.	Dry out or replace rotor blades.

Figure 15.6 (continued)

Thoroughly clean air filters when necessary. On oil bath-type filters, remove sludge when it has accumulated to a height of 1/4 of an inch in the sump. If a thickening of the oil is noted, change the oil even if excessive sludge is not present. Thickened oil prevents proper oil circulation in the filter. Clean air intake filters when the compressor is not operating. Remove the filter element and drain all oil. Thoroughly clean out all sludge accumulations. Clean the filter element by agitating it in a solvent or a hot water and detergent solution. Do not use gasoline, kerosene, or other low flash-point solvents, as an explosion within the compressor may result. When the filter element is dry, recharge with clean oil by dipping, brushing, or spraying. Use the quality of oil recommended the manufacturer.

Silencers
Semiannually inspect the silencers for the following:
- Externally—corrosion and peeling paint
- internally—corrosion and dirt accumulations
- Damaged gaskets

Semiannually, clean silencer internals for dirt accumulations. Replace defective gaskets as required.

Intercoolers and Aftercoolers
Avoid excess water flow rates, which cause erosion. Adjust water flow rates gradually to avoid sudden temperature changes. A sudden rise or drop in temperature induces stresses in the cooler part, which can loosen joints and result in leaks. If at any time evidence indicates an intercooler or aftercooler tube leak, shut down the compressor immediately and repair the leak. Excessive amounts of water drained from cooler shelves indicate a tube leak, which may result in large quantities of water carried over into succeeding compressor stages or into the distribution system. Air leaking into the water side of the cooler greatly reduces the cooler efficiency. A blowing water relief valve, although normal cooling water is indicated, suggests a tube failure, allowing air to build up the pressure in the water space.

Inspection: Check for the following at the noted intervals:

Daily
- Proper operation of the automatic controls and instruments
- Water leaks, temperature, and flow rate
- Deviations from normal temperature, or pressure drops across the cooler

Semiannually
- Corrosion and peeling paint
- Setting and proper operation of relief valves
- Leakage and corrosion of manual and automatic valves

Annually
- Corrosion and erosion of internal components, such as tubes, tube sheets, and baffles
- Leaking tubes
- Plug tubes
- Scale deposits and tubes or shelves

Maintenance: Clean tube interiors by flushing a stream of water through them. For more persistent deposits, brushes, rods, or other cleaning tools may be required. Clean tube exteriors by hosing with hot water. A stiff bristle brush aids in removing deposits from between tubes. Clean cooler interior without dismantling the unit by passing a chemical solution through it. The solution should dissolve the scale or other deposits without attacking the metal. Thoroughly wash out all chemicals from the cooler before returning it to service.

Seal off *leaky tubes* at both ends, using tapered hardwood plugs. The plug is inserted into the open end of the tube and seated with a hammer blow. Do not replace one tube at a time in a tube bundle. Keep tubes plugged until cooler performance is repaired, at which time the bundle can be retubed. Remove and replace tubes according to the manufacturer's instructions.

Separators

Drain separators regularly if automatic drainers are not provided. Frequency of draining is best determined by experience with the installation. Improperly drained separators result in moisture carryover into the air distribution system.

Inspect the separator once every six months for the following:

- Externally—rust, corrosion, and peeling paint
- Internally—corrosion and accumulations of dirt and oil
- Damaged gaskets
- Repaint separator exteriors as required.
- Semiannually clean separator interior.
- Replace defective gaskets.

Traps

Check for the following at the noted intervals:

Daily

Inspect the operation of the trap drains. Make sure the trap is draining properly and is not blowing air.

Annually

- Corrosion or erosion of parts
- Worn valves and seats
- Defective floats for buckets
- Loose or damaged linkages

Thoroughly clean trap interior at least once a year. Frequency of cleaning depends upon the condition of the system and whether or not a strainer is installed ahead of the trap. Remove all dirt accumulations from the trap body and mechanism, using a solvent if necessary. Clean valve seats using a small spiral brush.

Valves and Seats

Replace badly worn or grooved valves and seats in accordance with the manufacturer's instructions. Refer to Chapter 14 for additional information on valves.

Levers

Levers and linkages wear at pivot points. If excessive play on linkages is found, they should be replaced. Worn levers affect the bucket or float travel and result in a loss of capacity. Replace corroded or worn pins as required.

Buckets and Floats

Replace corroded or damaged buckets and floats in accordance with the manufacturer's instructions.

Air Receivers

Drain air receivers of accumulated condensation at least once each shift, if an automatic drainer is not provided.

Check automatic drainer for proper operation daily. Semiannually, check operation of relief valves and examine receiver for corrosion and peeling of paint. Annually, inspect the receiver internally once a year for corrosion and dirt accumulation.

- Annually clean the receivers.
- Semiannually calibrate the pressure gage.
- Repaint as required.

Air Dryers

Absorption-Type Dryers

Periodically check the pressure and temperature and drying towers of absorption-type dryers. Daily, check the operation of condensation traps on steam reactivated dryers. If reactivation is not under automatic control, reactivate drying towers at the intervals specified by the manufacturer.

Do not overrun the unit. Overrunning results in the tower becoming saturated. Then the moisture-laden air is carried over into distribution systems. In systems where oil carryover from the compressors is present, protect the desiccant bed of the dryer from becoming oil saturated. Oil deposits in the desiccant bed cause a decrease in drying efficiency and necessitate frequent placement of desiccant.

Inspection: Check for the following at the noted intervals:

Daily
- Proper operation of instruments and drain traps, where provided
- Air or steam leaks

Annually
- Relief valves for proper operation
- Dryer towers, piping, and valves for corrosion, rusting, and peeling paint.
- Desiccant bed for oil and dirt

Maintenance:
- Semiannually calibrate instruments.
- Lubricate and repack valves as required.
- Repair all leaks.
- Repair or replace all defective controls and instruments.
- Paint dryer towers and piping when necessary.
- Replace desiccant if it is dark brown or black or gummy. (A light brown color or slight discoloration does not indicate the need for replacement.)

Refrigeration-Type Dryers

Check the trap operation daily to make sure it is draining properly. Do not allow condensation to build up.

Inspection: Check for the following at the noted intervals:

Daily
- Proper operation of condensation drainer
- Air leaks

Quarterly
- Condition of filter cartridge
- Oil and dirt accumulations in the condensation collection chamber and condenser evaporator tube

Maintenance:
- Clean or replace filter element as required.
- Clean deposits from condensation collection chamber and condenser evaporator tubes with compressed air or steam.
- Lubricate and repack valves as required.
- Do not attempt to service the heat or refrigeration unit. If there is a malfunction, contact the manufacturer.

High-Pressure Air Systems

Reciprocating Compressors

Like other types of air systems, keep the compressor free of dust and dirt. For intervals and quality of lubrication, follow the manufacturer's instructions.

Daily, inspect for evidence of heating, which overloads the driver. Regularly service the air cleaner to keep excess dirt out of the air stream and away from the valves.

The first symptoms of trouble are low net air delivery and heating around the valve compartments. On the two-stage compressors, the intercooler pressure gages are used as a guide to locate defective valves. When low intercooler pressures occur, examine the valves on the low-pressure cylinders; when high intercooler pressures occur, examine those on the high-pressure cylinders. By feeling which valve cover plate gives the most heat, locate the defective valve. If high-pressure suction valves are leaking, the intercooler gage hand fluctuates above normal intercooler pressure, and the intercooler safety valve produces a popping noise. If the high-pressure discharge valves are leaking, the intercooler gage hand rises steadily, and pressure builds up in the intercooler until the intercooler safety valve releases it. When low-pressure suction valves leak, if the compressors are operated under load the air flows back through the suction line and the air cleaner. Leaking low-pressure discharge valves cause the intercooler pressure gage to fluctuate below normal intercooler pressure. Referred to the chapter on plumbing for further information on maintenance and repair of valves.

Piston Rings: If the compressor is losing efficiency, often the trouble is caused by the valves. However, if the valves are in good condition, the efficiency loss is due to piston and ring wear. Check piston rings by putting air pressure on top of the piston and listening or feeling for blow-by past the piston rings. To check a double-acting cylinder, remove a valve on one end and apply air at the other end, then check for blow-by on the end with the valve removed.

If blow-by is excessive, remove the pistons and check the piston-to-cylinder clearance and the rings for the amount of wear. Decide which parts need to be replaced. Scored cylinders may be rebored to a standard oversize and new pistons and rings fitted.

Bearings: Refer to Chapter 14 for information on bearings. Lubricate, maintain, repair, and replace bearings according to the manufacturer's instructions.

Intercoolers and Aftercoolers

Coolers are, in effect, condensors. Proper drainage of the moisture traps or compartments is the most important item of maintenance for these components. Condensation or moisture that is not drained regularly collects until water is carried over into the high-pressure cylinders and the intercoolers, and into the air receivers and the air lines of aftercoolers.

Frequent cleaning of outsides of cone sections or air-cooled intercoolers and radiators is essential. After dirt collects in the cone, the insulating effect causes less of the heat to be thrown off. Air blown in a direction opposite to the usual flow removes dust. If the dirt is contaminated with oil, apply a solvent and allow it to soak, after which it can be blown clean. Tube-type intercoolers and aftercoolers are subject to buildup from the mineral content in water. This scaled deposit has an insulating effect which retards cooling, and must be removed frequently.

Refer to the manufacturer's instructions for maintenance and repair of the loading and control systems.

V-Belts and Pulleys

For *drive pulleys*, inspect the V-belt for particles of rubber that cling to the grooves. Scrub with a cleaning solvent and a stiff brush.

Check the pulley grooves for roughness, cracks, and broken sections. Remove any roughness with a fine file and finish off the filed areas with a hand stone. Weld minor cracks around the perimeter of the pulley and dress the welds. Replace the pulley if the hub area is cracked, or if there are any broken sections. Check the inside of the pulley hub for scratches or burrs. Stone out any uneven areas.

High-Pressure Air Compressors: Problems and Solutions

Problem	Cause	Solution
Loss of pressure	Defective pressure gauge.	Replace gauge.
	Leak in pressure gauge line.	Repair or replace line.
	Leak in discharge line.	Repair or replace line.
	Worn or damaged piston rings.	Replace rings.
	Valve disks worn or not seating properly.	Repair or replace valve disks or springs.
	Valve seat worn.	Repair valve seat.
Excessive blowing of compressor pops safety valves	Broken valve in the following stage.	Replace valve.
	Safety valve out of adjustment.	Replace safety valve.
	Obstruction in following stage air inlet line.	Clean or replace inlet line.
Compressor air discharge temperature too high	Recompression due to leaky discharge valves.	Repair or replace valves.
	Increased cylinder discharge pressure due to obstructions in discharge passages.	Attempt to blow out obstructions by opening condensate trap, or remove section and clean.
	Insufficient cooling in intercoolers due to dirty fans.	Clean fans.
	Intercooler or aftercooler cores deteriorated from rust or corrosion.	Replace cores.
	Cores in intercooler or aftercooler leaking.	Replace cores.
	Fan belt slipping.	Adjust or replace fan belt.
Lubricating oil pressure too low	Broken oil line.	Replace oil line.
	Worn pump gears.	Replace gears.
	Dirty strainers or obstructions in line.	Clean strainer and line.
	Defective oil gauge.	Replace gauge.
	Main or connecting rod bearings too loose.	Adjust or replace bearings.
Excessive carbon on compressor valves	Too much oil in the crankcase.	Drain crankcase to proper level.
	Too great an oil flow from force-feed lubricator.	Adjust the force-feed lubricator to deliver the correct amount of oil.
	Worn piston rings.	Replace rings.
	Worn pistons.	Replace pistons and rings.
	Poor grade of oil used.	Drain crankcase and refill with proper grade of oil.
Compressor stage pressure too low	Worn piston rings.	Replace rings.
	Cracked pistons.	Replace piston.
	Cracked cylinder.	Replace cylinder.
	Sticking inlet valve.	Repair or replace valve.
Noisy fan	Fan pulley bearing loose or worn.	Adjust or replace fan bearings.
	Bent fan blades.	Straighten fan blades.
	Lower fan loose on pulley.	Tighten fan attaching cap screws.
	Bent fan guard or shroud.	Straighten guard or shroud.
	Fan belt too tight.	Adjust fan belt.

Figure 15.7

Check the fit of the shaft key and the shaft keyway and pulley hub keyway for looseness. If the key is loose in either keyway, measure the keyways in the key. If both keyways measure the same, replace the key. If one keyway is wider than the other, machine the smaller to the size of the larger, and by filing or machining provide key of proper dimensions.

Maintain air supply lines and filters, air receivers, and protective devices according to the manufacturer's instructions. Refer to Figure 15.7 for causes and solutions to problems with high-pressure air compressors.

Controls Prime Mover Controls

Adjust prime mover controls in order to regulate compressor output in accordance with the manufacturer's instructions. Periodically check liquid levels and hydraulic systems. Keep all electrical contact surfaces of components, such as rheostats, clean at all times and lightly greased, if required.

Inspection: Check for the following at the noted intervals:

Daily
Inspect the control system for proper operation.

Annually
- Inspect the control system for corrosion, wear, and dirt.
- Check all governor linkages, mechanisms, springs, and pins for mechanical defects and wear.
- Examine governor valves for erosion, corrosion, proper seating, and condition of packing and diaphragms.

Compressor operation noisy	V-belts too tight.	Adjust belt tension.
	Crankshaft end-play too great.	Adjust end-play.
	Connecting rod bearings too loose.	Inspect bearings and tighten or replace.
	Connecting rod bushings worn.	Replace bushings.
	Drive pulley loose.	Tighten drive pulley.
	Main bearings too loose.	Inspect bearings and tighten or replace.
	Valves loose.	Adjust valve setscrews.
	Valve springs broken.	Replace valve springs.
	Oil pressure too low.	Replace oil pump or oil regulator. Clean oil strainer.
Air leaks	Air leaking around valve covers.	Tighten valve cover cap screws or replace valve cover gasket.
	Air leak in inlet piping.	Tighten connection, replace gasket, or replace piping section.
	Air leak in discharge piping.	Tighten connection, replace gasket, or replace piping section.
	Air leak in intercooler-aftercooler assembly.	Replace intercooler or aftercooler as required.

Figure 15.7 (continued)

Maintenance: At each annual inspection, or as required, thoroughly clean all control components. Clean all sludge and accumulations from hydraulic systems with a solvent. Thoroughly flush out all control lines and refill systems with clean hydraulic fluid in accordance with the manufacturer's instructions. For frequency and quality of lubrication, refer to the manufacturer's instructions.

Repair or replace defective control system components as required. Consult the manufacturer's instructions for maintenance procedures. Perform the following maintenance work as needed.

- Replace defective diaphragms, bellows, and gaskets.
- Carefully inspect, clean, and test governor valves. Regrind or replace worn valve seats and discs. Replace badly worn valve stems, cylinder liners, piston rings, piston rods, and packing. Check valve positioner and spring adjustments.
- Repair or replace defective chains and sprockets.
- Replace badly worn linkage pins and pushings.
- Repaint piping as required. Repair all leaks.

Compressor Controls

Any malfunction in the compressor controls should be dealt with immediately. Thoroughly drain all moisture from the pneumatic control line and separators at least once each shift.

Inspection: Daily, inspect the compressor control system for proper operation of all control components. Annually, thoroughly inspect all control system components for wear, corrosion, dirt, and other defects.

Maintenance: Refer to the manufacturer's instructions for the maintenance procedures for each particular control component. Perform the following maintenance work as needed:

- Carefully inspect, clean, and test the operation of the onloaders, clearance control, and pilot valves. Regrind or replace worn valve seats and discs. Replace badly worn valve seats and defective diaphragms. Check spring tension adjustments.
- Thoroughly clean moisture separators and strainers. Remove all oil deposits. Replace defective strainer elements.

For quality and intervals of lubrication, refer to the manufacturer's instructions. Thoroughly clean and test for leaks all control system piping. Remove all dirt and oil from lines by blowing out with compressed air.

Steam Distribution Systems

The frequency of inspecting piping systems depends on the type of installation. Follow the maintenance schedules below to insure trouble-free service and piping system reliability for above ground systems:

Monthly
Check for the following:

- Leaks
- Damaged insulation
- Abnormal pressures and temperatures
- Abnormal pressure drops
- Vibration
- Correct operation of equipment, such as stream traps, pumps, pressure and temperature controllers, and strainers

Annually
Check for the following:

- Piping corrosion, leaks, and loose joints; damaged or missing supports
- Condition of insulation and protective jacket
- Setting or shifting in position of poles, hangers, or other supports
- Leaky, corroded, or defective valves; packing glands, hand wheels, body flanges, and gaskets

- Valve and meter pits for clogged vents, structural damage, missing covers, and accumulations of dust and dirt
- Condition of flanged fittings
  Note: Repair leaks as soon as possible to avoid wire drawing of flanges and damage to the insulation.
- Condition of expansion joints
- Condition of anchors, hangers, guides, and supports
- Condition of steam traps, strainers, and moisture separators
- Condition and calibration of pressure-reducing stations
- Setting of relief safety valves
- Condition and calibration of instrumentation
- Condensate return piping

Refer to the above-ground system checklists to inspect and maintain underground systems. In addition, inspect the following items:

- Leaks, monitored with a nitrogen monitoring system
- Condition of pipes
- Condition of tunnels and manholes
- Condition of any existing cathodic protection systems

Repairs

For main lines and conduits, make necessary repairs after periodic inspection as indicated in the preceding paragraphs. Repair all pipe leaks immediately by tightening loose connections, tightening or repacking valve glands and conduit-waled glands, replacing gaskets, and welding or replacing defective parts or sections.

Repair the coatings and coverings of protected metallic conduits as necessary. If the area to be repaired is small, use a standard asphalt blanket patch, cut to cover the required area. Heat the patch with a torch and place over the damaged area. For larger areas, a canvas blanket covered with hot asphalt is an effective repair procedure.

Replace damaged installation and repair leaks of metallic conduits by caulking, peening, or welding. If required, replace defective sections.

Provide and maintain good water drainage. Use proper gasket material of correct design for the flue pressure and temperature service conditions. Control corrosion of lines.

Repair or replace supports, hangers, guides, expansion joints, and auxiliary equipment as required.

Make repairs to manholes as required following periodic inspections. Stop water leaks in the floors and walls.

Other Distribution Systems

Maintenance procedures for *high-temperature water distribution systems* are essentially the same as those given for steam distribution systems.

The rest of this section address compressed air distribution systems. The frequency of inspecting piping systems depends on the type of installation. Use the following inspection schedules for average above ground installations.

Monthly
Check for the following:

- Leaks.
- Moisture and dirt—Check traps, strainers, and dehumidifiers for proper operation.
- Abnormal pressures—Insufficient pressure indicates excessive leakage, line obstructions preceding a using station, clogged strainers, defective compressor operation or controls, or improper operation of pressure-reducing stations. Excessive pressure indicates malfunction of the compressor control, improper setting or defective operation of the air receiver safety valve, or malfunction of the pressure-reducing stations.

- Excessive pressure drops—An overload of the pipe line or a line unduly obstructed results in an excessive pressure drop.
- Vibration—Inadequate or defective supports and improper anchorage causes vibration.
- Corrosion—Protect the external surfaces of air lines from corrosion by using corrosion-proof paint or adequate covering.

Annually

Check for the following:

- Piping corrosion, leaks, loose joints, damaged or missing supports
- Setting or shifting of poles, hangers, and other supports (determined by checking the grade of the lines)
- Leaky, corroded, or defective valves, packing glands, hand wheels, seats, bodies, flanges, and gaskets
- Valve pits for clogged vents, structural damage, missing covers, and accumulation of dust and dirt
- Condition of flange fittings
- Condition of expansion joints, if any
- Condition of hangers, guides, supports, and anchors
- Condition of traps, strainers, dehumidifiers, and moisture separators
- Condition and adjustment of pressure-reducing stations
- Setting of relief and safety valves
- Condition of air receivers
- Condition and calibration of instruments

Refer to the above-ground system checklists to inspect and maintain underground systems.

To repair compressed air distribution systems, refer to the section in this chapter on repairing steam distribution systems.

Instruments and Controls

Only authorized and trained personnel should repair, calibrate, or adjust instruments. Use the inspection schedules below for average installations and check for the following:

Daily check all accessible instruments and gages for defects such as leaks, cracks, and broken glass bed pointers

The following should be checked annually:

- Corrosion, deposits, binding, and mechanical defects of gages
- Ruptured or distorted pressure parts
- Incorrect calibration or adjustment
- Leaking pipes or meters
- Plugged internal passages
- Loose pointers
- Broken balance arm screws
- Broken or loosen linkages
- Broken or damaged adjustment assemblies
- Dust and dirt
- Binding of moving parts
- Dirty gage movements
- Linkage pins binding
- Defective gage glass gaskets
- Gage valve packing leaks
- Worn pins or bushings
- Mercury separation thermometers
- Temperature bulbs damaged by overheating
- Plugged piping or tubing and loose connections
- Mercury contamination
- Defective floats
- Operation of clockwork mechanism and integrator

Maintenance of Instruments

After the yearly inspection, repair or replace defective parts as necessary in accordance with the manufacturer's instructions. Below is a list of typical repairs.

- Replace bourdon tube assemblies distorted from over pressure.
- Replace temperature bulbs which are swollen as a result of excessive temperature.
- Replace all gage movement linkages, pins, and bushings which have lost motion due to wear.
- Replace or repair broken adjustment assemblies.
- Clean dirty mercury by straining through chamois cloth.
- Repair all leaks in piping or meters.
- Repair or replace defective floats.
- Replace defective gaskets.
- Rod out all plugged or partly plugged pipe line connections.
- Clean out the meter connecting piping.
- Lubricate according to the manufacturer's instructions. Use recommended lubricants.
- Replace all corroded, bent, or otherwise defective parts.

Inspection of Controls

Check for the following at the noted intervals.

Daily

- Observe operation of controls for proper operation.
- Inspect for leaks.
- Repair stuffing box leaks as soon as possible.

Annually

- Dismantle regulator valve and control mechanism; clean system components; and inspect for wear, corrosion, erosion, pitting, deposits, leaks, and mechanical defects.
- Inspect all safety devices and warning signals for defects and observe for correct operation.
- Check settings, adjustments, and operation of controls.
- Observe operation of safety devices.

Maintenance of Controls

After the yearly inspection, repair or replace defective parts as necessary in accordance with the manufacturer's instructions. Perform the following maintenance work as required.

- Examine the regulating valve stem and the valve plug and seat, and replace them if necessary. Change the valve to a smaller size if excessive cutting indicates valve is oversized. Check valve position and spring adjustment. Repack stuffing box.
- Clean, inspect, and test needle, pilot, poppet, and reducing and transfer valves. Replace defective parts.
- Observe condition of bellows and diaphragms. Replace if defective.
- Replace gaskets.
- Replace worn linkage pins and bushings.
- Clean air filter. Replace air filter cartridges.
- Repair all leaks. Rod out piping connections when necessary.
- Vent out air from liquid-filled systems.

Expansion and Construction Joints

No specific maintenance is required for expansion loops and bends other than inspection for alignment. General inspection procedures as used for ordinary piping include:

- Checking for leaks.
- Inspecting supports.
- Making certain that operating conditions do not exceed temperature and pressure ratings.

Annually, look for signs of erosion, corrosion, wear, deposits, and binding on *mechanical slip joints*. Repair or replace defective parts as required. Use the list below to inspect and maintain slip joints.

- **Alignment:** Check to be sure pipes are in alignment.
- **Packing:** Adjust or replace packing as required to prevent leaks and ensure free-working joint. Maintain the packing material and its installation in accordance with the manufacturer's instructions.
- **Travel:** Check the flange-to-flange distance of the joint, first when cold, then when hot. Compare the findings with the travel limit shown on the manufacturer's data. A change in slip travel may indicate a slippage in the anchorage and/or pipe guide. Locate and correct difficulty.
- **Lubrication:** Lubricate slip joints every six months or in accordance with the manufacturer's standards, utilizing a quality lubricant in the standards.

Annually inspect *bellow-type* joints for alignment, fatigue, corrosion, and erosion. Check the travel distance of the joint under both cold and hot conditions, and compare the findings with the manufacturer's instructions. In case of failure, the bellow sections must be replaced.

Usually no lubrication is required with *flexible ball joints*, as there is little friction or wear. The frequency of gasket replacement depends on operation conditions; replace gaskets as required in accordance with the manufacturer's instructions.

Anchors

Quarterly, inspect anchors for corrosion, breaks, and shifting. Make sure that the anchors are holding, and that the walls and footings near anchors do not show distress cracks. Make certain the bolts, turnbuckles, wire ropes, and other stressed members give no sign of possible early failure. If an anchor has shifted, take the line out of service and secure the anchor in place as required. Make necessary adjustments or repairs to the expansion joints and guides.

Hangers and Supports

Quarterly, inspect hangers and supports. Check for corrosion, wear, and failed parts. Make sure that all supports are in line with pipe and tracking tube, that supporting rollers turn freely, that each hanger carries its share of the load, and that there are no signs of possible early failure of any stressed member. Repair or replace defective parts as required. Line sags caused by misalignment can often be corrected by simple hanger adjustment.

Traps, Drains, and Vents

Test all traps for correct operation monthly. Annually dismantle and clean all traps and inspect for the following:

- Accumulation of foreign matter
- Plugging of orifices, valves, and vents
- Cracked, corroded, broken, loose, worn, or defective parts
- Excessive wear, grooving, and wire drawing of valves and seats
- Defective bells, buckets, or floats.
- Leaky vessels and pipes
- Defective bypass valves

Maintenance of Traps

After the yearly inspection, repair or replace defective gaskets, bellows, valves, valve seats, floats, buckets, linkages, and orifices as required in accordance with the manufacturer's instructions. Repair or replace leaking bypass valves and repack valve stems.

Maintenance of Air Vents

At least once a year, inspect and clean air vents. Remove vents from the line and allow to soak in a container of kerosene for about 24 hours to loosen rust and grease. However, do not soak vents using composition discs in kerosene, as the disc may be damaged. After the soaking period, place vent in vertical position, drain thoroughly, and allow to dry. Blow vent with air pressure. If vent performance cannot be corrected by cleaning, replace it. Dirt on the valve, failure of the thermostatic element, or a water pocket in the vent connection may result in water and steam spouting in the vent.

Air Conditioning

Refer to the checklist shown in Figure 15.8 when making an inspection of a building's air conditioning system.

Checklist for Inspection of Air Conditioning Systems

____ **Lubrication:** inadequate lubricating instructions, excessive bearing temperature; inadequate lubrication of bearings and moving parts, low oil level, poor oil condition.
Lubricate as required; add lubricant if grease dispenser or oil cup is less than half full; add oil to crankcase of refrigerant compressor if below correct level, and change if dirty; clean clogged oil lines.

____ **Rust and Corrosion:** Remove rust; paint bare spots and corroded areas.

____ **Motors, Drive Assemblies, and Fans:** dust, dirt, and graft accumulations; worn, loose, missing, or damaged connections and connectors; bent blades; worn or loose belts; unbalance, misalignment; excessive noise and vibration, end play of shafts; ineffective sound isolators; poor condition of motor windings and brushes.
Remove accumulations; tighten loose connections and parts; tighten loose belts, replace multiple belts in sets when one is worn; replace defective brushes; make other minor repairs and adjustments.

____ **Wiring and Electrical Controls:** loose connections; charred, broken, or wet insulation; short circuits; loose or weak contact springs; worn or pitted contacts; defective operation; wrong fuses, and other deficiencies.
Tighten loose connections and parts; replace cords with wet insulation or where broken in two or more places, and braid that is frayed more than 6 inches; replace or adjust contact springs; clean contacts; replace defective or improper fuses.

____ **Temperature and Humidity Controls:** improper setting; loose connections; worn, dirty, pitted, or misalignment of contacts; defective operation noted in observing operation through complete cycle; inaccuracy of thermostats found by comparing with mercury thermometer (dry-bulb type); inaccuracy of humidistats found by comparing with sling psychrometer and pyschrometric chart.
Adjust settings; tighten connections; clean contacts and adjust alignment.

Figure 15.8

Checklist for Inspection of Air Conditioning Systems
(continued)

—— **Steam and Hot Water Heating Units:** clogging; dirty heat-transfer surfaces; leaking; loose connections and parts; bent fins; misalignment; water hammer; air-binding; non-uniform heat spread; open bypass valves; below-normal temperature readings; defective valves, traps, and strainers.
Clean heat-transfer surfaces; tighten loose connections and parts; replace leaking valve packing; close bypass valves; clean valves, traps, and strainers.

—— **Electrical Heating Units:** burned, pitted, or dirty electrical contacts; short-circuited sections of elements; low voltage in electrical circuits; dirty reflective heat-transfer surfaces.
Clean contacts and heat-transfer surfaces; tighten loose connections; make minor repairs and adjustments.

—— **Air Ducts, Dampers, Registers, Grills, Louvers, and Bird and Insect Screens:** soot, dirt, dust, and other deposits; leaks; broken, loose, or missing connections and parts; excessive vibration; material defects; defective operation of movable parts; hinge parts failure; improper seasonal or operating settings of dampers; inadequate air distribution in branch circuits.
Remove deposits; tighten or replace defective connections and parts; caulk around flashings and make weathertight; adjust damper settings; make minor repairs and adjustments.

—— **Thermal Insulation and Vapor Barriers:** wet, damaged, or missing; broken tie-wires; loose bands; torn canvas jackets.
Repair or replace to restore insulating properties.

—— **Air Filters:** dust, grease, and other deposits; missing parts; improper fit.
Replace dirty throw-away type filters, and those missing or with improper fit; wash permanent-type filters in soap suds or solvents and rinse with hot water; restore viscous coating in accordance with the manufacturer's instructions.

—— **Guards, Casings, Hangers, Supports, Platforms, and Mounting Bolts:** loose, broken, or missing parts and connections; deformations; improper level; ineffective sound isolators.
Tighten loose connections and parts; adjust level; replace defective sound isolation; make minor repairs and replacements.

—— **Pump Units:** dust, dirt, and other deposits; leaks; noise; vibration; loose or missing connections or parts; defective operation.
Clean; repair leaks; make other minor repairs.

—— **Piping:** leaks; corrosion; deformations; material defects of fittings, copper tubing, and steel piping.
Repair leaks; make minor repairs.

—— **Water Sprays, Weirs, and Similar Devices:** external scale; leaks; defective valves, including float valve in sump; clogged nozzles or pipes; improper positioning of spray nozzles, baffles, or eliminators to control spray drift, to compensate for prevailing winds, or to provide cooling for entire coil; material defects.
Remove external scale; wipe external surfaces with cloth dipped in solvent; tighten leaky connections; replace defective gaskets; adjust float valve; clean clogged openings with wire; position nozzles and adjust baffles or eliminators; make other minor repairs.

—— **Compressors:** dirt and dust accumulations; leakage of oil, water or refrigerants; loose connections; loose or worn belts or parts; misalignment; excessive noise and vibration; incorrect suction and discharge pressure.

Figure 15.8 *(continued)*

433

Checklist for Inspection of Air Conditioning Systems
(continued)

Remove dirt, dust, and other accumulations; use Halide torch to check refrigerant leaks, and watch for flame color change indicating leak; tighten loose connections, belts and parts; replace worn belts; correct misalignment; make other minor repairs.

___ **Shell- and Tube-Type Condensers:** dust accumulations; leaks, including connection to cooling tower. (Internal scale indicated by small difference in temperature of refrigerant and cooling water when taken by mercury thermometer at inlets and outlets.) Remove dust accumulations.

___ **Self-Contained Evaporative Condensers:** leaks at pump or in piping; improper setting of float control device; improper overflow of solids contained in water; insufficient outdoor air flow; clogged nozzles; inadequate spread of water. Repair all leaks; adjust float controls; adjust outdoor air flow; clean clogged nozzles; adjust sprays.

___ **Air Cooled Condensers:** dust accumulations; leaks; excessive noise and vibration; loose, missing, or damaged parts. Vacuum or brush out dust; repair leaks; tighten loose connections; replace or repair missing or damaged parts.

___ **Liquid Receivers:** leaks; clogging; cracked gage glasses; damaged parts. Repair or replace defective refrigerant charging valves; make minor repairs.

___ **Refrigerant Driers, Strainers, Valves, Oil Traps, and Accessories:** inadequate operation. Clean out strainer baskets; replace missing or worn parts; repair leaks and other defects.

___ **Accessories:** In the fall, cover window-mounted air conditioners with insulated coverings (remove in the spring), close outdoor air-intake dampers on summer-use units, and adjust dampers to winter setting on summer-winter air conditioners; replace defective caulking around window units.

___ **Cooling Coils:** dust; leaks; bent fins; improper level, liquid flow not toward outlet of coil; obstructions to air flow; inadequate refrigerant output; excessive frosting; defective operation of direct expansion valve and automatic temperature controls. Clean out dust, particularly between fins, using vacuum cleaner or brush; straighten bent fins; correct level; remove obstructions from air flow; make adjustments and repairs.

Figure 15.8 (continued)

PART V

MANAGEMENT OF THE MAINTENANCE FUNCTION

Chapter 16 provides general guidelines for the evaluation of existing organizations or for structuring a new organization to carry out the function of maintenance. The chapter includes:

- Typical organization charts.
- Staffing and job descriptions.
- Size of maintenance crews.
- Methods of hiring, training, and evaluating in-house maintenance personnel.
- Advantages and disadvantages of doing in-house maintenance versus contracting it out.

Estimating serves as the foundation for most of the managerial tools used for directing the maintenance function. The effectiveness of any maintenance operation depends on how the estimator develops the estimate and follows through with it. The subject of estimating maintenance and repair is covered in Chapter 17.

Planning, scheduling, and cost control are basically fundamental processes which may seem confusing in relation to the total job. Chapter 18 is devoted to the presentation of actual methods of planning, scheduling, and cost control and includes examples of standard formats that can be utilized in part or in whole for the purposes of documenting cost data for future use.

CHAPTER 16 MAINTENANCE ORGANIZATION

This chapter provides general guidelines for evaluating existing maintenance organizations or for structuring a new or better organization to carry out the maintenance function.

Objectives of an Effective Organization

The first step is to establish objectives for the maintenance organization. The maintenance objectives must be consistent with those of the entire company. Below are examples of objectives for a maintenance organization.

- To maintain the plant and equipment in optimum condition in order to avoid repairs.
- To maintain and repair the plant and equipment economically and within the allocated budget.

Note that these are general objectives written in idealistic terms. These general objectives form the foundation of the organization.

The purpose of organization is to obtain the maximum benefit with the least effort and cost. Because budgeted dollars for manpower, equipment, and supplies frequently are not sufficient for the tasks to be performed, the maintenance manager is constantly challenged to find more efficient ways of doing things. Maintenance operations are managed more efficiently by effectively organizing two components of work: the work itself and the people who perform the work. These components are defined below.

- **The organization of work:** means establishing priorities so that critical tasks receive attention first, and scheduling the sequence of activities necessary to complete those tasks.
- **The organization of people:** includes an organizational chart, job descriptions for all employees, and the daily work hours available classified by skill.

The formation of a formal organization aids the maintenance manager in finding new ways to combine available manpower, materials, and equipment in order to increase the maintenance team's productivity. A formal structure enables the maintenance manager to match up the available manpower and materials with the workload. To realize the objectives of any organization, the maintenance organization must be formally organized and delegated specific responsibilities.

Organization of Maintenance

In order to function the most effectively, the maintenance organization must have as its foundation a formal structure that defines clearly the responsibilities at all levels in the company, as well as how those levels are interrelated (referred to as the hierarchy). The extent of detail of the formal hierarchy depends on the complexity of the maintenance organization. It may be as simple as a supervisor-employee relationship (a two-person department) or a department comprised of sub-departments having hundreds of employees.

The scope and activities of a maintenance organization (or department) depend on the factors listed below.

- Plant size
- Type of operation to be maintained
- Company policy
- Scope of operation in terms of geographic extensiveness

Setting up a maintenance organization involves organizing the type and amount of work and the hiring, scheduling, and training of qualified personnel. The basic needs of a maintenance organization are listed below.

- Schedule of daily and periodic activities
- Materials, tools, and equipment for necessary repairs
- Methods, techniques, and procedures for maintenance, repair, and housekeeping
- Trained personnel (and a training program)

The actual work of a maintenance organization can be grouped into the classes described below.

- Small jobs are frequent and routine, so standards for activities can be established, such as inspection and lubrication of equipment. These jobs can be assigned with definite instructions to specific personnel.
- Medium jobs require various skills, including the help of service groups, such as maintenance shop personnel. An example is changing filters in a piece of equipment.
- Major jobs require various skills, are lengthy, and possibly require the aid of outside consultants. An example is a major overhaul of equipment.
- Irregular jobs, such as the repair of a concrete beam, are done by outside consultants.

A typical maintenance organization establishes priorities; plans and schedules the various stages of work; establishes preventive maintenance schedules; establishes cost; and provides cost control, analysis, and follow-up.

Types of Maintenance Organizations

A maintenance department is organized according to a number of factors, including structure, objectives and goals, work orders, planning, and scheduling. A maintenance department may be functional or geographic, centralized or decentralized, or a combination.

A functional organization refers to the division of responsibility based on the types of services provided or the crafts involved. Each craft supervisor reports to the highest level of management in maintenance, usually the plant engineer. In a geographical organization, the division of responsibilities is based on plant layout, and is either centralized or decentralized.

Centralized Organization

In a central maintenance organization, people are assigned to work in any or all departments of the plant and report to the same head. The advantages of a central maintenance system are:

- Sufficient people are available to handle the work requirements of the plant.
- Considerable flexibility is available to assign members of different crafts to the various jobs.
- Emergency jobs, breakdowns and new work are handled quickly.
- The total number of men can be held reasonably level, minimizing hiring and layoffs.

- Specialists are utilized more effectively.
- Special maintenance equipment is used effectively.
- One individual is responsible for all maintenance.
- Accounting for maintenance costs is centralized.
- More control is obtained over capital or new work.
- People are better trained in their craft skills.
- A central group can justify more readily the need for trained personnel.

The central maintenance group provides excellent flexibility in getting maintenance work done in a plant. In industries or plants that normally need large crews to check equipment, the central maintenance group is generally used. For example, in oil refineries where cracking columns require periodic overhauling, or in steel industries for repair of blast furnaces. The entire maintenance cost of the plant, its equipment, buildings, and grounds are the responsibility of one person. Some of the disadvantages are:

- Mechanics are scattered all over the plant and not properly supervised.
- Time is lost in traveling to a job, obtaining tools, and getting instructions.
- Coordination or scheduling of several crafts to a job becomes difficult.
- More administrative controls are needed to function effectively.
- Different people are assigned to the same equipment, consequently, none really becomes responsible for and proficient in its repair.
- Intervals between initial job requests and completion for routine work are long.
- Priorities are given to the various production jobs by a maintenance supervisor, not a production supervisor.

The job priorities in this type of setup tend to be a headache. Every production or service department feels that their job is important. Backlog of work in a central group may be several weeks for effective planning of workers and material. This backlog is not appreciated by the production department. Weekly planning sessions are often needed with the production departments and plant manager to agree on the order of the work.

Decentralized Organization

Decentralized maintenance, sometimes referred to as area maintenance, is carried out within the confines of specific areas in a plant, and these specific areas, supervised by foremen or superintendents, report to a central location. The area may be defined geographically, by product or department. Some of the advantages of decentralized maintenance are:

- Reduced travel time to and from jobs.
- More intimate equipment knowledge by workers.
- Improved application to the job, due to closer alliance with the objectives of a smaller unit.
- Better preventive maintenance.
- Improved maintenance-production relationships.
- Time lag is minimized between the issuance of the work order and its completion.
- Mechanics are better supervised.
- Production line or process changeover is faster.
- Greater continuity from one shift to another is afforded.

Some of the disadvantages are:

- A tendency to overstaff the area.
- Major repairs for servicing jobs are difficult to handle.
- There are more problems and regulations pertaining to transferring, hiring, and working overtime.
- Special equipment is difficult to justify because usage may be limited.
- Duplication of equipment occurs in the area maintenance shops.
- More clerical help is needed if the area groups are large.
- A specialist is difficult to employ effectively.

Rather than assigning an arbitrary number of people to a decentralized facility, a study should be made of the types of services required to sustain production in the area under consideration.

Departmental Maintenance: Another type of organizational setup is the departmental maintenance organization, a form of area maintenance.

In the departmental maintenance organization the mechanics are assigned to a definite area or function and report to a production supervisor. Sometimes mechanics report to a maintenance foreman who in turn reports to a production supervisor. The advantages of the departmental group are similar to those in the decentralized maintenance category. The major difference is that the mechanics report to the production supervisor. The production supervisor has complete responsibility over an important factor affecting production and cost. There is no problem of priorities over other groups, and the work is scheduled by the supervisor responsible for new production quarters.

The disadvantages are also similar to decentralized maintenance, but may also include:

- Production supervisors are not qualified to direct a maintenance job.
- Production supervisors cannot give technical assistance to mechanics.
- Production supervisors may neglect maintenance in order to meet schedules.
- The maintenance responsibility of the plant is divided.
- The plant maintenance costs are harder to control.
- Personnel problems are more pronounced than in decentralized maintenance.

Centralized and Decentralized Combined

Management and plant personnel aware of the difficulty in balancing servicing and maintenance costs have attempted to resolve this problem by combining a central group with an area or department group. The advantages and disadvantages of the basic systems are still present to the extent that they are combined and modified by the basic factors that comprise the maintenance function. Some of the advantages for a large company may be:

- A group of central mechanics capable of handling the large projects and major repairs throughout the plant.
- Good control of maintenance costs.
- Area mechanics available to major service and production centers.
- Area mechanics familiar with key equipment in the production centers.

The major disadvantages are:

- Central mechanics assigned to work throughout the plant, with resulting high travel time and poor job supervision.
- Major job priorities determined by maintenance.
- Tendency to overstaff an area.
- Duplication of equipment.

In conclusion, in small plants a central maintenance plant is generally used. Large plants that manufacture one product or closely related products, may also have a central maintenance group of a thousand or more people. Plants with several different major products and/or service functions, such as research, will have a combination central and area department maintenance group. There are, of course, exceptions, and the selection of the best maintenance plan for a plant requires careful analysis of the maintenance function.

Examples of Maintenance Organizations

The following are some typical organizations that may be found relative to maintenance in different companies. Figure 16.1 depicts a medium-size company, geographically dispersed over about 100 acres and employing about 300 maintenance workers. Figure 16.2 shows a company similar to the previous figure, but one where the maintenance function is handled a little differently. Figure 16.3 shows the organization of another company, a small plant of 25 to 30 maintenance employees where the person in charge of maintenance is responsible for all types of work. Figure 16.4 is a large plant

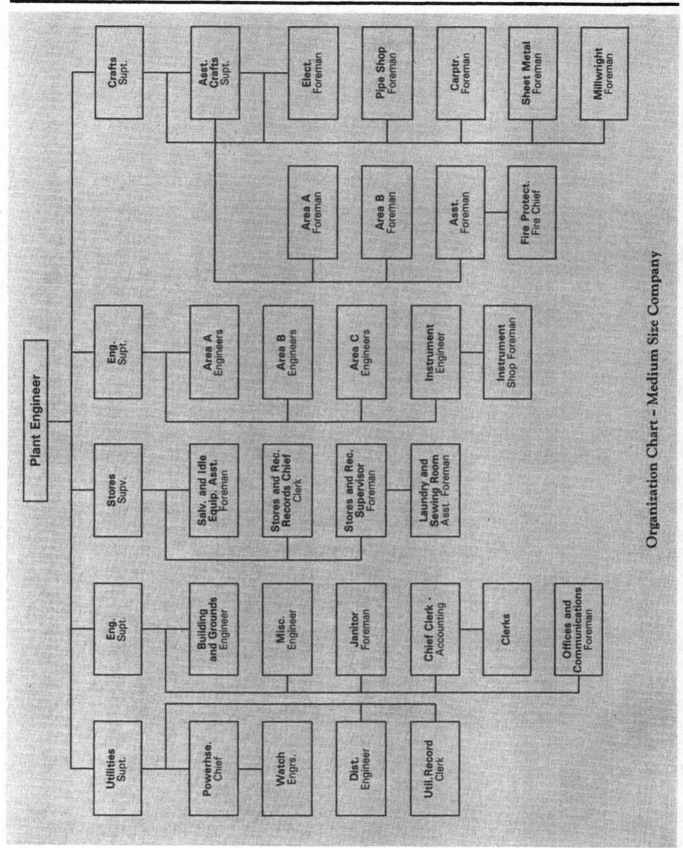

Figure 16.1

Organization Chart – Medium Size Company

441

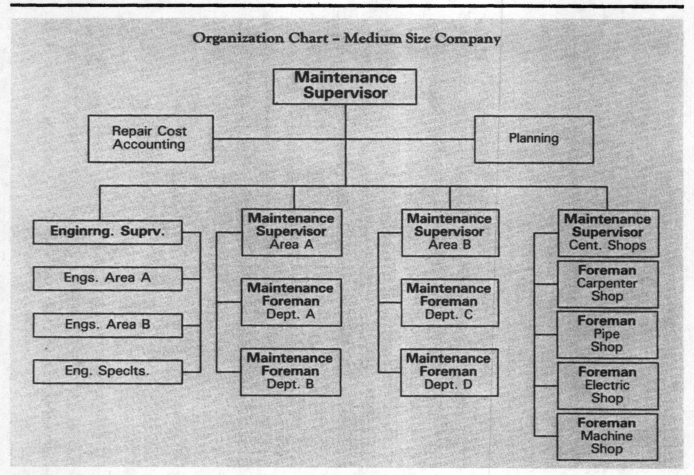

Figure 16.2

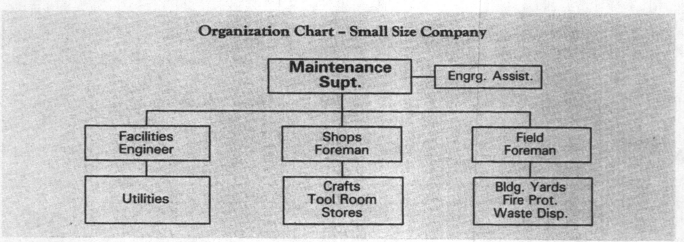

Figure 16.3

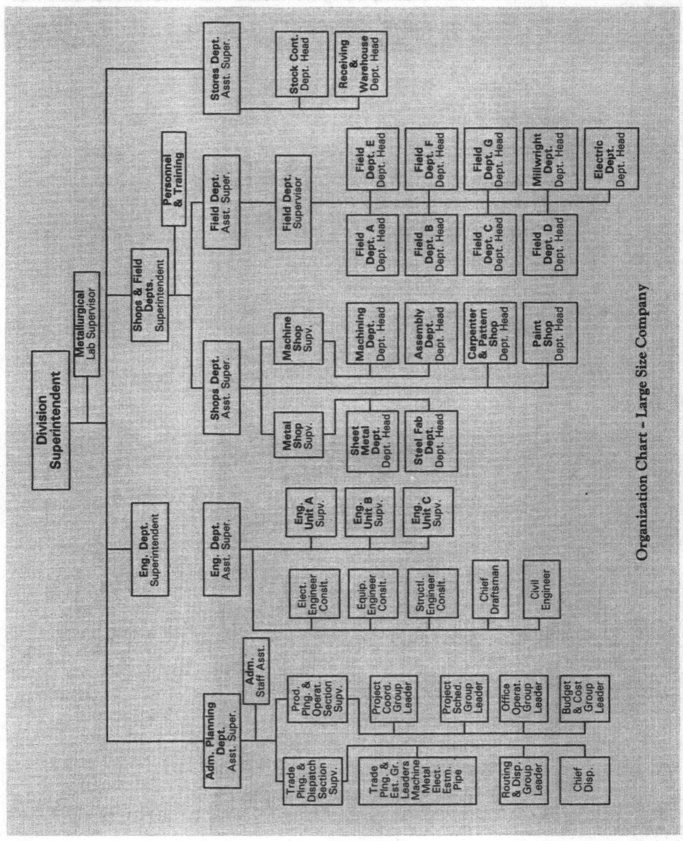

Organization Chart – Large Size Company

Figure 16.4

443

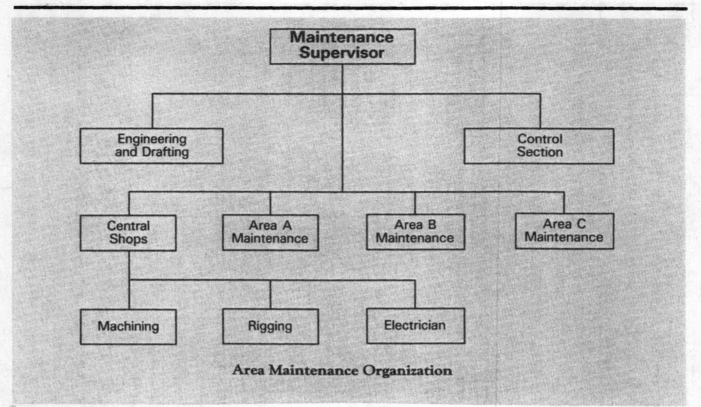

Area Maintenance Organization

Figure 16.5

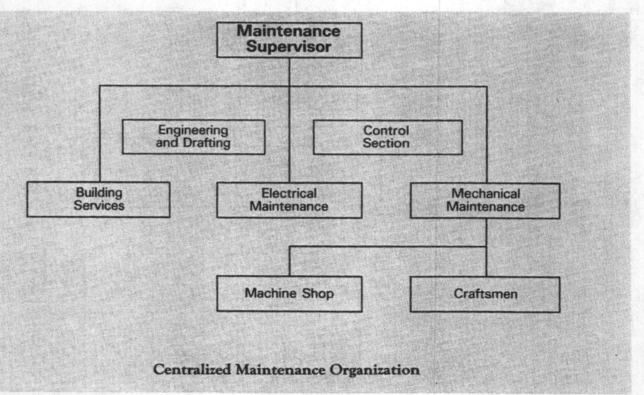

Centralized Maintenance Organization

Figure 16.6

with a working force in excess of 15,000 people and a maintenance force of 3,000 people. Figures 16.5 and 16.6 are comparisons of organizational charts for an area maintenance organization and a centralized maintenance organization.

Maintenance Policy Formation

Whatever type of maintenance organization is utilized, the goals on which it is founded should be clearly defined on paper and understood by all maintenance personnel. Each goal needs to be expanded into a policy statement. The policy statement defines the objectives, sets limits of authority and responsibilities, and serves to regulate and guide maintenance activities. In establishing an organization or revamping an existing one, all personnel affected by the policies should be involved. If this is done, the personnel will readily accept and abide by the policies. Once policies for the maintenance organization are developed, a workable plan to implement the policies is established. In the selection of a maintenance plan, top management's philosophy is important and the plan should be compatible with it. The organization of the maintenance department's policies and procedures must be stated in terms that are understandable and measurable. These include:

- A formal organization plan and chart.
- Maintenance practices, such as preventive maintenance plans, work schedules, and standardization of activities.
- A method of approval of work.
- Accounting procedures.
- Material requisition techniques.
- Training programs.

Example policy statements are listed below.

- Maintenance foreman will report to the maintenance head and shall be responsible for all assigned work as per their job description.
- The breakdown of any and all on-line machinery shall be given first priority for maintenance work assignments.
- Any single maintenance item requiring capital expenditure of more than $1,000 must have approval of the maintenance head.
- All new employees shall be given a ten-hour on-the-job training program relative to the responsibilities and duties spelled out in their job descriptions.

The type of organizational structure should be reviewed carefully, and the responsibilities of each group established. This may disclose such flaws as overlap of responsibilities. It is mandatory to keep the vertical lines of authority and responsibility as short as possible so as to facilitate maximum utilization of manpower toward productive work, and minimize the magnitude of effort expended in getting things done in an over-bureaucratic maze of formalities. For example, the number of assistants should be kept to a minimum. Also, the number of people reporting to any one supervisor should permit the accomplishment of the goals or sub-goals of the team. The "effective" number of people reporting to any one supervisor is known as span of control or span of management and is usually limited to between four and six persons.

Job Descriptions

Once the organizational chart has been developed, job descriptions should be prepared for all personnel, describing their responsibilities, authority, and duties. The job description should be written for each position and in language that is understandable, but more important, measurable.

The job description should begin with a definition or a general description of the position, such as assistant mechanical superintendent, general janitor or janitorial foreman. Once the position or job title has been established, the next thing is to clarify the area, department, or the location in the organization to which this position would be assigned. Then after this is done, the primary function of the person filling the position should be established. Following this, very specific duties and responsibilities of the position should be listed. A description should be developed for every job in the organization. Personnel should be selected, trained, and evaluated according to their job descriptions. An example of a job description for an area foreman follows.

> Department: Buildings and Grounds
> Responsible to: Plant Engineer
> Responsible for: 4 painters, 3 carpenters, 4 masons, 10 laborers
> Primary function: To supervise the performance of maintenance work
> in the field

Duties and Responsibilities

- Analyze repair orders as to required skill, materials, labor, equipment, tools, and methods, and prepare estimate.
- Requisition materials.
- Prepare daily schedules according to operating conditions, manpower, and availability of materials.
- Responsible for completion of all work.
- Responsible for quantity and quality of all work performed by personnel supervised.
- Account for time and charge for personnel supervised.
- Responsible for training of personnel supervised.
- Assist in the formulation of policies, the promotion of sound labor relations, and the settling of grievances. Attend meetings.
- Conduct meetings for own personnel—enforce use of all applicable safety regulations. Investigate any injuries affecting own group and prepare injury reports.
- Establish backlog.
- Establish priorities for all work done.

Hiring the Maintenance Manager

A formal advertisement and interview process should be used when hiring for the position of maintenance manager. Important factors to consider during the selection process include educational background, experience, age, and professional philosophy.

Experience is the most important factor in making the final decision. An alternative to hiring the maintenance department head after all policy has been formally established, would be to hire the person prior to making the policy and use him or her as a consultant in the process.

When a new facility is being constructed, the maintenance manager for that facility should be hired prior to the beginning of construction so he or she can observe the installation of all materials and equipment. If the facility is already in existence, a training or break-in period for the new head should be established.

The manager of a maintenance organization should follow the aims and objectives listed below.

- Plan and schedule all maintenance activities
- Establish control of quality, hours, and materials used for maintenance
- Provide effective maintenance service
- Keep to a minimum the amount of down time charged to maintenance
- Stay within the allocated budget established to carry out the planned maintenance activities

To achieve these objectives, certain basic management principles must be established, some of which are listed below.

- The maintenance manager should have the final say in the staffing of the maintenance positions.
- It is management's responsibility to plan for efficiency of labor.
- The amount of authority given to management position(s) should be commensurate with the amount of responsibility.

- There is a need for planning, scheduling, and cost control, and funds should be allocated for them.
- The actual day-to-day operations should be consistent with the formal organization established (as noted on the organization chart).
- The greatest production results when each worker understands the work he has to do, the manner in which he has to do it, and the time in which he should complete it.
- Formal and informal lines of communication must be kept open within all regions of the organization.

Hiring Maintenance Personnel

After top management has been hired and trained, the balance of the maintenance personnel in a new organization is hired. When recruiting personnel for in-plant staff, consideration should be given to the needs relative to the type of job, and the training and experience needed to accomplish the activity or job. The two major categories from which personnel can be drawn from are trained and untrained. Things to consider in each of these areas are as follows.

- **Untrained:** age, mechanical aptitude, manual dexterity, analytical ability, self-assurance, stability of character, and motivation.
- **Trained:** age, education, type and amount of experience.

The prospective employee must have at least a minimum amount of qualifications to fill the job. Otherwise, a complete training program would have to be established for this individual, which would mean additional costs charged against maintenance.

When reviewing the size of the maintenance force, it is advisable to plan for a slightly higher number of personnel in the beginning of the organization, not only to avoid future problems, but also to reduce the number of costly delayed startups. Do not lose sight of the fact that not only must personnel be hired to perform the actual line maintenance, but staff personnel will also be required to perform such functions as planning, scheduling, and estimating. In general, staff the in-plant force so that it can handle a workload slightly above what is required during slack periods.

New Employee Orientation

Once an employee has been hired, it is important that a company put him through a "new employee orientation." During this period the new employee should be introduced to the department or division head, his immediate supervisor, and fellow workers. He should be made aware of the work and organization in other departments. Along with his specific function in the organization, he should be told the performance standards expected and how he goes about securing supplies and equipment. He should also be informed of his work schedule, the number of hours he should work, any expected overtime, and periods of lunch and coffee breaks. He must also be informed of the pay period and how he will receive paychecks. If a company has any requirements for personnel preference such as no-smoking requirements on the job, parking areas, appearance, and grooming requirements, the respective employee should be informed of them at the time of the interview and again during his orientation. The employee should also be informed of vacations and holiday scheduling, sick and injury leave policies, and the responsibilities of the employee for notification when he is sick, injured, or late to the job.

Employee Records

There are many formats for employee applications. Many of these forms are commercially available at local office supply stores. Records should be established to keep track of the person in his progress and performance in the company. Therefore, a personnel file should be kept on every individual in the maintenance organization. This file should contain such information as present salary, benefits, record of tardiness, sickness, any training successfully completed, and any other information about the employee. It is important that a policy be instilled in an organization which allows the employee to have access to his records.

Employee Evaluation

Along with hiring and training of employees, evaluation of them is also very important. The evaluation system should be multifold. First, the immediate supervisor should hold an annual or semiannual evaluation of each of his employees. This evaluation should be in a written form outlining the major attributes of the employee and his points of weakness. This final product should be reviewed with the employee. It is important that evaluation be used for employee improvement and not as a threatening device. The total evaluation package should include a self-evaluation by the employee himself. This should be a short, written statement of the employee's attributes or the things that he did for the company during the last evaluation period, which he feels helped the company make a profit. The third item that could be part of the total evaluation package is an evaluation of the employee by the employee's peers.

Whatever evaluation methods are used, they should be formalized, understood by all involved, be in writing, and done on a regular (at least twice a year) basis. The evaluation should be open, honest, and directed toward the goal of employee improvement. Just as the craftsmen is evaluated, so should the supervisor, based on the same type of multi-phased evaluation process.

Maintenance Training Program

In order for the personnel in a maintenance organization to stay up-to-date on new maintenance, repair, and management techniques, an ongoing training program should be provided. A good training program will also keep an organization flexible and the personnel in it changeable. In general, training should be founded on the following methodology:

1. Determine needs of those to be trained.
2. Set goals and objectives for training by answering the following questions:
   —Why train?
   —What do you want to accomplish?
3. Determine subject areas to be included in the training program, and establish the prerequisites needed to begin working on a new subject.
4. Review the training program at regular intervals and make any necessary revisions.

It is important when considering the establishment of a training program that clear objectives be established. General objectives of a training program might be:

- To improve maintenance performance.
- To properly staff a maintenance department.

Prior to establishing a training program, a need must exist. Once the need is identified, the basic provisions must be established on paper, including the factors listed below.

- Selecting people to be trained
- Determining standards of performance of the students
- Establish a means of measuring the performance of students against standards of performance
- A means of solving individual trainee problems
- Determining the number of people to be trained at any one time
- Determining how, when, and where training should be given
- Establishing a main position of responsibility for carrying out the training program
- Setting up a formal process whereby classroom training can be correlated with shop and/or field training
- Developing evaluation methods to insure that the training program is always kept up-to-date
- Following up to see what happens to the student who passes or fails the program

- Determining how records should be kept for the purchase of supplies and textbooks for the program
- Establish a method by which the skills the student has acquired as a result of training are put to use in the field

Many union contracts will include provisions for a company-union sponsored training program. If this is the case, the specifics should be spelled out in writing as to who is responsible for what, and how the results will be measured and evaluated. Factors that influence the extensiveness of any training program are as follows:

- Size of plant.
- Attitude of the labor group.
- Availability of skilled workers.
- Management policies.

Type of Programs

The structure of the training curriculum will again depend upon the need of the specific maintenance organization. The curriculum could range from a two-day course in the review of the inspection of a boiler to a one-year program comprising many short and long courses. It is important that whatever the curriculum content is, it includes only the material necessary to enable the person to know how to do the job for which he is in training. The job description of each specific position is the best criterion on which to base the content of a curriculum.

Commercial curricula are available and, in part or in whole, can be adapted to any one company's needs. Listed below are the ones used the most.

- Bureau of Apprenticeship and Training Course—United States Department of Labor, Washington, D.C.
- Local union-sponsored programs
- TPC Training System—Barrington, Illinois

The danger in using standard curricula is that there may be superfluous information in them that would not be pertinent to a specific training program's needs. Therefore, prior to establishing one's own program based on evaluable curricula, review them carefully. To keep the cost down, this author suggests that current available programs be incorporated in whole or in part in any one training program to the company. It is ridiculous to "rediscover the wheel." Therefore, utilize what is available to meet your specific needs. However, if programs are not available, a company will be forced to develop their own in house programs or hire a professional.

Whatever type of curriculum is used or developed by the company, it should be based on specific objectives that are reasonable. Specific objectives usually come from the general objectives, noted earlier in this section. General objectives are hard to measure and are not task-oriented. However, specific objectives can be written in behavioral terms and measured as to whether or not they were attained. Once attained, one can say that the student has mastered and is ready to accept the responsibility of his new task. If he has not mastered or passed, his instructor will know that he must spend more time with the student to help him attain the objectives. Much literature is available today on how to write behavioral or task-oriented objectives. It is mandatory that if a program is not being developed in-plant, that objectives be written in a task-oriented sense. The person responsible for developing such a program should himself be trained in writing these objectives. Refer to the Appendix for a list of training programs.

Who Should Attend a Training Program?

It is the responsibility of management to select those who will attend the training program. Remember that the long-term result of the program should be lower maintenance costs.

The selection of who is to attend training programs may be accomplished by using a series of tests. Note, however, that tests should be only one tool in the selection process. In addition to general tests that measure how well a person is able to learn some, the types presently being used are:

- Interest tests, which indicate areas of activity in which a person shows a potential.
- Personality tests, which measure the pressures that produce behavior and may possibly help predict behavior.
- Achievement tests, which indicate the degree of retention of subject matter as compared with other people in the same subject field.
- Aptitude tests, which attempt to measure special individual abilities.

The major publishers of tests are:

- The Psychological Corporation, New York, New York.
- Educational Testing Service, Princeton, New Jersey.
- Harcourt Brace Jovanovich, Inc., New York, New York.
- California Test Bureau, Monterey, California.
- Science Research Associates, Chicago, Illinois.

Many personnel who need to be trained also need to be motivated to participate in the training experience. It is management's responsibility to make workers aware that by attending training programs they are in a better position to advance in the company and receive greater financial and personal rewards. An additional motivating factor for students is to hold the training program during working hours. It is more advantageous to hold classes in the morning, when people seem to be more awake, than later in the afternoon or end of a work day.

Who Should Conduct a Training Program?

The actual training can be conducted by in-plant personnel, such as shop supervisors, engineering personnel, or instructors from outside the company. Outside sources include local educational institutions such as junior/community colleges, vendor representatives, or training department personnel of local companies. Using in-house personnel to do the actual instruction has the advantage of making it possible to carry out the important principle that the responsibility for training must rest with maintenance supervision. It also provides opportunities for the supervisor and his workers to get to know each other, which is important for morale. It also insures that the workers will be taught the things they need to know to do their specific jobs in their specific departments. However, there are two disadvantages for using in-house supervisory personnel. They are:

- It is difficult to train most foremen or superintendents to be good instructors.
- They usually cannot take enough time from their other duties to prepare for their classes.

Whether using in-house personnel or consultants as instructors, it is important to select personnel who will relate well to the students and be well-trained not only in methods of teaching, but also in the subject matter they are teaching.

Because of this, probably the best instruction could be given by a full-time instructor with a degree in vocational education and some teaching experience. This person could be an employee of a maintenance organization or the training department of a company. In addition, a team approach would be advantageous, in which the maintenance supervisor would be part of the teaching team and be asked to instruct one or more of the lessons. Team teaching has many advantages such as the following:

- It breaks up the routine of one person.
- It allows the class participants to obtain a broader view of the topic.
- The training person knows how to teach, while the maintenance supervisor can furnish the expert knowledge relative to the specifics of the organization.

Classroom Technique

Methods of classroom techniques take the following forms:

- Program instruction or learning
- Group discussion and conference
- Role playing
- Case studies
- Sensitivity training
- On the job or laboratory training
- Job rotation
- Seminar and workshops put on by consultants outside the company
- Lectures
- Other special means

The actual classroom procedures should encompass as much discussion and hands-on problem-solving as possible, with the number of lectures minimized. The instructor, planner, developer, and management should realize that the goal of the classroom information is to aid the student obtain the objectives of the course, resulting in a more productive employee.

Program Instruction

The most individualized instruction is known as *program instruction*. Program instruction is the use of self-instructional courses to achieve an instructional objective on an individual basis. Audio-visual equipment is often used in program instruction. The process of program instruction is outlined below.

- Each student receives instruction individually and at his own pace.
- Each student responds continuously as he receives instruction. Instruction proceeds only as the student responds.
- Each student receives feedback immediately for each response he gives.
- A student should be able to enter and exit from such a program at any point during the program.

If more in-depth information or program instruction is desired, consult the professional literature on the market. Most companies have a junior/community college or university near their place of business, and these institutions can offer valuable insights in developing program instruction.

Employee Evaluation and Recognition

Once the program has been established and the methodology of the different class sessions worked out, the next important item to consider is evaluating classroom training. Evaluation of classroom training can take several forms, some of which are listed below.

- Periodic quizzes
- Major unit examinations
- Actual task-oriented jobs, such as lubrication of a piece of machinery in a hands-on laboratory situation
- Field training with observation

So that time is not wasted teaching information the student already knows, pre-testing should be considered. Pre-testing establishes whether or not and how much time, if any, need be spent on specific concepts. By giving a pretest before the new concept is presented, the instructor can ascertain to what degree the students are knowledgeable in a specific area. If the students are knowledgeable in the specific area, the instructor can go on to a new area, and thus save time and money on the program. However, if most of the students are not knowledgeable in the specific area, the instructor knows to spend more time on it. When working with a group of students, one is teaching to 'the average' and, therefore, means should be established whereby the student receives as much personal attention as possible. This can be done by keeping classes small, offering individualized instruction, utilizing audio-visual equipment, and employing instructors who are individual-oriented.

Those who participate in an educational endeavor or learning experience deserve to be recognized. The degree of recognition can be a simple handshake, the awarding of a certificate or, at the other extreme, a full-fledged graduation ceremony with a dinner inviting major company officers. Whatever way it is done, it is important that the student realize he is a more valuable person to the company as a result of his training, and this point should be recognized at the completion of the training program.

Contract Maintenance

The personnel required to staff a maintenance organization are either a part of that organization or hired from the outside. It is usually cheaper to staff internally. The following factors must be considered in making the determination:

- **Type of work to be done:** Is the internal staff trained to do this type of work?
- **Quantity of work involved:** Is there sufficient personnel internally to accomplish the work within the allocated budget?
- **Expediency** with which the work can be accomplished.

Whatever decision is made relative to in-plant versus contractual manpower, the optimum operation is one in which the in-plant work force is always occupied in accomplishing the organizational objectives. When the scope of the work, during peak periods, is such that not enough can be deferred to level the valleys (slack periods), outside contractual help should be considered. Contract maintenance is nothing more than the delegation to an outside organization the responsibility of maintenance in full or in part, while retaining full control over expenditures, quality, and quantity. It should be generally employed during periods of peak activity or in emergency situations, in which in-plant personnel are not available or do not have the experience to do the work.

Types of Contract Maintenance

There are five basic types of contract maintenance:

- The contractor acts solely as the supplier of workers who are directed by in-plant supervision.
- The contractor provides special equipment, skills, and possibly materials that are required only on rare occasions and possibly only on a one-time basis.
- The owner has a work force large enough to take care of normal needs and provide all the administration and supervision normally required and the contractor provides additional manpower for peak load periods as well as the supervisory and administrative personnel required.
- The owner and contractor have an arrangement under which each provides some of the technical and supervisory staff permanently required and the owner has a small work force of his own that is supplemented by the contractor as peak loads occur.
- The entire, or nearly entire, maintenance function is contracted out and the contractor assumes responsibility for supervision, technical assistance, purchasing, planning, scheduling and coordination of the work and the work force itself.

When selecting an outside contractor, the following qualifications should be considered:

- Availability
- Organization
- Character
- Experience
- Financial stability

In-plant people are responsible for the administration and supervision of the outside contractor, and they must be available to perform the functions listed below.

- Review specifications and drawings
- Provide inspection and quality control

Summary

No matter what type of maintenance organization is established, it must be flexible enough to accommodate the changing needs, responsibilities, and personnel of the organization and the company. Too rigid an organization results in a static situation, where innovation is minimized, and maximum efficiency and dollar return are never realized.

In order to accomplish the goals of the organization, carry out plans, and allow people to work their most efficiently and effectively, activities must be organized. Authority and responsibility must be established so that conflicts do not arise. The maintenance organization should not be a bureaucracy; it should be an understandable and workable operation both in practice and on paper.

The supervisor of the maintenance organization is responsible for the items listed below.

- Achieve control over quality of work, labor hours expended, and materials used.
- Provide prompt response to breakdowns and constantly improve the availability of all equipment.
- Operate the maintenance department at the lowest possible cost.
- Follow good management procedures in achieving the above.

No matter what the size of an organization, the administration and control of maintenance are not complicated. Four major factors are involved in developing an effective maintenance organization:

- A strong line organization.
- A well-trained work force.
- Good planning, scheduling, and cost control.
- Good human relations.

Refer to the flow chart shown in Figure 16.7.

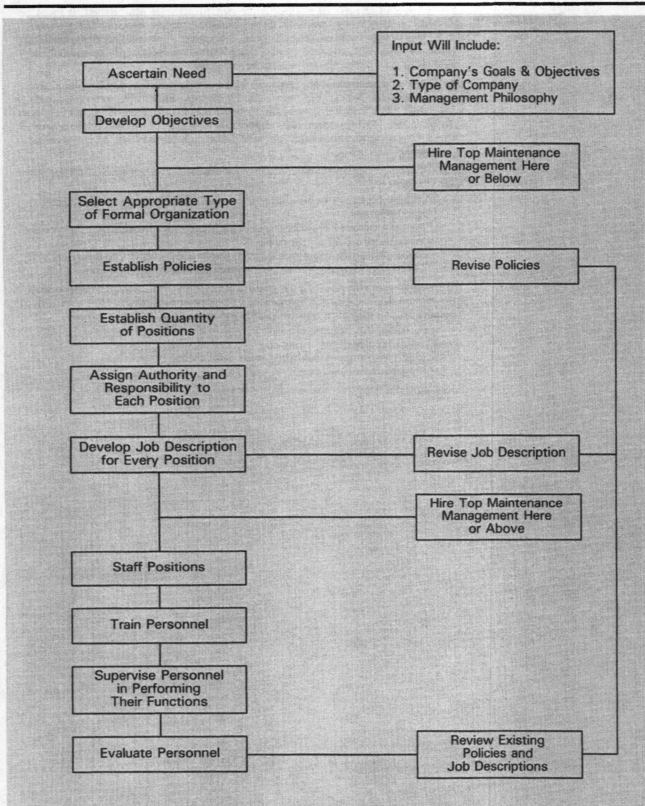

Flow Chart for Establishing, and Review of, an Effective, Flexible Maintenance Organization

Figure 16.7

CHAPTER 17

ESTIMATING MAINTENANCE AND REPAIR COSTS

Estimating serves as the foundation for most of the managerial tools used to direct the maintenance function. The effectiveness of any maintenance operation depends on how well the estimator develops the estimate and follows through with it. Some of the procedures of estimating, divided into major phases, are listed below.

- Compilation of the required types and quantities of materials needed to perform the task.
- Assignment of quantities of labor and equipment, and the amount of time needed to install or replace the material.
- Pricing of material, equipment, and labor.
- Assignment of indirect or overhead expenses, such as telephone calls; secretarial services and supervision; down time of effective operation; and profit, if outside contractors are used.
- Summation of all costs required to do the job.

In the field of maintenance and repair work, the two major categories of work when developing an estimate are repetitive jobs, such as window washing and changing oil, and nonrepetitive jobs, such as repair of a structural beam. Repetitive and nonrepetitive jobs are further subdivided into planned or emergency maintenance. The cost for any company is minimized by utilizing a preventive maintenance program (see Chapter 18). No matter what type of maintenance is being done, the same basic procedures are used to develop an estimate.

The extensiveness of an estimate depends on its use. Refer to Figure 17.1 for selecting the most useful estimating method.

No matter what method is used, the more exact an estimate, the better the input into the company's cost control program for future reference (see Chapter 18).

The Estimator

The estimate can be done at various levels of the organization. Usually, the more detailed or complete the job, the higher up in the organization it is prepared. Some of the people or departments who may prepare an estimate are the foreman, the engineering department, and the planning and scheduling department.

Use	Use of the Estimate			
	Approval, Backlog, and Forecasts	Individual Controls, Daily Schedules	Methods and Limited Schedule	Plant-Wide Controls
Emergencies Definite: none Indefinite: end result, work, methods	Judgment.	—	Judgment, historical records.	Historical records modified by work sampling.
Trouble shooting, breakdowns, construction without planning Definite: end result Indefinite: work, methods	Judgment.	Obtained elemental data.	Judgment, historical records, published data.	Time-study analysis, or ratios based on work sampling.
Overhauls, relocations, modifications, building repairs with drawings, repeat jobs without specific instructions Definite: end result some work, some methods Indefinite: some work, some methods	Judgment.	Obtained elemental data.	Judgment, published data.	Judgment based on detailed analysis, elemental obtained data.
Planned building modification, construction, spare parts, partially planned repairs Definite: end result work Indefinite: methods	Judgment.	Obtained elemental data.	Judgment, published data.	Judgment based on detailed analysis, obtained elemental data.
Fully planned repairs and maintenance Definite: end result work, methods (planned)	Sampling of elemental data.	Obtained elemental data.	Elemental obtained data, comparative based on published data obtained.	Elemental data.
Routine, repetitive PM and changeovers Definite: end result work, methods (repetitive)	Sampling of elemental data.	Obtained elemental data, comparative, time study.	Elemental obtained data, comparative based on published data obtained.	Elemental data, comparative, time study.

Figure 17.1

An effective estimator has many of the qualifications listed below.
- A knowledge of the details of performing the specific maintenance task
- Experience doing the particular task
- Ability to use an estimating manual that contains information such as material suppliers, labor hours required, and overhead items
- Sound judgment regarding work in different localities and complex jobs utilizing different classifications of labor
- A good method for preparing an estimate
- The ability to perform thorough, accurate work
- The ability to organize, clarify, collect, and evaluate data

Sources of Error in Estimating

Errors enter into an estimate in many ways. Some common estimating errors are listed below.
- Errors in mathematics (forgetting to check all computations and results)
- Errors in copying items from one paper to another
- Errors in estimating the length of time required to do certain items of work by different workmen and crews
- Errors in estimating wages
- Errors in estimating quantities of material
- Errors in estimating waste of material
- Errors in estimating prices of material, including the effect of rising, falling, and steady markets on prices
- Errors in estimating the cost of the transportation, handling, and storage of materials
- Omission of items, such as equipment charges and delays due to bad weather, labor troubles, lack of arrival of materials, and breakdown of machinery

Utilizing Standard Formats

When making an estimate, all figures should be recorded on standard formats. Separate format sheets should be used for the quantity takeoff, pricing of labor and equipment amounts, and summarizing. The most effective format is one that is developed by the specific organization for its specific needs. Standard format sheets can be classified into the categories listed below.
- Worksheets for recording quantities of all material and equipment
- Summary sheets to bring the quantities and prices for labor, material, and equipment together
- Recapitulation sheets to summarize final cost of labor, material, and equipment requirements

Preliminaries to Estimating

Prior to the preparation of an estimate, the estimator visits the site to study the area. It is helpful to use a standard format, such as the one shown in Figure 17.2 from *Means Forms for Building Construction Professionals*. After completing the site investigation report, the estimator is ready to go back into the office and begin the estimating procedure.

After visiting the job site the estimator must determine exactly what work is to be done along with how it is to be done. The major resources available to the estimator at this stage are:
- Manuals from manufacturers of the equipment and/or material
- Company job standards
- Recommendations of consultants
- Personal experience
- Combination of above

After determining what is to be done and how the job is to be performed, the next step is to develop a set of instructions, such as drawings and specifications. This can be done with the help of the engineering department for large jobs. For smaller jobs, a detailed list of what is to be done will suffice.

JOB SITE
ANALYSIS

SHEET NO.

PROJECT

BID DATE

LOCATION

NEAREST TOWN

ARCHITECT	ENGINEER	OWNER
Access, Highway	Surface	Capacity
Railroad Siding	Freight Station	Bus Station
Airport	Motels/Hotels	Hospital
Post Office	Communications	Police
Distance & Travel Time to Site		Dock Facilities
Water Source	Amount Available	Quality
Distance from Site	Pipe/Pump Required?	Tanks Required?
Owner	Price (MG)	Treatment Necessary?
Natural Water Availability		Amount
Power Availability	Location	Transformer
Distance	Amount Available	

Voltage	Phase	Cycle	KWH or HP Rate

Temporary Roads	Lengths & Widths
Bridges/Culverts	Number & Size
Drainage Problems	
Clearing Problems	
Grading Problems	
Fill Availability	Distance
Mobilization Time	Cost
Camps or Housing	Size of Work Force
Sewage Treatment	
Material Storage Area	Office & Shed Area
Labor Source	Union Affiliation
Common Labor Supply	Skilled Labor Supply
Local Wage Rates	Fringe Benefits
Travel Time	Per Diem

Taxes, Sales	Facilities	Equipment
Hauling	Transportation	Property
Other		

Material Availability: Aggregates		Cement
Ready Mix Concrete		
Reinforcing Steel	Structural Steel	
Brick & Block	Lumber & Plywood	
Building Supplies	Equipment Repair & Parts	

Demolition: Type	Number	
Size	Equip. Required	
Dump Site	Distance	Dump fees
Permits		

Page 1 of 2

Figure 17.2

🔔 Means Forms

Clearing: Area _____ Timber _____ Diameter _____ Species _____
  Brush Area _____ Burn on Site _____ Disposal Area _____
  Saleable Timber _____ Useable Timber _____ Haul _____
  Equipment Required _____

Weather: Mean Temperatures _____
  Highs _____ Lows _____
  Working Season Duration _____ Bad Weather Allowance _____
  Winter Construction _____
  Average Rainfall _____ Wet Season _____ Dry Season _____
  Stream or Tide Conditions _____
  Haul Road Problems _____
  Long Range Weather _____

Soils: Job Borings Adequate? _____ Test Pits _____
  Additional Borings Needed _____ Location _____ Extent _____
  Visible Rock _____
  U.S. Soil & Agriculture Maps _____
  Bureau of Mines Geological Data _____
  County/State Agriculture Agent _____
  Tests Required _____
  Ground Water _____

Construction Plant Required _____

  Alternate Method _____

  Equipment Available _____

  Rental Equipment _____ Location _____

Miscellaneous: Contractor Interest _____
  Sub Contractor Interest _____
  Material Fabricator Availability _____
  Possible Job Delays _____
  Political Situation _____
  Construction Money Availability _____
  Unusual Conditions _____

Summary _____

Figure 17.2 (continued)

QUANTITY SHEET

	SHEET NO.
PROJECT	ESTIMATE NO.
LOCATION	ARCHITECT	DATE
TAKE OFF BY	EXTENSIONS BY:	CHECKED BY:

DESCRIPTION	NO.	DIMENSIONS				UNIT		UNIT		UNIT		UNIT

Figure 17.3

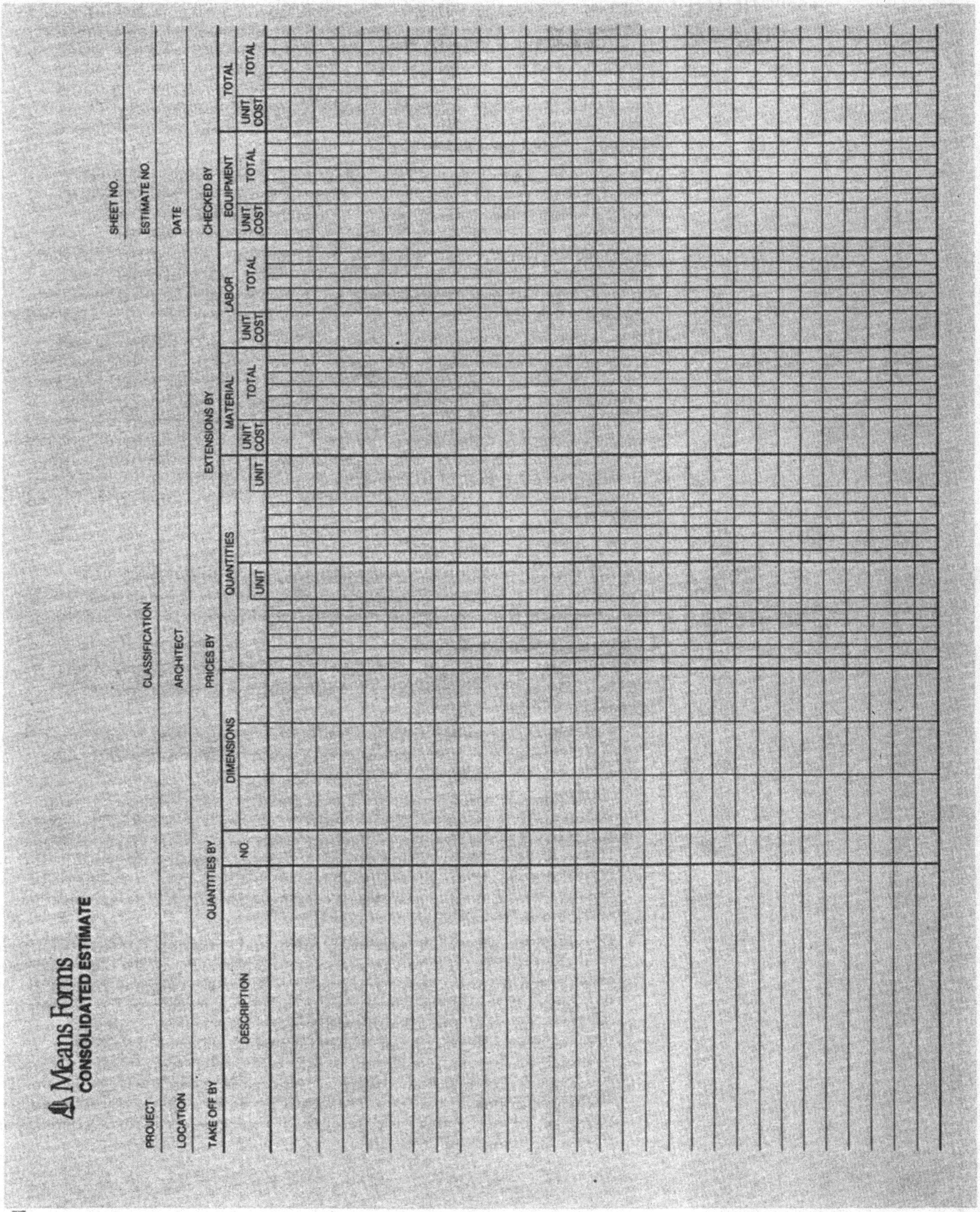

Figure 17.4

Estimating Material

Once a complete set of job instructions (or plans and specifications) is available, the estimator then proceeds to a quantity takeoff. Essentially the estimator takes off the various materials from the plans and specifications, or instructions, and tabulates them on a formal quantity takeoff sheet, an example of which is shown in Figure 17.3 from *Means Forms for Building Construction Professionals*. All like materials are kept together. The total quantities for the same items are brought forward in condensed form to a consolidation sheet (as shown in Figure 17.4 also from *Means Forms for Building Construction Professionals*).

Once the total quantities of each different type and grade of material are listed, the next step is to assign a cost to each one. The estimator must decide whether the materials are to be purchased from outside the organization or are already available in the inventory or will be manufactured by the company itself. It is, therefore, important that the estimator or estimating department have on hand a list of inventory or stock. Instead of ordering new material, it may be beneficial to substitute materials that are on hand, which can result in a lower cost to do the work.

If the material is not available in-house, the cost can be obtained, either through a formal bidding process or by quotes from local suppliers, a list of which should be at the disposal of the estimator. This list should be kept up-to-date. A review should be done at the end of the job to ascertain whether or not the suppliers actually fulfilled their obligations in terms of quality, quantity, and scheduled delivery of material. The estimator should not forget that the cost of freight, unloading, storage, inspection, testing, and insurance are part of the material cost. If an effective job was done in taking off material quantities, it can be used in whole or in part as the bill of materials.

Time Standards

At this particular point the estimator should keep in mind that if the company does not have elemental work time units or standards developed for estimating, this can be done during the performance of any specific job.

Types of Standards

The accuracy of a standard is a function of the degree of analysis of a job process. Some of the major types are listed below in increasing order of accuracy in terms of analysis:

1. **Published averages or ratios:** This method utilizes published cost data. The reader is referred to the latest edition of *Means Facilities Cost Data* for such data and related topics.

   This data is based on the total cost for doing work based on some unit such as the square foot for construction-type work or per ton of produced material for manufactured-type work. This type of estimate is an average output using skilled personnel. A word of caution: the reader should note that utilizing such a method may or may not be relevant to the specific working conditions of the job in question and should be modified to meet actual conditions.

2. **Specific company job standards (historical records):** For this particular method, the time, labor, materials, and equipment to do a specific job are noted from past experience and recorded as company historical records. When future jobs of a similar nature come up, the estimator simply refers to these historical records for estimated time and requirements for the labor, equipment, and material. The reader should note that these figures should be updated from time to time for reasons of inflation, change in wage rates and material prices, and changes in production rates. Also, if one or more aspects of the total similar job are somewhat different from the one on record, corrections should be made appropriately.

Average Cleaning Times

FLOOR OPERATIONS

Sweeping — 1000 sq. ft.
- Halls & Corridors — 15 min.
- General Rooms — 30 min.

Mopping — 1000 sq. ft.
- Dust Mop (unobstructed) — 10 min.
- Dust Mop (obstructed) — 15 min.
- Damp Mop (unobstructed) — 20 min.
- Damp Mop (obstructed) — 40 min.
- Wet Mop and Rinse (obstructed) — 100 min.

Scrubbing — 1000 sq. ft.
- Hand Scrub 12" brush — 300 min.
- Deck Scrub — 100 min.

Machine Scrub — 1000 sq. ft.
- Machine Scrub 12" Diameter — 50 min.
- Machine Scrub 14" Diameter — 40 min.
- Machine Scrub 16" Diameter — 35 min.
- Machine Scrub 18" Diameter — 30.5 min.
- Machine Scrub 19" Diameter — 28 min.
- Machine Scrub 21" Diameter — 25 min.
- Machine Scrub 23" Diameter — 23 min.
- Machine Scrub 24" Diameter — 20 min.
- Machine Scrub 32" Diameter — 18 min.
- Machine Scrub 36" Diameter — 15 min.
- Automatic Scrub Machine (24") — 5 min.

Vacuum Pick Up — 1000 sq. ft.
- Vacuum Pick Up (unobstructed) — 20 min.
- Vacuum Pick Up (obstructed) — 30 min.

Wax — 1000 sq. ft.
- Wax — 30 min.
- Machine Polish (19" machine) — 15 min.
- Rectangular Machine (48" plate) — 5 min.
- Buff with Steel Wool — 20 min.
- Strip and Rewax (1 man) — 150 min.
- Dry Strip and Rewax (1 man) — 120 min.

Spray Buffing — 1000 sq. ft.
- Spray Buffing (unobstructed) — 30 min.
- Spray Buffing (obstructed) — 45 min.

Carpeting — 1000 sq. ft.
- Vacuuming (unobstructed) — 20 min.
- Vacuuming (obstructed) — 30 min.
- Spot Vacuuming — 15 min.
- Shampoo (dry-foam) — 60 min.
- Pile Lift — 30 min.
- Lockers — 20 min.
- Radiators — 30 min.
- Tables (medium) — 50 min.
- Telephones — 15 min.
- Towel Dispensers — .12 min.
- Towel Disposal Cans — .40 min.
- Typewriter & Stand — .50 min.
- Washbasin (Office) — .60 min.
- Wastebasket — .50 min.
- Windowsill — .20 min.
- Venetian Blinds (std. size) — 3.50 min.

WASHROOMS

Cleaning
- Cleaning Commode — 4. min.
- Door (spot wash both sides) — 1 min.
- Mirrors — 1 min.
- Sanitary Napkin Dispenser — .50 min.
- Urinals — 3.00 min.
- Washbasin — Soap Dispenser — 3.00 min.

General Washroom Cleaning
- General Cleaning per 1000 sq. ft. — 120 min.

MISCELLANEOUS OPERATIONS

Wall Washing — 1000 sq. ft.
- Painted Walls (manual) — 240 min.
- Painted Walls (machine) — 150 min.
- Marble Walls (manual) — 90 min.

Ceiling Washing — 1000 sq. ft.
- Ceiling Washing (manual) — 300 min.
- Ceiling Washing (machine) — 180 min.

Window Washing — 1000 sq. ft.
- Single Pane — 125 min.
- Multi-pane — 170 min.
- Frosted Single Pane — 190 min.
- Opaque Glass — 50 min.
- Plate Glass — 35 min.
- Office Partitions (glass) — 110 min.

Dusting Lamps & Light Fixtures
- Wall Fluorescent Fixtures — .15 min.
- Desk Fluorescent Lamp — .30 min.
- Table Lamp & Shade — .60 min.
- Floor Lamp & Shade — .60 min.

Washing Fluorescent Light Fixtures
- Ceiling Fixture (egg crate) 4' ea. — 9 min.
- Ceiling Fixture (egg crate) 8' ea. — 12 min.

Dusting — 1000 sq. ft.
- Air Conditioners — .30 min.
- Ash Trays (desk) — .25 min.
- Book Cases (3-tier sect.) — .30 min.
- Chairs — .30 min.
- Cigarette Stands — .40 min.
- Couch — .25 min.
- Desks — .80 min.
- Desk Trays — .15 min.
- File Cabinets (4 drawer) — .40 min.

Fabric Upholstery Cleaning
- Whisk or Vacuum Armless Chair — .50 min.
- Whisk or Vacuum Armchair — 1 min.
- Shampooing Armless Chair — 4 min.
- Shampooing Armchair — 7 min.
- Shampooing Couch — 20 min.

Stairway Cleaning
- Sweep & Dust 1 flight 15 steps — 6 min.
- Damp Mop 1 flight 15 steps — 5 min.
- Scrubbing (hand) — 20 min.

Figure 17.5

3. **Activity or task standards:** Every job can be broken down into a series of activities (as discussed in the chapter on cost control). The activities can further be divided into elements. It is possible through such means as time and motion studies (including time-lapse photography) and methods-time measurement to obtain times for each one of these elements. The elements, along with the times, can then be recorded in a standard format book from which the estimator can draw for any and all jobs. For instance, the element erecting scaffolding could be used in many different types of jobs such as painting a wall, changing light bulbs, and changing air conditioning filters. When an estimator has a job to estimate, he can first divide the job down into its activities and then each activity into its element. Using his recorded elemental times, he can then assign times to each of the elements for the specific job in question. These times can be summed to give total times for activities, and the activity times summed to give the total job time. The reader is referred to Figures 17.5 and 17.6 for examples of times.

Typical Production Rates

Activity	Number of Men	Travel Allow (Hrs.)	Labor Time (Hrs.)	Total (Hrs.)
Carpentry				
Repair door surface closer	1	1/2	1	1½
Repair concealed door closer	2	1/2	1½	4
Repair door damage at shop	1	1	3	4
Repair & replace screens	1	1/2	1/2	1
Repair broken glass	1	1/2	1	1½
Repair & replace ceiling tile	1	1/2	1½	2
Repair small dry-wall damage	1	1/2	1	1½
Repair & replace sash balance	2	1/2	1½	4
Change lock on door	1	1/2	1	1½
Painting				
Repaint 20′ x 15′ room, 1 coat	2	1		34
Repaint small bathroom, 1 cost	1	1/2		11½
Repaint exterior window	1	1/2		2½
Plumbing & Steamfitting				
Clear stopped water closet	2	1/2		3
Clear stopped basin	1	1/2		1½
Replace leaking radiator valve	2	1/2		6
Clear external sewer stoppage	2	1/2		9
Install replacement valve, faucet, or trap	2	1/2		10
Electrical				
Replace fluorescent lamp ballast	1	1/2		2
Reset fire or security alarm				
Replace fractional HP motor	2	1		10
Replace brown fuse or reset cir. bkr.	1	1/2		1½
Repair exterior light damage	2	1/2		4
Control circuit problems	2	1		5

Figure 17.6

TIME AND MOTION STUDY

PROJECT		OPERATION											SHEET NO.				
START TIMING													DATE				
END TIMING		OPERATOR											OBSERVER				
ELEMENT			CYCLES, TIME IN MIN.												SUMMARY		
			1	2	3	4	5	6	7	8	9	10			ΣT		T̄
START																	
FILL DIPPER	T																
	R																
SWING	T																
	R																
DUMP	T																
	R																
RETURN	T																
	R																
DELAY	T																
	R																
CYCLE	T																

ΣT = sum of element times   T̄ = average time for element = ΣT/N

Figure 17.7

465

List of Engineering Performance Standards Publications

MAVFAC	Title and Contents
P-700.0	Engineers Manual
P-700.1	Planner and Estimator's Workbook, Instructor's Manual
P-700.2	Planner and Estimator's Workbook
P-701.0	General Handbook
P-701.1	General Formulas
	Travel
	Job Preparation
	Allowances
	Material Handling
	Instructions
P-701.2	Elemental Standard Time Data Handbook
P-701.3	Elemental Standard Time Data Element Analysis
P-702.1	Carpentry Formulas
	Carpentry
	Roofing
	Floor Covering
	Millroom
	Glazing
	Furniture Repair
P-703.0	Electrical, Electronic Handbook
P-703.1	Electrical, Electronic Formulas
	Electrical
	Electronic
P-704.0	Heating, Cooling and Ventilating Handbook
P-704.1	Heating, Cooling and Ventilating Formulas
	Heating, Cooling, and Ventilating
P-705.0	Emergency/Service Handbook
P-706.0	Janitorial Handbook
P-706.1	Janitorial Formulas
	Janitorial
P-707.0	Machine Shop, Machine Repair Handbook
P-707.1	Machine Shop, Machine Repair Formulas
	Machine Shop
	Machine Repair
P-708.0	Masonry Handbook
P-708.1	Masonry Formulas
	Masonry
	Plaster
P-709.0	Moving and Rigging Handbook
P-709.1	Moving and Rigging Formulas
	Moving and Rigging
P-710.0	Paint Handbook
P-710.1	Paint Formulas
	Paint
P-711.0	Pipefitting and Plumbing Handbook
P-711.1	Pipefitting and Plumbing Formulas
	Pipefitting and Plumbing
P-712.0	Roads, Grounds, Pest Control Handbook
P-712.1	Roads, Grounds, Pest Control Formulas
	Pest Control
	Grounds
	Roads and Paving
	Refuse Collection and Disposal
P-713.0	Sheetmetal, Structural Iron and Welding Handbook
P-713.1	Sheetmetal, Structural Iron and Welding Formulas
	Sheetmetal
	Welding
P-714.0	Trackage Handbook
P-714.1	Trackage Formulas
	Trackage

Figure 17.8

Time and Motion Studies

Time study techniques have been developed and standardized over the years. There are many excellent resources in terms of textbooks and manufactured literature which describe in detail how to conduct such studies, how to level for performance, and what allowances to make for personal time, rest, or delay. Time study techniques are being considered to obtain elemental or activity time, the first thing to do is to define the maintenance operation as specifically as possible. Each element to be timed should be noted individually.

Time study observers should be familiar with the operations being studied. The actual timing is done in the field using a stopwatch or time-lapse photography. The elements and times must be recorded on standard format sheets as noted in Figure 17.7. The total operational time is a sum of the elemental times attributed to the operation. The reader should note that all interruptions in non-productive times should be noted clearly on the time study sheet and not be included in operational times. Unavoidable delays are noted and a delayed time factor must be included in the total operational time to take these interruptions into consideration.

When operations are performed in sequence, each operation must be evaluated and recorded separately. Exactly how many observations must be taken depends on the degree of accuracy required for the elemental time. The amount of time used or assigned to the element is an average of the different observations of the particular element.

Method-Time-Measurement (MTM)

Method time measurement is a method of developing standard times for any operation or job by assigning predetermined standards to elemental motions such as reach or move, position, turn, kneel, sit, and walk. Much work in this area has been established by the MTM Association for Standards and Research. The reader is referred to this association for more information and data relative to its publications. Another one of the widely used systems of MTM are Engineering Performance Standards. These have been developed by the United States Department of Navy, Naval Facilities Engineering Command, and Public Works for Maintenance. These publications are available at a nominal charge from the Superintendent of Documents, United States Government Printing Office, Washington, DC 20401. A list of Engineering Performance Standards available is contained in Figure 17.8.

Estimating Labor

When labor estimates are prepared, allowances must be made for variations in wages, the length of time required to do certain items of work, working conditions, and in the classes (skilled and unskilled) of labor required for different kinds of work. Some allowance should also be made in regard to the probable labor supply and demand, especially if one must go out of the plant to obtain the labor.

The length of time required to do a certain piece of work may vary considerably owing to the following items:
- The person, his skill, and mental attitude
- The working environment
- The crew in which he works
- The equipment which is used

In some localities, union rules and regulations tend to limit the hourly output of the workers. The estimator should keep in mind that when work is plentiful, labor is scarce, and jobs are easy to get, the time required for a laborer to do a certain unit of work will probably be more than the average and vice versa.

General Office Cleaning Frequencies

Daily

Sweep office floors.
Empty wastepaper baskets.
Empty and wipe ash trays and smoking stands.
Dust tops of desks, tables, file cabinets.
Empty pencil sharpeners.
Clean chalkboards and erasers.
Clean drinking fountains.
Spot-clean doors, partition glass, walls.
Sweep and wet-mop rest room floors.
Replenish soap, towel, tissue, and napkin dispensers.
Clean washbasins.
Clean toilet bowls.
Clean urinals.
Clean plumbing supply lines, drainpipes.
Clean mirrors.
Clean supply closet.
Empty napkin disposal receptacles.
Vacuum carpets and remove spots.

Weekly

Machine-scrub and buff rest room floors and high traffic office
  aisles.
Dust all furniture.
Dust window sills, ledges, picture frames.
Wash interior glass, display cases, shelves, and doors.
Wash ash trays and smoking stands.
Vacuum air vents.
Wash toilet booth partitions.
Acid-clean toilet bowls and urinals.
Paraffin-treat toilet seats.
Spray rest rooms for insect control.

Monthly

Wash rest room walls.
Machine scrub and refinish rest room floors and high-traffic office
  aisles.

Once every 3 months

Wash and polish wood furniture.
Wash all desks, tables, chairs, file cabinets.
Vacuum drapes.

Once every 6 months

Wash light fixtures.
Vacuum and/or wash venetian blinds.
Wash wastepaper baskets.
Machine-scrub and refinish office floors.
Wash exterior glass.

Once a year

Wash all walls and ceilings.
Machine shampoo carpeting. Wash and polish wood furniture
Wash all desks, tables, chairs, file cabinets
Vacuum drapes.

Figure 17.9

Establishing a Crew Size

The following points must be considered when establishing crew size:

- Physical limitations of work in terms of handling of material and equipment
- Safety limitations in terms of whether or not a worker will need a helper to perform the job safely
- Urgency of completing the job
- Demands of the job

Initially, the size of the crew needed to perform a job is likely based on judgment and previous experience. However, for repetitive-type jobs, more sophisticated methods such as time and motion studies should be utilized in establishing the most productive and efficient crew size.

Frequencies

Some types of repetitive jobs need to be performed at a regular interval or frequency. A list of figures should be developed by each company based on their own experience or in conjunction with the experience of others such as:

- Recommendation of manufacturers of equipment
- Recommendation of manufacturers of building materials
- Published data contained in periodicals and literature of trade associations

Examples of some frequencies for cleaning are shown in Figure 17.9.

Assigning Labor Wage Rates

When estimating wages, consideration must be given to the different kinds of work involved, to the different classes of labor needed, and to the prevailing hourly wages. Hourly wages of various classes of labor vary considerably in different parts of the country and between union and open shop jobs. Local customs or union regulations often require the use of certain classes of labor for certain types of work. A list of prevailing wage rates should be kept on record by the estimator and updated as needed.

Estimating Equipment

The equipment for any job will include all the necessary tools, machinery, and all temporary buildings and structures. The estimator must determine which of the items above are already available in-house and which must be rented or purchased from outside sources. The machines or tools which will only be used for one job could probably be rented more economically than purchased for the job. Therefore, consideration should be given in the estimate to the estimated life of any particular machine or tool, the length of time it will be used on the job, the amount of work to be done using this specific tool or machinery, and its potential use in the future for other jobs. The estimator should have on hand an inventory of equipment and tools, along with a list of acceptable suppliers.

Equipment costs include costs of ownership (or rentals), transportation, erection, moving, dismantling, and operation. These costs may also include the labor costs of machine operators. Overhead such as supervision may also be included here.

Estimating Overhead

Overhead is a cost which is not directly attributed to a specific job. Overhead costs are divided into two types: *general* and *job*.

General overhead costs include all costs that cannot be charged directly to any one specific job. They include such items as office rent, office equipment and supplies, utilities, telephone, general insurance, taxes, interest, legal expenses, traveling, plans, and specifications. General overhead costs may also include the salaries or parts of salaries of general officers, managers, estimators, engineers, secretaries, clerks, craftsmen, etc., that cannot be readily charged to a particular job.

Means Forms

ESTIMATE SUMMARY

		SHEET NO.	
PROJECT		ESTIMATE NO.	
LOCATION	TOTAL AREA/VOLUME	DATE	
ARCHITECT	COST PER S.F./C.F.	NO. OF STORIES	
PRICES BY:	EXTENSIONS BY:	CHECKED BY:	

DIV.	DESCRIPTION	MATERIAL	LABOR	EQUIPMENT	SUBCONTRACT	TOTAL
1.0	**General Requirements**					
	Insurance, Taxes, Bonds					
	Equipment & Tools					
	Design, Engineering, Supervision					
2.0	**Site Work**					
	Site Preparation, Demolition					
	Earthwork					
	Caissons & Piling					
	Drainage & Utilities					
	Paving & Surfacing					
	Site Improvements, Landscaping					
3.0	**Concrete**					
	Formwork					
	Reinforcing Steel & Mesh					
	Foundations					
	Superstructure					
	Precast Concrete					
4.0	**Masonry**					
	Mortar & Reinforcing					
	Brick, Block, Stonework					
5.0	**Metal**					
	Structural Steel					
	Open-Web Joists					
	Steel Deck					
	Misc. & Ornamental Metals					
	Fasteners, Rough Hardware					
6.0	**Carpentry**					
	Rough					
	Finish					
	Architectural Woodwork					
7.0	**Moisture & Thermal Protection**					
	Water & Dampproofing					
	Insulation & Fireproofing					
	Roofing & Sheet Metal					
	Siding					
	Roof Accessories					
8.0	**Doors, Windows, Glass**					
	Doors & Frames					
	Windows					
	Finish Hardware					
	Glass & Glazing					
	Curtain Wall & Entrances					
	PAGE TOTALS					

Page 1 of 2

Figure 17.10

DIV	DESCRIPTION	MATERIAL	LABOR	EQUIPMENT	SUBCONTRACT	TOTAL
	Totals Brought Forward					
9.0	**Finishes**					
	Studs & Furring					
	Lath, Plaster & Stucco					
	Drywall					
	Tile, Terrazzo, Etc.					
	Acoustical Treatment					
	Floor Covering					
	Painting & Wall Coverings.					
10.0	**Specialties**					
	Bathroom Accessories					
	Lockers					
	Partitions					
	Signs & Bulletin Boards					
11.0	**Equipment**					
	Appliances					
	Dock					
	Kitchen					
12.0	**Furnishings**					
	Blinds					
	Seating					
13.0	**Special Construction**					
	Integrated Ceilings					
	Pedestal Floors					
	Pre Fab Rooms & Bldgs.					
14.0	**Conveying Systems**					
	Elevators, Escalators					
	Pneumatic Tube Systems					
15.0	**Mechanical**					
	Pipe & Fittings					
	Plumbing Fixtures & Appliances					
	Fire Protection					
	Heating					
	Air Conditioning & Ventilation					
16.0	**Electrical**					
	Raceways					
	Conductors & Grounding					
	Boxes & Wiring Devices					
	Starters, Boards & Switches					
	Transformers & Bus Duct					
	Lighting					
	Special Systems					
	Subtotals					
	Sales Tax %					
	Overhead %					
	Subtotal					
	Profit %					
	Adjustments/Contingency					
	TOTAL BID					

Figure 17.10 (continued)

Means Forms

CONSOLIDATED ESTIMATE

SHEET NO.

PROJECT

ESTIMATE NO.

ARCHITECT

DATE

TAKE OFF BY	QUANTITIES BY	PRICES BY	EXTENSIONS BY	CHECKED BY		
DESCRIPTION	NO.	DIMENSIONS	QUANTITIES		MATERIAL	LABOR
			UNIT	UNIT	UNIT COST TOTAL	UNIT COST TOTAL

Figure 17.11

472

Means Forms
CONSOLIDATED ESTIMATE

			SHEET NO.
PROJECT			ESTIMATE NO.
ARCHITECT			DATE

TAKE OFF BY		QUANTITIES BY		PRICES BY		EXTENSIONS BY		CHECKED BY

DESCRIPTION	NO.	DIMENSIONS				QUANTITIES			UNIT		MATERIAL			LABOR			EQ./TOTAL	
							UNIT			UNIT	UNIT COST	TOTAL		UNIT COST	TOTAL		UNIT COST	TOTAL

Figure 17.12

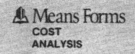

Means Forms

COST
ANALYSIS

	SHEET NO
PROJECT:	ESTIMATE NO.
ARCHITECT	DATE
TAKE OFF BY: QUANTITIES BY: PRICES BY: EXTENSIONS BY:	CHECKED BY:

DESCRIPTION	SOURCE/DIMENSIONS			QUANTITY	UNIT	MATERIAL		LABOR	
						UNIT COST	TOTAL	UNIT COST	TOTAL

Figure 17.13

MAINTENANCE COST DATA SHEET

WORK ORDER NO. _____   JOB _____

ACCOUNT _____   PREPARED BY _____

LABOR				MATERIAL		
EMP. NO.	HOURS	DATE	COST	DESCRIPTION	DATE	COST
	TOTAL LABOR COST $				TOTAL MATERIAL COST $	

COMPLETED DATE _____   TOTAL JOB COST $ _____

Figure 17.14

Job overhead costs include all costs which can be charged directly to a job, but which cannot be charged to materials, labor, or equipment. Some such items are low production time, job insurance, public liability insurance, job telephone, extra sets of plans and specifications, surveys, bonds, permits, inspection costs, architects and engineers fees, part of wages of superintendent or general foreman. Some estimators include payroll taxes and insurance in overhead costs, but most estimators add these items directly to labor costs. These items usually include Social Security (FICA), Medicare, Federal unemployment insurance, State unemployment insurance, Workmen's Compensation insurance, any type of liability insurance, and health and welfare charges along with any profit sharing or other similar fringe benefits available in the company.

To account for all the costs of operating a maintenance department, some means of applying these general costs to individual jobs is adopted. Usually this is accomplished by dividing the total operating overhead costs for a given period of time by the total maintenance labor hours charged to specific jobs and thus establishing an overhead rate per maintenance direct labor hour.

The Final Estimate

The total material, labor, equipment, and overhead costs contained on the consolidation sheets are then listed on a summary sheet, and finally summed to arrive at a total job cost. The reader is referred to Figure 17.10 for an example of an Estimate Summary Sheet From *Means Forms for Building Construction Professionals*.

Refer to Figures 17.11 through 17.14 for other examples of forms (from *Means Forms for Building Construction Professionals*) that can be utilized in the estimating process. Also refer to Chapter 18 for methods of completing format sheets.

Estimating the Cost of Deferring Maintenance

The cost of lost production and possible damage to equipment or product because of deferring maintenance generally can be estimated by judgment based on basic guide figures. Out-of-pocket costs will be most realistic and useful, and detailed accuracy is, therefore, not necessary. Estimation of material or product losses can be developed by utilizing average ratios of labor and material costs. For example, if direct labor for a product line averages half the material costs and 20 units are produced per day by four people at $5.00 per hour the approximate labor cost is $8.00 per unit, and the material cost is $16.00. These figures could then be used to compare against the cost of doing or not doing the maintenance work.

Formal engineering economy analysis can be applied for determining exactly whether or not maintenance and/or repair work should be deferred to a future date.

CHAPTER 18

PLANNING, SCHEDULING, AND COST CONTROL

The subjects of planning, scheduling, and cost control are often problem areas, because each is a separate process that exists only in relation to the total job. This chapter presents methods of planning, scheduling, and cost control, including examples of standard formats that can be used for documenting cost data for future projects.

Planning

Planning is defined as the selection of future courses of action through the analysis and evaluation of possible alternatives. Planning of maintenance operations is the attempt to establish the duration of time that a specified amount of work will take, having the necessary resources available.

Effective planning exposes likely difficulties, facilitates the maintenance department to overcome them, and minimizes the unproductive time of both manpower and equipment. Planning focuses on the cost of production by labor and equipment, and is useful as a control device against which progress can be measured during and after the completion of the project. Whatever type of planning is done, it must be sufficiently accurate to use for forecasting material, labor, and equipment needs.

In the planning stage, get feedback from everyone involved in the maintenance activity, including not only supervisory personnel, but also key tradespeople, such as carpenters, electricians, and millwrights. Any type of labor-saving device aids in insuring the completion of the job within the time and cost limitations. Not involving all personnel who have the potential of giving productive information may result in delayed time and increased costs. A feeling of teamwork results among the entire staff, both office and field personnel, when working toward common goals. This minimizes time delays due to work, method, and personnel problems, and jurisdictional disputes.

Effective planning facilitates the establishment of minimum-time, optimum-cost work methods with the resources available. The planning team considers all the parameters involved in doing the job, along with the sequence in which these parameters come into play. There is no measuring device available to indicate exactly how long and how much money to spend on planning a job. Experience is the best measuring device as to how much time and money can be devoted to the planning of any maintenance activity.

The stages of maintenance planning are long-range, yearly, daily, and emergency.

Long-Range Planning

The main goal of long-range planning is to ensure that the objectives, policies, and procedures of the maintenance organization are the same as those of the company. This type of planning is dependent on forecasts of future sales and production. Two of the most important factors to consider at this planning stage are possible changes in production equipment and in maintenance equipment and facilities.

Long-range planning focuses on the items listed below.
- A general plan for improving the maintenance function within the company.
- The possibility of establishing a training program for crafts and supervisory personnel.
- Establishing work-methods-improvement objectives.
- Identifying future capital needs for which additional funding will be required.
- An equipment replacement schedule.
- Development of a schedule of equipment and systems overhaul.

Annual Planning

This stage of planning is devoted to the establishment of an annual maintenance plan, consisting of the items listed below.
- Schedule of labor distribution throughout the plant.
- Schedule of major production shutdowns such as overhaul of major pieces of machinery.
- Determining the availability of materials and equipment to perform the maintenance function during the year.
- Establishment of a schedule for other maintenance responsibilities such as changing all incandescent light bulbs or complete sealing of concrete floor areas.

The annual maintenance plan assumes that manpower, equipment, and material will be available to perform any maintenance function during the coming year and allows for the need for lead time to acquire these if not available within the plant.

Daily Planning

The last stage of planning is done on a daily basis and does not include emergencies, which are unanticipated type maintenance activities. Daily job planning includes the following:
- Scheduling time for the job.
- Determination of a completion date.
- Establishing a list of material and equipment, including the writing of specifications for such material.
- Estimating the amount of labor required.
- Delivery of job instructions to workers.
- Identifying unusual safety hazards which may be involved in doing the job.
- Preparing all necessary work orders.

Emergency Planning

Much of the work done in maintenance today is done on an emergency priority; all breakdown maintenance factors cannot be anticipated. Therefore, planned work does not include emergencies. The best one can do for planning relative to an emergency would be to:

1. Develop a list, incorporating feedback from supervisory personnel, of emergencies typical to all areas in the plant.
2. Establish a written procedure to handle emergencies that cannot be handled incorporating normal knowledge and skill, within a reasonable time frame.
3. Make provisions for training personnel in emergency procedures.
4. Anticipate equipment and material needs and have them in stock within reason.

Preventive Maintenance (PM)

One very important way to minimize emergency maintenance is to have an effective maintenance program, which is nothing more than planned, or preventive maintenance. The objective of a preventive maintenance program is to increase productivity of the company and lower the maintenance cost of doing business. Preventive maintenance includes periodic inspection of plant and equipment to discover conditions which may lead to production or material breakdown. It also insures the upkeep of the plant by correcting defects while they are still in a minor stage. The installation of a preventive maintenance program is an investment that needs extra cash. All size plants can benefit from a preventive maintenance program. The larger the plant, the more detailed will be the paper work and supervision required to administer such a program. The major returns on an investment of installing a preventive maintenance program include:

- Less production downtime.
- Less overtime pay for maintenance personnel for emergency maintenance situations.
- Fewer large scale and repetitive repairs.
- Less repair cost for simple repairs located before larger break-downs.
- Fewer product rejections.
- Better quality control.
- Less standby equipment needed.
- Lower unit cost of manufacture.
- Better spare parts control.
- A decreased quantity of breakdown maintenance.
- Fewer safety hazards.
- Increased life expectancy of equipment and buildings.
- Proper lubrication on a systematic basis insuring less breakdowns due to improper lubrication schedule.
- Reduction of power requirement and utility expenses.

To ascertain the estimated dollar savings of installing a preventive maintenance program:

1. Review records for the past year for machine breakdown.
2. List the total cost of breakdown repairs.
3. List what each breakdown cost in idle time of operator, spoilage, and rework.
4. List other possible costs in terms of dollar amount such as overhead.
5. Estimate what the repairs would have cost if performed before the breakdown.
6. The difference is what might be spent on a preventive maintenance program.

Establishing a Preventive Maintenance Program

When establishing a preventive maintenance program do so in terms of the type of company and the philosophy of management. Be sure to include all personnel in this development and periodic review.

In setting up a preventive maintenance program, attempt to follow the principles noted below:

- Do not initiate a total preventive maintenance program company-wide; begin in one or two departments or areas and build from there.
- Begin with the simple objective of minimizing breakdowns through periodic inspection to discover and correct unfavorable conditions.
- Develop a list of items to be inspected and the duration of time between inspections. Things to be included in that list would be motors, production machinery, controls, building, material handling equipment, process equipment, lighting, HVAC systems, and plumbing systems.

In determining or analyzing what should be and what should not be included on the inspection first consider the following:

- Is this a critical item in terms of causing major shutdowns or costly damage or harm to an employee?
- Is standby equipment available in case of failure? If available, the need for preventive maintenance is then dependent on other factors such as the cost of breakdown maintenance.
- Does the cost of preventive maintenance exceed the cost of downtime and cost of repairs or replacement? If it costs no less to tear down a machine to repair a repetitive wear part, than the overall cost of the repair itself, the value of preventive maintenance is questionable.
- Does the normal life of equipment without preventive maintenance exceed manufacture's needs? If obsolescence is expected sooner than deterioration preventive maintenance may be a waste of dollars.

For non-operating equipment, if failure in upkeep or adjustment of the elements harms either production or the employee, or wastes plant's assets, the item in question should not be excluded from the preventive maintenance inspection practice.

As to what physical parts of each piece of equipment to inspect or tend to one must depend on:

- Experience of plant personnel.
- Service manuals provided with equipment.

Once a comprehensive list is made of what to inspect, a checklist should be developed highlighting all the points to be checked on any one piece or type of equipment or property. Checklists should be complete and allow for the inclusions of general information such as day of inspection, name of inspector, and room for specific general comments.

There are no readily available 'always' answers as to how often to inspect. This information must be developed by the company itself. The age of plant, kind of equipment, environment, type of operation, and other similar factors must be considered. To develop a timetable consult with:

- Historical records of the plant.
- Manufacturer of the equipment.
- Manufacturer's associations.
- Employees of the company.
- Insurance agent.
- Trade associations.

Keep in mind that once a figure has been established, it should be reviewed on a regular basis for possible alteration.

Scheduling Preventive Maintenance Functions

Again the most effective and efficient schedule is one that fits into the company operations resulting in a minimum of down time for production.

Then are three categories of preventive maintenance functions that should be scheduled. These are:

- Routine upkeep.
- Periodic inspections and replacement of parts.
- Contingent work which includes work at indefinite intervals where equipment is down for other reasons.

In scheduling preventive maintenance functions for the first two categories above, the following objectives should be kept in mind:

- Handle them on the day shift if possible to minimize overtime.
- Distribute these as evenly as possible over the year. Make use of slack seasons wherever possible.
- Keep production down time to a minimum.

All scheduling should begin with a formal recording system. The two types being utilized today are the following:

- The overall chart.
- The individual cards.

The overall chart is a list of every piece of equipment in the plant, department or service function. The other method is an individual card for each piece of equipment or service function.

Once this has been accomplished, the total scheduling for a preventive maintenance program means stipulating a definite day and time along with the materials, tools and equipment, and class of labor to perform the preventive maintenance function. The exact routine depends upon the type of company. However, once the scheduling is established it should be somewhat flexible so as to allow for emergency maintenance. As for every other maintenance function the personnel performing an inspection must be trained and experienced in this area.

Paper Work

Probably the biggest argument against establishing a preventive maintenance program is the expected quantity of paper work that will result. Some paper work is needed to control the system; however, the following things should be kept in mind:

- Minimize the number of forms and entries. If the data to be recorded is of no immediate or future use, do not record it. All cost data needs to be included.
- Attempt to integrate the preventive maintenance system within other maintenance paper work procedures.
- Provide for a periodic review check of preventive maintenance performance in terms of summary formats.

A minimum of five forms is recommended for preventive maintenance programs. These are:

- Equipment record (see Figure 18.1).
- Checklist (see Figure 18.2).
- Inspection schedule (see Figure 18.3).
- Inspection report (see Figure 18.4).
- Equipment maintenance preventive maintenance cost record (see Figure 18.5).

Once a preventive maintenance program is in operation and assumed to be operating effectively, it should be evaluated on a periodic basis through cost comparisons of estimated versus actual cost. Some ways of maximizing the program efficiency are listed below:

- Incorporate better materials which need less maintenance into present equipment in the plant.
- Redesign components in systems.
- Provide ongoing training programs for operators and maintenance personnel.
- Perform time and motion studies to be used in a regular job analysis.
- Develop a standard preventive maintenance manual or overall maintenance manual for the company.
- Standardize all tools, methods, equipment and other specialties.
- Establish and maintain a maintenance storeroom.
- Extend equipment life by applying protective coatings.
- Perform regular job analysis to measure the effectiveness of the program.

Much has been written on preventive maintenance programs and the reader is referred to the bibliography at the end of the book for resource information.

It is not the purpose of this book to explain preventive maintenance programs in depth, but only to highlight their important factors.

Responsibility for Planning

The responsibility of doing the actual planning may be in different parts of the formal company structure depending upon the size and type of company. Larger companies may have their own planning section or department while smaller ones may authorize their foremen to prepare all plans. No matter who performs this function, two objectives of planning to keep in mind are:

- Do only necessary work.
- Accomplish the work in the most effective way.

EQUIPMENT RECORD

DESCRIPTION Model FY-280 Automatic Thread Grinder MAKER   SERIAL NO. 12345

VENDOR   ADDRESS

ORDERED 10-01-78	WIDTH 12'	WEIGHT 12,000 lbs.		WATER No
RECEIVED 12-08-78	DEPTH 5'	EXHAUST 2-6" sq. exhaust ducts		GAS No
INSTALLED 12-16-78	HEIGHT 5'	AIR Air gun by machine		STEAM No

DETAILS Uses refrigerating unit to cool grinding oil

LOCATION Shop No. 1 NOW IN DEPT. #42	TRANSFERS	DATE 12/05	TO Dept. 80	DATE	TO	DATE	TO

LUBRICATION

SPINDLE BEARINGS 3 Drip feed-daily
BASE 1 Reservoir-(W) 6 mos.
GEAR CHANGE BOX 1 Reservoir-(W) 6 mos.
GRINDING 5 Reservoir-(W) 4 mos.
ELEC. MOTOR BEAR. 12 Grease gun-4 mos.

Ball thrust bearings used
5207K
3207 Internal attachment bearings
1-4" x 24" lg. endless flat belt

MOTOR RECORD

MFG	HP	RPM	TYPE	FRAME	STYLE	SERIAL NO.	MODEL NO.	VOLTS	AMPS
	1.5	1000	B	422		BT2144	1T442C864	220/440	4.98/2.49
	5	1150/3450	T	428		FL712	14462B232	220	10.8
	1/8	1725	XP	4524		OV5327		230	.7

PHASE	CYCLE	CONTROL EQUIP.	DRIVES	BELTS	TYPE BEARING	BEARING DATA	NO.
3	60	Main cont. bx.	Direct		Ball bearing	1305W	8448
DC	DC	Main cont. bx.	Direct	A-56		207MF	9294
DC	DC	Main cont. bx.	Direct		Ball bearing	424	9284

Figure 18.1

TO BE CHECKED	EQUIPMENT NUMBER									
	DEFECT	OK	DEFECT	OK	DEFECT	OK	DEFECT	OK	DEFECT	OK
BEARINGS	D			*		*		*	D	
BRAKE		*	D			*		*		*
CLUTCH		*		*		*	D			*
CUTTER		*	D			*		*		*
GEARS	D		D		D		D		D	
GUARDS		*		*	D			*		*
LUBRICATION	D		D		D			*	D	
MOTOR		*		*		*		*		*
PIPING		*		*		*		*		*
ROLLS		*		*		*		*		*

INSPECTION CHECKLIST

Figure 18.2

INSPECTION
SCHEDULE   PREPARED BY   DATE

WEEKLY MAINTENANCE SCHEDULE
AIR CONDITIONING SYSTEMS

C – CLEAN   O – OVERHAUL   FC – FUNCTIONAL CHECK

SYSTEM NUMBER	1	2	3	4	5	6	7	8	9	10	11	12	13	14	15	46	47	48	49	50	51	52
EAST BUILDING								C														
UNIT		C	O		C			FC			C			C		FC	C			C		
FILTERS	C							C								C						
FANS	C FC															C						
CONTROL		C		FC	C			C FC			C	FC		C		FC	C			C FC		
TOOL ROOM UNIT				FC			C O			FC					C FC					FC		
FILTERS	C	C	C	C	C	C	C	C	C	C	C	C	C	C	C	C	C	C	C	C	C	C
FANS			C								C									C		
CONTROL				C FC					C					C				FC				
STORE ROOM UNIT						O														FC		
FILTERS	C			C				C								C				C		
FANS																						
CONTROL	FC							FC								FC						

Figure 18.3

484

INSPECTION REPORT

INSPECTION ENGINEER'S REPORT TO DEPT. HEAD	TO DEPARTMENT HEAD B. White	DEPT. 25
1. MACHINE CONDITION	MACHINE & NO. 9876 wire drawing	EQUIP. NO.
2. RECOMMENDATION FOR REPAIR	SIGNED, INSPECTION ENGINEER	DATE Dec. 8

ITEM NO.	MACHINE CONDITION REPORT	CORRECTION	SAFETY	EMERGENCY REPAIR	ENGINEER RECOMMENDATION
1.	Main roll bearings worn - all shims removed at last overhaul	Replace bearings on all four roll shafts			X
2.	Oil pump shaft scored and worn	Install new oil pump - rework damaged pump for spare		X	
3.	Capstan bearing worn and housing damaged	Rebore casting and fabricate sleeve to adapt new bearing			
4.	Bolt guard damaged	Repair guard and paint	X		X
5.	Noisy drawing roll drive gears	Replace gears			X

ESTIMATED REPAIR COSTS			ENGINEER'S REMARKS	DEPARTMENT HEAD DISPOSITION	
	TOTAL ITEM COSTS			APPROVED ITEM REPAIRS	REMARKS
COST DISTRIBUTION	ALL LISTED	RECOMMENDED		ALL	
MATERIAL	276.00	162.00		SAFETY	
LABOR	173.00	134.00		EMERGENCY REPAIR	
OVERHEAD	152.60	109.20		ENGINEER'S RECOMMENDATIONS X	
TOTAL	601.60	405.20		OTHER (SEE REMARKS)	
				NONE (SEE REMARKS)	
				REPAIR ORDER ISSUED X	SIGNATURE DATE

Figure 18.4

EQUIPMENT MAINTENANCE COST RECORD

| MACHINE NO. | | TYPE MACHINE | | CAT. NO. | | SERIAL NO. | |

| VENDOR | | DATE PURCHASED | | | DATE INSTALLED | | |

| PURCHASE COST | INSTALLATION COST | TOTAL COST | | DEPRECIATON | YEAR | PER YEAR | |

DEPRECIATION SCHEDULE					MECHANICAL EQUIPMENT		
YEAR	DEPRE-CIATION	ADDED TO VALUE	TO BE AUTHORIZED	MAINTENANCE COST			
19							
19							
19							
19							
19							
19							
19							
19							
19							
19							
19					ELECTRICAL EQUIPMENT		
19					EQUIPMENT		
19					MAKE		
19					SERIAL NO.		
19					TYPE FRAME		
19					VOLTAGE		
19					PHASE		
19					AMPERS		
19					HORSEPOWER		
19					R.P.M.		
19					DRIVE		
19					CIRCUIT		
19					DATE INSTALLED		

Figure 18.5

Work Orders

The need for planning maintenance is initiated with a form known as a *work order*. A work order is a written authorization for a specific maintenance activity. It is a part of a total work order system. A complete work order system includes the following activities:

1. Establish a cost accounting code system.
2. Develop a formal format for a work order.
3. Estimate the maintenance activity.
4. Prepare a schedule.
5. Follow up during and after the job by applying effective cost control parameters.

Cost Accounting Code Systems

The first step in establishing a work order system is to develop a system of time charges to provide positive identity for every activity to be done by maintenance. An effective time charge system must allow a cost accumulation of maintenance and materials on a day basis by charging every labor hour and material dollar against a designated job or account number. These account numbers should not only provide for charging time and materials against each job, but should provide an accurate breakdown to demonstrate how costs are distributed within each project by material, labor, and format. This information can, in turn, be used to compare the estimated to the actual cost of doing work, and also serve as historical data for future use in preparing estimates for similar projects.

Maintenance personnel must keep in mind that every hour and dollar of maintenance time and money must be charged against a valid, active maintenance account number. If the entire cost numbering system is not used appropriately and accurately by field personnel, the information drawn from this system will be of no use. It is important that all company personnel who are involved in using this system, in whole or in part, be trained in its application.

A cost accounting code is nothing other than establishing a numbering system for the purpose of identifying the various job activities and tasks connected to each activity. There are various systems in use today and the best system is one that fits the individual company's needs.

There are various types of coded numbering systems. The first utilizes all numbers in some sequence to designate a meaningful choice. An example would be 1920.5 where 1900 is changing incandescent lamps; 20 is all office areas and .5 signifies that the electrical part of the maintenance department would do the work.

Another system is utilizing numbers and alphabet letters. An example would be 125A; where 100 would indicate cleaning cooling towers; 25 indicates a specific cooling tower; and A indicates the job is to be done by maintenance group A. The above systems could be enlarged to include a number to indicate the priority given to the work order. The reader is also referred to Figure 18.6 for an example of a list of maintenance work accounting codes for a small maintenance group.

Work Order Format

The format of a work order should give the workers all the information needed to do the job, and also let the employee report back how well the job actually went. A complete work order format should include the following information:

- Originator's name.
- Authorization.
- Department in which job is to be done.
- Identification of the equipment that is to be maintained.
- Completion date.
- Description of work.
- Estimate of labor, material, equipment, and overhead or labor burden required.
- Time actually spent in doing the job.
- Parts and materials used.
- Job classification by number of code.
- Cost comparison.
- Personnel actually doing the job.

There are two types of work orders. The first type is a specific work order issued for the completion of a specific maintenance job. The second type is the "open work order," which covers the activity to be done on a regular basis such as waxing office floors once a week. An open work order is usually written for a year's duration. Its issuance saves time in not having to write a new work order every time the floor is to be waxed. Maintenance management must determine the type of maintenance work that can be covered in an open work order. Refer to Figure 18.7 for an example of a work order format.

Establishing Work Priorities

Once a work order has been initiated, one of the first activities that maintenance management needs to undertake is to assign a priority to the job. This, in turn, requires establishing a formal work order priority system. As for the work order, priority systems should be established by each individual company to meet its own internal needs. Whatever type of system is developed, it should be simple and easy to understand and issue. An example would be a three-point priority system as noted below:

Priority # 1: Tasks which are mandatory to successful operation of the company's facilities. This priority would take precedence over all other maintenance work.

Priority # 2: Tasks which need to be completed as soon as practicable.

Priority # 3: Tasks which are desirable, but which may be completed when convenient. These last tasks make up the backlog of a maintenance organization. Backlog is used to level out peaks and valleys in the maintenance organization's daily schedule.

In addition to the above direct in-line maintenance tasks, there are also staff support functions such as clerical work. These tasks can be given a separate priority number or designation.

Again, it is up to the company and its management to establish a list of which tasks fall into what priority classification: what might be priority number one for one company might not be for another. The reader should also keep in mind that emergencies, unplanned-for maintenance jobs, cannot be given a priority classification prior to the emergency happening. At best, a list of the various types of possible anticipated emergencies could be included in the company's priority system. Whatever type of system is used, a general list of all priorities should be kept. The list will serve as a summary of what work is ahead and what work is in the process of being done.

Maintenance Work Accounting Codes

Site
- 10 Grounds
- 11 Parking Lots and Driveways
- 12 Lawn Sprinkler Systems
- 13 Parking Lot Lights
- 14 Interior Planters and Plants
- 15 Painting-Curbs and Roadway

Building—Structural
- 20 Roofs
- 21 Painting—Inside
- 22 Painting—Outside
- 23 Floors
- 24 Walls
- 25 Windows and Blinds
- 26 Doors
- 27 Ceilings

Building—Electrical
- 30 Electrical
- 31 Power System
- 32 Telephone, and Communications Systems
- 33 Instrumentation Cabling and Duct

Buildings—Mechanical
- 40 Refrigeration System
- 41 Heating System
- 42 Air-Handling system
- 43 Plumbing System—Including Sewers and Fire Sprinklers
- 44 Fuel Storage and Distribution
- 45 Steam Storage and Distribution
- 46 Sheet Metal Work

Building Services
- 50 Janitorial Services
- 51 Fire and Safety Devices
- 52 Material and Equipment Handling Stripes

Supply Operations
- 60 Storeroom Operations
- 61 Receiving Operations

Production Machinery and Equipment
- 70 Scheduled Lube and Check
- 71 Mechanical Repairs
- 72 Electrical Repairs
- 73 Installations, Mechanical and Electrical Equipment
- 74 Inspection and Calibration
- 75 Painting
- 76 Furniture Repair Service Call

Special Services
- 80 Requests for installing or relocating blackboards, bookcases, etc.
- 81 Moves

Figure 18.6

SPECIFIC WORK ORDER

MAINTENANCE WORK ORDER			JOB NO. 4-6-79		BORN 638	
TITLE Equipment for Toll Control Station			REVISION		DATE	
PURPOSE Needed to Complete Job Improvement Request			ENG. REL. NO.		PRIORITY 3	
			DELIVER TO DEPT. 6-32		REQ. COMPLETE DATE 5/1/	

	SIGNATURE	DATE		ESTIMATED COST	ACTUAL COST
REQUESTED	S. F. Sims	4/1/	LABOR	380.00	
PREPARED	E. Blue	4/4/	BURDEN	57.00	
APPROVED	Joe Sims	4/6/	MATL. ON HAND	417.00	
APPROVED			MATL. TO PURCHASE		
APPROVED			FREIGHT		
EXPENSE	X		TAX		
CAPITAL			OUTSIDE CONTRACT		
CHARGE TO ACCOUNT			TOTAL	854.00	

ITEM	DESCRIPTION	ESTIMATED		ACTUAL	
		LABOR	MATERIAL	LABOR	MATERIAL
1.	Dept. 19-31 Construct one (1) foreman's desk per Std 151-248 (18 hours @ $4.50)	81.00	82.00		
2.	Dept. 20-30 Fabricate two (2) steel racks per Std Dwg 151-286 (68 hours @ $4.00)	272.00	320.00		
3.	Dept. 20-32 Paint the above described desk and steel racks std colors (6 hours @ $4.50)	27.00	15.00		
4.	Dept. 26-82 Deliver the above desk and steel racks to Dept. 6-32 Attention J. A. Mikes, foreman.				

PERSONNEL ASSIGNED TO JOB:	DUTIES:
Joe James	Carpenter
Mike James	Painter
Jim Wright	Delivery

Figure 18.7

490

Follow-up Action

Once a work order is received, the persons doing the planning should actually go to the field to check the job out. The planner's next responsibility is to insure that everything necessary for the job is available when needed. The planner should never assume that a critical item is on hand. When an item that is needed is located, it should be tagged with the work order number or otherwise designated for the particular job. The materials to be purchased for the job should be on hand before the planner releases the work order to the shop. While this will require that some work order be held for a period of time, the shop doesn't want the work order until the work can actually be done.

The next job is to prepare any drawings or sketches that may be needed. In conjunction with many of the items above, an estimate must also be made as to how long the job will take, along with the required amount of materials and equipment. This is the estimator's job in a large company, or the foreman's job in a small company. Once all this information is accumulated and put together the work order is ready to be released, along with a materials list and any required sketches to the workers in the field doing the repair or maintenance item. The reader is referred to Figure 18.8 for a planning checklist.

Figure 18.8

Planning Checklist

_____ How much work will be done in-house and how much will be contracted?
_____ Site visitation
_____ Review specifications and drawings
_____ Site access
_____ Availability of utilities
_____ Disposal of salvaged material
_____ Labor availability
_____ Local weather conditions (if exterior maintenance)
_____ Availability of outside contractors
_____ Staff to be used on job
_____ Personnel experience
_____ Equipment availability
_____ Probable completion date
_____ Preliminary planning-bar chart
_____ Activity job analysis
_____ Selection of planning team
_____ Optimization of the methods used a they apply to the particular job
_____ Budget
_____ Purchasing of materials
_____ Bill of material

Planning Material

In planning material requirements, consideration should be given to prices, price trends, time to buy, amount to buy, dealer or manufacturers, transportation, deliveries, amount to be kept on hand, checking, testing, and insuring. Figure 18.9 is an example of a material organization chart.

Some general points to keep in mind when planning material requirements are listed below.

- When a shipment is received, make sure someone is available to receive it and ensure that what has been ordered is what has been delivered.
- Make sure the entire quantity has been delivered and that it is in good condition.
- After the material is received, make sure you know where it will be stored and that is is protected physically from loss, damage, or theft as reasonably as possible.
- Whether it is delivered directly to the job for immediate use, or is stored in the yard, warehouse, or stock room, be sure a written record is kept of who it was delivered to and for what, before you release it. The person to whom it was delivered should do the same.

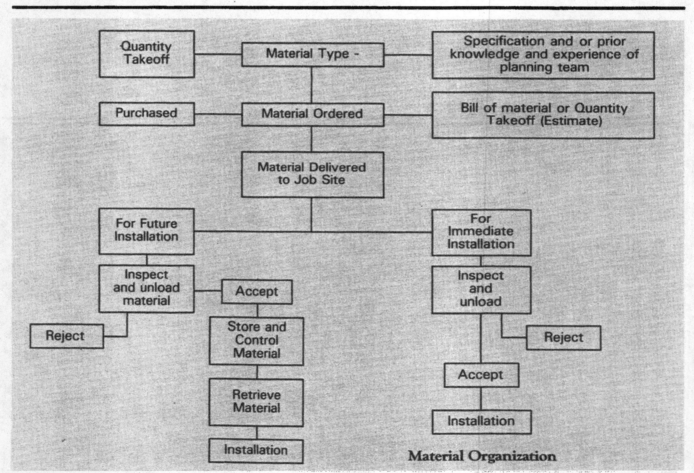

Figure 18.9

492

Planning Tools

Some of the major considerations in planning the use of tools for any project are listed below.

- Establish complete list of tools required for job.
- Establish who provides which tools.
- Establish tool standards for tools to be purchased.
- Establish well controlled and organized tool storage area, to make tools readily available to the workmen with minimum delay.
- Establish and maintain a tool inventory.
- Instruct in safe use of tools.
- Stamp and identify tools.
- Maintain spare parts.
- Use the proper tools for the specific job.

Planning Labor

When planning labor some of the major considerations are:

- Kinds and amounts will vary with different jobs.
- The amounts of skilled and unskilled labor should be determined.
- Consideration should be given to the men in the permanent organization, old employees who may be available, labor that may be available locally, and labor that may need to be imported from another locality.
- Watch labor trends and wages. In general, when there is a larger amount of work in any locality, the labor supply will be low, the wages will be high, and the labor efficiency will be low. Labor wages depend not only upon the amount of work and the labor supply available, but also upon union rules and regulations, government regulations, general prosperity of the country or locality, work available in other fields, and inflation.
- Time estimated for doing each kind of work and the amount of each kind of work.
- Need to know date and time that each type of work is to be performed.
- Make any needed reassignments.
- Be sure everyone knows what to do and where to go.
- Make sure your people have their work properly completed in preparation for an outside contractor.

Planning Equipment

The costs of owning and operating equipment are as follows:

- Operation cost
  —Fuel
  —Oil
- Repairs
  —Preventive Maintenance
  —Crash Maintenance
- Operator.
- Downtime due to equipment itself or other equipment.
- Moving in and moving out (transportation).
- Set-up and take-down time.
- Inspection.
- Housing and/or storage of equipment, equipment parts and maintenance facilities.
- Other minor costs.

Some of the major considerations the planner should keep in mind when planning for equipment utilization are as follows:

- Various pieces of major equipment.
- Kinds of work they are to do.
- Date and time each piece of equipment is to be available for use on each kind of work.
- Selection will depend on what is available and on what can readily be procured for the job.
- Equipment layout is important so that the various pieces will work most efficiently.
- See that all equipment is on hand when necessary to ensure productivity.
- Be sure each laborer has proper tools to perform with efficiency.

Job Analysis

The secret of effective planning is to be able to break down any job into single activities of which the job is comprised. Also it is the ability to use such resources as one's own experience, the experience of others, historical data and available, reliable technical information to assign times, labor, material, and equipment to accomplish the job.

One of the goals of effective planning is to be able to perform an analysis of the job by breaking it down into tasks or activities and combining the tasks. This is done in such a manner as to result in performing the job in a minimum amount of time and thus for the lowest cost. A basic example of this would be the job of changing a light bulb. The following are only two ways the job can be accomplished, relative to breaking it into activities:

Method One	Method Two
Obtain ladder	Obtain ladder
Set up ladder	Obtain new light bulb
Climb ladder	Set up ladder
Remove burned-out light bulb	Climb up ladder
Climb down ladder	Remove burned-out light bulb
Discard old light bulb	Install new light bulb
Procure new light bulb	Climb down ladder
Climb up ladder	Take down ladder
Install new light bulb	Put ladder away
Climb down ladder	
Take down ladder	
Put ladder away	

As the reader can see, the second method would require less time and thus use less of the appropriated maintenance budget to complete the job.

Job analysis begins with a brief written description of the job to be done, followed by answering specific questions noted below:
- What is being done?
- What is the purpose of what is being done?
- Why is the work done?
- What would happen if the work were not done at all?
- Is every part of the job necessary?
- Who does the work?
- Who could do it better?
- Where is the work done?
- Could it be done somewhere else more economically?
- When is the work done?
- Would it be better to do it some other time?
- How is the work done?

From the answers to these questions a detailed list of activities is developed. Estimated time, manpower, and equipment requirements are then assigned to each activity. Once this is done, the activities should be combined, separated, or changed, and put together in such a manner as to result in doing the job in the shortest amount of time and with the least cost possible.

The amount of time spent in performing job analysis must be proportional to the available planning resources and complexity of the job to be performed. However, the reader should keep in mind that some planning is better than no planning at all.

Job Analysis Format

Without using a specific standard format, job analysis can become very confusing. If the result of the pre-job analysis is formatted, it can serve as a control device after the job is completed. By recording how the actual job is performed; along with the time and resources used, management can obtain accurate data on the effectiveness of its planning and scheduling process. Also, the information can be utilized, as historical data, in determining ways the job could be more effectively done in the future.

One format that can be used is the job analysis chart noted in Figure 18.10. The following method is used to devise such a chart:

1. Under the activities column, list the activities comprising the job. And opposite each one assign a number in numerical order (1, 2, etc). When listing the activities do not forget any types of delays such as not obtaining a full 60 minutes from the workmen.
2. For each activity note the estimated distance for all activities which have travel time (such as moving to a job).
3. Assign the estimated time to each activity.
4. Along the top of the chart list the labor classifications and equipment needed to do the job.
5. Under each labor classification and/or piece of equipment note the activity number that the personnel and equipment are involved in during the respective activity in the right-hand column. For example, referring to Figure 18.10, while activity five, inspection, is taking place, the foreman will be inspecting; the pump operator and hopperman will be making ready tools (involving activity one); the laborers and vibrator men will be setting up the machines (involving activity four); and the pump and fittings are on the job, but not being used (a delay for activity nine). For each activity, come up with the total cost of performing it. Finally, calculate the total cost and time for doing the job as noted on the example in Figure 18.10.

Once this is accomplished, an analysis of the sequence of activities and estimated costs can be made. The objective is to:

- eliminate unnecessary detail.
- minimize travel and anticipated delay time.
- rearrange for better sequencing.
- simplify tasks where possible.
- suggest the use of more efficient tools, material, and equipment.

Ask the questions noted earlier in this section to see if a more efficient method can be used to do the job.

After devoting as much time to the process as possible and coming up with the method to be utilized for doing the job, copy the final activity chart and make it available to field personnel. Also, keep a copy in the office for future use.

As the job progresses, a new job analysis chart should be made outlining the actual activities, labor, and equipment utilized, along with the actual time spent doing each activity. Once completed, the actual and estimated activity charts can be compared to ascertain how effective the planning process was, following up with adjustments to be made in the future. This information will then be utilized as historical data for similar future jobs.

Scheduling

While planning insures smooth processes of carrying out the maintenance function, scheduling is the formality of carrying out the planning function by assigning exact times for the work to be done. There should be as much scheduling as needed for maximum overall efficiency as long as the system costs less than the cost of operating without it. However, the extent and sophistication of the scheduling system should be dependent upon how well it is adhered to. Another point to consider is how effective the scheduling system has been relative to increasing maintenance efficiency. The essence of scheduling is to maintain a proper balance between work capacity and work load.

JOB ANALYSIS CHART

DWG. NO. 2854	CHART NO. 1
SUBJECT Concrete Pumping-Beam Repair	LOCATION Loading Dock
DATE 4/6/87	CONT. NO. 500	CHARTERED BY RWL
WORK CYCLE Place conc. by pipe line	UNIT Cu. Yd.
EST. QUANTITY 180 C.Y.	EST. COST 10.00/C.Y.	ACTUAL COST
PLANT EQUIPMENT Contractor Provides	PRESENT
SUPT.	FOREMAN	PROPOSED

HOURLY RATES

NO.	ACTIVITY	DIST.	TIME	15.00	12.00	11.00	10.00	10.00	10.00	10.00	11.00	11.00	50.00	25.00	COST
1.	Make ready-tools.	200	60	1		2	1	1	1	1	1	1			88.50
2.	Move to job site.	150	15	2	3	2	2	2	2	2	2	2			33.00
3.	Pump mach. to job.	2 Mi.	45	3									3	3	70.50
4.	Setup mach. time.		90	4	4	4	4	4	4	4	4	4	4		287.00
5.	Inspection.		30	5	1	1	4	4	4	4	4	4	9		15.00
6.	Mach. adjusting.		30	6	6	6	4	4	4	4	4	4	9		19.00
7.	Delay-ready-mix.		20	7	7	7	7	7	7	7	7	7	7		50.00
8.	Pump conc. to forms.	115	360	8	8	8	8	8	8	8	8	8	8		923.50
9.	50 min. hr. normal delay.		80	9	9	9	9	9	9	9	9	9	9		200.00
10.	Dismantle pump.		45		10	10	10	8	8	10	10	8	10		90.50
11.	Tools to shed.	150	15		10	10	11	11	11	11	11	11	10		15.50
12.	Return pump.	2 Mi.	30	5	12	12	13	13	13	13	13	13	12	12	49.00
13.	Final cleanup.		10												10.33
14.	Crew check out.		5				14	14	14	14	14	14			5.17
15.															
16.															
17.															
18.															

Vibrators and Tools in Plant Setup.

													TOTAL COST	1,857.00

TOTAL TIME	MINUTES	760	715	775	780	780	780	780	780	780	745	75	
	HOURS	12.67	11.92	12.92	13	13	13	13	13	13	12.42	1.25	1,857.00

Figure 18.10

Factors that must be considered in establishing a schedule are as follows:

- Unit of labor, such as man hours or man days, on which the estimates will be based and measured.
- Size of job or jobs to be scheduled, which will be dependent upon the company.
- Percent of total work load to be scheduled versus that which cannot be scheduled. A percentage of the work crew needs to be available for emergency or breakdown maintenance. Between 10 and 20 percent of the man power should be available for this purpose.
- Necessary lead time for planning and scheduling.

There are two general types of scheduling; master and detail.

Master Scheduling

A master schedule should be set up for a long period of time such as one week, 90 days, or one year. It needs not only to include preventive maintenance functions, but also all items that are a part of future maintenance activity in the company. The master schedule should be flexible, not fixed.

The master schedule should indicate the nature and magnitude of each repair and construction task segment of maintenance for a specified amount of time. The master schedule will serve as the foundation for more detailed scheduling.

The entire master schedule should be reviewed on a monthly basis and revised accordingly. Figure 18.11 is an example of a master schedule.

Detail Scheduling

Detail scheduling is nothing more than breaking down the time span allocated in the master schedule. This is done to properly sequence the various phases of the jobs and thus to insure that each task will be done in the most efficient manner. The master schedule is based on established priorities and availability of equipment and labor within a company. The detail schedule indicates what work is to be done, task time, and special tools and/or equipment needed. As for master scheduling, detail scheduling must also be flexible. The most effective and carefully prepared detail schedule may be offset suddenly by unforeseen emergencies. When this occurs, scheduling personnel must be able to improvise or reschedule rapidly to meet the new conditions.

A detail schedule is done on a daily basis. The first jobs to be included in the schedule are those which are carryovers from the previous day. Next, priority jobs one and two are added to the schedule. Finally, priority three jobs are added until all available man hours have been scheduled. The reader is referred to Figure 18.12 for an example of a daily work schedule, one type of a detail schedule.

When scheduling minor jobs or maintenance activities on small pieces of machinery, it is customary to prepare a comprehensive maintenance schedule covering all trades required for the maintenance item. On a larger job and/or machines, each schedule should then be split up into individual trade schedules. No matter how it is done, each schedule should indicate the data computation and be given an issue number starting from one and counting upward.

Not only should a schedule be prepared for every day's work in the department, but each job should also be scheduled.

Types of Schedules

The two most commonly used types of schedules are the Gantt Bar Chart and Network Diagram such as the Critical Path Method (CPM).

MASTER SCHEDULE

SCHEDULE DURATION

DATE COMPLETED   SCHEDULED BY

WORK ORDER	DEPT.	OPERATIONS	MEN	EST. HRS.	PLANNED DATE START	PLANNED DATE COMPLETE

Figure 18.11

DAILY SCHEDULE

DATE _____ PREPARED BY _____

PERTINENT INFORMATION _____

NAME	EMPLOYEE	ACTUAL TIME	WORK ORDER NO. / JOB COST NO.					
			TOTAL					

Figure 18.12

Gantt Bar Chart

Bar charts can be utilized for both short- and long-term jobs. However, they are essentially advantageous when being used for jobs having a short duration. This type of chart provides management with a means of determining the following:

- The overall time required, through the use of a logical method rather than a calculated guess.
- A means to review each phase of the job in detail to see that items such as material and equipment are covered.
- Coordination requirements between crafts.
- Comparing alternate methods of performing the work.

One can develop a bar chart by following these steps:

1. List all the activities that will be done to perform the job. The steps should be listed in sequential order of occurrence. Once the list is compiled, the overall job can be reviewed to determine if any steps can be handled simultaneously.
2. Assign times required to perform each activity.
3. Duration of the job is broken down into some type of constant time units which are determined by the total estimated time of the job. These time units are placed at the top of the chart. The following general rule applies as to what time units to use:

Job Duration	Time Units
Less than a week	Hours
One week to one year	Days or months
Over a year	Weeks or months

4. Using a solid line for the estimated time, draw in one for each activity. The line or bar is drawn to scale of the time unit chart above. It begins on the anticipated beginning date and ends at the estimated completion date.
5. At any specific time during the job, such as on a daily, weekly, monthly basis or at the completion of the job, draw in a dashed line which indicates the actual progress, including the actual starting time, and the percent completed to date.

This type of schedule can be used as a control device to ascertain whether or not the job is on schedule. If for one reason or another the job is not on schedule, immediate action can be taken in the office and/or field to correct the problem at hand to insure the particular activity, and thus the job, is completed on schedule. It might be that the estimate was done incorrectly and the schedule will have to be redone.

The bar chart is only another tool to be used in ascertaining whether or not the job is on schedule and is being carried out as planned and estimated. If any changes are be needed to insure that the job is completed as budgeted, the time to know this comes during the job, and not after it is completed. The bar chart will aid in this investigation. The reader is referred to Figure 18.13 for an example of a bar chart for a job that is in progress, and Figure 18.14 for a bar chart prepared at the planning stage.

Network Diagram—the Critical Path Method

Critical Path Method (CPM) is a more sophisticated method of scheduling than the Gantt Bar Chart. CPM is also a management tool whereby the many aspects of a project can be made immediately apparent. It is a road map on which a person can see many things, including material or engineering bottle necks, field problems, completion dates of various job phases, and overall completion time. This method is usually utilized for jobs which are complex and take a relatively large span of time to complete.

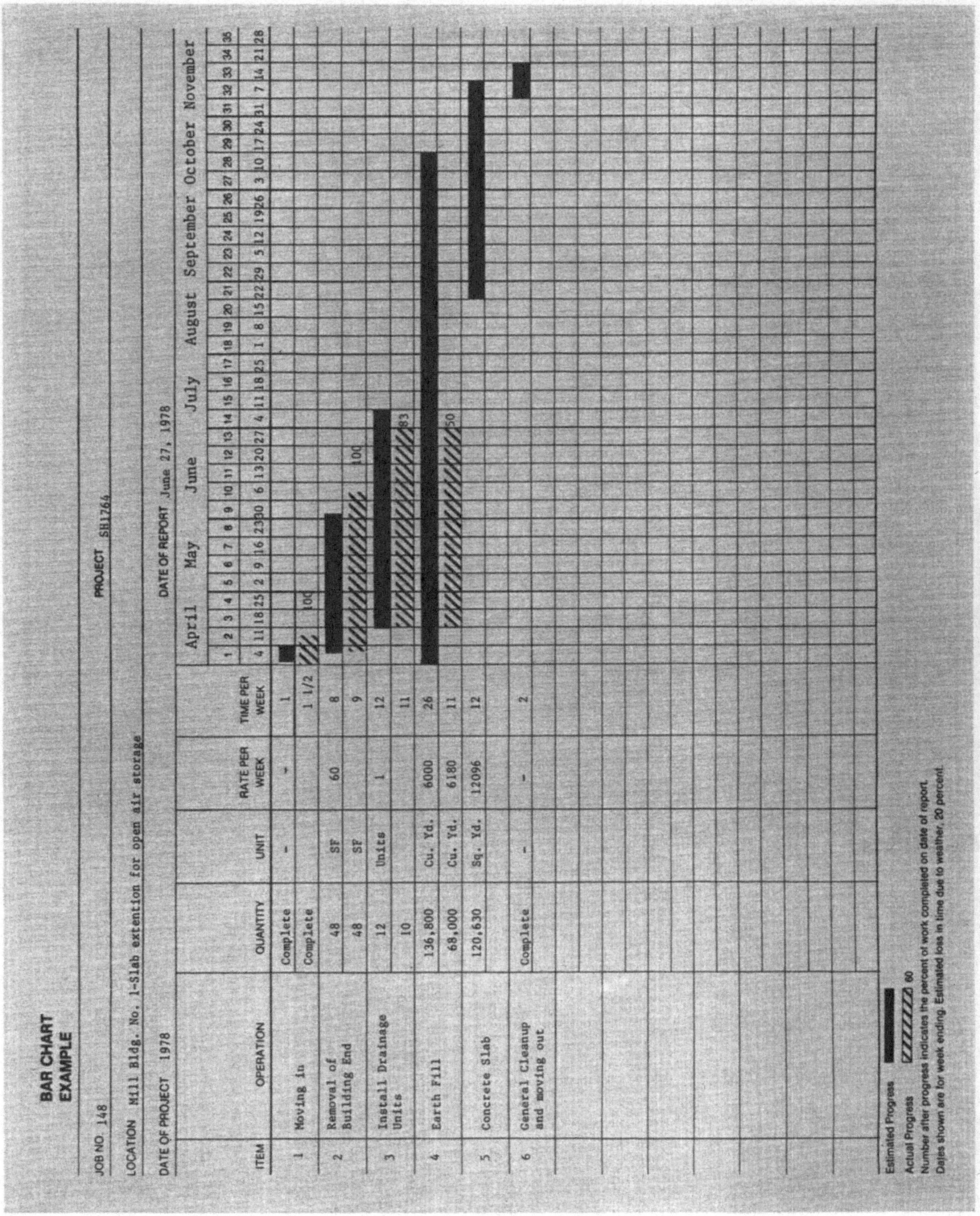

BAR CHART EXAMPLE

JOB NO. 148
PROJECT SH1764
LOCATION Mill Bldg. No. 1-Slab extention for open air storage
DATE OF PROJECT 1978
DATE OF REPORT June 27, 1978

ITEM	OPERATION	QUANTITY	UNIT	RATE PER WEEK	TIME PER WEEK
1	Moving in	Complete	-	-	1
2	Removal of Building End	Complete	-	-	1 1/2
		48	SF	60	8
		48	SF		9
3	Install Drainage Units	12	Units	1	12
		10			11
4	Earth Fill	136,800	Cu. Yd.	6000	26
		68,000	Cu. Yd.	6180	11
5	Concrete Slab	120,630	Sq. Yd.	12096	12
6	General Cleanup and moving out	Complete	-	-	2

Months: April | May | June | July | August | September | October | November
Week endings: 4 11 18 25 | 2 9 16 23 30 | 6 13 20 27 | 4 11 18 25 | 1 8 15 22 29 | 5 12 19 26 | 3 10 17 24 31 | 7 14 21 28

Progress percentages shown: 100, 100, 100, 83, 50

Legend:
Estimated Progress (solid bar)
Actual Progress (hatched bar) 60
Number after progress indicates the percent of work completed on date of report.
Dates shown are for week ending. Estimated loss in time due to weather, 20 percent.

Figure 18.13

Bar Chart Example

Installation of New Bearings and Gears

Task						
Prepare rigging for removal of equipment cover	▭					
Remove oil	▭					
Remove bolts from bearing covers	▭					
Remove top half of cover and bearing cover	▭					
Remove old bearings	▭ 4 mech.					
Remove old gears		▭				
Remove couplings		▭				
Install new couplings		5 mech.	▭ 2 mech.			
Clean gear bar			▭ 3 janitors			
Install new gears				▭		
Install new bearings				▭ 5 mech.		
Check gear contact and bearing tolerances					▭	
Install reducer and bearing covers					4 mech.	▭
Fill reservoir						▭ 4 mech.

Figure 18.14

Arrow Diagramming: The basis of the Critical Path Method is a network representation of the project, known as an arrow diagram. In summary, the arrow diagram graphically describes not only the sequence of activities, but also their interrelationship. On an arrow diagram, both arrows and circles are used to describe the sequence of work. An arrow stands for an activity, while a circle represents an event. An event is the starting or ending point of an activity and occurs only when all the activities preceding it (which means all the arrows leading to the circle) have been completed.

When constructing an arrow diagram, the following items must be kept in mind:

- Everything in the diagram has a significant meaning.
- An activity has a single definite starting point and a single definite ending point.
- The arrow diagram describes dependency relationships rather than time relationships.

Arrow diagrams proceed from left to right and are developed by locating the arrows in accordance with the following three criteria:

- What jobs come before?
- What jobs come after?
- What jobs can be done concurrently?

For example:

Job y follows job x and therefore cannot start until job x is completed. Conversely, job x comes before job y.

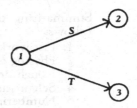

In the above example, job s and t can be done concurrently and are therefore independent of each other.

In drawing arrow diagrams, it is helpful to remember that you are planning the next job and not trying to represent the way a job has always been done in the past. How detailed in the actual breakdown should one get? Is it better to start with the simple diagram and expand?

Remember that when an arrow is placed into the diagram, two questions are asked about it:

- What activities must be completed before this one can start? This determines the event that starts the activity.
- What activities cannot start until this one has been completed? The answer to this question indicates at what event the activity ends.

Since the number of arrows in a diagram for any particular job can be hundreds or even thousands, a specific method of identification is required for each activity. One method is to number the event circles. The number at the tail of the arrow should always be smaller than the number at the head of the arrow. This is mandatory, especially when computerizing the Critical Path Method. Each activity should have its own set of unique numbers. These numbers are sometimes referred to as the i-j numbers; i at the tail of the arrow and j at the head of the arrow.

As mentioned above, arrows cannot have the same number on both the head and the tail. Example: if two jobs start and end at the same time, they cannot be diagrammed as follows:

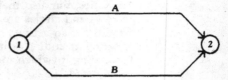

Instead, dummy arrows are used to avoid the situation as noted below:

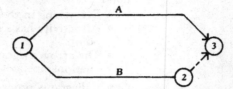

Dummy arrows are also used to indicate dependence. In the following example T depends on the completion of A while E depends on the completion of A and C. T and E are independent.

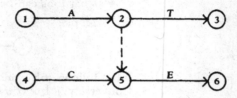

Summarizing, the steps in developing a network logic diagram are as follows:

1. List work items.
2. Phasing of work items.
3. Rough draft of network.
4. Schematic of network finalized.
5. Numbering the events.

An example of constructing a network diagram is given below, based on the following list of activities and dependencies:

1. A and D start up the origin.
2. J follows F but precedes K.
3. C follows A but precedes G.
4. H follows D but precedes L & K.
5. B follows A but precedes F & E.
6. K, L and E are terminal activities.
7. K follows G and H.
8. E follows B and C.
9. F is independent of C.
10. L is independent of J.

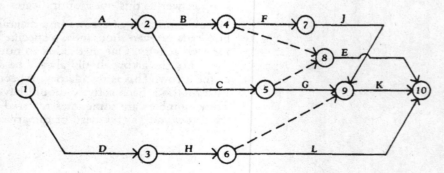

Figure 18.15

Activity Durations: Once the network diagram has been completed, the next step is to assign durations to each activity. The durations are nothing more than the estimated time each activity would take. Factors affecting duration of an activity are as follows:

- Man-days or man-hours required to perform each activity from the estimate.
- Crew size.
- Interference by other crafts on concurrent work.
- Availability of work area.
- Nominal delays resulting from lead or lag time and partial completions.
- Allowance for weekends and holidays when the duration is expressed in calendar days.

The unit selected for activity duration is dependent on the scope of the job and the amount of refinement that the person drawing the diagram desires. For example, the unit selected for a one-year construction project might be days or weeks while the most desirable unit for a four-hour maintenance job is probably minutes. Dummy activities have an elapsed time of zero.

Once the duration of each activity is established it is placed on the arrow in the network diagram. Figure 18.15 can be used in recording labor and equipment durations.

By definition the critical path is the longest path, in time, through the network. Since these critical activities are added to determine the total duration of the project, any delay of one of the activities will proportionately delay the job. Conversely, any speed-up will decrease the total duration time. The reader is referred to Figure 18.16. This figure is a network diagram that contains four paths. A first path, activities B, E, F, and H, takes a total of 20 days to complete. A second path, comprising of activities A, D, F, and H, takes a total of 19 days to complete. The third path, comprising activities C, G, and H, takes a total of seven days to complete and the fourth path of activities C and I takes 17 days to complete.

The longest path in time through the network, activities B, E, F, and H, are the critical activities. This is the path controlling the project length. If it is desired or required that the amount of time needed to complete the project be shortened, these are the activities upon which to concentrate. Non-critical activities are strictly dependent upon the completion of the critical items, so speeding up non-critical activities is of no value in terms of project duration. Generally, a very small number of activities make up the critical path. This means that a large percentage of the activities in a project have extra time available since they are, in a sense, waiting for the critical items to be completed. A scheduler or manager can adjust his non-critical activities to take best advantage of manpower and equipment availabilities, weather conditions and other similar items, without delaying the project. An important point to keep in mind when using the Critical Path Method is that there is not necessarily only one critical path. There may be several. Also, remember that if a critical path is shortened, a new parallel path will most likely become critical.

When producing a schedule, the Critical Path Method user is interested in the following items of information about each activity in the job:

- **Early start date:** the first day upon which all work preceding the activity has been completed.
- **Early finish date:** the first day upon which no work will be done on an activity.
- **Latest start date:** the latest allowable day by which work must start on an activity if the project is not to be delayed.
- **Latest finish date:** the deadline date for finishing the activity if the project is to be completed on time.
- **Status:** whether the activity is on the critical path.
- **Float:** the amount of extra time available for non-critical activities.

The calculation of each of these items can best be described through illustration. The arrow diagram in Figure 18.16 describes a project for which a schedule is desired. The first step necessary is the calculation of the earliest occurrence or start time for each event. The earliest occurrence time for each event will be placed in a box by that event (at the tail of the activity arrow).

The earliest event time for event number one is generally assumed to be time zero. Therefore, a zero should be placed in a box by event one. Zero means that the job will start at the end of day zero or at the beginning of day one.

In Figure 18.17, activity D cannot start until activity A has been completed. It takes seven days to complete activity A, therefore, the earliest start time of activity D would be the earliest start time of the preceding activity A, which is zero plus the duration of the preceding activity, which is seven. Therefore, the earliest start time of activity D would be seven. Which number would go in a box by event two?

Continuing with our example, activity E cannot begin until activity B has been completed and therefore the earliest start time of activity E would be at the end of the sixth day. Therefore the earliest start time of activity E is six and this is placed in a box by event number three.

When more than one arrow (activity) enters into a circle (event), calculations must be made for each arrow relative to the early start time of the next activity leaving the event. Turning to Figure 18.18, activity F cannot start until activities E and D are completed. Activity E will not be done until the eleventh day of the job while activity D will be done at the end of the tenth day of the job. Even though activity D will be done before activity E, activity F cannot start until the end of the eleventh day or the beginning of the twelfth day of the job. Therefore the earliest start time of activity F is eleven and this is placed in a small box adjacent to event number five.

The balance of this example is completed using the same logic as presented above and results shown in Figure 18.19.

Once the forward path has been completed, the next step in the schedule production is the calculation of the latest finishing time of each activity. This is done by doing a backward pass, which is starting at the end of the network diagram and working toward the front or beginning of the job. In this particular example, the latest finish times will be placed in hexagons adjacent to the ending events of each activity.

The latest finish time of the job is the summation of the longest path. This number would be placed in the hexagon at event number seven. In this case it is 20. After the latest finish time for the last event has been calculated, the Critical Path Method user would go to the event number that is one less than the ending event. In Figure 18.20 this would be event number six. At event six the question would be asked how late can this event occur so that any activity that starts with it will be finished by time 20. Since there is only one activity that starts with event six and it has a duration of five days, it is apparent that event six can occur as late as time 15 (the latest finish time of an event is equal to the latest finish time of the following event minus the duration of the event in question). The number in the hexagon at event six is, therefore, 15.

At event five, the same question is asked. How late can event five occur so that any activity that starts with event five will be finished by time 15? Again, there is only one activity to consider, so event six can occur as late as time eleven (15–4). Therefore, eleven is placed in the hexagon adjacent to five, as noted in Figure 18.21.

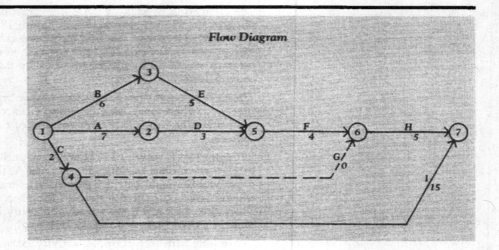

Figure 18.16

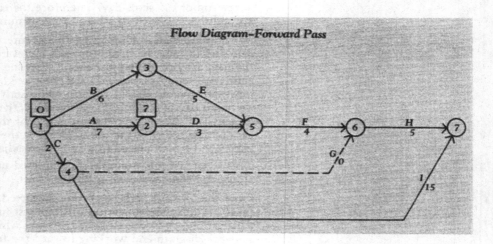

Figure 18.17

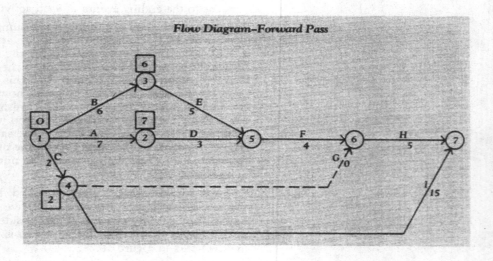

Figure 18.18

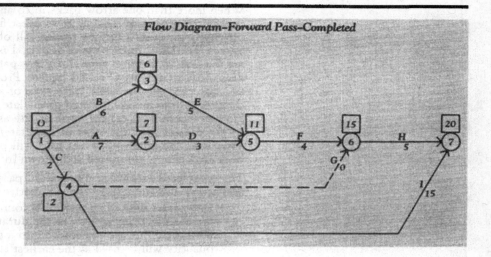

Figure 18.19

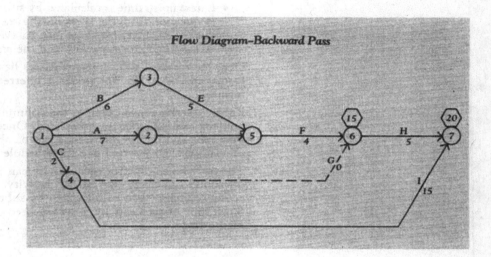

Figure 18.20

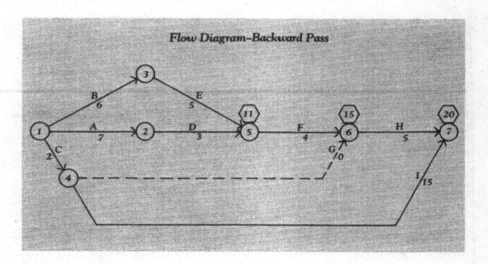

Figure 18.21

509

When more than one arrow backs in to an event, each one must be considered separately relative to its latest finish time. Therefore, considering activity C (or event four) the reader will observe that activities G and I back into it. The latest finish time of activity I is 20. Subtracting 15 from 20 we see that the latest finish time along that path would indicate that the latest finish time of activity C would be five. Proceeding backwards along path G, one observes that the latest finish time of activity G, 15 minus zero (the duration of the dummy), would give a late finish time of 15 for activity C. However, activity C cannot finish as late as the fifteenth day without holding up the project since the other path dictates a late finish time of five days. Therefore, the number at event four will be five. The balance of the backward pass is completed and shown in Figure 18.22.

The completed forward and backward pass results have been summarized and shown in Figure 18.23. To summarize the definitions:

- **Early start time** of an activity is found by adding the early start time of the the preceding activity to the duration of the activity in question. If more than one activity comes into a starting event, the largest resulting quantity will be used as the earliest start time of the activity in question.
- **Latest finish time** is calculated by subtracting the duration of the activity in question from the latest finish time of the preceding activity. If more than one activity backs up into an event, separate late finish times must be calculated for each path and the smallest one used.

Once the forward and backward passes have been completed, a schedule format can be set up. The reader is referred to Figure 18.24 for such a format.

Relative to the above example, the columns headed activity, (the i-j numbers) and *duration* are completed. Once this has been done, the early start time and the late finish time can be abstracted from the network calculations and inserted into the schedule format.

The early finish time for each activity can then be calculated by adding the duration to the early start of each activity. The late start time of each activity can also then be calculated by subtracting the duration from the late finish time of each activity. This has been done in the example shown in Figure 18.25.

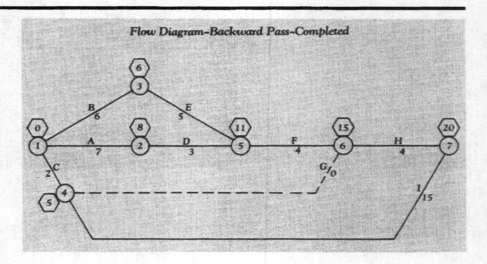

Figure 18.22

510

Once these columns have been completed the remaining information to be found is *total float*, and *status*. The total float is the amount of time that the job can be delayed without affecting the critical path or overall completion time. It is calculated by subtracting either the earliest finish from the latest finish or the early start from the latest start. Any activity with a total float of zero will be on the critical path. Thus, activities B, E, F, and H are on the critical path. The status of an activity is whether or not it is on the critical path.

As the reader can see, the calculations are not complicated and can be done manually for jobs with any number of arrows. However, on large jobs involving thousands of arrows, the calculations become extremely tedious. For this reason, computer programs have been developed to make them. It is only necessary to estimate the elapsed time for each activity and inform the computer of the dependency relationships. Programs have been designed to report starts and finishes keyed to calendar dates or the shifts of a day to list the critical path items, to report upcoming activities, to produce bar charts and network diagrams.

Updating procedures, as the job progresses, involves only a few entries with revised estimates that are run through the computer to yield another schedule. A manpower leveling program also has been developed which considers the crafts involved and levels demanded by shifting those activities with float time. The intent is to develop a schedule with the most practical allocation of crafts by desired time increments such as shifts, days, weeks, and months, throughout the entire job. The reports are compiled, showing the total manpower by crafts and the manpower on the near critical and critical activities for each time increment. A report is then developed which is based on the leveling and shows the recommended start and finish for each job. If the manpower leveling reports show more time from any craft than is available, the arrow diagram plan is revised and the calculations run again.

An example of the Critical Path Method applied to maintenance is shown in Figures 18.26a and 18.26b.

Figure 18.23

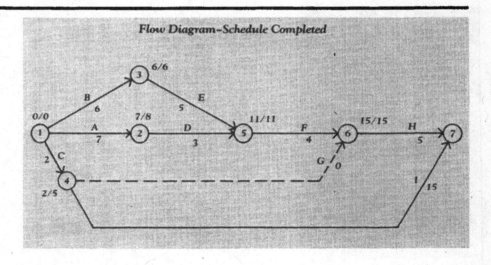

ACTIVITY		DURATION	DESCRIPTION	EARLIEST		LATEST		TOTAL FLOAT
i	j			START	FINISH	START	FINISH	

CPM SCHEDULE FORMAT — FINAL SCHEDULE   PREPARED BY   DATE

Figure 18.24

| ACTIVITY | | DURATION | DESCRIPTION | EARLIEST | | LATEST | | TOTAL FLOAT |
i	j			START	FINISH	START	FINISH	
1	2	7	A	0	7	1	8	1
1	3	6	B	0	6	0	6	0
1	4	2	C	0	2	3	5	3
2	5	3	D	7	10	8	11	1
3	5	5	E	6	11	6	11	0
4	6	0	G	2	2	15	15	13
5	6	4	F	11	15	11	15	0
4	7	15	I	2	17	5	20	3
6	7	5	H	15	20	15	20	0

CPM SCHEDULE — EXAMPLE   PREPARED BY   DATE

Figure 18.25

Cost Control

Cost control can be defined as the regulation of expenditures to within the authorized estimated cost, while providing the desired quality of work within the time limit specified.

The major purposes of having a cost control system are as follows:

- To create historical cost data for future use in estimating.
- To check actual cost against estimated cost.
- To motivate the desire for lower cost.
- To place the responsibility and credit where they belong.
- To enable corrective action for high cost during the progress of the work, not after the job has been completed at a higher cost than estimated.
- To become a foundation for work-simplification, cost-reduction studies.
- To enable the estimator to check outside contractor bids.

Relative to factors affecting job cost control, some of the major ones are noted below:

- Job location
- Procurement of labor and materials
- Lead time
- Equipment utilization
- Management personnel
- Financial requirements
- Purchasing
- Expediting
- Accounting and record keeping
- Cost control procedure during job
- Estimating and cost accounting
- Cost control records for material, labor, and equipment

Even though control of job costs can begin as early as the point in which the job is initiated through the development of correct, concise and clear designs, drawings, specifications, and labor instructions, the basis for the cost control program is the estimate.

Initially this means that the work classification, as estimated, must be in the same terms as measured in the field either during (for long-duration jobs) or at the end of the job (for short-duration jobs). The units of labor costs as estimated must be the same as those secured from the payroll in order to be directly comparable. The same reasoning holds for material and equipment control.

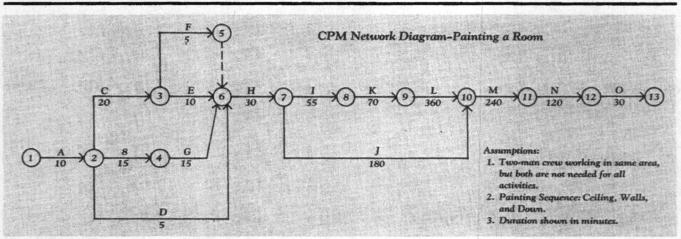

Figure 18.26a

CPM SCHEDULE — Painting a Room   PREPARED BY   DATE

ACTIVITY		DURATION (IN MINUTES)	DESCRIPTION	EARLIEST		LATEST		TOTAL FLOAT
i	j			START	FINISH	START	FINISH	
1	2	10	Obtain Work Order	0	10	0	10	0
2	4	15	Check Tool room for rollers, Pan, ect.	10	25	10	25	0
2	3	20	Estimate Paint Needed	10	30	10	30	0
2	6	5	Order Ladder, drop cloth, ect.	10	15	35	40	25
3	6	10	Order Paint from Warehouse	30	40	30	40	0
3	5	5	Request No. of Needed Painters	30	35	35	40	5
4	6	15	Obtain Tools from Tool Room	25	40	25	40	0
5	6	0	Dummy	35	35	40	40	5
6	7	30	Stir Paint and Setup Equipment	40	70	40	70	0
7	8	55	Dust & Roller Paint Ceiling-1st Coat	70	125	70	125	0
7	10	180	Wash Doors and Screens	70	250	375	555	305
8	9	70	Roller Paint Ceiling-2nd Coat	125	195	125	195	0
9	10	360	Roller Paint Sidewalls	195	555	195	555	0
10	11	240	Brush Paint Doors	555	795	555	795	0
11	12	120	Wash Desks and Chairs	795	915	795	915	0
12	13	30	Clean Floors, Rollers & Turn in Equipment	915	945	915	945	0

Figure 18.26b

To insure consistency between what is estimated and what is measured in the field, an in-depth cost code numbering system needs to be developed. The system should be:

- As logical and understandable as possible.
- Simple.
- Founded on the idea that it will be the basis of the cost control system, therefore complete enough not to have to charge items to a "miscellaneous" account. In fact there should not be an account number for "miscellaneous."

The specific cost account numbers will be utilized for the specific item throughout the job to keep a continuous record of the actual cost of doing the particular item. The reader is referred to an earlier discussion in this chapter for more specific information and examples of cost accounting coding system.

Developing a Field Reporting System

Once a cost coding system has been established and its importance and use understood by everyone affected, the next step is to develop an effective field reporting system.

The objective of the field reporting system is to gather information of the actual work progress in terms of the following:

- Labor time expended.
- Equipment time expended.
- Quantities of material used.

The gathered information must be in terms of the same units as measured for the estimate. For instance, if a concrete beam repair job is estimated in terms of cubic yards of concrete in place, the field information must also be recorded in terms of cubic yards of concrete placed over a specified duration of time. And furthermore, the estimated and actual work should be referenced to the same job code number.

The information obtained in the field must be put or recorded on standard formats. The type of format must meet the need of the specific organization. Some of the main forms that should be developed are listed below. The reader should remember that it is not the goal of a cost control program to create more paper work. Therefore, the information obtained from the separate forms listed below could possibly be obtained from combining two or more of the forms into a single form or by utilizing a specific type of form giving the same types of information.

Types of Forms

- Labor time cards
- Equipment utilization forms
- Material installation reports
- Labor summary reports
- Equipment summary reports
- Material summary reports

The first three items listed above should be completed on a daily basis. This information is usually completed by the foreman (superintendent) or the workers themselves (if they are so trained). This information is then sent to the main office or main department for summarization on a weekly or monthly basis. Items 4, 5, and 6 are therefore completed on a weekly or monthly basis.

A word of caution—no matter how many or how few forms one uses and how detailed the type of information one deems necessary from the field, if they are not measured and recorded accurately they will be of little use in the cost control program. It is mandatory that if the system is to produce effective-results, all personnel should be trained not only in how to use the system, but more important how to measure and report units of labor, material, and equipment utilization on the standard forms. The lack of personal motivation to take the system seriously or the lack of knowledge on the importance of the system and how it is used to help the worker is the cause of many cost control program failures in industry today. Many companies have found that by sharing with the employees any saving that result from using the control system, the system will be utilized by the employee more effectively and taken more seriously.

Once summary reports are produced in the office, the tabulated unit cost of doing the actual work is compared with the estimated unit cost.

The entire cost control process, including the standard forms, can be as simple or sophisticated as the company deems necessary to better control its costs. The following is an example, with forms to be utilized in a cost control system for three activities of a repair project.

The estimating department will work up the estimate to do the work. In this particular example three activities will be examined. They are listed below:

- Excavate soil by bulldozer.
- Set anchor bolts.
- Strip and clean forms.

Based on the estimator's experience and knowledge of the scope of the work in terms of sizes, shapes, location, etc., and using the drawings and specifications, the following quantities are taken off:

- Set anchor bolts—100 anchor bolts
- Excavation of soil by dozer—370.4 cubic yards
- Strip and clean forms—2000 square feet contact area

The information developed above is developed on a standard format known as a Work Sheet, Figure 18.27.

Once the basic quantities are known, the next step is to apply material cost, labor production factors, labor wage rates, and equipment production factors to obtain a final item cost. This information can be obtained from:

- Experience.
- Company records.
- Historical data.
- Commercially published technical data. (Information is to be used only when nothing else is available.)

For this particular hypothetical company historical records or data are utilized. Sample headings for these forms are noted below.

For labor production:

Activity	Crew Size List Quantity of Each Trade	Crew or Individual Production (Quanity Per Unit)	Units of Production	Remarks
Strip and Clean Forms	2 Laborers	125 S.F./Hr	Square Feet	
Set Anchor Bolts-Column	1 Carpenter 1 Laborer	4 Bolts/Hr	Each	

For equipment production:

Activity	Equipment	Unit of Production	Rate of Production	Auxiliary Equipment or Labor
Excavation	Bulldozer	Cubic Yards	60 C.Y./Hr	1 Operator
Hauling— Less Than Two Miles	5 Cubic Yards Truck Rear Dump	Cubic Yards Loose	50 C.Y./Hr	1 Driver

Means Forms
QUANTITY SHEET

PROJECT	SHEET NO.
	ESTIMATE NO.
LOCATION	ARCHITECT	DATE
TAKE OFF BY	EXTENSIONS BY:	CHECKED BY:

DESCRIPTION	NO.	DIMENSIONS L	W	H		UNIT	Volume	UNIT	Area	UNIT	Misc.	UNIT
Set Anchor Bolts-3/4"	100	1									100	Ea.
Excavation Dozer		100'	200'	0.5'	10000	CF	371	CF				
Strip & Clean Forms-3/4" Ply		200'		10'					2000	SF		

Figure 18.27

Apply this information to the quantities. Placing all work on a standard format sheet known as a Summary Sheet (see Figure 18.28) the final costs are calculated as follows:

- Set anchor bolts—$900.
- Excavate soil—$209.81.
- Strip and clean forms—$160.

Based on the total quantity from the work sheet and the total cost given above, the cost per unit can be calculated and results in the following:

- Set anchor bolts—$11.25 each
- Excavate soil—$.62 per cubic yard
- Strip and clean forms—$.144 per square foot

The above are known as unit costs, and will be the basis on which the actual unit costs will be compared against later in this example.

The final cost for each item or activity is listed on a separate format known as a Recapitulation Sheet (Figure 18.29) to be used for summary review of job costs.

Once the job begins, it is mandatory to put one's field reporting system into operation on a daily basis to acquire actual work quantities done in the field. Specifically, the type of information one needs is:

- What each worker did.
- How many hours spent doing it.
- How much material is used.
- Types and length of times equipment was used.

The amount of labor time expenses along with what each laborer did would be reported on a daily time card. An example of a time card for the labor included in the above-noted three activities is shown in Figure 18.30. The reader will note that the minimum types of information needed on a time card are as follows:

- Name of worker
- Job classification of worker
- Date
- Times
- Activities or jobs involved in and hours charged against each job, along with the code number it should be charged against.
- Other types of information deemed needed by the company.

For excavating soil, not only is a time required for the operator of the dozer, but a daily equipment report is also required (see Figure 18.31). For stripping and cleaning forms, not only are time cards required of the workers, but also a quantity report of the amount of cleaning that was done is needed (see Figure 18.32). The same holds true for placing anchor bolts (see Figures 18.33 and 18.34). This information, completed daily, is submitted to a central department or office and is summarized on a weekly or monthly basis. In this example it is summarized on a weekly basis as noted in Figures 18.35 through 18.37. Not only will the daily information acquired from the field be placed on the summary formats, but also spaces will be made available to summarize hours, amounts of dollars, and other types of information that may be required.

Once the summary sheets have been completed, it is then possible to get a total actual cost of labor, material and equipment. After this information is accumulated, a weekly total cost summary can be prepared. On this particular form (refer to Figure 18.38), the cost code, work description, estimated quantity cost, and unit cost are first inserted. Once this has been completed, all actual costs to date and of a particular date, are noted in the columns provided. Following this, all work done for this particular week of the cost summary statement is then inserted in the form. Again the information required is the quantity installed, the total cost, and unit cost. Once this has been completed, the analysis of each activity is done, based on a comparison of estimated unit costs to actual unit costs.

COST ANALYSIS

					SHEET NO.	
PROJECT					ESTIMATE NO.	
ARCHITECT					DATE	
TAKE OFF BY:	QUANTITIES BY:	PRICES BY:	EXTENSIONS BY:	CHECKED BY:		

DESCRIPTION	SOURCE/DIMENSIONS			QUANTITY	UNIT	MATERIAL		LABOR		EQ./TOTAL	
						UNIT COST	TOTAL	UNIT COST	TOTAL	UNIT COST	TOTAL
Set Anchor Bolts				100	Ea	6	600				
4 Bolts/4r 1 Carpenter				25	Hr			12–	300		
1 Laborer				25	Hr			9–	225		
Excavation											
Dozer 60 CY/Hr				6.2	Hr					25	154
Operator				6.2	Hr			12–	74		
Strip & Clean											
Forms											
2 Laborers				32	Hr			9–	288		
125 SF/Hr											

Figure 18.28

Means Forms

CONDENSED ESTIMATE SUMMARY

PROJECT				SHEET NO.
LOCATION		TOTAL AREA/VOLUME		DATE
ARCHITECT		COST PER S.F./C.F.		NO. OF STORIES
PRICES BY:		EXTENSIONS BY:		CHECKED BY:

		Material	Labor	Equipment	Total	Unit Cost
100 Ea.	Set Anchor Bolts	600	525	---	11 25	11 25
371 CY.	Excavate Soil	---	74	154	228	.62
2000 SF.	Strip & Clean Forms	---	288	---	288	.15

Figure 18.29

TIME CARD — Excavating

NAME John Jones, Jr., I   CLASSIFICATION Operator

MACHINE AND NUMBER Dozer-64

DATE October 13, 19	START 8:00 A.M.	TOTAL HOURS 8
WEATHER Warm-Clear	QUIT 4:30 P.M.	

JOBS AND ACCOUNT NO. Excavate   HOURS 8

3010

REPAIR TIME ———   STAND BY TIME ———

REMARKS ———

Figure 18.30

DAILY EQUIPMENT REPORT — Soil Excavation

FOREMAN'S DAILY EQUIPMENT REPORT

DATE 10/13/   JOB Acct. No. 3010   SHIFT 1st   FOREMAN

MATERIALS HANDLED Topsoil

SANDY SOIL ☐   CLAY ☐   LOAM ☒   CLAY LOAM ☐   GRAVELLY OR STONEY SOIL ☐   ROCK ☐   OTHER

CONDITION OF HAUL ROAD   ROUGH ☐   SMOOTH ☐   DRY ☐   DAMP ☐   MUDDY ☐   SOFT ☐   STICKY OR SLIPPERY ☐

GRADE   OPERATOR

UNIT	UNIT NO.	ATTACH- MENTS	COST CODE	HOURS WORKED	PAY HOURS	L.T. MECH.	L.T. SERV.	HOUR RATE	EARNINGS
AIR COMPRESSOR									
AIR TOOL									
AIR TOOL									
BIT. PLANT									
BOILER									
BUCKET									
BATCHER, CONCRETE									
BIN, AGGREGATE									
CART, CONCRETE									
CEMENT GUN									
CEMENT PLANT									
CONCRETE MIXER									
CONCRETE PAVER									
CRANE, CRAWLER									
CRANE, TRUCK									
CRUSHER									
DRAG LINE									
ENGINE, GAS									
GRADER									
GRADER									
HOIST									
PUMP, GAS									
PUSHER									
PUSHER									
PUSHER									
ROLLER, ROAD									
ROLLER, COMPACT									
SAW, CHAIN									
SAW, TABLE									
SAW, RADIAL									
SCRAPER									
SCRAPER									
SHOVEL, GAS									
SHOVEL, ATTACH.									
TRACTOR, CRAWLER	1	Blade	640	8	8	--	--	25.00	150.00
TRACTOR, RUBBER									
TRACTOR, DOZER									
TRUCK, DUMP									
TRUCK, FLAT									
TRUCK, SERVICE									
TOWER, HOIST									
WAGON, DUMP									

Figure 18.31

QUANTITY REPORT — Stripping and Cleaning Forms

DAY OR WEEK ENDING   October 17   19   REPORTED BY

ACCOUNT NO.	TOTAL LAST REPORT	QUANTITY FOR						TOTAL TO WEEK	TOTAL TO DATE	ITEM NO.
		MONDAY	TUESDAY	WEDNESDAY	THURSDAY	FRIDAY	SATURDAY			
4260	500 s.f.					1000		1000 s.f.	1500 s.f.	

Figure 18.32

FOREMAN'S QUANTITY REPORT

DATE 10/17/

J.O. NO. 1 UNIT Anchor Blt. CRAFT Laborer, Carpenter ACCOUNT NO. 2834

DESCRIPTION OF WORK Set Anchor Bolts

ORIGINAL BUDGET QUANTITY	100
BUDGET INCL. AUTH. CHANGE NO'S TO QUANTITY	---
CURRENT ESTIMATED COMPLETION QUANTITY	100
UNITS PREVIOUSLY REPORTED	90
UNITS COMPLETED TODAY	10
UNITS COMPLETED TO DATE	100
FOREMAN'S ESTIMATED UNITS TO COMPLETE	---

TOTAL MAN-HOURS WORKED TODAY SIGNED FOREMAN SUPERVISOR

FIELD ENGINEER'S PROGRESS MEASUREMENT

FIELD ENGINEER'S REMARKS All Bolts Set

FIELD ENGINEER DATE 10/18/

Figure 18.33

QUANTITY REPORT — ANCHOR BOLTS

MATERIAL COST SUMMARY

COMMODITY   Anchor Bolts   ACCT. NO.   2834

VENDOR	P.O. NO.	DESCRIPTION	QUANTITY	AMOUNT	MONTHLY TOTAL
Sam Snow	1111	1/2" x 36" anchor bolts	100	$600.00	$600.00

Figure 18.34

WEEKLY SUMMARY — Set Anchor Bolts

LABOR ANALYSIS & SUMMARY

ACCT. NO. 2834   SHEET NO. 1

CLASS OF WORK Set Anchor Bolts   WEEK ENDING 10-17-87   JOB NO. 1

CLASSIFICATION				15th	16th	17th		HOURS	RATE	AMOUNT	
1 Laborer				8	8	4		20	9.00	180	00
2 Carpenter				8	8	4		20	12.00	240	00
TOTAL				16	16	8		40	--	420	00

	QTY. WORK IN PLACE	PAYROLL COSTS	LABOR AVERAGE UNIT COST	AVERAGE QTY. PER 8 HR. DAY	QTY. WORK IN PLACE	PAYROLL COSTS	LABOR AVERAGE UNIT COST	AVERAGE QTY. PER 8 HR. DAY
PREVIOUS	----	----	----	----				
THIS WEEK	100	420.00	4.20	40				
TOTAL	100	420.00	4.20	40				

Figure 18.35

WEEKLY SUMMARY — Excavation

LABOR ANALYSIS & SUMMARY

ACCT. NO. 3010   SHEET NO. 2

CLASS OF WORK Excavation   WEEK ENDING 10–17–87   JOB NO. 1

CLASSIFICATION		13th	14th	15th	16th	17th		HOURS	RATE	AMOUNT	
1 Operator		8	2					10	12.00	120	00
TOTAL		8	2					10	12.00	120	00

	QTY. WORK IN PLACE	PAYROLL COSTS	LABOR AVERAGE UNIT COST	AVERAGE QTY. PER 8 HR. DAY	QTY. WORK IN PLACE	PAYROLL COSTS	LABOR AVERAGE UNIT COST	AVERAGE QTY. PER 8 HR. DAY
PREVIOUS	----	----	----	----				
THIS WEEK	400.00	120.00	.30	400 c.y.				
TOTAL	400.00	120.00	.30	400 c.y.				

Figure 18.36

WEEKLY SUMMARY — Strip & Clean Forms

LABOR ANALYSIS & SUMMARY

ACCT. NO. 4260   SHEET NO. 3

CLASS OF WORK Strip&Clean Forms   WEEK ENDING 10-17-87   JOB NO. 1

CLASSIFICATION		13th	14th	15th	16th	17th		HOURS	RATE	AMOUNT	
1 Laborer					6	8		14	9.00	126	00
2 Laborer					6	8		14	9.00	126	00
TOTAL					12	16		28	---	252	00

	QTY. WORK IN PLACE	PAYROLL COSTS	LABOR AVERAGE UNIT COST	AVERAGE QTY. PER 8 HR. DAY	QTY. WORK IN PLACE		PAYROLL COSTS	LABOR AVERAGE UNIT COST	AVERAGE QTY PER 8 HR. DAY
PREVIOUS	500 s.f.	$ 72.00	.144	1000					
THIS WEEK	1000 s.f.	$252.00	.252	1000					
TOTAL	1500 s.f.	$324.00	.216	1000					

Figure 18.37

For the activity, "set anchor bolts", the reader will note that the estimated cost is $11.25 per bolt as compared to the actual cost of $10.20 per bolt. This resulted in a total savings of $105.00. This savings is noted in the column on the form. Since this activity has been completed there will be no projected savings or loss. However, comparing the unit cost, one will note that the bolts were placed at $1.05 each less than estimated. Examining this point a little further, one or more of the following conclusions could be made:

- The estimator estimated the cost of the material higher than it actually cost.
- The estimator used a higher labor production rate than actually occurred.
- Field personnel performed better than expected (as estimated).

In general, this activity would not be a concern to the manager.

The reader will note that the "excavation" was actually done for $.93 a cubic yard which is approximately $.31 per cubic yard more than estimated. By further examining the figures, the reader will also note that 400 cubic yards of dirt were excavated instead of the 340.4 estimated. Therefore, it would be safe to say the estimator was a little low in his calculations. Some of the other conclusions that could be reached:

- Equipment rates were estimated lower than they actually were.
- Labor production rates were lower than estimated.
- Field personnel did not perform as estimated.

In the end, $141.71 was lost on this activity. Since this activity is also complete there is no immediate need for any follow-up action; however, when reviewing this job at its completion it would be a good idea to consider revising the labor and equipment production figures.

For the last item, "clean and strip forms", it is noted that prior to this week's report, the work was being done at exactly the estimated unit cost ($.144 per square foot). However, this week it is noted that the unit cost for doing the work was $.252 per square foot. By subtracting $.144 from $.252 one will note that it is costing the company $.108 per square foot more to do the work. Therefore, to date (including figures from this week) a present loss of $108.00 is indicated. Looking at the long term, if the work continues to be done for $.252 per square foot there will be a projected loss of $90.00 (which is obtained from multiplying the difference, $.108 per square foot, times the amount of work that is done at $.252 per square foot which would be 1,500 per square feet). At this point, since this activity is not complete, it is up to management to decide how much time and manpower they would like to devote to attempting to bring the unit price back to the estimated one (if possible). This is one of the beautiful things about a cost control program. As the reader will note, there are 500 square feet more formwork to be stripped and cleaned. If management steps in and takes some immediate action in the field, they will possibly be able to reduce the present per square foot cost to some lower figure approaching the estimate and thus save money. Even though the example involves very small quantities and costs, these costs will add up for most jobs. In the past it has been management's feeling that they fully realize they will lose money on some activities and make money on other activities and, hopefully, at the end of the job the loses will be balanced by the gains. This is a dangerous assumption and without a cost control program, management cannot ascertain exactly where they stand relative to the actual cost of doing the job until it is completed, and then it is too late to do anything about it.

WEEKLY COST SUMMARY

JOB _____ JOB NO. 1 PAGE 1 ____ OF —

WEEK ENDING 10-17- PREPARED BY _____

COST CODE	WORK DESCRIPTION	UNIT	ESTIMATED			TO DATE			THIS WEEK			TO DATE		PROJECTED	
			QTY.	TOTAL COST	UNIT COST	QTY.	TOTAL COST	UNIT COST	QTY.	TOTAL COST	UNIT COST	SAVING	LOSS	SAVING	LOSS
2834	Set Anchor Bolts	EA	100	1125	11.25	----	----	----	100	1020	10.20	105.00		----	----
4260	Excavate	CY	370.4	228.29	.62	----	----	----	400	370	.93		141.71	----	----
3010	Strip & Clean Forms	SF	2000	288.00	.144	500	72	.144	1000	252.00	.144	108.00			162.0

Figure 18.38

Once the job is done, it is then important that company personnel review the actual unit cost of doing the work, along with the actual production and equipment rates. From these new rates, management will be able to revise its labor and equipment production figures which are used in developing future estimates. In this particular job, such an analysis was made and the conclusions are noted below:

Anchor Bolts

Material	600.00
Labor	180.00
	240.00
	1020.00

Strip and Stockpile

Materials	= None	
Quantity	= 400 C.Y.	
Equipment	= 25 × 8 =	\$250.00
Operator	=	120.00
		\$370.00

Analysis:

1 Carpenter and 1
Laborer = Crew Size
20 Hours for 100
Anchor Bolts = 5/Hr

Analysis:

Equipment = 400/E = 50 C.Y./Hr
400/10Hrs = 40C.Y./Hr

Strip and Clean Forms (Assuming activity continues to be performed at: .252cts/S.F. and total quantity actually done is 2,000 S.F.)

$$\text{Job done for } (.252 \times 1,500) + (.144 \times 500) = \$450$$

$$\text{Revised labor production } \frac{2,000 \text{ S.F.} \times \$18/\text{Hr}}{\$450} = 80 \text{ S.F./Hr}$$

The foregoing example can best be shown by a flow chart as noted in Figure 18.39. This flow chart will help management not only obtain, but retain, better control of its cost of doing work.

Follow-up Action

Control problems can vary depending on such items as:

- Type of job
- Kinds of equipment involved
- Material types and variations
- Classifications and experience of laborer
- Inventory requirements

The reader is referred to Figure 18.40 for different types of control problems and recommended follow-up action.

As noted earlier in this section, one of the main objectives of a cost control program is to obtain data of actual labor and equipment production rates. The cost of a specific quantity and type of material is more or less stipulated by the material supplier and therefore the amount of control is minimal in relation to a factor affecting total cost of a particular job. Productivity of equipment can be controlled by knowing the actual cost of operating and owning a specific type of equipment by studying such things as:

- Original cost
- Cost of fuel and oil
- Cost of maintenance and repair
- Storage and insurance cost
- Interest (worth the present dollar)
- Obsolescence
- Depreciation
- Downtime
- Salvage

Forms	Flow Chart	Input	Source
1. Worksheets	Quantity Takeoff	Drawings Specifications	Entire Department
2. Labor Production Summary Form Equipment Operator Summary Form	Calculate Labor and Equipment Production Figures	Average Labor Output Average Equipment Operation and Production	Company Records
3. Summary Sheet	Assign Costs to Quantities to Obtain Cost/Item	Material Cost Labor Cost Equipment Cost	Suppliers Prevailing Wage Rates Rental or In-Company Rates Previous Work Sheets & Production Figures
4. Recapitulation Sheet	Summarize	Total Quantities and Costs	Summary Sheets
5. Time Sheets Equipment Utilization Sheets Material Installation Sheets	Monitor Job	Field Reports on Quantities of Material Hours of Labor and Equipment Utilization	Field
6. Summary Sheets Labor Equipment Material	Comparison Estimate With Actual Costs	Summary Sheets of Information Gathered During Monitoring Stage	Office
	Abstract Actual Cost and Production Figures into History Records	Summary Sheets	Office

Cost Control Flow Chart

Figure 18.39

533

Control Problems

Control Problem	Recommended Types of Corrective Action
1. High off-specification production and high waste, reworking, and downgrading	a. Check processing procedures b. Check material quality c. Check equipment cleaning and changeover procedures d. Check control instruments
2. High labor costs	a. Check for avoidable overtime b. Check production scheduling for economic labor scheduling c. Check absenteeism and lateness records for chronic offenders
3. High maintenance costs	a. Review preventive and routine maintenance procedures b. Check that maintenance is performed by qualified personnel c. Emphasize maintenance scheduling against waiting for emergencies d. Displace equipment requiring excessive maintenance
4. High utility costs	a. Check for proper control and eliminate leaks b. Check that utilities are shut off when not required c. Check utility metering for accuracy
5. Low production rates	a. Check for excessive rework b. Cheek maintenance down time and causes c. Check for excessive changeover and downtime d. Check processing procedures
6. Low yields-material	a. Check usage directions; see usage too high that they are being followed b. Check material-weighing and metering devices c. Check material quality d. Check for leaks, spills, even theft

Time and motion studies have indicated that a typical maintenance worker a day is broken down into the following tasks:

1. Receiving instructions.
2. Reaching for materials, tools and equipment.
3. Searching for materials, tools and equipment.
4. Walking with materials tools and equipment.
5. Work of his trade.
6. Work other than his trade.
7. Held up by other tradesmen.
8. Waiting for materials, tools and equipment.
9. Walking and loading.
10. Idle for no apparent reason.
11. Talk other than his work.
12. Personal delays.
13. Planning, discussing and lay-out.

Figure 18.40

This information can be readily obtained and recorded for future use for any type of equipment. As a result one can then use more accurate figures when estimating future jobs utilizing the same equipment.

Controlling Labor Costs

The hardest aspect of any job to control is labor. Only by improving labor productivity, however, will larger savings and lower maintenance costs result. Many studies have shown that maintenance personnel are only 35 to 40 percent efficient. This means that in any one typical hour of any one typical day only 35 to 40 percent of the time is spent in doing the task assigned. The balance (60 to 65 percent) of the time is devoted to nonproductive or indirect tasks and thus is a waste of maintenance dollars.

To improve productivity, management needs to minimize the time that the worker is not being productive; that is item numbers 1, 2, 3, 4, 6, 7, 8, 9, 10, 11 and 12 in Figure 18.40.

Studies have also shown that when a worker does not do a reasonable day's work it is because of one of the following reasons:

- Lack of information.
- Lack of work to do or interferences.
- Lack of tools.
- Lack of motivation.
- Lack of ability.
- Lack of desire.

The first three items above can be corrected by an effective program of planning, scheduling and supervision.

The fourth and fifth items, above, can be corrected by effective leadership and training. Motivation and desire can be corrected by instilling in the worker incentive, integrity, initiative, and loyalty.

Some specific ways the supervisor can maximize production are as follows:

On a Weekly Basis

- Plan how the work is to be done.
- Get any doubtful points cleared up.
- Make any necessary sketches or drawings that may help in performing the work (include as part of the work order).
- Make a complete, detailed list of all materials required for the work (include as part of the work order).
- Select the proper size work group.
- Check delivery of all special materials.
- Check availability of all tools required.
- Request specialists as needed.
- Check job progress against estimated time.
- On a periodic basis, make sure that the job is proceeding efficiently and according to schedule.
- Get prepared for the next week's work.

On a Daily Basis

- Have all material for the entire day's work assembled at the work location.
- Have all tools and equipment required available at the work location.
- Give detailed instructions to the workers so they know exactly what to do. This may include showing them.
- Be available to answer any questions from the men.
- Keep schedules up to date.
- Mark up any drawing or sketches showing previous day's work completed.

- Police lost time especially during the eight "danger periods." These danger periods are as follows:
  - —First 15 minutes in the morning
  - —15 minutes before break
  - —Breaks
  - —15 minutes after breaks
  - —15 minutes before lunch
  - —15 minutes after lunch
  - —15 minutes after rest period
  - —15 minutes prior to clean up
- If the worker is not doing his job, let him know about it immediately.

In summary, labor costs can be minimized by:
- Careful planning of work so that each man knows what he is to do.
- Proper scheduling of men so that they know who is to do it, how many it takes to do it, and when it should be done.

Therefore, to aid in reducing maintenance and repair cost:
- Apply better methods.
- Utilize better and more efficient tools.
- Use motion and time study method analysis.
- Constantly train and retrain maintenance personnel.

APPENDICES

TRAINING PROGRAMS

Electricity

Allen-Bradley, Systems Div.
747 Alpha Dr.
Highland Heights, OH 44143

Eaton Corp.
I.D.D. Dynamatic Plant
3122 14th Ave.
Kenosha, WI 53140

Biddle Technical School
James G. Biddle Co.
Township Line & Jolly Roads
Plymouth Meeting, PA 19462

General Electric Co.
641 Lexington Ave.
New York, NY 10022

Reliance Electric Co.
Cleveland, OH 44117

General Electric Co.
641 Lexington Ave.
New York, NY 10022

Training Center-Industry Services Div.
Westinghouse Electric Corp.
875 Greentree Rd.
Pittsburgh, PA 15220

General Electric Lighting Institute
Nela Park
Cleveland, OH 44112

Transformer Consultants Div. of
S.D. Myers, Inc.
P.O. Box 3575
Akron, OH 44310

Eaton Corp.
2445 Maryland Rd.
Willow Grove, PA 19090

E.I.L. Instruments, Inc.
Electrical Testing Div.
18030 York Rd.
Timonium, MD 21093

Power Transmission

Parker Hannifin Corp.
17325 Euclid Ave.
Cleveland, OH 44112

Taylor Machine Works, Inc.
P.O. Box 150
Louisville, MS 39339

CR Industries
(Chicago Rawhide Mfg.
900 N. State
Elgin, IL 60120

Sperry Vickers
P.O. Box 302
Troy, MI 48084

SEPCO Corp.
P.O. Box 10846
Birmingham, AL 35202

General Electric Co.
Building 600, One River Road
Schenectady, NY 12345

Solar Div., International Harvester
200 Pacific Hwy., P.O. Box 80966
San Diego, CA 92138

Bentley Nevada Corp.
P.O. Box 157
Minden, NV 8923

Air Conditioning, Ventilation, and Refrigeration

George Koch Sons, Inc.
P.O. Box 358
Evansville, IN 47744

Westinghouse Electric Corp.
Box 2510
Staunton, VA 24401

Carrier Air Conditioning Co.
Carrier Parkway
Syracuse, NY 13201

United Sheet Metal Div.
United McGill Corp.
200 E. Broadway
Westerville, OH 43081

Heating

Sunbeam Equipment Corp.
180 Mercer St.
Meadville, PA 16335

Dearborn Chemical (U.S.)
300 Genesee
Lake Zurich, IL 60047

Eclipse Combustion Div.
1100 Buchanan St.
Rockford, IL 61101

Maxon Corp.
P.O. Box 2068
Muncie, IN 47302

Sam Dick Industries
P.O. Box 70498
Seattle, WA 98107

Piping and Valves

Ric-Wil, Inc.
10100 Brecksville Rd.
Brecksville, OH 44141

Aeroquip Corp.
c/o Training Center
300 S. East Ave.
Jackson, MI 49203

General Cable Corp.
Apparatus Div.
P.O. Box 666

Westminster, CO 80030
Yarway Corp.
Blue Bell, PA 19422

Pumps and Compressors

Marketing Manager
SEPCO Corp.
P.O. Box 10846
Birmingham, AL 35202

PULSA feeder/INTERPACE Corp.
77 Ridgeland Rd.
Rochester, NY 14482

Durametallic Corp.
2104 Factory St.
Kalamazoo, MI 49001

Chem Pump Div.
Crane Co.
175 Titus Ave.
Warrington, PA 18901

Maintenance

Airco Technical Institute
121 Kane St.
Baltimore, MD 21224

Manager Manufacturers Div.
Huntington Laboratories, Inc.
P.O. Box 710
Huntington, IN 46750

Marketing Manager
SEPCO Corp.
P.O. Box 10846
Birmingham, AL 35202

Environeers Inc.
P.O. Box 374
Ridgefield, CT 06877

Loctite
North Mountain Road
Newington, CT 06011

Magnaflux Corp.
7300 W. Lawrence Ave.
Chicago, Il 60656

Taylor Machine Works, Inc.
P.O. Box 150
Lousiville, MS 39339

Plastico
P.O. Box 12183
32 Flicker St.
Memphis, TN 38112

The Seaman Nuclear Corp.
3946 W. Wisconsin Ave.
Milwaukee, WI 53208

General Cable Corp.
Apparatus Div.
P.O. Box 666
Westminster, CO 80030

Paints and Coatings

Speeflo, Inc.
4631 Winfield Rd.
Houston, TX 77039

Binks Manufacturing Co.
9201 Belmont
Franklin Park, IL 60131

Standard Dry Wall Products
7800 NW 38th St.
Miami, FL 33166

Environmental Control

Bruner Div.
Calgon Corp.
4767 N. 32nd St.
Milwaukee, WI 53209

EIMCO BSP Services
One David Dr.
Belmont, CA 94002

Fischer & Porter
County Line Rd.
Warminster, PA 18974

Instruments and Controls

Eagle Signal Div.
736 Federal St.
Davenport, IA 52803

Educational Services
The Foxboro Co.
Foxboro, MA 02035

General Electric Co.
Building 600, One River Road
Schenectady, NY 12345

Bailey Training Center
Bailey Meter Co.
Wickliffe, OH 44092

Fisher Controls Co.
Box 190
Marshalltown, IA 50158

Fischer & Porter
County Line Rd.
Warminster, PA 18974

E.F. Johnson Co.
216 2nd Ave.
Waseca, MN 56093

General Electric Co.
1501 Roanoke Blvd.
Salem, VA 24153

Sybron/Taylor
95 Ames St.
Rochester, NY 14601

General Electric Co.
P.O. Box 2830 Terminal Annex
Los Angeles, CA 90051

Telemotive Div., Dynascan Corp.
6460 W. Cortland St.
Chicago, IL 60635

Spectral Dynamics Corp. of San
Diego
P.O. Box 671
San Diego, CA 92112

Fairbanks Weighing Div.
722 E. St. Johnsburg Rd.
St. Johnsburg, VT 05819

Combustion Engineering, Inc.
1000 Prospect Hill Rd.
Windsor, CT 06095

APPENDIX

B   TRADE ASSOCIATIONS

Acoustical Society of America, Woodbury, New York
Air Conditioning and Refrigeration Institute, Arlington, Virginia
Aluminum Association, The, Washington, D.C.
American Architectural Manufacturers Association, Des Plaines, Illinois
American Association of Healthcare Consultants, Arlington, Virginia
American Boiler Manufacturers Association, Arlington, Virginia
American Chemical Society, Washington, D.C.
American Concrete Institute, Detroit, Michigan
American Concrete Pavement Association, Arlington Heights, Illinois
American Concrete Pipe Association, Vienna, Virginia
American Foundrymen's Society, Des Plaines, Illinois
American Hot Dip Galvanizers Association, Washington, D.C.
American Institute of Chemical Engineers, New York, New York
American Institute of Industrial Engineers, Norcross, Georgia
American Institute of Plant Engineers, Cincinnati, Ohio
American Institute of Steel Construction, Chicago, Illinois
American Institute of Timber Construction, Englewood, Colorado
American Iron and Steel Institute, Washington, D.C.
American Petroleum Institute, Washington, D.C.
American Public Power Association, Washington, D.C.
American Society of Civil Engineers, New York, New York
American Society of Gas Engineers, Tinley Park, Illinois
American Society of Heating, Refrigerating, and Air Conditioning Engineers,
    Atlanta, Georgia
American Society of Lubrication Engineers, Park Ridge, Illinois
American Society of Mechanical Engineers, New York, New York
American Society for Metals, Metals Park, Ohio
American Society for Testing and Materials, Philadelphia, Pennsylvania
American Water Works Association, Denver, Colorado
American Welding Society, Inc., Miami, Florida
American Wood Preserver Institute, Vienna, Virginia
Architectural Woodwork Institute, Arlington, Virginia
Asphalt Institute, College Park, Maryland
Asphalt Roofing Manufacturers Association, Rockville, Maryland
Association of Physical Plant Administrators of Universities and Colleges,
    Alexandria, Virginia
Brick Institute of America, Reston, Virginia
Builders Hardware Manufacturers Association, New York, New York
Building Owners and Managers Association International, Washington, D.C.
Building Service Contractors Association, Fairfax, Virginia
Carpet and Rug Institute, Dalton, Georgia
Ceramics Arts Foundation Int., Anaheim, California

Cleaning Management Institute, Irvine, California
Copper Development Association, Inc., Greenwich, Connecticut
Compressed Air and Gas Institute, Cleveland, Ohio
Compressed Gas Association, Inc., Arlington, Virginia
Computer Society of the Institute of Electrical and Electronics Engineers, Inc.,
    Washington, D.C.
Construction Specifications Institute, Alexandria, Virginia
Edison Electrical Institute, Washington, D.C.
Electrical Apparatus Service Association, Inc., St. Louis, Missouri
Engine Manufacturers Association, Chicago, Illinois
Environmental Management Association, Largo, Florida
Expansion Joint Manufacturers Association, Inc., Tarrytown, New York
Flat Glass Marketing Association, Topeka, Kansas
Fluid Controls Institute, Inc., Morristown, New Jersey
Fluid Sealing Association, Philadelphia, Pennsylvania
Hardwood Plywood Manufacturers Association, Reston, Virginia
Hydraulic Institute, Cleveland, Ohio
Illuminating Engineering Society, New York, New York
Industrial Heating Equipment Association, Arlington, Virginia
Institute of Environmental Sciences, Mt. Prospect, Illinois
Instrument Society of America, Research Triangle Park, North Carolina
International Masonry Institute, Washington, D.C.
International Sanitary Supply Association, Chicago, Illinois
Lead Industries Association, New York, New York
Lighting Protection Institute, Harvard, Illinois
Manufacturer Standardization Society of the Valve and Fittings Industry,
    Vienna, Virginia
Maple Flooring Manufacturers Association, Northbrook, Illinois
Mechanical Contractors Association of America, Inc., Bethesda, Maryland
Metal Treating Institute, Inc., Jacksonville Beach, Florida
National Asphalt Pavement Association, Riverdale, Maryland
National Association of Architectural Metal Manufacturers, Chicago, Illinois
National Association of Corrosion Engineers, Houston, Texas
National Association of Homebuilders, Washington, D.C.
National Association of Plumbing-Heating-Cooling Contractors,
    Falls Church, Virginia
National Association of Power Engineers, Inc., Des Plaines, Illinois
National Clay Pipe Institute, Alexandria, Virginia
National Concrete Masonry Association, Herndon, Virginia
National Council of Acoustical Consultants, Inc., Springfield, New Jersey
National Electrical Contractors Association, Inc., Bethesda, Maryland
National Electrical Manufacturers Association, Washington, D.C.
National Environmental Balancing Bureau, Inc., Vienna, Virginia
National Executive Housekeepers Association, Westerville, Ohio
National Fire Protection Association, Quincy, Massachusetts
National Fluid Power Association, Milwaukee, Wisconsin
National Forest Products Association, Washington, D.C.
National Insulation Contractors Association, Washington, D.C.
National Kitchen and Bath Association, Hackettstown, New Jersey
National LP-Gas Association, Oak Brook, Illinois
National Oak Flooring Manufacturers Association, Memphis, Tennessee
National Paint and Coatings Association, Washington, D.C.
National Ready Mix Concrete Association, Silver Springs, Maryland
National Roofing Contractors Association, Des Plaines, Illinois
National Society of Professional Engineers, Alexandria, Virginia
Pipe Fabrication Institute, Springdale, Pennsylvania
Plastics Pipe Institute, New York, New York
Plumbing-Heating-Cooling Information Bureau, Chicago, Illinois
Portland Cement Association, Skokie, Illinois
Pressure Vessel Manufacturers Association, Chicago, Illinois
Prestress Concrete Institute, Chicago, Illinois
Process Equipment Manufacturers Association, Falls Church, Virginia
Red Cedar Shingle and Hard Split Shake Bureau, Bellevue, Washington

Refrigeration Engineers and Technicians Association, Chicago, Illinois
Refrigeration Service Engineers Society, Des Plaines, Illinois
Sheet Metal and Air Conditioning Contractors National Association, Inc.,
    Vienna, Virginia
Society of American Value Engineers, Chicago, Illinois
Society of Plastics Engineers, Inc., Brookfield Center, Connecticut
Solar Energy Industries Association, Arlington, Virginia
Steel Structures Painting Council, Pittsburgh, Pennsylvania
Steel Window Institute, Cleveland, Ohio
Sump Pump Manufacturers Association, Chicago, Illinois
Thermal Insulation Manufacturers Association, Inc., Mt. Kisco, New York
Underwriters Laboratories, Inc., Northbrook, Illinois
Valve Manufacturers Association, Washington, D.C.
Welded Steel Tube Institute, Cleveland, Ohio
Wood and Synthetic Flooring Institute, Hillside, Illinois

FACILITIES MAINTENANCE MAN-HOURS

This section lists minimum and maximum cleaning times per unit. For more information, see chapters listed.

Facilities Maintenance Man-Hours		
	Unit	
	Minimum	Maximum
Blast Cleaning (See Chapter 6)		
White-metal	100 S.F./Hr.	
Near-white	175 S.F./Hr.	
Commercial	370 S.F./Hr.	
Brush-off	870 S.F./Hr.	
Paint Application (See Chapter 6, and Figures 6.29 and 6.30)		
Brushing	125 S.F./Hr.	
Rolling	125 S.F./Hr.	
Spraying	500 S.F./Hr.	
Plaster Cleaning (See Chapter 7)		
Wall Dusting	2 sec./S.F.	3 sec./S.F.
Vacuuming	4 sec./S.F.	5 sec. /S.F.
Spot Washing	125 S.F./Hr.	175 S.F./Hr.
Thorough cleaning	275 S.F./Hr.	
Plaster Repair (See Chapter 7)		
Gypsum and Lime repair	5 S.Y./Hr.	10 S.Y./Hr.
Ceramic Tile Repair (See Chapter 7)		
General	7 S.F./Hr.	10 S.F./Hr.
Adhesive Tile Setting	9 S.F./Hr.	12 S.F./Hr.
Pointing Tile Joints	10 S.F./Hr.	15 S.F./Hr.
Floor Cleaning (See Chapter 8)		
Manual Sweeping	10 min./1000 S.F.	25 min./1000 S.F.
Dust Mopping	5 min./1000 S.F.	20 min./1000 S.F.
Buffing	15 min./1000 S.F.	40 min./1000 S.F.
Spray Buffing	20 min./1000 S.F.	50 min./1000 S.F.
Damp Mopping	15 min./1000 S.F.	30 min./1000 S.F.
Wet Mopping	30 min./1000 S.F.	50 min./1000 S.F.
Scrubbing	50 min./1000 S.F.	140 min./1000 S.F.
Scrubbing using electric floor machine		
General	15 min./1000 S.F.	30 min./1000 S.F.
Stripping	100 min./1000 S.F.	200 min./1000 S.F.

Facilities Maintenance Man-Hours (continued)

	Unit	
	Minimum	**Maximum**
Waxing and Buffing using power machine (See Chapter 8)		
Rewaxing	15 min./1000 S.F.	30 min./1000 S.F.
Stripping and Rewaxing (two coats)	100 min./1000 S.F.	300 min.1000 S.F.
Waxing and Buffing (one coat)	30 min./1000 S.F.	70 min./1000 S.F.
Carpets (See Chapter 8)		
Dry Vacuuming	15 min./1000 S.F.	40 min./1000 S.F.
Wet Vacuuming	30 min./1000 S.F.	50 min./1000 S.F.
Carpet Mopping	20 min./1000 S.F.	40 min./1000 S.F.
Shampooing	175 min./1000 S.F.	250 min./1000 S.F.
Resilient Floor Repair (See Chapter 8)		
Grinding	50 S.F./Hr.	80 S.F./Hr.
Floor Replacement (See Chapter 8)		
Removal (by hand)		
Tiles	100 S.F./Hr.	130 S.F./Hr.
Sheet	120 S.F./Hr.	160 S.F./Hr.
Hardwood	40 S.F./Hr.	60 S.F./Hr.
Replacement		
Ceramic	10 S.F./Hr.	20 S.F./Hr.
Resilient	40 S.F./Hr.	70 S.F./Hr.
Hardwood	25 S.F./Hr.	35 S.F./Hr.
Add for related items:		
Replace wood subfloor	80 S.F./Hr.	100 S.F./Hr.
Replace underlayment	75 S.F./Hr.	90 S.F./Hr.
Replace floor moulding	10 S.F./Hr.	30 S.F./Hr.
Seamless Floor Repair (See Chapter 8)		
	5 S.F./Hr.	15 S.F./Hr.
Wood Floors (See Chapter 8)		
Sanding	40 S.F./Hr.	60 S.F./Hr.
Sealing	200 S.F./Hr.	300 S.F./Hr.
Waxing	*See waxing and buffing using electric power machine.*	
Wood Floor Repair (See Chapter 8)		
Loose boards or tiles	50 S.F./Hr.	250 S.F./Hr.
Wood strip floor replacement	30 S.F./Hr.	60 S.F./Hr.
Wood block floor replacement	20 S.F./Hr.	60 S.F./Hr.
Window Washing (See Chapter 10)		
	300 S.F./Hr.	450 S.F./Hr.
Venetian Blinds, cleaning (See Chapter 10)		
	15 min./set of blinds	

This section lists average cleaning times per each, or square foot; for more information, see Chapter 17.

Facilities Maintenance Man-Hours	
	Average
Floor operations	
Sweeping	
Halls and Corridors	15 min./1000 S.F.
General Rooms	30 min./1000 S.F.
Dust Mop (unobstructed)	10 min./1000 S.F.
Dust Mop (obstructed)	15 min./1000 S.F.
Damp Mop (unobstructed)	20 min./1000 S.F.
Damp Mop (obstructed)	40 min./1000 S.F.
Wet Mop and Rinse	100 min./1000 S.F.
Hand Scrubbing 12" brush	300 min./1000 S.F.
Deck Scrubbing	100 min./1000 S.F.
Machine Scrubbing	
12" diameter	50 min./1000 S.F.
14" diameter	40 min./1000 S.F.
16" diameter	35 min./1000 S.F.
18" diameter	30.5 min./1000 S.F.
19" diameter	28 min./1000 S.F.
21" diameter	25 min./1000 S.F.
23" diameter	23 min./1000 S.F.
24" diameter	20 min./1000 S.F.
32" diameter	18 min./1000 S.F.
36" diameter	15 min./1000 S.F.
Automatic Scrub Machine (24")	5 min./1000 S.F.
Vacuum (unobstructed)	20 min./1000 S.F.
Vacuum (obstructed)	30 min./1000 S.F.
Waxing	30 min./1000 S.F.
Machine Polish (19" machine)	15 min./1000 S.F.
Rectangular Machine (48" plate)	5 min./1000 S.F.
Buff with steel wool	20 min./1000 S.F.
Strip and Rewax	150 min./1000 S.F.
Dry Strip and Rewax	120 min./1000 S.F.
Spray Buffing (unobstructed)	30 min./1000 S.F.
Spray Buffing (obstructed)	45 min./1000 S.F.
Carpeting	
Vacuuming (unobstructed)	20 min./1000 S.F.
Vacuuming (obstructed)	30 min./1000 S.F.
Spot Vacuuming	15 min./1000 S.F.
Shampoo (dry foam)	60 min./1000 S.F.
Pile Lift	30 min./1000 S.F.
Lockers	.20 min.
Radiators	.30 min.
Tables (medium)	.50 min.
Telephones	.15 min.
Towel Dispensers	.12 min.
Towel Disposal Cans	.40 min.
Typewriter and Stand	.50 min.
Wash Basin (office)	.60 min.
Waste Basin (office)	.50 min.
Window Sill	.20 min.
Venetian Blinds std. size	3.50 min.
Washrooms	
Cleaning Commode	4 min.
Door (spot wash both sides)	1 min.
Mirrors	1 min.
Sanitary Napkin Dispenser	.50 min.
Urinals	3 min.
Wash Basin-soap Dispenser	3 min.
General Cleaning	120 min./1000 S.F.

Facilities Maintenance Man-Hours (continued)

	Average
Wall Washing	
Painted Walls (manual)	240 min./1000 S.F.
Painted Walls (machine)	150 min./1000 S.F.
Marble Walls (manual)	90 min./1000 S.F.
Ceiling Washing	
Ceiling Washing (manual)	300 min./1000 S.F.
Ceiling Washing (machine)	180 min./1000 S.F.
Window Washing	
Single Pane	125 min./1000 S.F.
Multi-pane	170 min./1000 S.F.
Frosted Single Pane	190 min./1000 S.F.
Opaque Glass	50 min./1000 S.F.
Plate Glass	35 min./1000 S.F.
Office Partitions (glass)	110 min./1000 S.F.
Dusting Lamps and Lighting Fixtures	
Wall Fluorescent Fixtures	.15 min.
Desk Fluorescent Lamp	.30 min.
Table Lamp and Shade	.60 min.
Floor Lamp and Shade	.60 min.
Washing Fluorescent Light Fixtures	
Ceiling Fixtures (egg crate) 4' ea.	9 min.
Ceiling Fixture (egg crate) 8' ea.	12 min.
Dusting	
Air Conditioners	.30 min.
Ash Trays (desk)	.25 min.
Book Cases (3-tier set)	.30 min.
Chairs	.30 min.
Cigarette Stands	.40 min.
Couch	.25 min.
Desks	.80 min.
Desk Trays	.15 min.
File Cabinets (4 drawer)	.40 min.
Fabric Upholstery Cleaning	
Whisk or vacuum armless chair	.50 min.
Armchair	1 min.
Couch	2 min.
Shampooing armless chair	4 min.
Armchair	7 min.
Couch	20 min.
Stairway Cleaning	
Sweep and dust, 1 flight, 15 steps	6 min.
Damp mop, 1 flight, 15 steps	5 min.
Scrubbing (hand)	20 min.
Carpentry	
Repair door surface closer	1-1/2 Hrs.
Repair concealed door closer	4 Hrs.
Repair door damage at shop	4 Hrs.
Repair and replace screens	1 Hr.
Repair broken glass	1-1/2 Hrs.
Repair and replace ceiling tile	2 Hrs.
Repair small drywall damage	4 Hrs.
Repair and replace sash balance	4 Hrs.
Change lock on door	1-1/2 Hrs.
Painting	
Repaint 20' x 15' room, 1 coat	34 Hrs.
Repaint small bathroom, 1 coat	11-1/2 Hrs.
Repaint exterior window	2-1/2 Hrs.

Facilities Maintenance Man-Hours (continued)

	Average
Plumbing and Steamfitting	
Clear stopped water closet	3 Hrs.
Clear stopped basin	1-1/2 Hrs.
Replace leaking radiator valve	6 Hrs.
Clear external sewer stoppage	9 Hrs.
Install replacement valve, faucet, or trap	10 Hrs.
Electrical	
Replace flourescent lamp ballast	2 Hrs.
Replace fractional HP motor	10 Hrs.
Replace blown fuse or reset circuit breaker	1-1/2 Hrs.
Repair exterior light damage	4 Hrs.
Control circuit problems	5 Hrs.

BIBLIOGRAPHY

Books

Allen, R.T.I., The Repair of Concrete Structures, London, England: Cement and Concrete Association, 1976.

Ambrose, J.E., Building Structure Primer, New York: John Wiley and Sons, Inc., 1967.

Asphalt Institute, Asphalt Maintenance Manual, College Park, Maryland, 1965.

Berkeley, Bernard, Floors: Selection and Maintenance, Chicago: American Library Association, 1968.

Blanchard, B.S. and Lowery, E.E., Maintainability: Principles and Practice, McGraw-Hill, 1969.

Bond, Horatio, ed., N.F.P.A. Inspection Manual, Boston: National Fire Protection Association, 1970.

Bosich, Joseph F., Corrosion Prevention for Practicing Engineers, New York: Barnes and Noble, 1970.

Callender, J.H., ed., Timer-Saver Standards for Architectural Design Data, 5th ed., New York: McGraw-Hill Book Company, 1974.

Carpet Selection and Care, American Institute of Maintenance, Glendale, California, 1975.

Champion, S., Failure and Repair of Concrete Structures, New York: John Wiley and Sons, 1961.

Clement, Edward, Plant Maintenance Manual, New York: Conover-Mast Publications, 1952.

Computerized Maintenance Control Systems, Cos Cob, Connecticut: Cleworth Publishing Company.

Conover, H.S., Grounds Maintenance Handbook, 2nd ed., New York: McGraw-Hill Book Company, 1964.

Cooling, W.C., Low Cost Maintenance Control, New York: AMACOM, 1973.

Corder, A.S., Maintenance Management Techniques, New York: McGraw-Hill, 1976.

Cotz, Victor J., Plant Engineers Manual and Guide, New York: Prentice Hall, Inc., 1967.

Diamant, R.M.E., Industrialized Building, Series 3. London: Iliffe, Book Ltd., 1968.

Electrical Maintenance and Improvement Products, Cos Cob, Connecticut: Cleworth Publishing Company.

Emerick, Robert H., Heating Handbook, New York: McGraw-Hill Book Company, 1964.

Establishing and Conducting Maintenance Training Programs, Cos Cob, Connecticut: Cleworth Publishing Company.

Fancutt and Hudson, Protective Practices of Structural Steel, London: Chancen and Hall LTD, 1957.

Feld, Jacob, Construction Failures, New York: John Wiley and Sons, Inc., 1968.

Feldman, Edwin B., Building Design for Maintainability, New York: McGraw-Hill Book Company, 1975.

Feldman, Edwin B., Housekeeping Handbook for Institutions, Businesses and Industry, New York: Frederick Fell, Inc., 1969.

Floor Care Guide, American Institute of Maintenance, Glendale, California, 1975.

Groner, John C., Programmed Cleaning and Environmental Sanitation, New York: Soap Detergent Association.

Godfrey, R.S., Building Construction Cost Data, Duxbury, Massachusetts: Robert Snow Means Company, Inc., latest edition.

Goldman, A.S. and T.B. Slatter, Maintainability: A Major Element of System Effectiveness, New York: John Wiley & Son, Inc., 1964.

Gordon, Frank, Maintenance Engineering Organization and Management, New York: John Wiley and Son, 1973.

Greathouse, Glenn A., Deterioration of Materials, New York: Reinhold Publishing Corporation, 1954.

Guide and Data Book, American Society of Heating, Refrigerating and Air Conditioning Engineers, 1972.

Haines, John E., Automatic Control of Heating and Air Conditioning, 2nd ed., New York: McGraw-Hill Book Company, Inc., 1961.

Handy Maintenance Tips American Institute of Maintenance, Glendale, California, 1975.

Harroun, Jack T., ed., Good School Maintenance, A Manual of Programs and Procedures for Buildings, Grounds and Equipment, Illinois Association of School Boards, 1976.

Heintzelman, John E., The Complete Handbook of Maintenance Management, New York: Prentice-Hall, Inc., New York, 1977.

Hewes, L.I. and C.H. Oglesby, Highway Engineering, New York: John Wiley and Sons, Inc., Latest edition.

Johnson, Sidney M., Deterioration, Maintenance and Repair of Structures, New York: McGraw-Hill Book Co., 1965.

Knowles, Asa., Handbook of College and University Administration, Vols. I and II. New York: McGraw-Hill Book Company, Inc., 1970.

Lewis, Bernard, Controlling Maintenance Costs, Waterford, Connecticut: National Foreman Institute, 1964.

Lewis, B.T. and Marron, J.P., Facilities and Plant Engineering Handbook, New York: McGraw-Hill Book Company, 1975.

Lewis, B.T. and Pearson, W.W., Maintenance Management, New York: John F. Rider Publishers, 1963.

Lubricating Maintenance and Improvement Procedures, Cleworth Publishing Company, Cos Cob, Connecticut.

Maintenance as a Management Function, Cos Cob, Connecticut: Cleworth Publishing Company.

Maintenance Guide for Commercial Buildings, Cedar Rapids, Iowa: Stamats Publishing Company, 1970.

Maintenance Hints, Trafford, Pennsylvania: Westinghouse Electric Corporation, Printing Division.

McKaig, Thomas H., Field Inspection of Building Construction, New York: McGraw-Hill Book Company, 1958.

McLaughlin, P., The Cleaning Hygiene and Maintenance Handbook, Englewood Cliffs, New Jersey: Prentice-Hall, Inc. 1973.

Mechanical Maintenance and Improvement Procedures, Cos Cob, Connecticut: Cleworth Publishing Company.

Merrit, Frederick, ed., Building Construction Handbook, 3rd ed., New York: McGraw-Hill Book Company, Latest edition.

Miller, J., Blood, Jerome, et al, Modern Maintenance Management, New York: American Management Association, 1963.

Modern Plastics Encyclopedia, New York: Modern Plastics, latest edition.

Morrow, L.C., Maintenance Engineering Handbook, 2nd ed., New York: McGraw-Hill Book Company, 1966.

Muther, Richard, Practical Plant Layout, Hightstown, New Jersey: McGraw-Hill Book Company, Inc., 1955.

Newbrough, E.T., and et al, Effective Maintenance Management, Organization, Motivation and Control in Industrial Maintenance, New York: McGraw-Hill, 1967.

O'Brien, James J., Construction Inspection Handbook, New York: Van Nostrand Reinhold Co., 1974.

Perkins, Phillip, Concrete Structure.- Repair, Waterproofing and Protection, New York: John Wiley and Sons, 1977.

Plastics Engineering Handbook, Society of the Plastics Industry, Inc., New York: Reinhold Publishing Corporation, 2nd Edition, 1954.

Preventive Maintenance Procedures, Cos Cob, Connecticut: Cleworth Publishing Company.

Price, Seymour G., Air Conditioning for Building Engineers and Managers, Operation and Maintenance, New York: Industrial Press, Inc., 1970.

Proceedings of New England Plant Engineering and Maintenance Conference, New England Region American Institute of Plant Engineers, 1969.

Productive Maintenance, General Electric Company.

Programmed Cleaning and Environmental Sanitation for Buildings, Plants, Offices and Institutions, New York: The Soap and Detergent Association, 1971.

Ramsey, C.G. Architectural Graphic Standards, New York: John Wiley and Sons, Inc., 1970.

Readings on Maintenance Management, edited by Lewis, Bernard and Leonard, Tom, Boston: Cahner Books, 1973.

Reed, Ruddle, Plant Location, Layout and Maintenance, Homewood, Illinois: Richard D. Irwin, Inc., 1967.

Sack, Thomas F., A Complete Guide to Building and Plant Maintenance, Englewood Cliffs, New Jersey: Prentice-Hall, Inc., 1971.

Schiff, M. and Schirger, Joseph, Control of Maintenance Cost, New York: National Association of Accountants, 1964.

Schmidt, John L. et. al, Construction Principles, Materials and Methods, Chicago: American Savings and Load Institute Press, 1970.

Selecting Proper Floor Care Materials, American Institute of Maintenance, 1974, Glendale, California.

Simpson, J.W. and Horrobin, P.J., ed., The Weathering and Performance of Building Materials, New York: Halsted Press, 1970.

Small Plant Maintenance and Improvement Procedures, Cos Cob, Connecticut: Cleworth Publishing Company.

Smeaton, Robert W., Motor Application and Maintenance Handbook, New York: McGraw-Hill, 1971.

Speller, F.N., Corrosion Causes and Prevention, New York: McGraw-Hill Book Company, Inc., 1951.

Standard Handbook for Civil Engineers, edited by Fredrick S. Merritt, New York: McGraw-Hill Book Company.

Standard Handbook for Electrical Engineers, New York: McGraw-Hill Book Company, Inc., 1971.

Standard Plant Operations, New York: McGraw-Hill, 1975,

Staniar, William, ed., Plant Engineering Handbook, New York: McGraw-Hill, 1959.

Steel, Ernest W., Water Supply and Sewerage, New York: McGraw-Hill Book Company, Inc., 1947.

Steel Structures Painting Manual, Joseph Bigos, Steel Structures Painting Council, Pittsburgh: Volume 1, Good Painting Practice, 1954; Volume 2 Systems and Specifications, 1955.

Techniques of Electrical Construction Design, Volume 8, Maintaining Electrical Systems, New York: McGraw-Hill, 1975.

Terry, Harry, Mechanical-Electrical Equipment Handbook for School Buildings: Installation, Maintenance and Use, New York: John Wiley and Sons, Inc., 1960.

Urquhart, Leonard Church, Civil Engineering Handbook, New York: McGraw-Hill Book Company, Inc., latest edition.

Watt, John H. and Summers, Wilford, eds., NFPA Handbook of the National Electrical Code, 4th ed., New York: McGraw-Hill, 1975.

Weber, G.O., ed, A Basic Manual for Physical Plant Administration, Washington, D.C.: The Association of Physical Plant Administration of Universities and Colleges, 1974.

Wertenberger, Morris Jr., Floor Maintenance Manual, Milwaukee, Wisconsin: Trade Press Publishing Company, 1972.

Wilson, C.L. and Oatis, J.A., Corrosion and the Maintenance Engineer, New York, Hart Publishing Co., Inc., 1968.

Magazines

It is recommended that the reader refer to the "Readers Guide to Periodical Literature" (technical or general) for current specific articles contained in the magazines noted herein.

American Institute of Concrete Journal Publications and Committee Reports, Detroit, Michigan: American Concrete Institute.

American School and University, The Buildings Magazine. Philadelphia, Pennsylvania: Educational Communications Division of North American Publishing Company.

Building Operating Management Milwaukee, Wisconsin: Trade Press Publishing Company.

Building Research, Journal of the BRAB. Washington, D.C.: Building Research Institute, BRAB Building, National Research Council.

Building Services Contractor, New York: MacNair-Dorland Company, Inc.

Buildings, The Constructor and Building Management Journal, Cedar Rapids, Iowa: Stamats Publishing Company.

Canadian Building Digest, Ottawa, Canada: Division of Building Research, National Research Council.

Civil Engineering, New York: American Society of Civil Engineers.

College and University Business, New York: McGraw-Hill Publications Company.

Concrete Construction, Addison, Illinois: Concrete Construction Publications.

Electrical Construction and Maintenance, Hightstown, New Jersey: McGraw-Hill Company.

Engineering Journal, American Institute of Steel Construction.

Engineering News-Record, New York: McGraw-Hill Publishing Company.

Iron and Steel Engineer, New York: American Iron and Steel Institute.

Journal of the Prestress Concrete Institute, Chicago, Illinois: Prestress Concrete Institute. Maintenance Supplies, New York: MacNair-Dorland Company.

Maintenance Viewpoint, Cos Cob, Connecticut: Cleworth Publishers.

Plant Engineering and Research Management Information, Barrington, Illinois: Technical Publishing Company.

Professional Sanitation Management, Clearwater, Florida: The Environmental Management Association.

Sweets Catalogue, New York: McGraw-Hill Publishing Co.

Technical Notes on Brick Construction, McLean, Virginia: Brick Institute of America.

TEK Service, McLean, Virginia: National Concrete Masonry Association.

Technical Notes, McClean, Virginia: Structural Clay Products Institute.

United States Governmental Publications

(United States Governmental Printing Office, Pueblo, Colorado)

U.S. Department of Agriculture, Series of Pamphlets published by Forest Products Laboratory, Forest Service, U.S. Department of Agriculture on Wood.

Concrete Manual, A Water Resources Technical Publication, 8th Edition, U.S. Department of Interior, Bureau of Reclamation.

"Maintenance of Grounds," Department of Navy, Facilities Engineering Command, MO-100, 1963.

"Maintenance of Miscellaneous Ground Structures," Department of Navy, Naval Facilities Engineering Command, MO-101, 1963.

"Maintenance of Pavements," Department of Navy, Naval Facilities Engineering Command, MO-102, 1970.

"Maintenance of Trackage," Department of Navy, Naval Facilities Engineering Command, MO-103, 1974.

"Maintenance of Waterfront Facilities," Department of Navy, Naval Facilities Engineering Command, MO-104, 1963.

"Paints and Protective Coatings," Department of Navy, Naval Facilities Engineering Command, MO-110, 1969.

"Management of Maintenance Painting of Facilities," Department of Navy, Naval Facilities Engineering Command, MO-110.1, 1963.

"Building Maintenance-Structural," Department of Navy, Naval Facilities Engineering Command, MO-111, 1963.

"Maintenance and Repair of Roofs," Department of Navy, Naval Facilities Engineering Command, MO-113, 1974.

"Building Maintenance—Plumbing, Heating, and Ventilation," Department of Navy, Naval Facilities Engineering Command, MO-114, 1964.

"Building Maintenance—Air Conditioning and Refrigeration," Department of Navy, Naval Facilities Engineering Command, MO-115, 1962.

"Electrical Interior Facilities," Department of Navy, Naval Facilities Engineering Command, MO-116, 1972.

"Fire Alarm and Sprinkler Maintenance," Department of Navy, Naval Facilities Engineering Command, MO-117, 1968.

"Building Maintenance-Galley Equipment," Department of Navy, Naval Facilities Engineering Command, MO-119, 1963.

"Building Maintenance—Furniture and Furnishings," Department of Navy Naval Facilities Engineering Command, MO-120, 1964.

"Military Custodial Services Manual," Department of Navy, Naval Facilities Engineering Command, MO-125, 1969.

"Electric Power Distribution Systems Maintenance," Department of Navy, Naval Facilities Engineering Command, MO-200, 1962.

"Operation of Electric Power Distribution Systems," Department of Navy, Naval Facilities Engineering Command, MO-201, 1963.

"Overhead Powerlines Electromagnetic Interference Handbook," Department of Navy, Naval Facilities Engineering Command, MO-202, 1968.

"Central Heating and Stem Electrical Generating Plants," Department of Navy, Naval Facilities Engineering Command, MO-205, 1964, 1968.

"Operation and Maintenance of Air Compressor Plants," Department of Navy, Naval Facilities Engineering Command, MO-206, 1964.

"Operation and Maintenance of Internal Combustion Engines," Department of Navy, Naval Facilities Engineering Command, MO-207. 1966.

"Maintenance of Steam, Hot Water, and Compressed Air Distribution Systems," Department of Navy, Naval Facilities Engineering Command, MO-209, 1966.

"Maintenance and Operation of Water Supply Systems," Department of Navy, Naval Facilities Engineering Command, MO-210, 1964.

"Sewerage and Industrial Waste Systems (Interim)," Department of Navy, Naval Facilities Engineering Command, MO-212, 1962.

"Maintenance and Operation of Gas Systems," Department of Navy, Naval Facilities Engineering Command, MO-220, 1970.

"Corrosion Prevention and Control," Department of Navy, Naval Facilities Engineering Command, MO-306, 1964.

"Corrosion Control by Catholic Protection," Department of Navy, Naval Facilities Engineering Command, MO-307, 1964.

"Wood Preservation," Department of Navy, Naval Facilities Engineering Command, MO-312, 1968.

"Maintenance Management of Public Works and Public Utilities," Department of Navy, Naval Facilities Engineering Command, MO-321, 1975.

"Maintenance Management of Public Works and Public Utilities—For Small Activities," Department of Navy, Naval Facilities Engineering Command, MO-321.1, 1972.

"Inspection for Maintenance of Public Works and Public Utilities, Vol. 1," Department of Navy, Naval Facilities Engineering Command, MO-322, 1974.

"Inspection for Maintenance of Public Works and Public Utilities, Vol. 2," Department of Navy, Naval Facilities Engineering Command, MO-322, 1975,

"Inspection for Maintenance of Public Works and Public Utilities, Vol. 3," Department of Navy, Naval Facilities Engineering Command, MO-322, 1975.

"Work Simplification for Maintenance of Public Works and Public Utilities," Department of Navy, Naval Facilities Engineering Command, MO-325, 1969.

"Construction Inspection Guide," Department of Navy, Naval Facilities Engineering Command, P-456, 1974.

"General Handbook," Department of Navy, Naval Facilities Engineering Command, P-701.0, 1964.

"Elemental Standard Time Data Handbook," Department of Navy, Naval Facilities Engineering Command, P-701.2, 1973.

"Elemental Standard Time Data Element Analysis," Department of Navy, Naval Facilities Engineering Command, P-701.3, 1973.

"EPS Craft Formulas," Department of Navy, Naval Facilities Engineering Command, P-701.4, 1975.

"Carpentry Handbook," Department of Navy, Naval Facilities Engineering Command, P-702.0, 1962.

"Carpentry Formulas," Department of Navy, Naval Facilities Engineering Command, P-702.1, 1962.

"Electric, Electronic Handbook," Department of Navy, Naval Facilities Engineering Command, P-703.0, 1966.

"Electric, Electronic Formulas," Department of Navy, Naval Facilities Engineering Command, P-703.1, 1966.

"Heating, Cooling, and Ventilating Handbook," Department of Navy, Naval Facilities Engineering Command, P-704.0, 1963.

"Heating, Cooling, and Ventilating Formulas," Department of Navy, Naval Facilities Engineering Command, P-704.1, 1963.

"Janitorial Handbook," Department of Navy, Naval Facilities Engineering Command, P-706.0, 1962.

"Machine Shop, Machine Repair Handbook," Department of Navy, Naval Facilities Engineering Command, P-707.0, 1966.

"Machine Shop, Machine Repair Formulas," Department of Navy, Naval Facilities Engineering Command, P-707.1, 1966.

"Masonry Handbook," Department of Navy, Naval Facilities Engineering Command, P-708.0, 1963.

"Masonry Formulas," Department of Navy, Naval Facilities Engineering Command, P-708.1, 1963.

"Moving, Rigging Handbook," Department of Navy, Naval Facilities Engineering Command, P-709.0, 1962.

"Moving, Rigging Formulas," Department of Navy, Naval Facilities Engineering Command, P-709.1, 1962.

"Paint Handbook," Department of Navy, Naval Facilities Engineering Command, P-710.0, 1963.

"Paint Formulas," Department of Navy, Naval Facilities Engineering Command, P-710.1, 1963.

"Pipefitting, Plumbing Handbook," Department of Navy, Naval Facilities Engineering Command, P-711.0, 1966.

"Pipefitting, Plumbing Formulas," Department of Navy, Naval Facilities Engineering Command, P-711.1, 1966.

"Roads, Grounds, Pest Control, Refuse Collection Handbook," Department of Navy, Naval Facilities Engineering Command, P-712.0, 1969.

"Roads, Grounds, Pest Control, Refuse Collection Formulas," Department of Navy, Naval Facilities Engineering Command, P-712.1, 1963.

"Sheetmetal, Structural Iron, and Welding Handbook," Department of Navy, Naval Facilities Engineering Command, P-713.0, 1965.

"Sheet metal, Structural Iron, and Welding Formulas," Department of Navy, Naval Facilities Engineering Command, P-713.1, 1965.

"Trackage Handbook," Department of Navy, Naval Facilities Engineering Command, P-714.0, 1963.

"Trackage Formulas," Department of Navy, Naval Facilities Engineering Command, P-714.1, 1963.

"Wharfbuilding Handbook," Department of Navy, Naval Facilities Engineering Command, P-715.0, 1963.

567